International Series in Operations Research & Management Science

Volume 170

Series Editor:
Frederick S. Hillier
Stanford University, CA, USA

Special Editorial Consultant:
Camille C. Price
Stephen F. Austin State University, TX, USA

For further volumes
www.springer.com/series/6161

John N. Hooker

Integrated Methods
for Optimization

Second Edition

 Springer

John N. Hooker
Carnegie Mellon University
Tepper School of Business
Pittsburgh, Pennsylvania
USA
john@hooker.tepper.cmu.edu

ISSN 0884-8289
ISBN 978-1-4899-8994-9   ISBN 978-1-4614-1900-6 (eBook)
DOI 10.1007/978-1-4614-1900-6
Springer New York Dordrecht Heidelberg London

Springer is part of Springer Science+Business Media (www.springer.com)

Preface

Optimization has become a versatile tool in a wide array of application areas, ranging from manufacturing and information technology to the social sciences. Methods for solving optimization problems are equally numerous and provide a large reservoir of problem-solving technology. In fact, there is such a variety of methods that it is difficult to take full advantage of them. They are described in different technical languages and are implemented in different software packages. Many are not implemented at all. It is hard to tell which one is best for a given problem, and there is too seldom an opportunity to combine techniques that have complementary strengths.

The ideal would be to bring these methods under one roof, so that they and their combinations are all available to solve a problem. As it turns out, many of them share, at some level, a common problem-solving strategy. This opens the door to integration—to the design of a modeling and algorithmic framework within which different techniques can work together in a principled way.

This book undertakes such a project. It deals primarily with the unification of mathematical programming and constraint programming, since this has been the focus of most recent research on integrated methods. Mathematical programming brings to the table its sophisticated relaxation techniques and concepts of duality. Constraint programming contributes its inference and propagation methods, along with a powerful modeling approach. It is possible to have all of these advantages at once, rather than being forced to choose between them. Continuous global optimization and heuristic methods can also be brought into the framework.

The book is intended for those who wish to learn about optimization from an integrated point of view, including researchers, software developers, and practitioners. It is also for postgraduate students interested in a unified treatment of the field. It is written as an advanced textbook, with exercises, that develops optimization concepts from the ground up. It takes an interdisciplinary approach that presupposes mathematical sophistication but no specific knowledge of either mathematical programming or constraint programming.

The choice of topics is guided by what is relevant to understanding the principles behind popular linear, mixed integer, and constraint programming solvers—and more importantly, integrated solvers of the present and foreseeable future. On the mathematical programming side, it presents the basic theory of linear and integer programming, cutting planes, Lagrangean and other types of duality, mixed integer modeling, and polyhedral relaxations for a wide range of combinatorial constraints. On the constraint programming side it discusses constraint propagation, domain filtering, consistency, global constraints, and modeling techniques. The material ranges from the classical to the very recent, with some results presented here for the first time.

The ideas are tied together by a search-infer-and-relax algorithmic framework, an underlying theory of inference and relaxation duality, and the use of metaconstraints (a generalization of global constraints) for modeling.

The first edition of the book was published only four years ago, but the field has moved ahead. This second edition expands, reorganizes, and updates the earlier edition in several ways. The examples that began the first book now occupy a separate chapter, followed by two new chapters. A chapter on optimization basics makes the book more nearly self-contained. A second new chapter on duality presents a stronger case for its centrality and provides conceptual background for the search chapter that follows, which is much expanded. The chapter on inference covers additional global constraints, and the dictionary of metaconstraints in the final chapter has been enlarged. The material throughout has been updated and elaborated where appropriate, resulting in some 170 new references.

I would like to acknowledge the many collaborators and former students from whom I have learned much about integrated problem solving. They include Henrik Andersen, Kim Allen Andersen, Ionut Aron, David Bergman, Alexander Bockmayr, Endre Boros, Srinivas Bollapragada, Jonathan Cagan, Vijay Chandru, Andre Cire, Elvin

Coban, Milind Dawande, Giorgio Gallo, Latife Genç Kaya, Omar Ghattas, Ignacio Grossmann, Tarik Hadzic, Peter Hammer, Samid Hoda, Willem-Jan van Hoeve, Hak-Jin Kim, Maria Auxilio Osorio, Barry O'Sullivan, Greger Ottosson, Gabriella Rago, Ramesh Raman, Erlendur Thorsteinsson, Peter Tiedemann, H. Paul Williams, Hong Yan, and Tallys Yunes.

Contents

Chapter 1
Introduction

Optimization calls for a multifaceted approach. Some classes of optimization problems, such as linear programming models, can be solved by an all-purpose method. However, most problems require individual attention. The secret to solving a problem is to take advantage of its particular structure.

The result is a proliferation of optimization methods. Thousands of journal articles address narrowly defined problems, as they must if the problems are to be solved in a reasonable amount of time. Not only this, but the articles are distributed across several literatures that speak different technical languages. Chief among these are the literatures of mathematical programming, constraint programming, continuous global optimization, and heuristic methods. This imposes an obvious burden on anyone who seeks the right method and software to solve a given problem.

Some of the individual disciplines have made significant progress toward developing robust, general-purpose solvers. Each generation of mathematical programming software solves a wider range of problems, and similarly for constraint programming and global optimization software. Yet even these solvers fail to exploit most problem-specific methods—with the partial exception of constraint programming, whose concept of a global constraint provides a clue to how to overcome this weakness. Also, the four disciplines continue to move in largely separate orbits. This not only imposes the inconvenience of becoming familiar with multiple solvers, but it also passes up the advantages of integrated problem solving.

Recent research has shown that there is much to be gained by exploiting the complementary strengths of different approaches to optimization. Mathematical programmers are expert at relaxation techniques and polyhedral analysis. Constraint programming is distinguished by its inference techniques and modeling power. Continuous global optimization is known for its convexification methods, and heuristic methods for their search strategies. Rather than choose between these, one would like to have them all available to attack a given problem. Some problems submit to a single approach, but others benefit from the more flexible modeling and orders-of-magnitude computational speedups that can result when ideas from different fields are combined.

The advantages of integrated methods are being demonstrated in a growing literature but, as a result, they themselves are multiplying. If there are many solution methods that might be combined, there are even more ways to combine them. The problem of proliferation seems only to be compounded by efforts to integrate.

A change of perspective can bring order into this chaos. Rather than look for ways to combine methods, one can look for what the methods already have in common. Perhaps there is a general problem-solving strategy that the various communities have arrived at independently, albeit from different directions and with different emphases, because it is a strategy that works.

This book takes such a perspective. It develops an algorithmic framework in which the different optimization methods, and more importantly their many combinations, are variations on a theme. It proposes a modeling practice that can bring problem-specific methods into the scheme in a natural way. In short, it seeks an underlying unity in optimization methods.

The book emphasizes the integration of mathematical programming and constraint programming in particular, since this is where most of the research on integrated methods has been focused to date. Nonetheless, some attempt is made to show how global optimization and heuristic methods fit into the same framework.

Unification is good for learning as well as practice. Students of operations research or computer science who confine themselves to their own field miss the insights of the other, as well as an overarching perspective on optimization that helps make sense of it all. This book is therefore designed as a graduate-level optimization text that belongs to neither field but tries to construct a coherent body of material from both.

Neither this nor any other text covers all concepts related to optimization. There is nothing here about stochastic optimization, multiobjective programming, semidefinite programming, or approximation methods, and only a little about nonlinear programming. Some important ideas of combinatorial analysis are left out, as are whole areas of constraint programming. A broad selection of topics is nonetheless presented, guided primarily by what is relevant to the major general-purpose solvers of linear and mixed-integer programming, constraint programming, and to some extent global optimization. The emphasis, of course, is on how these topics form more than just a miscellaneous collection but are parts of an integrated approach to optimization.

Although not comprehensive, the material presented here is more than adequate to provide a substantial grounding in optimization. It is a starting point from which the student can explore other subdisciplines from an ecumenical point of view.

1.1 A Unifying Framework

Optimization methods tend to employ three strategies that interact in specific ways: *search*, *inference*, and *relaxation*. Search is the enumeration of problem restrictions, while inference and relaxation make the search more intelligent. Inference draws out implicit information about where the solution might lie, so that less search is necessary. Relaxation replaces the problem with a simpler one whose solution may point the way to a solution of the original one.

Search is fundamental because the optimal solution of a problem lies somewhere in a solution space, and one must find it. The solution space is the set of solutions that satisfy the constraints of the problem. Some solutions are more desirable than others, and the objective is to find the best one, or at least a good one. In many practical problems, the solution space tends to be huge and multidimensional. Its boundaries may be highly irregular, and if there are discrete as well as continuous variables, it is full of gaps and holes.

Whether searching for lost keys or a fugitive from the law, a common strategy is to divide up the territory and focus on one region at a time. Optimization methods almost invariably do the same. A region is the solution set for a *restriction* of the problem, or a problem to which additional constraints have been added. A sequence of restrictions are solved, and the best solution found is selected. If the search is exhaustive, meaning that the entire search space is covered, the best solution found is optimal.

The most prevalent scheme for exhaustive search is *branching*: splitting the territory (say) in half, splitting it again if either half is still too large to manage, and so forth. Another basic strategy is *constraint-directed search*: whenever a region has been searched, a constraint or *nogood* is created that excludes that part of the search space, and perhaps other parts that are unpromising for the same reason. The next examined region must satisfy the constraints so far accumulated.

Branching methods include the popular branch-and-cut methods of mixed-integer programming and branch-and-infer methods of constraint programming, on which all the major commercial solvers are based. Constraint-directed methods include Benders decomposition in mathematical programming and such nogood-based methods as branching with clause learning for propositional satisfiability problems, and partial-order dynamic backtracking. Continuous global optimizers use a branching scheme to divide space into multidimensional boxes. Local search or heuristic methods likewise enumerate a sequence of problem restrictions, represented by a sequence of neighborhoods. The distinction of branching and constraint-directed search carries over to heuristic methods: a greedy adaptive search procedure, for example, is an incomplete form of branching search, and tabu search is an incomplete constraint-directed search.

Inference is a way of learning more about the search space, so as not to waste time looking in the wrong places. Police might deduce from a street map that a suspect would never frequent certain neighborhoods, just as a problem solver might deduce from a constraint that certain variables would never take certain values in an optimal solution. This not only shrinks the region to be searched, but also, sometimes more importantly, reduces its dimensionality. Mathematical programming systems use inference methods in the form of cutting planes and preprocessing techniques, while constraint programming systems rely heavily on domain filtering and bounds propagation.

Relaxation enlarges the search space in a way that makes it easier to examine. A common strategy is to replace the current problem restriction with a continuous, linear relaxation. This replaces the corresponding region of the search space with a polyhedron that contains it, thus smoothing out the boundary and filling in the holes, and simplifying the search for an optimal solution. Relaxation can help in several ways.

The solution of the relaxation may happen to lie inside the original search space, in which case it solves the current problem restriction. If not, the optimal solution of the relaxation may be no better than the best feasible solution found so far, in which case one can immediately move on to another region. Even if this does not occur, the optimal solution of the relaxation may provide a clue as to where the optimal solution of the original problem lies.

Search, inference, and relaxation reinforce one another. Restricting the problem in a search process allows one to draw more inferences and formulate relaxations that are closer to the original. Inference accelerates the search by excluding part of the search space, as when filtering reduces domains in constraint programming, when logic-based Benders cuts (a form of nogood) are generated in Benders decomposition, or when items are added to the tabu list in local search. It deduces constraints that can strengthen the relaxation, such as cutting planes in integer programming.

Relaxation abbreviates the search by providing bounds, as in the branch-and-relax and branch-and-cut methods of integer programming. The solution of the relaxation, even when infeasible in the original problem, can provide information about which problem restriction to examine next. In branch-and-cut methods, for example, one branches on a variable that has a fractional value in the solution of the relaxation. In constraint-directed search, the set of accumulated nogoods is, in effect, a relaxation whose solution defines the next problem restriction to be solved. Less obviously, the solution of the relaxation can direct inference, since one can give priority to deducing constraints that exclude this solution. The separating cuts of integer programming provide an example of this.

This search-infer-and-relax scheme provides a framework within which one can mix elements from different traditions. As a search scheme one might select branching or constraint-based search. For inference one might apply domain filtering to some constraints, generate cutting planes for others, or for others use some of the interval propagation and variable fixing methods characteristic of global optimization. Some constraints might be given relaxations based on integer programming models, others given relaxations that are tailor-made for global constraints, others relaxed with the factorization methods used in continuous global solvers, and still others left out of the relaxation altogether.

1.2 Modeling to Reveal Problem Structure

Search-infer-and-relax methods can succeed only to the extent that they exploit problem-specific structure. Experience teaches that inference and relaxation can be blunt instruments unless sharpened with specific knowledge of the problem class being solved. Yet, it is impractical to invent specialized techniques for every new type of problem that comes along.

The answer is to analyze constraints rather than problems. Although every problem is different, certain patterns tend to recur in the constraints. Many scheduling problems require, for example, that jobs run sequentially without overlapping. Other problems require that employees be assigned to work a specified number of days in a row. It is not hard to identify structured subsets of constraints that keep coming up. Each subset can be represented by a single *metaconstraint*, and specialized inference and relaxation methods can be designed for each metaconstraint. The modeler's choice of metaconstraints can then communicate much about the problem structure to the solver. In particular, it can dictate which inference and relaxation techniques are used in the solution process.

This might be called *constraint-based control*, which can extend to the search procedure as well as the choice of inference and relaxation methods. The user begins by choosing the overall search algorithm, perhaps branching or constraint-directed search, and perhaps an exhaustive or heuristic version of it. The choice of metaconstraints determines the rest. In a branching framework, for example, the search branches when the solution of the current relaxation violates one or more constraints. A priority list designates the violated constraint on which to branch. The constraint "knows" how to branch when it is violated, and the search proceeds accordingly.

This scheme can work only if the modeler uses constraints that are rich enough to capture substantial islands of structure in the problem. This requires a change from the traditional practice of mathematical programming, which is to build models with a small vocabulary of primitive constraints such as inequalities and equations. It recommends something closer to the constraint programmer's practice of using *global constraints*, so-named because each constraint stands for a collection of more elementary constraints whose global structure is exploited by the solver.

This book therefore advocates modeling with metaconstraints, which generalize the idea of global constraints. A metaconstraint may consist of a set of inequalities of a certain type, a set of constraints to be activated under certain conditions, or a global constraint familiar to the constraint programming world—to mention only a few possibilities. The advantages of metaconstraints are twofold. They not only reveal the problem structure to the solver, which may lead to faster solution, but they also allow one to write more concise models that are easier to follow and easier to debug.

Modeling with metaconstraints immediately raises the issue of what to do when the metaconstraints begin to proliferate, much as the special-purpose algorithms that were mentioned earlier. New problems often require new metaconstraints to capture a substructure that did not occur in previous problems. Yet, this is not the stumbling block that it may seem to be.

The lexicon of metaconstraints can grow large, but there are still many fewer constraints than problems, just as there are many fewer words than sentences. In any field of endeavor, people tend to settle on a limited number of terms that prove adequate over time for expressing the key ideas. There is no alternative, since most of us can master only a limited vocabulary. This is true of technical domains in particular, since a limited number of technical terms tend to evolve and prove adequate for most situations. Sailors must know about halyards, booms, mizzens, and much else, but the nautical vocabulary is finite and learnable.

The same applies to modeling. In any given domain, practitioners are likely to develop a limited stock of metaconstraints that frequently arise. There might be one stock for project management, one for process scheduling, one for supply chains, and so forth, with much overlap between them. In fact, this has already happened in some domains for which specialized software has developed, such as project scheduling. There will be many metaconstraints overall, just as there are many technical terms in the world. But no one is obliged to know more than a small fraction of them.

Computer-based modeling systems can ease the task further. There is no need to write models in a formal modeling language in which one must get the syntax right or generate error messages. An intelligent user interface can provide menus of constraints, conveniently organized by application domain or along other dimensions. Selecting a constraint activates a window that allows one to import data, set parameters, and choose options for inference, relaxation, and search. The window contains links to related constraints that may be more suitable. The system prompts the user with checklists or queries to guide the modeling process. The solver keeps updating the solution of a small problem as the modeling proceeds, so that the modeler can see when the solution begins to look reasonable.

1.3 The Role of Duality

Duality is a perennial theme of optimization. It occurs in such forms as linear programming duality, Lagrangean duality, surrogate duality, and superadditive duality. It is also a unifying theme for this book for two reasons. These various duals turn out to be closely related, a fact that helps to unify optimization theory. They can all be classified as *inference duals* or *relaxation duals*, and in most cases as both.

Secondly, the two types of duals help elucidate how search, inference, and relaxation relate to one another: inference duality is a duality of search and inference, while relaxation duality is a duality of search and relaxation. Successful solution methods for combinatorial problems tend to be *primal–dual* methods, which move toward a solution of the original (primal) problem and a dual problem simultaneously. This book sees integrated methods as *primal–dual–dual* methods, which at least partially solve both the inference and relaxation duals as they solve the primal.

Inference duality arises as follows. An optimization problem can be seen as the problem of finding a set of values for the problem variables that minimize the objective function. But it can also be seen as the problem of inferring from the constraint set the tightest possible lower bound on the value of the objective function. In the first case, one searches over values of the variables, and in the second case, one searches over proofs. The problem of finding the proof that yields the best bound is the inference dual.

The precise nature of the inference dual depends on what inference method one uses to derive bounds. If the inference method is complete for the problem class in question, the inference dual has the same optimal value as the original problem. One particular inference method—nonnegative linear combination—yields the classical linear programming dual for linear problems and the surrogate dual for general inequality-constrained problems. A slightly different inference method gives rise to the all-important Lagrangean dual. In fact, the close connection between surrogate and Lagrangean duals, which are superficially unrelated, becomes evident when one regards them as inference duals. Still other inference methods yield superadditive and branching duals, which arise in integer programming and other combinatorial problems.

Inference duality is a unifying concept because, first of all, it can be defined for any optimization problem, not just the inequality-constrained problems traditionally studied in mathematical programming. Secondly, it can serve as a general basis for sensitivity analysis, which examines the sensitivity of the optimal solution to perturbations in the problem data, thus revealing which data must be accurate to get a meaningful solution. Most importantly, the proof that solves the inference dual is the source of nogoods or logic-based Benders cuts in a constraint-directed search method. Methods as disparate as Benders decomposition—a classic technique of operations research—and state-of-the-art propositional satisfiability methods with clause learning are closely related because they both generate nogoods by solving an inference dual. In principle, any inference dual can give rise to a nogood-based algorithm. The use of new inference duals in decomposition methods has resulted in computational speedups of several orders of magnitude.

Relaxation duality is a duality of search and relaxation, or more precisely a duality of restriction and relaxation. A motivation for solving a sequence of restrictions is that the restrictions are easier to solve than the original. Since relaxations are also designed to be easier than the original, one might ask whether a problem can be addressed by solving a sequence of relaxations. It can, if the relaxations are parameterized by *dual variables*, which allow one to search the space of relaxations by enumerating values of the dual variables. The solution of each relaxation provides a bound on the optimal value, and the problem of finding the best bound is the relaxation dual. In general, an enumeration of relaxations does not solve the problem, as does an enumeration of restrictions, because the best bound may not be equal to the optimal value of the original problem. The bound may nonetheless be useful, as in the surrogate and particularly Lagrangean duals, which were originally conceived as relaxation duals rather than inference duals.

1.4 Advantages of Integrated Methods

The academic literature tends to emphasize computation speed when evaluating a new approach to problem solving, perhaps because it is easily measured. Practitioners know, however, that model development time is often at least as important as solution time. This argues for the convenience of having all the modeling and algorithmic resources available in a single integrated system. One can try several approaches to a problem without having to learn several systems and port data between them. Metaconstraints can also be a significant time saver, as they lead to simpler models that are easier to build and maintain.

The computational advantages are there as well. Certainly, one need never pay a computational price for using an integrated system, because the traditional techniques can always be available as one of the options. However, experience confirms that a more broad-based strategy can substantially speed computation. This is borne out by a sampling of results from the literature. The focus here is on methods that integrate constraint programming (CP) and mixed-integer/linear programming (MILP), which can be classified roughly by the type of integration they use.

Many integrated methods combine CP with linear relaxations from MILP. One study, for example, combined a CP algorithm with an assignment problem relaxation (Section 7.11.2) and reduced-cost variable fixing (Sections 6.2.3, 6.3) to solve lesson timetabling problems 2 to 50 times faster than CP [202]. Another study, based on [413] and reimplemented in the integrated solver SIMPL [517], combined CP with convex hull relaxations to solve production planning problems with piecewise linear costs (Section 2.4) 20 to 120 times faster than state-of-the-art MILP software, generating search trees that were 1000–8000 times smaller. Linear relaxations combined with logic processing (Section 6.4) solved a boat party scheduling problem in five minutes that MILP could not solve in twelve hours, and solved flow shop instances four times faster than MILP [295].

Experiments have been conducted with other kinds of relaxations as well. A combination of CP and Lagrangean relaxation (Section 4.5) solved automatic digital recording problems one to ten times faster than MILP, which was faster than CP [448]. Another CP–Lagrangean hybrid achieved order-of-magnitude speedups on a radiation therapy problem [115]. CP assisted by semidefinite programming relaxations has been successfuly used for stable set problems and clustering problems on bipartite graphs [250, 485]. Logic processing and linear quasi-relaxations solved nonlinear structural design problems up to 600 times faster than MILP and solved two of the problems in less than six minutes when MILP could not solve them in 20 hours [104].

Branch-and-price integer programming methods (Section 5.1.6) have also been combined with CP processing, particularly in the area of airline and transit crew scheduling (such a problem is discussed in Section 5.1.7). In one study [519], this approach solved significantly larger urban transit crew management problems than traditional branch and price could solve. A CP-based branch-and-price method was the first to solve the eight-team traveling tournament problem [184].

Perhaps the greatest speedups have been achieved by integrating CP and MILP through generalized forms of Benders decomposition (Section 5.2.2). One study [300] solved minimum-cost machine allocation and scheduling problems 20 to 1000 times faster than CP or MILP. A subsequent study [473] improved upon these results by an additional factor of ten. SIMPL [517] brought further improvements, solving some problems in less than a second that were intractable for CP and MILP. A logic-based Benders approach solved single-machine scheduling problems with much longer time horizons than were tractable for either CP or MILP, even though the problem does not obviously decompose (the time horizon was broken into segments) [138, 139].

Other work has extended the applicability of logic-based Benders methods. One industrial implementation [475] solved, in ten minutes, polypropylene batch scheduling problems at BASF that were previously insoluble. A CP/MILP hybrid solved twice as many call center scheduling problems as traditional Benders [77]. A different CP/MILP hybrid solved planning and scheduling problems, with resource-constrained scheduling, 100 to 1000 times faster than CP or MILP when minimizing cost or makespan [283], 10 to 1000 times faster when minimizing the number of late jobs, and about 10 times faster (with much better solutions when optimality was not obtained) when minimizing total tardiness [285] (Section 2.8 shows how to solve a simplified minimum-makespan problem with a logic-based Benders technique). Finally, a hybrid Benders approach was applied [410] to obtain speedups of several orders of magnitude relative to the state of the art in sports scheduling.

It is important to bear in mind that none of these results were achieved with the full resources of integration. They are also preliminary results obtained with experimental codes. Integrated solution software will doubtless improve over time. The chief advantage of integrated methods, however, may be that they encourage a broader perspective on problem solving. This may inspire developments that would not have been possible inside individual disciplines.

1.5 Some Applications

Integrated methods have been successfully applied in a wide variety of contexts. The literature can again be roughly organized according to the type of integration used.

Applications that combine MILP/cutting planes with CP or logic-based methods include the orthogonal Latin squares problem [18], truss structure design [104], processing network design [245, 295], single-vehicle routing [432], resource-constrained scheduling [83, 170], multiple machine scheduling [102], shuttle transit routing [402], boat party scheduling [295], and the multidimensional knapsack problem [379]. Cutting planes for disjunctions of linear systems (Section 7.4.5) have been applied to factory retrofit planning, strip packing, and zero-wait job shop scheduling [435].

Convex hull relaxations for disjunctions of linear and nonlinear systems (Sections 7.4.1 and 7.5.1) have been used to solve several chemical process engineering problems [336, 337, 409, 436, 492]. Convex hull relaxations of piecewise linear constraints have been used in a CP context to solve fixed-charge problems and transportation problems with piecewise linear costs [413], as well as production planning problems with piecewise linear costs [380, 381].

Applications that combine CP with reduced-cost variable fixing include the traveling salesman problem with time windows [358], product configuration [358], fixed-charge network flows [317], and lesson timetabling [202].

CP-based Lagrangean methods have been applied to network design [153], automatic digital recording [448], traveling tournament problems [78], the resource-constrained shortest-path problem [219], the delivery of radiation therapy [115], and the general problem of filtering domains [313].

The most popular application of CP-based branch-and-price methods is to airline crew assignment and crew rostering [125, 190, 308, 321, 450]. Other applications include transit bus crew scheduling [519], aircraft scheduling [244], vehicle routing [431], network design [126], employee timetabling [171], physician scheduling [222], radiation therapy delivery [115], and the traveling tournament problem [183, 184, 481].

Benders methods that combine MILP with CP or logic-based methods have been developed for circuit verification problems [298], integer programming [131, 296] and the propositional satisfiability problem [279, 296]. A series of papers have described applications to planning and scheduling [132, 258, 279, 283, 285, 300]. Other applications include dispatching of automated guided vehicles [149], steel production scheduling [257], batch scheduling in a chemical plant [345], and polypropylene batch scheduling in particular [475]. CP-based Benders methods have also been applied to scheduling of computer processors [75, 76, 114], location-allocation problems [194], and traffic diversion problems [512]. Other applications of logic-based Benders include transportation network design [394], queuing design and control [471], and sports scheduling [129, 411].

1.6 Software

Methods that combine constraint programming and optimization have found their way into a number of software packages. Because software rapidly evolves, only a brief summary is given here. A more detailed survey can be found in [516].

The earliest integrated solver was the Prolog-based constraint logic programming system ECL^{i}PS^e [19, 428, 497]. It is now an open-source project, and the software continues to be used and maintained. It is organized around the concept of cooperating solvers. Constraint and LP solvers, for example, exchange information that reduce variable domains. In addition, column generation, Benders decomposition, and Lagrangean relaxation have been implemented in ECL^{i}PS^e.

IBM/ILOG's OPL Studio [327, 328, 382] provides a modeling language that invokes CP and MILP solvers. Models can be written for either type of solver, and a versatile script language allows users to run solvers iteratively and transfer information between them. This is convenient for logic-based Benders methods, for example. The CP modeling language is enhanced for scheduling applications and includes such novel features as interval-valued and conditional variables to represent operations that can be present or absent from the schedule.

Xpress-Mosel [142, 143] is a modeling language that allows low-level control of cooperating solvers in FICO's Xpress suite of mathematical modeling and optimization tools. These include CP, MILP, and nonlinear programming solvers. Mosel can implement a variety of hybrid methods (e.g., [102, 261, 432]) but, like OPL Studio, may require detailed coding to do so.

NICTA's G12 system [466] includes a solver-independent modeling language, Zinc; a mapping language, Cadmium; and an algorithmic language, Mercury (all elements of group 12 in the Periodic Table). Cadmium is used to generate a Mercury script from the Zinc model. The solution strategy can invoke a variety of solvers, including constraint programming, linear programming, and satisfiability (SAT) solvers. The user can experiment with different complete, local, or hybrid methods to solve the Zinc model.

The noncommercial system SIMPL [23, 517] provides a high-level modeling language that invokes integrated methods at the micro level, using the idea of constraint-based control. The modeling language contains meta-constraints that allow the solver to exploit problem substructure. The solver is organized around a single search-infer-and-relax loop that integrates inference and relaxation methods from CP and OR to suit the problem at hand, in an approach similar to that described in this book. The system therefore supports both branching and constraint-based search, including nogood generation and logic-based Benders methods.

SCIP [4, 5], developed at Zuse-Institut Berlin, accommodates constraint propagation within an efficient MILP solver ("constraint integer programming"). Its "constraint handler" and "plugin"-based architecture allows the user to supply filtering algorithms for specific constraint types. The system also supports nogood generation, with built-in conflict analysis for MILP.

There has been some investigation of integrated problem solving beyond the CP/MILP interface. The global optimization package BARON [468, 469] combines nonlinear (as well as linear) integer programming with CP-style domain reduction, although it uses a modeling system (AIMMS) that does not support CP-style constraints. The modeling language in the Comet system [267], which evolved from an earlier system, Localizer [357], allows CP and MILP constraints as well as high-level constraint-based specifications of local search. SALSA [329] is a language that can be used to design hybrids of global and local search algorithms. ToOLS [166] is a search specification language with a facility to invoke and combine existing algorithms. Numberjack [260] is a Python-based language for embedding CP, MILP and SAT technology into larger applications.

None of the existing systems fully integrate solution methods. Yet the necessary concepts and technology have reached a stage where a seamlessly integrated system is within reach. Perhaps the discussion to follow will help encourage efforts in this direction.

1.7 Plan of the Book

The remainder of the book begins by introducing the main ideas of the book in a series of examples (Chapter 2). Seven problems illustrate how models are constructed and how search, inference, and relaxation interact to solve them. One can get a very good idea of what integrated problem solving is all about by reading this chapter alone.

Chapter 3 provides necessary background in the basics of optimization, with brief treatments of linear programming, network flows, matching problems, optimality conditions for nonlinear programming, and deterministic dynamic programming. Chapter 4 then moves into the theory of integrated methods with an introduction to inference and relaxation duality, and their relation to problem solving and sensitivity analysis. Several specific types of duality are presented, including linear programming, surrogate, Lagrangean, subadditive, and branching duality.

The three longest chapters of the book—Chapters 5, 6, and 7—correspond to the three parts of the search-infer-and-relax framework. Chapter 5, on search, focuses successively on branching search, constraint-directed search, and local search. The section on branching search covers node and variable selection, cost-based branching, primal heuristics, and branch-and-price methods. Constraint-directed search includes logic-based Benders decomposition, nogood-directed branching in MILP and elsewhere, and partial-order dynamic backtracking. Particular attention is paid to conflict clause generation in algorithms for the propositional satisfiability problem.

Chapter 6, on inference, begins with some basic concepts of constraint programming, including k-consistency, domain consistency, and bounds consistency. It develops a theory of inference for inequality-constrained problems and propositional logic. The remainder of the chapter presents filtering methods for some popular global constraints, specifically element, all-different, cardinality, nvalues, among, sequence, stretch, regular, and circuit constraints. It concludes with bounds reduction algorithms for disjunctive and cumulative scheduling, which have contributed much to the success of constraint programming.

Chapter 7, on relaxation, has a stronger flavor of mathematical programming due to its use of mixed-integer modeling and cutting-plane theory. The chapter provides continuous relaxations for disjunctions of linear and nonlinear inequality systems, as well as for several global constraints. There is a certain parallelism between Chapters 6 and 7, in that both take the reader through linear and integer inequalities, propositional logic, and roughly the same set of global constraints. The main difference, of course, is that Chapter 6 presents inference methods for each of these constraint types, and Chapter 7 presents relaxation methods.

Chapter 8 is something of an appendix. It lists 48 metaconstraints, including a number of global constraints from the CP community, as a starting point for a menu of constraints in an integrated solver. When possible, it says something about usage, inference methods, relaxation methods, and related constraints, providing pointers to the literature and to relevant sections of the book.

1.8 Bibliographic Notes

Integrated methods have developed over the last 20 years or so in both the constraint programming (CP) and operations research (OR) communities. While a fuller history can be found elsewhere [281], a very brief synopsis might go as follows.

On the OR side, it is interesting that implicit enumeration [216], an early 1960s technique for integer programming, can be seen as anticipating the use of constraint propagation in an integer programming context. Constraint programming is explicitly mentioned in the OR literature as early as 1989 [111], but integrated methods were yet to develop.

The CP community began to investigate integrated methods in a serious way during the 1990s. They were initially conceived as double-modeling approaches, in which some constraints receive both CP and MILP formulations that exchange domain reduction and/or infeasibility information [340]. This mechanism was implemented in the constraint logic programming system ECL^{i}PS^e [428, 497]. The constraints community also began to recognize the parallel between constraint solvers and mixed-integer solvers, as evidenced by [101].

In later work, such OR ideas as reduced-cost variable fixing, linear relaxations of global constraints, and convex hull relaxations of piecewise linear functions were brought into CP-based algorithms [202, 204, 205, 380, 413, 414]. ILOG's OPL Studio [382] provided a modeling language that invokes CP and MILP solvers.

While this research was underway in CP, the OR community introduced hybrid methods as generalizations of branch and cut or a logic-based form of Benders decomposition. Integer variables were replaced with logical disjunctions and their relaxations as early as 1990 [50]. A series of papers appearing the 1990s integrated CP and logic-based methods with branch and cut [245, 276, 279, 295]. The logic-based Benders approach was developed during the same period, initially for circuit verification [298] and later as a general method [279, 296]. A Benders method that joins MILP and CP was proposed [279] and successfully implemented [300]. CP-based branch and price, a very different approach, was also developed [308, 518].

The First International Joint Workshop on AI and OR was organized in 1995 to provide an early forum for discussion of integrated methods. The idea was revived in 1999 with the annual CP-AI-OR workshop (Integration of AI and OR Techniques in CP for Combinatorial Optimization), now an annual conference series with published proceedings. Papers on hybrid methods regularly appear in CP and OR conferences.

Chapter 2
Examples

This chapter shows how an integrated approach to optimization plays itself out in some concrete cases. It presents several small examples that cover a wide range of application areas. It each case, it formulates a model that is appropriate for integrateld solution, and in most cases it carries the solution procedure to completion.

The first example—a simple freight transfer problem—demonstrates how inference methods from CP and relaxation methods from MILP can work together to accelerate branching search. A production planning example then shows how discrete choices can be formulated in an integrated modeling environment to result in much faster solution. An employee scheduling example demonstrates the modeling power of metaconstraints.

A fourth example shows how inference and relaxation can be profitably combined in continuous global optimization. A product configuration problem illustrates how high-level modeling can tell the solver how to combine inference and relaxation. A machine scheduling problem shows how integer programming and constraint programming can be integrated in a framework of logic-based Benders decomposition. A final and more complex example deals with communications network routing and frequency assignment. It combines elements of network flow programming and constraint programming using another form of decomposition.

The chapter begins with some basic definitions, followed a general description of the search-infer-and-relax solution process that is illustrated by the examples.

2.1 Basic Definitions

For the purposes of this book, an optimization problem can be written

$$\min \text{ (or max) } f(x)$$
$$\mathcal{C}(x) \tag{2.1}$$
$$x \in D$$

where $f(x)$ is a real-valued function of variable x and D is the *domain* of x. The function $f(x)$ is to be minimized (or maximized) subject to a set $\mathcal{C}$ of constraints, each of which is either satisfied or violated by any given $x \in D$. Generally, x is a tuple $(x_1, \ldots, x_n)$ and D is a Cartesian product $D_1 \times \cdots \times D_n$, where each $x_j \in D_j$. The notation $\mathcal{C}(x)$ means that x satisfies all the constraints in $\mathcal{C}$.

Adopting terminology from mathematical programming, any $x \in D$ is a *solution* of (2.1). Solution x is a *feasible* solution if $\mathcal{C}(x)$, and the *feasible set* of (2.1) is the set of feasible solutions. A feasible solution x^* is *optimal* if $f(x^*) \leq f(x)$ for all feasible x. An *infeasible* problem is one with no feasible solution. If (2.1) is infeasible, it is convenient to say that it has optimal value ∞ (or $-\infty$ in the case of a maximization problem). The problem is *unbounded* if there is no lower bound on $f(x)$ for feasible values of x, in which case the optimal value is $-\infty$ (or ∞ for maximization).

It is assumed throughout this book that (2.1) is either infeasible, unbounded, or has a finite optimal value. Thus, such problems as minimizing x subject to $x > 0$ are not considered. An optimization problem is considered to be *solved* when an optimal solution is found, or when the problem is shown to be unbounded or infeasible. In incomplete search methods that do not guarantee an optimal solution, the problem is solved when a solution is found that is acceptable in some sense, or when the problem is shown to be unbounded or infeasible.

A constraint G can be *inferred* from $\mathcal{C}$ if any $x \in D$ that satisfies $\mathcal{C}(x)$ also satisfies G. Equivalently, $\mathcal{C}$ *implies* G, or G is *valid* for $\mathcal{C}$. A *relaxation* R of the minimization problem (2.1) is obtained by dropping constraints and/or replacing the objective function $f(x)$ with a lower bounding function $f'(x)$. That is, any $x \in D$ that is feasible in (2.1) is feasible in R and satisfies $f'(x) \leq f(x)$. When one is minimizing, the optimal value of a relaxation is always a lower bound on the optimal value of the original problem.

2.2 The Solution Process

Search, inference, and relaxation interact to provide a general scheme for solving (2.1). The search procedure solves a series of *restrictions* or special cases of the problem. The best solution of a restriction is accepted as a solution of the original problem. Inference and relaxation provide opportunities to exploit problem structure, a key element of any successful approach to solving combinatorial problems. These ideas are more formally developed in Chapters 5, 6, and 7. Only a brief overview is given here.

Restrictions are obtained by adding constraints to the problem. The rationale for searching over restrictions is that they may be easier to solve than the original problem, even when there are many of them. Branching search methods, for example, divide the feasible set into smaller and smaller subsets by branching on alternatives. Each subset corresponds to a restriction of the original problem. Well-known examples of branching search include the branch-and-cut algorithms of MILP solvers and the branch-and-infer methods of CP solvers. Benders decomposition also enumerates restrictions by solving a sequence of subproblems in which certain variables are fixed. Local search methods examine a sequence of neighborhoods, each of which defines a restriction of the problem.

Search can often be accelerated by *inference*—that is, by inferring new constraints from the constraint set. The new constraints are then added to the problem, which may ease solution by describing the feasible set more explicitly. Common forms of inference are domain filtering in CP, cutting-plane generation in MILP, and Benders cuts in Benders decomposition. Inferred constraints may also result in a stronger relaxation, as in the case of cutting planes and Benders cuts.

Inference is most effective when it exploits problem structure. When a group of constraints has special characteristics, the model can indicate this by combining the constraints into a single *metaconstraint*, known as a *global constraint* in constraint programming. This allows inference algorithms to exploit the global structure of the group when filtering domains or generating valid cuts. Developing special-purpose inference procedures has been a major theme of research in CP.

Relaxation is the third element of the solution scheme. Like a restriction, a relaxation of the problem may be easier to solve than the original. For instance, an MILP problem becomes much easier to solve if one relaxes it by allowing the integer-valued variables to take any real value.

Solving a relaxation can be useful in several ways. Its solution may happen to be a solution of the original problem. Even if not, its solution may provide a clue to where one might find a solution of the original problem and therefore help guide the search. In addition, the optimal value of the relaxation provides a lower bound on the optimal value of the original problem (when one is minimizing). This is particularly useful in branching search, where bounds can often help prune the search tree.

Relaxation also provides a valuable opportunity to exploit special structure in individual constraints or groups of constraints, and this has been a perennial theme of the optimization literature. For example, strong cutting planes can be inferred from sets of inequalities that define certain types of polyhedra. One can use metaconstraints to indicate which inequalities have special structure.

2.3 Freight Transfer

A simple freight transfer problem illustrates, at an elementary level, how inference and relaxation can interact with branching search. Forty-two tons of freight must be conveyed overland. The shipper has a fleet of trucks in four sizes, with three vehicles of each size (Table 2.1). The eight available loading docks must accommodate all the trucks used, because the trucks must be loaded simultaneously. Due to the shape of the loading area, three loading docks must be allocated to the largest trucks even if only one or two of them are loaded. The problem is to select trucks to carry the freight at minimum cost.

Table 2.1 Data for a small instance of a freight transfer problem.

Truck type	Number available	Capacity (tons)	Cost per truck
1	3	7	90
2	3	5	60
3	3	4	50
4	3	3	40

2.3.1 Formulating the Problem

If variable x_i indicates the number of trucks of type i loaded, the requirement that 42 tons be transported can be written

$$7x_1 + 5x_2 + 4x_3 + 3x_4 \geq 42 \tag{2.2}$$

The loading dock constraints can be written

$$x_1 + x_2 + x_3 + x_4 \leq 8$$
$$(1 \leq x_1 \leq 2) \Rightarrow (x_2 + x_3 + x_4 \leq 5) \tag{2.3}$$

where $\Rightarrow$ means "implies." The problem can therefore be formulated

$$\text{integerLinear:} \begin{cases} \min\ 90x_1 + 60x_2 + 50x_3 + 40x_4 \\ 7x_1 + 5x_2 + 4x_3 + 3x_4 \geq 42 & (a) \\ x_1 + x_2 + x_3 + x_4 \leq 8 & (b) \end{cases} \tag{2.4}$$

$$\text{conditional: } (1 \leq x_1 \leq 2) \Rightarrow (x_2 + x_3 + x_4 \leq 5)$$

$$\text{domains: } x_i \in \{0, 1, 2, 3\},\ \ i = 1, \ldots, 4$$

In accord with the spirit of revealing problem structure, the constraints are grouped by type to provide the solver guidance as to how to process them. Since the objective function becomes an integer linear inequality whenever it is bounded, it is grouped with the integer linear constraints to form an integer linear metaconstraint. An optimal solution is $(x_1, \ldots, x_4) = (3, 2, 2, 1)$, with a minimum cost of 530.

The problem may be solved by a branching algorithm that uses inference and relaxation. The following sections first show how to carry out the inference and relaxation steps and then how to conduct the search.

2.3.2 Inference: Bounds Propagation

Domain filtering removes values from a variable domain when the variable cannot take those values in any feasible solution. The reduced domains inferred from one constraint can be used as a starting point for domain filtering in another constraint (a form of *constraint propagation*). One can, in principle, cycle through the constraints in this fashion until no further filtering is possible. This allows one to draw some inferences that do not follow from any single constraint, even though the constraints are processed individually.

A type of domain filtering known as *bounds propagation* is useful for the freight transport problem. Focusing first on constraint (2.2), one can "solve" it for variable x_1, for example, to obtain

$$x_1 \geq \frac{42 - 5x_2 - 4x_3 - 3x_4}{7} \geq \frac{42 - 5 \cdot 3 - 4 \cdot 3 - 3 \cdot 3}{7} = \frac{6}{7}$$

where the second inequality is due to the fact that each $x_j \in \{0, 1, 2, 3\}$. Because x_1 must be integral, one can round up the bound to obtain $x_1 \geq 1$. The domain of x_1 is therefore reduced from $\{0, 1, 2, 3\}$ to $\{1, 2, 3\}$. The same procedure can be applied to the other variables, but as it happens, no further domain reductions are possible. The smaller domain for x_1 can now be propagated to the second constraint (2.3), which by similar reasoning implies that $x_1 \leq 8$ and $x_j \leq 7$ for $j = 2, 3, 4$. Unfortunately, this does not further reduce any of the domains. If it did, one could cycle back to the first constraint and repeat the procedure.

If the reduced domain of x_1 were a subset of $\{1, 2\}$, one could infer from the implication constraint that $x_2 + x_3 + x_4 \leq 5$ and perhaps reduce domains further. This condition is not initially satisfied, but it may become satisfied during the search.

2.3.3 Inference: Valid Inequalities

In addition to deducing smaller domains, one can deduce *valid inequalities* or *cutting planes* from the knapsack constraints. The inferred inequalities are added to the original constraint set in order to produce a stronger continuous relaxation.

One type of cutting plane, known as a *general integer knapsack cut*, can be inferred from constraints like (2.2). Note that the first two terms $7x_1, 5x_2$ of (2.2) cannot by themselves satisfy the inequality, even if x_1 and x_2 are set to their largest possible value of 3. To satisfy the inequality, one must have

$$4x_3 + 3x_4 \geq 42 - 7 \cdot 3 - 5 \cdot 3$$

which implies

$$x_3 + x_4 \geq \left\lceil \frac{42 - 7 \cdot 3 - 5 \cdot 3}{4} \right\rceil = \left\lceil \frac{6}{4} \right\rceil = 2$$

The inequality $x_3 + x_4 \geq 2$ is an integer knapsack cut.

Because the first two terms of (2.2) cannot satisfy the inequality by themselves, the index set $\{1, 2\}$ is a *packing*. In fact, it is a *maximal packing*, because no proper superset is a packing. There are four maximal packings for (2.2), each of which gives rise to an integer knapsack cut:

$$
\begin{array}{lll}
\{1, 2\}: & x_3 + x_4 \geq 2 & \\
\{1, 3\}: & x_2 + x_4 \geq 2 & \\
\{1, 4\}: & x_2 + x_3 \geq 3 & (2.5) \\
\{2, 3, 4\}: & x_1 \geq 1 &
\end{array}
$$

These cuts are implied by (2.2) because any integral $x = (x_1, \ldots, x_4)$ that satisfies (2.2) must satisfy (2.5). Nonmaximal packings also give rise to integer knapsack cuts, and they may be nonredundant. For example, the packing $\{2, 3\}$ produces the cut $x_1 + x_4 \geq 3$, which is not redundant of (2.5) or any other knapsack cut.

Knapsack cuts can sometimes be strengethened using domains. For example, the packing $\{2\}$ gives rise to the cut $x_1 + x_3 + x_4 \geq 4$, which can be strengthened to $x_1 + x_3 + x_4 \geq 5$. This is because $x_1 \leq 3$, and as a result the largest possible left-hand side of (2.2) when $x_1 + x_3 + x_4 = 4$ is 40. It is therefore neessary to have $x_1 + x_3 + x_4 \geq 5$.

The fourth cut in (2.5) duplicates the bound $x_1 \geq 1$ already obtained from bounds propagation. In fact, it is easy to see that any propagated bound $x_j \geq L$ can be obtained from the knapsack cut corresponding to the maximal packing that consists of the remaining variables. Knapsack cuts corresponding to maximal packings therefore dominate propagated bounds. However, it is much more costly to generate all maximal packings than to propagate bounds.

2.3.4 Relaxation: Linear Programming

A continuous relaxation of the problem instance (2.4) can be obtained from the inequality constraints by allowing each variable x_i to take any real value in the range between its lowest and highest value. One can also add the knapsack cuts (2.5) before relaxing the problem. This yields the relaxation below:

$$
\text{linear:} \begin{cases}
\min \; 90x_1 + 60x_2 + 50x_3 + 40x_4 \\
7x_1 + 5x_2 + 4x_3 + 3x_4 \geq 42 & (a) \\
x_1 + x_2 + x_3 + x_4 \leq 8 & (b) \\
x_3 + x_4 \geq 2 & (c) \\
x_2 + x_4 \geq 2 & (d) \\
x_2 + x_3 \geq 3 & (e) \\
L_i \leq x_i \leq U_i, \;\; i = 1, \ldots, 4
\end{cases} \quad (2.6)
$$

$$
\text{domains:} \; x_i \in \mathbb{R}, \; i = 1, \ldots, 4
$$

where initially $(L_1, U_1) = (1, 3)$ and $(L_i, U_i) = (0, 3)$ for $i = 2, 3, 4$. This relaxed problem can be easily solved by linear programming. An optimal solution is $x = (x_1, \ldots, x_4) = (2\frac{1}{3}, 3, 2\frac{2}{3}, 0)$, which has cost $523\frac{1}{3}$. This solution is not feasible in the original problem because it is not integral, but it provides a lower bound on the optimal cost. The freight cannot be transported for less than $523\frac{1}{3}$.

The knapsack cuts (c)–(e) in (2.6) make the relaxation tighter because they "cut off" solutions that satisfy the other constraints. For example, the solution $x = (3, 3, 1.5, 0)$ satisfies the other constraints but violates (c). As it happens, these particular knapsack cuts do not improve the lower bound, because the optimal value of the relaxation is $523\frac{1}{3}$ without them. Yet if the knapsack cut $x_1 + x_4 \geq 3$ had been included in the relaxation, it would have cut off the solution $x = (2\frac{1}{3}, 3, 2\frac{2}{3}, 0)$ and provided the tighter bound of 525.

2.3.5 Branching Search

A search tree for problem instance (2.4) appears in Figure 2.1. Each node of the tree below the root corresponds to a restriction of the original problem. A restriction is processed by applying inference methods (bounds propagation and cut generation), then solving a continuous relaxation, and finally branching if necessary.

Maximal packings are generated only at the root node, due to the cost of identifying them at every node. Thus the same three knapsack cuts appear in the relaxation at every node of the tree.

Bounds propagation applied to the original problem at the root node reduces the domain of x_1 to $\{1, 2, 3\}$, as described earlier. Figure 2.1 shows the resulting domains in braces. Next, three knapsack cuts (2.5) are generated (note that the knapsack cut $x_1 \geq 1$ is already implicit

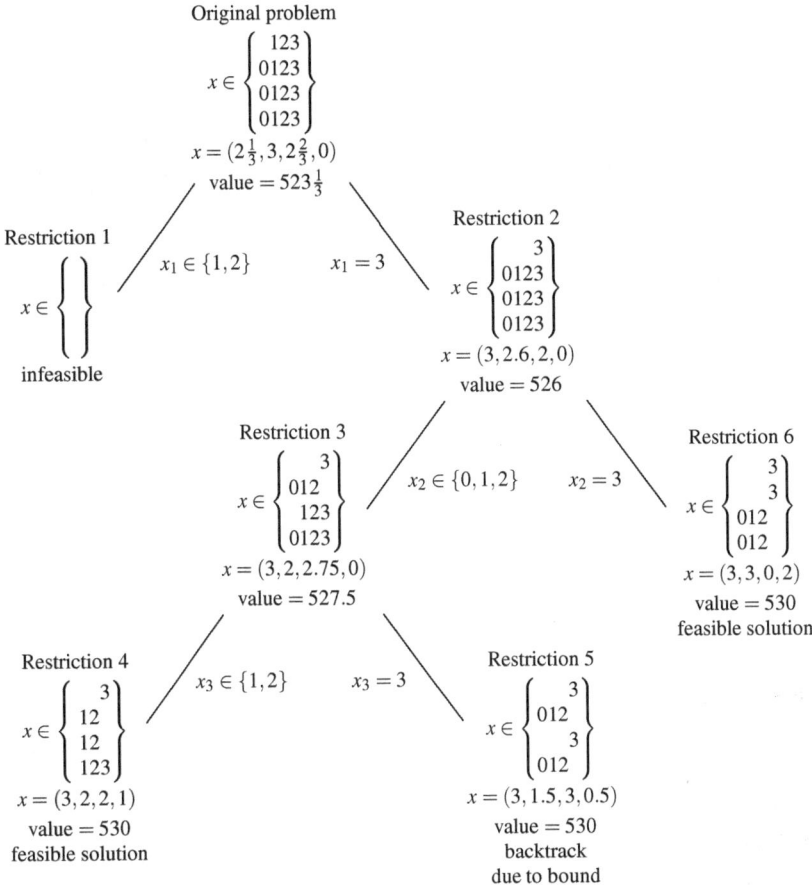

Fig. 2.1 Branch-and-relax tree for the freight transfer problem. Each node of the tree shows, in braces, the filtered domains after domain reduction. The rows inside the braces correspond to $x_1, \ldots, x_4$. The solution of the continuous relaxation appears immediately below the domains.

in x_1's domain). They are added to the constraint set in order to obtain the continuous relaxation (2.6), whose optimal solution $x = (2\frac{1}{3}, 3, 2\frac{2}{3}, 0)$ is shown at the root node.

This solution is infeasible in the original problem because x_1 and x_3 do not belong to their respective domains (they are nonintegral). It is therefore necessary to branch on one of the domain constraints $x_1 \in \{1, 2, 3\}$, $x_3 \in \{0, 1, 2, 3\}$. Branching on the first splits the domain into $\{1, 2\}$ and $\{3\}$, as the solution value $2\frac{1}{3}$ of x_1 lies between 2 and 3. This generates restrictions 1 and 2.

The tree can be traversed in a depth-first manner. Moving first to restriction 1, where x_1's domain is $\{1, 2\}$, bounds propagation applied to the original inequality constraint yields domain $\{2, 3\}$ for x_2 and $\{1, 2, 3\}$ for x_3 and x_4. Because the domain of x_1 activates the conditional constraint, one can infer $x_2 + x_3 + x_4 \leq 5$ and further reduce the domains of x_3 and x_4 to $\{1, 2\}$. One more round of propagation on these two inequalities reduces all domains to the empty set. Restriction 1 is therefore infeasible, and there is no need to solve the relaxation or to branch.

Moving now to restriction 2, no further domain filtering is possible, and solution of the relaxation (2.6) yields $x = (3, 2.6, 2, 0)$. Branching on x_2 creates restrictions 3 and 6.

Continuing in a depth-first manner, restriction 3 is processed next. Inference yields the domains shown, and branching on x_3 produces restrictions 4 and 5. The continuous relaxation of restriction 4 has the integral solution $x = (3, 2, 2, 1)$, which is feasible in the restriction. It becomes the *incumbent* solution (the best feasible solution so far), and no branching is necessary.

Restriction 5 is processed next. Here, the relaxation has a nonintegral solution, but its optimal value 530 is no better than the value of the incumbent solution. Since 530 is a lower bound on the optimal value of any further restriction of restriction 5, there is no need to branch. The tree is therefore "pruned" at restriction 5, and the search proceeds to restriction 6. A branching search in which the tree is pruned in this manner is called *branch and bound*.

The continuous relaxation of restriction 6 has an integral solution, and there is no need to branch, thus completing the search. Because the solution is no better than the incumbent (in fact it is equally good), it and the incumbent solution $x = (3, 2, 2, 1)$ are optimal. The minimum cost of transporting the freight is 530.

Inference and relaxation work together in this example to reduce the solution time. Solutions of relaxations help to guide the branching. At restriction 1, inference alone proves infeasibility, and there is no need to branch. As it happens, solving the relaxation at this node would also prove infeasibility, but due to inference there is no need to incur the greater overhead of solving a relaxation. At restrictions 4 and 6, the relaxation (with the help of prior domain reduction) obtains feasible solutions, and no branching is necessary. If inference alone were used, one would be able to find feasible solutions only by branching until all the domains are singletons. At restriction 5, the relaxation provides a bound that again obviates the necessity of further branching.

Exercises

2.1. Apply domain propagation to the inequalities

$$5x_1 + 4x_2 + 3x_3 \geq 18$$
$$2x_1 + 3x_2 + 4x_3 \leq 10$$

with initial domains $D_i = \{0, 1, 2\}$ for $i = 1, 2, 3$. Cycle through the inequalities until no further domain reduction is possible.

2.2. Identify all packings for inequality (2.2) using domains $x_j \in \{0, 1, 2, 3\}$, and write the corresponding knapsack cuts. *Hints.* Cuts corresponding to maximal packings appear in (2.5), but there are also nonmaximal packings. In addition, some of the cuts can be strengthened using domains.

2.3. Solve the problem of minimizing $3x_1 + 4x_2$ subject to $2x_1 + 3x_2 \geq 10$ and $x_1, x_2 \in \{0, 1, 2, 3\}$ using branch and relax without bounds propagation. Now, solve it using branch and infer without relaxation. Finally, solve it using branch and relax with propagation. Which results in the smallest search tree? (When using branch and infer, follow the common constraint programming practice of branching on the variable with the smallest domain.)

2.4. Write two integer linear inequalities (with initial domains specified) for which bounds propagation reduces domains more than minimizing and maximizing each variable subject to a continuous relaxation of the constraint set.

2.5. Write two integer linear inequalities (with domains specified) for which minimizing and maximizing each variable subject to a continuous relaxation of the constraint set reduces domains more than bounds propagation.

2.4 Production Planning

A very simple production planning problem illustrates how logical and continuous variables can interact. A manufacturing plant has three operating modes, each of which imposes different constraints. The objective is to decide in which mode to run the plant, and how much of each of two products to make, so as to maximize net income.

2.4.1 Formulating the Problem

Let x_A, x_B be the production levels of products A and B, respectively. In mode 0 the plant is shut down, and $x_A = x_B = 0$. In mode 1, it incurs a fixed cost of 35, and the production levels must satisfy the constraint $2x_A + x_B \leq 10$. Mode 2 incurs a fixed cost of 45 and the constraint $x_A + 2x_B \leq 10$ is imposed. The company earns a net income of 5 for each unit of product A manufactured, and 3 for each unit of product B.

A natural modeling approach is to let a Boolean variable δ_k be true when the plant runs in mode k. Then, the model is immediate:

$$\text{linear: max } 5x_A + 3x_B - f$$

$$\text{logic: } \delta_0 \vee \delta_1 \vee \delta_2$$

$$\text{conditional: } \begin{cases} \delta_0 \Rightarrow (x_A = x_B = f = 0) \\ \delta_1 \Rightarrow (2x_A + x_B \leq 10, \ f = 35) \\ \delta_2 \Rightarrow (x_A + 2x_B \leq 10, \ f = 45) \end{cases} \tag{2.7}$$

$$\text{domains: } x_A, x_B \geq 0, \ \delta_k \in \{\text{true}, \text{false}\}, \ k = 0, 1, 2$$

where variable f represents the fixed cost. The logic constraint means that at least one of the three Boolean variables must be true.

2.4.2 Relaxation

The model has an interesting relaxation because each production mode k enforces constraints that define a polyhedron in the space of the continuous variables x_A, x_B, f. So the projection of the feasible set onto this space is the union of the three polyhedra, illustrated in Figure 2.2. The best possible continuous relaxation of this set is its *convex hull*, which is itself a polyhedron and is also shown in the figure. The convex hull of a set is union of all line segments connecting any two points of the set.

There is a general procedure, given in Section 7.4.1, for writing a linear constraint set that describes the convex hull of a disjunction of linear systems. In this case, the convex hull description is

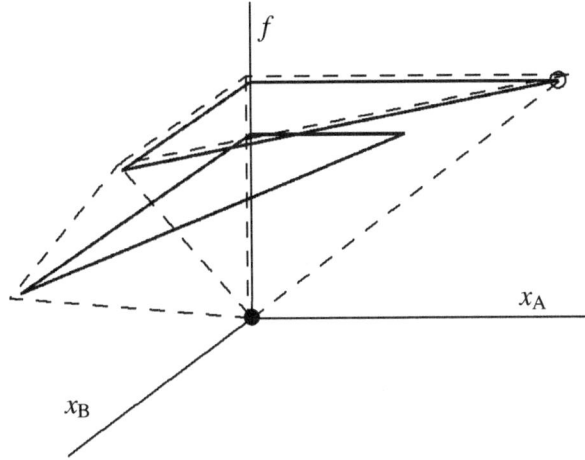

Fig. 2.2 Feasible set (two triangular areas and black circle) of a production planning problem, projected onto the space of the continuous variables. The convex hull is the volume inside the dashed polyhedron. The open circle marks the optimal solution.

$$2x_{A1} + x_{B1} \leq 10y_1$$
$$x_{A2} + 2x_{B2} \leq 10y_2$$
$$f \geq 35y_1 + 45y_2$$
$$x_A = x_{A1} + x_{A2}, \quad x_B = x_{B1} + x_{B2} \tag{2.8}$$
$$y_0 + y_1 + y_2 = 1, \quad y_k \geq 0, \ k = 0, 1, 2$$
$$x_{Ak} \geq 0, \ x_{Bk} \geq 0, \ k = 1, 2$$

where the variables y_k correspond to the Boolean variables δ_k, and fixing $\delta_k =$ true is equivalent to setting $y_k = 1$. Since there are new variables x_{Ak}, x_{Bk}, y_k in the relaxation, one should, strictly speaking, say that it describes a set whose projection onto x_A, x_B, f is the convex hull of (2.7)'s feasible set. A lower bound on the optimal value can now be obtained by minimizing $5x_A + 3x_B - f$ subject to (2.8).

To take advantage of this relaxation, the solver must somehow recognize that the feasible set in continuous space is a union of polyhedra. This can be done by collecting the constraints of (2.7) into a single *linear disjunction*, resulting in the model:

linear: max $5x_A + 3x_B - f$

linearDisjunction:

$$\begin{bmatrix} \delta_0 \\ x_A = x_B = 0 \\ f = 0 \end{bmatrix} \vee \begin{bmatrix} \delta_1 \\ 2x_A + x_B \le 10 \\ f = 35 \end{bmatrix} \vee \begin{bmatrix} \delta_2 \\ x_A + 2x_B \le 10 \\ f = 45 \end{bmatrix}$$

domains: $x_A, x_B \ge 0$, $\delta_k \in \{\text{true}, \text{false}\}$, $k = 0, 1, 2$

The linear disjunction has precisely the same meaning as the logic and conditional constraints of (2.7). Its presence tells the solver to formulate a relaxation for the union of polyhedra described by its three disjuncts. In general, the user should be alert for metaconstraints that can exploit problem structure, although in this case an intelligent modeling system could automatically rewrite (2.7) as (2.8).

2.4.3 Branching Search

The problem can now be solved by branching on the Boolean variables. The relaxation of the problem at the root node has solution $(y_1, y_2) = (0, 1)$, with $(x_A, x_B, f) = (10, 0, 35)$. Because this solution is feasible in (2.7) with $(\delta_0, \delta_1, \delta_2) = (\text{false}, \text{false}, \text{true})$, there is no need to branch. One should operate the plant in mode 2 and make product A only.

In this problem, there is no branching because (2.8) is a convex hull relaxation not only for the corresponding linear disjunction but for the entire problem. In more complex problems, an auxiliary variable y_k may take a fractional value, in which case one can branch by setting $\delta_k = \text{true}$ and $\delta_k = \text{false}$.

2.4.4 Inference

Inference plays a role in such problems when the logical conditions become more complicated. Suppose, for example, that plant 1 can operate in two modes (indicated by Boolean variables δ_{1k} for $k = 0, 1$) and plant 2 in three modes (indicated by variables δ_{2k}). There is also a rule that if plant 1 operates in mode 1, then plant 2 cannot operate in mode 2. The model might be as follows (where $\neg$ means *not*).

linear: max cx

logic: $\begin{cases} \delta_{10} \vee \delta_{11} \\ \delta_{20} \vee \delta_{21} \vee \delta_{22} \\ \delta_{11} \rightarrow \neg\delta_{22} \end{cases}$

conditional: $\delta_{ik} \Rightarrow A^{ik}x \geq b^{ik}$, all i, k

domains: $x \geq 0$, $\delta_{ik} \in \{\text{true, false}\}$

To extract as many linear disjunctions as possible, one can use the *resolution* algorithm (discussed in Section 6.4.2) to compute the *prime implications* of the logical formulas in the above model:

$$
\begin{array}{ll}
\delta_{10} \vee \delta_{11} & \delta_{10} \vee \neg\delta_{22} \\
\delta_{20} \vee \delta_{21} \vee \delta_{22} & \neg\delta_{11} \vee \delta_{20} \vee \delta_{21} \\
\neg\delta_{11} \vee \neg\delta_{22} & \delta_{10} \vee \delta_{20} \vee \delta_{21}
\end{array}
$$

These are the undominated disjunctions implied by the logical formulas. Three of the prime implications contain no negative terms, and they provide the basis for linear disjunctions. The model becomes:

linear: max cx

logic: $\begin{cases} \neg\delta_{11} \vee \neg\delta_{22} \\ \delta_{10} \vee \neg\delta_{22} \\ \neg\delta_{11} \vee \delta_{20} \vee \delta_{21} \end{cases}$

linear disjunction:

$$
\bigvee_{k \in \{0,1\}} \begin{bmatrix} \delta_{1k} \\ A^{1k}x \geq b^{1k} \end{bmatrix}
$$

$$
\bigvee_{k \in \{0,1,2\}} \begin{bmatrix} \delta_{2k} \\ A^{2k}x \geq b^{2k} \end{bmatrix}
$$

$$
\begin{bmatrix} \delta_{10} \\ A^{10}x \geq b^{10} \end{bmatrix} \vee \begin{bmatrix} \delta_{20} \\ A^{20}x \geq b^{20} \end{bmatrix} \vee \begin{bmatrix} \delta_{21} \\ A^{21}x \geq b^{21} \end{bmatrix}
$$

domains: $x \geq 0$, $\delta_{ik} \in \{\text{true, false}\}$

Again, an intelligent modeling system can carry out this process automatically.

Inference plays a further role as branching proceeds. When branching fixes a Boolean variable to true or false, it may be possible to deduce that some other variables must be true or false, thus reducing their domains to a singleton. For example, if δ_{10} is fixed to false, then δ_{11} must be true and δ_{22} false. The resolution method draws all such inferences.

Exercises

2.6. Formulate the following problem with the help of conditional constraints. A lumber operation wishes to build temporary roads from forests 1 and 2 to sawmills A and B. Due to topography and environmental regulations, roads can be built only in certain combinations: (a) from 1 to A, 1 to B, and 2 to B; (b) from 1 to A, 1 to B, and 2 to A; (c) from 1 to A, 2 to A, and 2 to B. Each sawmill j requires d_j units of timber, the unit cost of shipping timber over a road from forest i to sawmill j is c_{ij}, and the fixed cost of a road from i to j is f_{ij}. Choose which roads to build, and how much timber to transport over each road, so as to minimize cost while meeting the sawmill demands. Now, formulate the problem without conditional constraints and with a disjunctive linear constraint.

2.7. Consider the problem

$$\text{linear: min } cx$$
$$\text{logic:} \begin{cases} y_1 \vee y_2 \\ y_1 \rightarrow y_3 \end{cases}$$
$$\text{conditional: } y_k \Rightarrow \left(A^k x \geq b^k \right), \ k = 1, 2, 3$$

where each y_k is a Boolean variable. Write the problem without conditional constraints and using as many linear disjunctions as possible. *Hint:* The formula $y_1 \rightarrow y_3$ is equivalent to $\neg y_1 \vee y_3$.

2.5 Employee Scheduling

An employee scheduling problem is useful for introducing several of the modeling and inference techniques that have been developed in the constraint programming field. Since the objective is to find a feasible rather than an optimal solution, relaxation bounds do not play a role. Nonetheless, the integration of technologies is a key ingredient of the solution method, partly because the filtering algorithms rely on such optimization techniques as maximum flow algorithms and dynamic programming.

A certain hospital ward requires that a head nurse be on duty seven days a week, twenty-four hours a day. There are 3 eight-hour shifts, and on a given day each shift must be staffed by a different nurse. The schedule must be the same every week. Four nurses (denoted A, B, C, and D) are available, all of whom must work at least five days a week. Since there are 21 eight-hour periods a week, this implies that three nurses will work five days and one will work six days. For continuity, no shift should be staffed by more than two different nurses during the week.

Two additional rules reduce the burden of adjusting to shift changes. No employee is asked to work different shifts on two consecutive days; there must be at least one day off in between. Also, an employee who works shift 2 or 3 must do so at least two days in a row. Shift 1 is the daytime shift and requires less adjustment.

2.5.1 Formulating the Problem

There are two natural ways to think about an employee schedule. One, shown in Table 2.2, is to view it as assigning a nurse to each shift on each day. Another is to see it as assigning a shift (or day off) to each nurse on each day; this is illustrated by Table 2.3, in which shift 0 represents a day off. Either viewpoint gives rise to a different problem formulation, neither of which is convenient for expressing all the con-

Table 2.2 Employee scheduling viewed as assigning workers to shifts.

	Sun	Mon	Tue	Wed	Thu	Fri	Sat
Shift 1	A	B	A	A	A	A	A
Shift 2	C	C	C	B	B	B	B
Shift 3	D	D	D	D	C	C	D

Table 2.3 Employee scheduling viewed as assigning shifts to workers.

	Sun	Mon	Tue	Wed	Thu	Fri	Sat
Worker A	1	0	1	1	1	1	1
Worker B	0	1	0	2	2	2	2
Worker C	2	2	2	0	3	3	0
Worker D	3	3	3	3	0	0	3

straints of the problem. Fortunately, there is no need to choose between them. One can use both formulations in the same model and connect them with *channeling constraints*. This not only accommodates all the constraints but also makes propagation more effective.

To proceed with the first formulation, let variable w_{sd} be the nurse that is assigned to shift s on day d. Three of the scheduling requirements are readily expressed in this formulation, using metaconstraints that are well known to constraint programmers. One is the *all-different* constraint, alldiff(X), where X denotes a set $\{x_1, \ldots, x_n\}$ of variables. It simply requires that $x_1, \ldots, x_n$ take different values. It can express the requirement that three different nurses be scheduled each day:

$$\text{alldiff}(w_{\cdot d}), \text{ all } d$$

The notation $w_{\cdot d}$ refers to $\{w_{1d}, w_{2d}, w_{3d}\}$.

The *cardinality* constraint can be used to require that every nurse be assigned at least five days of work. The constraint in general is written

$$\text{cardinality}(X \mid v, \ell, u)$$

where $X = \{x_1, \ldots, x_n\}$ and v is an m-tuple $(v_1, \ldots, v_m)$ of values. $\ell = (\ell_1, \ldots, \ell_m)$ and $u = (u_1, \ldots, u_m)$ contain lower and upper bounds respectively. The vertical bar in the argument list indicates that everything before the bar is a variable and everything after is a parameter. The constraint requires, for each $i = 1, \ldots, m$, that at least ℓ_i and at most u_i of the variables in X take the value v_i. To require that each nurse work at least five and at most six days, one can write

$$\text{cardinality}(w_{\cdot\cdot} \mid (A, B, C, D), (5, 5, 5, 5), (6, 6, 6, 6))$$

where $w_{\cdot\cdot}$ refers to the set of all the variables w_{sd}.

The *nvalues* constraint is written

$$\text{nvalues}(X \mid \ell, u)$$

and requires that the variables in $X = \{x_1, \ldots, x_n\}$ take at least ℓ and at most u different values. To require that at most two nurses work any given shift, one can write

$$\text{nvalues}(w_{s\cdot} \mid 1, 2), \text{ all } s$$

Both nvalues and cardinality generalize the alldiff constraint, but one should use alldiff when possible, since it invokes a more efficient filtering algorithm.

The remaining constraints are not easily expressed in the notation developed so far, because they relate to the pattern of shifts worked by a given nurse. For this reason, it is useful to move to the formulation suggested by Table 2.3. Let y_{id} be the shift assigned to nurse i on day d, where shift 0 denotes a day off. It is first necessary to ensure that all three shifts be assigned for each day. The alldiff constraint serves the purpose:

$$\text{alldiff}(y_{\cdot d}), \text{ all } d$$

This condition is implicit in the constraints already written, but it is generally good practice to write redundant constraints when one is aware of them, in order to strengthen propagation.

The *stretch* constraint was expressly developed to impose conditions on stretches or contiguous sequences of shift assignments in employee scheduling problems. Given a tuple $x = (x_1, \ldots, x_n)$ of variables, a stretch is a maximal sequence of consecutive variables that take the same value. Thus, if $x = (x_1, x_2, x_3) = (a, a, b)$, x contains a stretch of value a having length 2 and a stretch of b having length 1, but it does not contain a stretch of a having length 1. The stretch constraint is written

$$\text{stretch}(x \mid v, \ell, u, P)$$

where $x = (x_1, \ldots, x_n)$, while v, ℓ, and u are defined as in the cardinality constraint. P is a set of *patterns*, each of which is a pair (v, v') of distinct values. The stretch constraint requires, for each v_i in v, that every stretch of value v_i have length at least ℓ_i and at most u_i. It also requires that whenever a stretch of value v comes immediately before a stretch of value v', the pair (v, v') must occur in the pattern set P. The requirements concerning consecutive nursing shifts can now be written

$$\text{stretchCycle}(y_{i\cdot} \mid (2,3), (2,2), (6,6), P), \text{ all } i$$

where P contains all patterns that include a day off:

$$P = \{(s, 0), (0, s) \mid s = 1, 2, 3\}$$

A cyclic version of the constraint is necessary because every week must have the same schedule. The cyclic version treats the week as a cycle and allows a single stretch to extend across the weekend.

Finally, the two formulations must be forced to have the same solution. This is accomplished with *channeling constraints*, which in this case take the form

$$w_{y_{id}d} = i, \text{ all } i, d \tag{2.9}$$

and

$$y_{w_{sd}s} = s, \text{ all } s, d \tag{2.10}$$

Constraint (2.9) says that on any given day d, the nurse assigned to the shift to which nurse i is assigned must be nurse i, and similarly for (2.10). The subscripts y_{id} in (2.9) and w_{sd} in (2.10) are *variable indices* or *variable subscripts*.

The model is now complete. It can be written with five metaconstraints, plus domains:

alldiff: $\left\{ \begin{matrix} (w_{\cdot d}) \\ (y_{\cdot d}) \end{matrix} \right\}$, all d

cardinality: $(w_{\cdot \cdot} \mid (A, B, C, D), (5, 5, 5, 5), (6, 6, 6, 6))$

nvalues: $(w_{s\cdot} \mid 1, 2)$, all s

stretchCycle: $(y_{i\cdot} \mid (2, 3), (2, 2), (6, 6), P)$, all i $\qquad$ (2.11)

linear: $\left\{ \begin{matrix} w_{y_{id}d} = i, \text{ all } i \\ y_{w_{sd}d} = s, \text{ all } s \end{matrix} \right\}$, all d

domains: $\left\{ \begin{matrix} w_{sd} \in \{A, B, C, D\}, \ s = 1, 2, 3 \\ y_{id} \in \{0, 1, 2, 3\}, \ i = A, B, C, D \end{matrix} \right\}$, all d

The linear constraints are so classified because they are linear aside from the variable indices, which will be eliminated as described below. One can appreciate the convenience of metaconstraints by attempting to formulate this problem with, say, an MILP model.

2.5.2 Inference: Domain Filtering

The alldiff constraint poses an interesting filtering problem that, fortunately, can be easily solved. Suppose, for example, that the current domains of assignment variables w_{s1} are:

$$w_{11} \in \{A, B\}, \ w_{21} \in \{A, B\}, w_{31} \in \{A, B, C, D\}$$

The days are numbered $1, \ldots, 7$ so that the subscript 1 in w_{s1} refers to Sunday. Thus, only nurse A or B can be assigned to shift 1 or 2 on Sunday, while any nurse can be assigned to shift 3. Since

$$\text{alldiff}(w_{11}, w_{21}, w_{31})$$

must be enforced, one can immediately deduce that neither A nor B can be assigned to shift 3. Thus, the domain of w_{31} can be reduced to $\{C, D\}$. This type of reasoning can be generalized by viewing the solution of alldiff as a matching problem on a bipartite graph, for which there are verys fast algorithms. Optimality conditions for maximum cardinality matching allow one to identify values that can be removed from domains. These ideas are presented in Section 6.8. In a similar fashion, network flow models can provide filtering for the cardinality and nvalues constraints.

Filtering for the stretch constraint is more complicated, but nonetheless tractable. Suppose, for example, that the domains of y_{Ad} contain the values listed beneath each variable:

y_{A1}	y_{A2}	y_{A3}	y_{A4}	y_{A5}	y_{A6}	y_{A7}
0	0		0		0	0
1			1	1	1	1
2	2	2		2		2
3	3	3	3		3	3

$$(2.12)$$

This means, for instance, that nurse A must work either shift 2 or shift 3 on Tuesday. After some thought, one can deduce from the stretch constraint in (2.11) that several values can be removed, resulting in the following domains:

y_{A1}	y_{A2}	y_{A3}	y_{A4}	y_{A5}	y_{A6}	y_{A7}
0			0		0	0
				1	1	1
2	2	2				2
3	3	3				3

$$(2.13)$$

Note that two variables are fixed. A polynomial-time dynamic programming algorithm can remove all infeasible values for any given stretch constraint. It is described in Section 6.11.

2.5.3 Inference for Variable Indices

The variably indexed expression $w_{y_{id}d}$ in (2.11) can be processed with the help of an *element* constraint. Domain reduction algorithms for this constraint are well known in the constraint programming community, and elementary polyhedral theory provides a continuous relaxation for the constraint.

The type of element constraint required here has the form

$$\text{element}(y, x, z) \tag{2.14}$$

where y is an integer-valued variable, x is a tuple $(x_1, \ldots, x_m)$ of variables, and z is a variable. The constraint requires that z be equal to the yth variable in the list $x_1, \ldots, x_m$. One can therefore deal with a variably indexed expression like x_y by replacing it with z and adding the element constraint (2.14). In particular, the constraint $w_{y_{id}d} = i$ is parsed by replacing it with $z_{id} = i$ and generating the constraint

$$\text{element}\,(y_{id}, (w_{0d}, \ldots, w_{3d}), z_{id})$$

Similarly, $y_{w_{sd}d} = s$ is replaced with $\bar{z}_{sd} = s$ and

$$\text{element}\,(w_{sd}, (y_{Ad}, \ldots, y_{Dd}), \bar{z}_{sd})$$

Filtering for element is a simple matter. An example will illustrate this and show how propagation of channeling constraints between two models can reduce domains. Focusing on Sunday (day 1), suppose the domains of w_{s1} and y_{i1} contain the elements listed beneath each variable:

w_{01}	w_{11}	w_{21}	w_{31}	y_{A1}	y_{B1}	y_{C1}	y_{D1}
A		A	A	0	0		0
B	B		B	1	1	1	
C	C	C			2	2	2
	D	D	D	3		3	3

Suppose that no further propagation is possible among the constraints involving only the variables w_{sd}, and similarly for the constraints involving only the variables y_{id}. Nonetheless, the channeling constraints can yield further domain reduction. For instance, since y_{A1} has domain $\{0, 1\}$, the constraint

$$\text{element}\,(y_{A1}, (w_{01}, \ldots, w_{31}), z_{A1})$$

implies that y_{A1} must select either w_{01} or w_{11} to be equated with z_{A1}. But $z_{A1} = A$, and only w_{01}'s domain contains A. Thus, y_{A1} must select

w_{01}, and y_{A1}'s domain can be reduced to $\{0\}$. Similarly, the constraints $\bar{z}_{01} = 0$ and

$$\text{element}\,(w_{01}, (y_{A1}, \ldots, y_{D1}), \bar{z}_{01})$$

remove C from the domain $\{A, B, C\}$ of w_{01}. Deductions of this kind reduce the domains to:

w_{01}	w_{11}	w_{21}	w_{31}	y_{A1}	y_{B1}	y_{C1}	y_{D1}
A			A	0	0		
B	B				1	1	
	C	C				2	2
	D	D		3			3

Exercises

2.8. Formulate the following problem using appropriate global constraints (alldiff, cardinality, nvalues, stretchCycle) and channeling constraints. There are six security guards and four stations. Each station must be staffed by exactly one guard each night of the week. A guard must be on duty four or five nights a week. No station should be staffed by more than three different guards during the week. A guard must never staff the same station two or more nights in a row and must staff at least three different stations during the week. A guard must not staff stations 1 and 2 on consecutive nights (in either order), and similarly for stations 3 and 4. Every week will have the same schedule. *Hints.* Let w_{sd} be the guard at station s on night d, and let y_{id} be the station assigned to guard i on night d. Because two guards are unassigned on a given night, two dummy stations are necessary to make the channeling constraints work.

2.9. Write all feasible solutions of the stretchCycle constraint in (2.11) to confirm that the domains in (2.12) can be reduced to (2.13) and cannot be reduced further.

2.10. Formulate this lot-sizing problem using a combination of variable indices, linear constraints, and stretch constraints. In each period t, there is a demand d_{it} for product i, and this demand must be met from the stock of product i at the end of period t. At most, one product can be manufactured in each period t, represented by variable y_t. If nothing is manufactured, $y_t = 0$, where product 0 is a dummy product. The quantity of product i manufactured in any period must be either q_i or zero. When product i is manufactured, its manufacture must continue no fewer than ℓ_i and no more than

u_i periods in a row. The manufacture of product i in any period must be followed in the next period by the manufacture of one of the products in S_i (one may assume $0, i \in S_i$). The unit holding cost per period for product i is h_i, and the unit manufacturing cost is g_i. The setup cost of making a transition from product i in one period to product j in the next is c_{ij} (where possibly i and/or j is 0). Minimize total manufacturing, holding, and setup costs over an n-period horizon while meeting demand. After formulating the problem, indicate how to replace the variable indices with element constraints. *Hints:* Let variable x_{ij} represent the quantity of product i manufactured in period t, and s_{it} the stock at the end of the period. The element(y, x, z) constraint also has a form element$(y, z \,|\, a)$, where a is a tuple of constants. It sets z equal to a_y.

2.11. A possible difficulty with the model of the previous exercise is that the setup cost after an idle period is always the same, regardless of which (nondummy) product was manufactured last. How can the model be modified to allow setup cost to depend on the last nondummy product? *Hint:* Define several dummy products. For example, let $y_t = -i$ indicate that no product is manufactured in peropd t, and the last product manufactured is i.

2.12. Formulate the following problem using linear disjunctions, variable indices, and stretchCycle constraints. The week is divided into n periods, and in each period t, d_t megawatts of electricity must be generated. Each power plant i generates at most q_i megawatts while operating. Once plant i is started, it must run for at least ℓ_i periods, and once shut down, it must remain idle for at least ℓ'_i periods. The cost of producing one megawatt of power for one period at plant i is g_i, and the cost of starting up the plant is c_i. Determine which plants to operate in each period to minimize cost while meeting demand. The schedule must follow a weekly cycle. Indicate how to replace the variable indices with element constraints. *Hint:* Let $y_{it} = 1$ if plant i operates in period t, and let the setup cost incurred by plant i in period t be $c_{iy_{i,t-1}y_{it}}$, where $c_{i01} = c_i$ and $c_{i00} = c_{i01} = c_{i11} = 0$.

2.6 Continuous Global Optimization

Optimization problems need not contain discrete variables to be combinatorial in nature. A continuous optimization problem may have a large number of locally optimal solutions, which are solutions that are optimal in a neighborhood about them. Nonlinear programming solvers, highly developed as they are, are often geared to finding only a local optimum. To identify a global optimum, there may be no alternative but to examine the entire solution space, at least implicitly.

The most popular and effective global solvers use a branch-and-bound search that is analogous to the one presented for the freight transfer problem. The main difference is that because the variable domains are continuous intervals rather than finite sets, the algorithm branches on a variable by splitting an interval into two or more intervals. This sort of branching divides continuous space into increasingly smaller boxes until a global solution can be isolated. Constraint propagation and relaxation are also somewhat different than in discrete problems, because they employ techniques that are specialized to the nonlinear functions that typically occur in continuous problems.

Global optimization is best illustrated with a very simple example that is chosen to highlight some of the simpler techniques rather than to represent a practical application. The problem is:

$$\text{linear:} \begin{cases} \max x_1 + x_2 \\ 2x_1 + x_2 \leq 2 \end{cases}$$

$$\text{bilinear:} \ 4x_1x_2 = 1$$

$$\text{domains:} \ x_1 \in [0, 1], \ x_2 \in [0, 2]$$

(2.15)

The feasible set is illustrated in Figure 2.3. There are two locally optimal solutions,

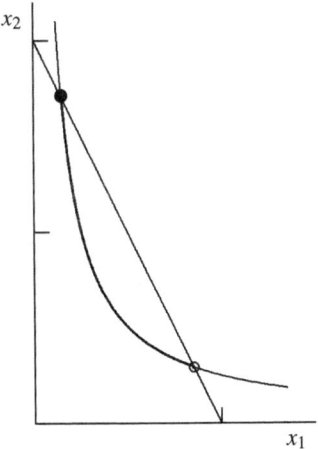

Fig. 2.3 Illustration of a continuous global optimization problem. The heavy curve represents the feasible set. The solid circle marks the optimal solution, and the open circle marks a second locally optimal solution.

$$(x_1, x_2) = (\tfrac{1}{2} + \tfrac{1}{4}\sqrt{2}, 1 - \tfrac{1}{2}\sqrt{2}) \approx (0.853553, 0.292893)$$
$$(x_1, x_2) = (\tfrac{1}{2} - \tfrac{1}{4}\sqrt{2}, 1 + \tfrac{1}{2}\sqrt{2}) \approx (0.146447, 1.707107)$$

the latter of which is globally optimal.

2.6.1 Inference: Bounds Propagation

Bounds propagation can be applied to nonlinear constraints as well as linear inequalities. This is easiest when a constraint can be "solved" for a bound on each variable in terms of a monotone function of the other variables, as is possible for linear inequalities. To obtain a new bound for a given variable, one need only substitute the smallest or largest possible values for the other variables.

In the example, $4x_1 x_2 = 1$ can be solved for $x_1 = \tfrac{1}{4} x_2$. By substituting the upper bound of 2 for x_2, one can update the lower bound on x_1 to $\tfrac{1}{8}$, resulting in the domain $[0.125, 1]$. Similarly, the domain for x_2 can be reduced to $[0.25, 2]$. The linear constraint $2x_1 + x_2 \leq 2$ allows further reduction of the domains to $[0.125, 0.875]$ and $[0.25, 1.75]$, respectively. By cycling back through the two constraints repeatedly, one asymptotically converges to domains of approximately $[0.146, 0.854]$ for x_1 and $[0.293, 1.707]$ for x_2. Global solvers typically truncate this process quite early, however, because the marginal gains of further iterations are not worth the time investment. The computations to follow reflect the results of cycling through the constraints only once.

2.6.2 Relaxation: Factored Functions

Linear relaxations can often be created for nonlinear constraints by *factoring* the functions involved into more elementary functions for which linear relaxations are known. For instance, the constraint $x_1 x_2 / x_3 \leq 1$ can be written $y_1 \leq 1$ by setting $y_1 = y_2 / x_3$ and $y_2 = x_1 x_2$. This factors the function $x_1 x_2 / x_3$ into the elementary operations of multiplication and division, for which tight linear relaxations have been derived.

The constraint $4x_1 x_2 = 1$ in the example (2.15) can be written $4y = 1$ by setting $y = x_1 x_2$. The product $y = x_1 x_2$ has the well-known relaxation

$$L_2x_1 + L_1x_2 - L_1L_2 \le y \le L_2x_1 + U_1x_2 - U_1L_2$$
$$U_2x_1 + U_1x_2 - U_1U_2 \le y \le U_2x_1 + L_1x_2 - L_1U_2$$

(2.16)

where $[L_i, U_i]$ is the current interval domain of x_i.

The example (2.15) therefore has the relaxation

$$
\text{linear:} \begin{cases}
\max\ x_1 + x_2 \\
4y = 1 \\
2x_1 + x_2 \le 2 \\
L_2x_1 + L_1x_2 - L_1L_2 \le y \le L_2x_1 + U_1x_2 - U_1L_2 \\
U_2x_1 + U_1x_2 - U_1U_2 \le y \le U_2x_1 + L_1x_2 - L_1U_2 \\
L_i \le x_i \le U_i, \quad i = 1,2
\end{cases}
$$

(2.17)

The constraint $4y = 1$ can be eliminated by substituting $1/4$ for y in the other constraints. Initially, $[L_1, U_1] = [0.125, 0.875]$ and $[L_2, U_2] = [0.25, 1.75]$, because these domains result from the bounds propagation described in the previous section.

2.6.3 Branching Search

At the root node of a branching tree (Figure 2.4), the initial linear relaxation (2.17) is solved to obtain the solution $(x_1, x_2) = (\frac{1}{7}, \frac{41}{24}) \approx (0.143, 1.708)$. This solution is infeasible in the original problem (2.15), because $4x_1x_2 = 1$ is not satisfied. It is therefore necessary to branch by splitting the domain of a variable.

The choice of branching heuristic can be crucial to fast solution, but for purposes of illustration one can simply split the domain $[0.25, 1.75]$ of x_2 into two equal parts. Restriction 1 in Figure 2.4 corresponds to the lower branch, $x_2 \in [0.25, 1]$. The domains reduce as shown. The solution $(x_1, x_2) = (0.25, 1)$ of the relaxation is feasible and becomes the incumbent solution. The objective function value is 1.25, which provides a lower bound on the optimal value of the problem. No further branching on the left side is necessary. This, in effect, rules out the locally optimal point $(0.854, 0.293)$ as a global optimum, and it remains only to find the other locally optimal point to a desired degree of accuracy.

Moving to restriction 2, the solution of the relaxation is $(x_1, x_2) = (\frac{41}{280}, \frac{239}{140}) \approx (0.146429, 1.707143)$. This is slightly infeasible, as $4x_1x_2$ is 0.999898 rather than 1. One could declare this solution to be feasible

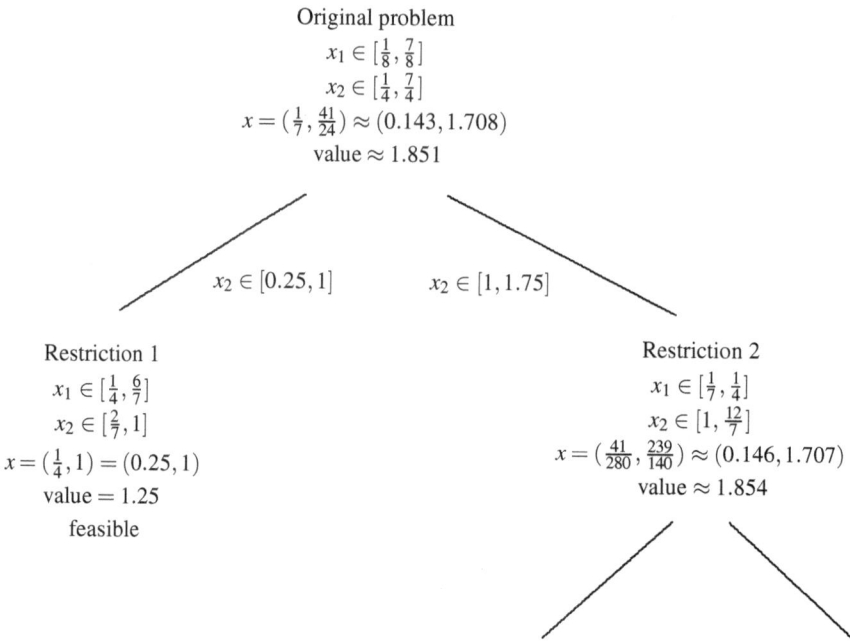

Fig. 2.4 A portion of a search tree for a global optimization problem. The reduced domains and solution of the relaxation are shown at each node.

within tolerances and terminate the search. Alternatively, one could continue branching to find a strictly feasible solution that is closer to the optimum than the incumbent.

Before branching, however, it may be possible to reduce domains by using Lagrange multipliers and the lower bound 1.25 on the optimal value. The constraints $2x_1 + x_2 \leq 2$ and $\frac{1}{4} \leq U_2 x_1 + L_1 x_2 - L_1 U_2$, respectively, have Lagrange multipliers 1.1 and 0.7 in the solution of the relaxation (2.17) (the remaining constraints have vanishing multipliers since they are not satisfied as equalities). The multiplier 1.1 indicates that any reduction Δ in the right-hand side of $2x_1 + x_2 \leq 2$ reduces the optimal value of the relaxation, currently 1.854, by at least 1.1Δ. Thus any reduction Δ in the left-hand side of the constraint, currently equal to 2, has the same effect. But the value of the relaxation should not be reduced below 1.25, because the optimal value of the original problem is at least 1.25. So one can impose the bound $1.1\Delta \leq 1.854 - 1.25$, which can be written

$$1.1[2 - (2x_1 + x_2)] \leq 1.854 - 1.25$$

This implies $2x_1 + x_2 \geq 1.451$, an inequality that must be satisfied by any optimal solution found in the subtree rooted at the current node. Propagation of this inequality reduces the current domain $[1, 1.714]$ of x_2 to $[1.166, 1.714]$. Similar treatment of the inequality

$$\frac{1}{4} \leq U_2 x_1 + L_1 x_2 - L_1 U_2$$

which currently is $\frac{12}{7}x_1 + \frac{1}{7}x_2 \geq \frac{97}{196}$, has no effect on the domains.

Exercises

2.13. Derive a linear relaxation for $y = x^2$ from (2.16), and derive a linear relaxation for $y = a^x$ when $a > 0$. Now, write a factored relaxation for the problem

$$\max\ a^{x_1} + b^{x_2}$$
$$(x_1 + 2x_2)^2 \leq 1, \quad L_j \leq x_j \leq U_j, \quad j = 1, 2$$

when $a, b > 0$.

2.14. Consider the problem:

$$\max\ -x_1^2 - x_2^2$$
$$-3x_1 - x_2 \leq -15$$
$$x_1, x_2 \geq 0 \text{ and integral}$$

Suppose that the best known feasible solution for the problem is $(x_1, x_2) = (4, 3)$. The continuous relaxation has the optimal solution $(x_1, x_2) = (4.5, 1.5)$ with a Lagrange multiplier of 3 corresponding to the first constraint. Derive an inequality that must be satisfied by any optimal solution. (Round up the right-hand side, because the coefficients and variables are integral.)

2.7 Product Configuration

A product configuration example shows how inference and relaxation combine in a slightly more complex problem. It also introduces indexed linear constraints, which occur when the coefficients of a linear constraint are indexed by variables.

The problem is to decide what type of power supply, disk drive, and memory chip to install in a laptop computer, and how many of each. The objective is to minimize total weight while meeting the computer's

Table 2.4 Data for a small instance of the product configuration problem.

Component i	Type k	Net power generation A_{i1k}	Disk space A_{i2k}	Memory capacity A_{i3k}	Weight A_{i4k}	Max number used
1. Power supply	A	70	0	0	200	1
	B	100	0	0	250	
	C	150	0	0	350	
2. Disk drive	A	−30	500	0	140	3
	B	−50	800	0	300	
3. Memory chip	A	−20	0	250	20	3
	B	−25	0	300	25	
	C	−30	0	400	30	
Lower bound L_j		0	700	850	0	
Upper bound U_j		∞	∞	∞	∞	
Unit cost c_j		0	0	0	1	

requirements—upper and lower bounds on memory, disk space, net power supplied, and weight. Only one type of each component may be installed. One power supply, at most three disk drives, and at most three memory chips may be used. Data for a small instance of the problem appear in Table 2.4.

2.7.1 Formulating the Problem

To formulate the problem as an optimization model, let variable t_i be the type of component i that is chosen, and q_i the quantity. It is convenient to let variable r_j denote the total amount of attribute j that is produced, where $j = 1, 2, 3, 4$ correspond, respectively, to net power, disk space, memory, and weight. The computer requires a minimum L_j and maximum U_j of each attribute j.

In general, the objective in such a problem is to minimize the cost of supplying the attributes, where c_j is the unit cost of attribute j. In this case, only the weight is of concern, so that $(c_1, \ldots, c_4) = (0, 0, 0, 1)$.

The interesting aspect of the problem is modeling the connection between the amount r_j of attribute j supplied and the configuration chosen. The component characteristics can be arranged in a three-dimensional array A, in which A_{ijk} is the net amount of attribute j produced by type k of component i. A natural way to write r_j is to sum the contributions of each component

$$r_j = \sum_i A_{ijt_i} q_i$$

where A_{ijt_i} is the amount contributed by the selected type of component i. This leads immediately to the optimization model

$$\text{linear:} \begin{cases} \min \sum_j c_j r_j & (a) \\[2mm] r_j = \sum_i A_{ijt_i} q_i, \quad \text{all } j & (b) \end{cases}$$

$$\text{domains:} \begin{cases} t_1 \in \{\text{A,B,C}\}, t_2 \in \{\text{A,B}\}, t_3 \in \{\text{A,B,C}\} \\ r_j \in [L_j, U_j], \quad \text{all } j \\ q_1 \in \{1\}, \quad q_2, q_3 \in \{1, 2, 3\} \end{cases}$$

(2.18)

The optimal solution of this model is $t = (t_1, t_2, t_3) = (\text{C, A, B})$ and $q = (1, 2, 3)$, which results in weight 705. This calls for installation of the largest power supply (type C), two of the smaller disk drives (type A), and three of the medium-sized memory chips (type B).

Constraints (b) of (2.18) are linear equations in which the coefficients have variable indices t_i. For each j, (b) is processed as an *indexed linear* metaconstraint:

$$\text{indexedLinear:} \begin{cases} r_j = \sum_i z_{ij} \\[2mm] \text{elementCoeff: } (t_i, q_i, z_{ij} \mid (A_{ij1}, \ldots, A_{ijm})), \quad \text{all } i \end{cases}$$

(2.19)

The *element coefficient* constraint is a specially structured element constraint. It sets z_{ij} equal to the t_i^{th} tuple in the list $A_{ij1} q_i, \ldots, A_{ijm} q_i$. The solution software should generate this reformulation automatically.

2.7.2 Inference: Indexed Linear Constraint

Domain reduction for the linear constraints in (2.18) can be achieved by bounds propagation, as discussed earlier. It remains to propagate the indexed linear metaconstraint.

Domain filtering for the element coefficient constraint is similar to that of an element constraint but takes advantage of its special form. The details are presented in Sections 6.7.1–6.7.2. Additional filtering may be derived from the fact that the indexed linear constraint implies integer knapsack inequalities. If $[L_j, U_j]$ is the current domain of r_j, then

$$L_j \leq \sum_i A_{ijt_i} q_i \leq U_j$$

This in turn yields the integer knapsack inequalities

$$L_j \leq \sum_i \max_{k \in D_{t_i}} \{A_{ijk}\} q_i$$

$$\sum_i \min_{k \in D_{t_i}} \{A_{ijk}\} q_i \leq U_j$$

which can be used to reduce the domains of the variables q_i.

Three of the attributes in the example problem (power, disk space, and memory), respectively, yield the following knapsack inequalities:

$$0 \leq \max\{70, 100, 150\} q_1 + \max\{-30, -50\} q_2 +$$
$$\max\{-20, -25, -30\} q_3$$
$$700 \leq \max\{0, 0, 0\} q_1 + \max\{500, 800\} q_2 + \max\{0, 0, 0\} q_3$$
$$850 \leq \max\{0, 0, 0\} q_1 + \max\{0, 0\} q_2 + \max\{250, 300, 400\} q_3$$
$$(2.20)$$

which simplify to

$$0 \leq 150q_1 - 30q_2 - 20q_3, \quad 700 \leq 800q_2, \quad 850 \leq 400q_3 \quad (2.21)$$

Weight yields no useful inequality as its lower bound is zero. Propagation of these inequalities reduces the domain of q_3 to $\{3\}$ but does not affect the other domains. When all the constraints of (2.18) are propagated, the domains are reduced to the following:

$$
\begin{aligned}
& q_1 \in \{1\} && q_2 \in \{1,2\} && q_3 \in \{3\} \\
& t_1 \in \{C\} && t_2 \in \{A,B\} && t_3 \in \{B,C\} \\
& z_{11} \in [150,150] && z_{21} \in [-75,-30] && z_{31} \in [-90,-75] \\
& z_{12} \in [0,0] && z_{22} \in [700,1600] && z_{32} \in [0,0] \\
& z_{13} \in [0,0] && z_{23} \in [0,0] && z_{33} \in [900,1200] \\
& z_{14} \in [350,350] && z_{24} \in [140,600] && z_{34} \in [75,90] \\
& r_1 \in [0,45] && r_2 \in [700,1600] && r_3 \in [900,1200] \\
& r_4 \in [565,1040]
\end{aligned}
\tag{2.22}
$$

2.7.3 Relaxation: Indexed Linear Constraint

Problem (2.18) is linear except for the indexed linear constraints and the integrality restriction on q_i. It can therefore be relaxed by dropping the integrality condition and finding a linear relaxation for the indexed linear constraints.

The relaxation consists of two parts: (i) a relaxation for the knapsack inequalities (2.21) and (ii) a relaxation the element coefficient constraint. The knapsack inequalities themselves provide a linear relaxation, which can be strengthened with knapsack cuts, Gomory cuts (Section 7.3.2), and other cutting planes for integer linear inequalities.

The key to relaxing an element constraint is to note that it implies a disjunction. In particular, the element coefficient constraint in (2.19) implies the disjunction of linear equations:

$$
\bigvee_{k \in D_{t_i}} (z_{ij} = A_{ijk} q_i)
\tag{2.23}
$$

This says that at least one of the equations $z_{ij} = A_{ijk} q_i$ must hold. Relaxations for disjunctions of linear systems in general are discussed in Section 7.4. The particular disjunction (2.23) can be relaxed as follows:

$$
z_{ij} = \sum_{k \in D_{t_i}} A_{ijk} q_{ik}, \quad q_i = \sum_{k \in D_{t_i}} q_{ik}
$$

where $q_{ik} \geq 0$ are new variables. This is a convex hull relaxation, which is the tightest possible linear relaxation. It can be tightened further by using known bounds on the variables, say $L_j \leq z_{ij} \leq U_j$ and $L_{q_i} \leq q_i \leq U_{q_i}$, by adding the constraints

$$L_j \delta_k \leq A_{ijk} q_{ik} \leq U_j \delta_k, \quad \text{all } k \in D_{t_i}$$
$$L_{q_i} \delta_k \leq q_{ik} \leq U_{q_i} \delta_k, \quad \delta_k \geq 0, \quad \text{all } k \in D_{t_i}$$
$$\sum_{k \in D_{t_i}} \delta_k = 1$$

A solution of the relaxation is feasible for (2.23) if exactly one of the δ_k, say δ_{k^*}, is equal to 1, which indicates that $q_{ik^*} = q_i$ and $z_{ij} = A_{ijk^*} q_i$.

Based on this idea, the relaxation of (2.18) becomes:

$$\text{linear:} \begin{cases} \min \sum_j c_j r_j & (a) \\[2mm] r_j = \sum_i \sum_{k \in D_{t_i}} A_{ijk} q_{ik}, \quad \text{all } j & (b) \\[2mm] q_i = \sum_{k \in D_{t_i}} q_{ik}, \quad \text{all } i & (c) \\[2mm] L_j \delta_k \leq A_{ijk} q_{ik} \leq U_j \delta_k, \quad k \in D_{t_i}, \text{ all } i & (d) \\[2mm] L_{q_i} \delta_k \leq q_{ik} \leq U_{q_i} \delta_k, \quad \delta_k \geq 0, \quad k \in D_{t_i}, \text{ all } i & (e) \\[2mm] \sum_{k \in D_{t_i}} \delta_k = 1 & (f) \\[2mm] L_j \leq r_j \leq U_j, \quad \text{all } j & (g) \\[2mm] L_{q_i} \leq q_i \leq U_{q_i}, \quad \text{all } i & (h) \\[2mm] L_j \leq \sum_i \max_{k \in D_{t_i}} \{A_{ijk} q_i\}, \quad \text{all } j & (i) \\[2mm] \sum_i \min_{k \in D_{t_i}} \{A_{ijk} q_i\} \leq U_j, \quad \text{all } j & (j) \\[2mm] q_{ik} \geq 0, \quad \text{all } i, k & (k) \end{cases} \quad (2.24)$$

domains: $q_i, q_{ik}, r_j, \delta_k \in \mathbb{R}$, all i, j, k

The relaxation can be strengthened with knapsack cuts for (i) and (j) and other cutting planes, such as Gomory cuts (Section 7.3.2).

The solution of the relaxation (2.24), given the reduced domains in (2.22), is $q_1 = q_{13} = 1$, $q_2 = q_{2A} = 2$, and $q_3 = q_{3B} = 3$, with the other q_{ik}'s equal to zero. Because the q_i's and δ_k's are integer, the solution is feasible and no branching is necessary. The δ_k's indicate that $(t_1, t_2, t_3) = (C, A, B)$. Also $(r_1, \ldots, r_4) = (15, 1000, 900, 705)$, which indicates that the minimum weight of the computer is $r_4 = 705$.

2.7.4 Branching Search

A solution of the relaxation (2.24) is feasible for the original problem (2.18) unless it violates the integrality constraint for some q_i or fails to satisfy the indexed linear constraint. In the former case, branching follows the usual pattern. If $\hat{q}_i$ is a noninteger value in the solution of the relaxation, the domain $\{L_{q_i}, \ldots, U_{q_i}\}$ of q_i is split into $\{L_{q_i}, \ldots, \lfloor \hat{q}_i \rfloor\}$ and $\{\lceil \hat{q}_i \rceil, \ldots, U_{q_i}\}$. At either branch, the smaller domains are propagated and the resulting relaxation solved.

The solution of the relaxation violates the indexed linear constraint for some i when at least two q_{ik}s are positive, say q_{ik_1} and q_{ik_2}. In this case, the search can branch by splitting the domain of t_i into two sets, one excluding k_1 and the other excluding k_2.

Exercises

2.15. What knapsack covering inequality can be inferred from the metaconstraint (2.19)? Assume each $x_i \geq 0$.

2.16. A farmer wishes to apply fertilizer to each of several plots. The additional crop yield from plot i per unit of type k fertilizer applied is a_{ik}, and the runoff into streams is c_{ik}. There are storage facilities on the farm for at most K different kinds of fertilizer. Use variable indices and the nvalues constraint to formulate the problem of identifying which fertilizer to apply to each plot, and how much, to maximize total additional yield subject to an upper limit U on the amount of runoff. Now, reformulate the problem using indexed linear constraints in place of variable indices.

2.8 Planning and Scheduling

An elementary planning and scheduling problem illustrates *logic-based Benders decomposition*, a form of *constraint-directed search*. It is formally similar to branching in that it enumerates a sequence of problem restrictions, but the restrictions are created differently. Relaxations also help guide the search, but again in a different way.

The restrictions are formed by enumerating assignments to a subset of the variables (the *search variables*) and, for each assignment, solving a restricted problem (*subproblem*) for the remaining variables.

The solution of the subproblem yields a constraint (*Benders cut*) that encodes information about the solution. The next assignment enumerated must satisfy all the Benders cuts so far generated. The object is to avoid assignments that are similar to those already tried, so that a better one might be found. The search is constraint-directed in that it is directed by Benders cuts. Relaxation also helps direct the search, because the current set of Benders cuts is in effect a relaxation of the original problem.

Classical Benders decomposition requires that the subproblem be a linear or nonlinear programming problem, because the Benders cuts rely on Lagrange multipliers obtained from the subproblem. However, these multipliers can be more generally viewed as the solution of an *inference dual* of the subproblem. Because an inference dual can be defined for a subproblem of any form, this leads to a generalization of Benders decomposition. The generalization is "logic-based" in the sense that the inference dual is solved by constructing a proof of optimality for the subproblem solution.

The planning and scheduling problem presented here represents an important class of problems that frequently occur in manufacturing, supply chain management, and elsewhere. The goal is to assign tasks to facilities and then schedule the tasks. The facilities might be factories, machines in a factory, transport modes, delivery vehicles, or computer processors.

In the problem instance at hand, five jobs are to be allocated to two machines, named A and B, and scheduled on them. Each job j has a release time r_j and a deadline d_j. The time required to process job j on machine i is p_{ij}. The specific problem data appear in Table 2.5. Note that machine A is faster than machine B. The objective is to minimize makespan; that is, to minimize the finish time of the last job to finish.

2.8.1 Formulating the Problem

It is convenient to use a metaconstraint noOverlap to represent the scheduling portion of the problem. The constraint may, in general, be written

$$\text{noOverlap}(s \mid p)$$

where $s = (s_1, \dots, s_n)$ are the start times of the jobs to be scheduled, and $p = (p_1, \dots, p_n)$ are the processing times. The constraint is

Table 2.5 Data for a simple two-machine machine scheduling problem.

Job j	Release time r_j	Dead-line d_j	Processing time p_{Aj}	p_{Bj}
1	0	10	1	5
2	0	10	3	6
3	2	7	3	7
4	2	10	4	6
5	4	7	2	5

satisfied when the jobs do not overlap. That is,

$$s_j + p_j \leq s_k \text{ or } s_k + p_k \leq s_j, \text{ all jobs } j, k \text{ with } j \neq k$$

Only two types of decision variables are needed to formulate the problem—the start time s_j already mentioned, and the machine x_j to which job j is assigned. If there are n jobs and m machines, the formulation is

linear: min M

subproblem: $\begin{cases} \{x_1, \ldots, x_n\} \\ \text{linear:} \begin{cases} M \geq s_j + p_{x_j j}, & \text{all } j \\ s_j + p_{x_j j} \leq d_j, & \text{all } j \end{cases} \\ \text{noOverlap: } ((s_j \mid x_j = i) \mid (p_{ij} \mid x_j = i)), \quad \text{all } i \end{cases}$

domains: $s_j \in [r_j, \infty)$, $x_j \in \{1, \ldots, m\}$, all j

In the objective function, M represents the makespan. The inequality constraints define the makespan and enforce the deadlines. A noOverlap constraint is imposed for each machine. In these constraints, the notation $(s_j \mid x_j = i)$ denotes the tuple of start times s_j for jobs assigned to machine i, and similarly for the processing times.

The *subproblem* constraint makes it possible to write a noOverlap constraint whose argument list would otherwise depend on the values of the assignment variables x_j. The subproblem constraint begins with a set of variables, in this case $\{x_1, \ldots, x_n\}$, that are treated as constants inside the scope of the constraint. The noOverlap argument list is therefore fixed. In general, the solver must have at hand one or more decomposition methods that can deal with a subproblem constraint. In this case, Benders decomposition is used.

A natural decomposition distinguishes the assignment portion from the scheduling portion. One can search over various assignments of jobs to machines and, for each, try to find a feasible schedule for the jobs assigned to each machine. The assignment variables x_j are therefore the search variables, and each subproblem is a scheduling problem that decouples into separate scheduling problems for the individual machines.

The inequality constraints could in principle be written outside the subproblem constraint, where they would be parsed as indexed linear constraints, as in the product configuration problem of the previous section. However, the decomposition used here places them in the subproblem, where they regulate the scheduling of assigned jobs.

2.8.2 Relaxation: The Master Problem

The master problem (relaxation) minimizes makespan subject to the Benders cuts generated so far. It can be solved by whatever method is most suitable for its structure. One option is to solve it as an MILP problem, because it is naturally expressed in this form. For this purpose, the variables x_i can be replaced with 0-1 variables x_{ij}, where $x_{ij} = 1$ when job i is assigned to machine j.

The master problem has the form

$$
\begin{aligned}
&\text{min } M \\
&\text{Benders cuts} \\
&\text{relaxation of subproblem (optional)} \\
&x_{ij} \in \{0, 1\}, \text{ all } i, j
\end{aligned}
\tag{2.25}
$$

The Benders cuts are described in the next section. The subproblem relaxation allows the master problem to select reasonable assignments before many Benders cuts have been accumulated.

A simple subproblem relaxation observes, for example, that the jobs assigned to a machine must fit within the earliest release time and latest deadline of those jobs. In fact, this is true of any subset of the jobs assigned to a given machine. To formulate this condition, let $J(t_1, t_2)$ be the set of jobs whose time windows lie in the interval $[t_1, t_2]$. So

$$
J(t_1, t_2) = \{j \mid [r_j, d_j] \subset [t_1, t_2]\}
$$

The total processing times of the jobs in $J_i(t_1, t_2)$ that are assigned to a given machine i must not exceed $t_2 - t_1$:

$$\sum_{j \in J(t_1, t_2)} p_{ij} x_{ij} \leq t_2 - t_1 \tag{2.26}$$

It suffices to consider release times for t_1 and deadlines for t_2. In the problem instance at hand,

$$
\begin{aligned}
J(0, 7) &= \{3, 5\} & J(2, 10) &= \{3, 4, 5\} \\
J(0, 10) &= \{1, 2, 3, 4, 5\} & J(4, 7) &= \{5\} \\
J(2, 7) &= \{3, 5\} & J(4, 10) &= \{5\}
\end{aligned}
$$

Some of these sets give rise to vacuous or redundant inequalities (2.26). For instance, the inequality for $J(0, 7)$ is $p_{i3}x_{i3} + p_{i5}x_{i5} \leq 7$, which is $3x_{A3} + 2x_{A5} \leq 7$ for $i = A$ and $7x_{B3} + 5x_{B5} \leq 7$ for $i = B$. The former is obviously redundant since $x_{ij} \in \{0, 1\}$. The latter is dominated by another inequality for machine B (namely, $7x_{B3} + 5x_{B5} \leq 5$). The nonredundant inequalities for machine A are

$$
\begin{aligned}
J(0, 10): &\quad \sum_{j \in \{1,2,3,4,5\}} p_{Aj} x_{Aj} \leq 10 \\
J(2, 10): &\quad \sum_{j \in \{3,4,5\}} p_{Aj} x_{Aj} \leq 8
\end{aligned}
\tag{2.27}
$$

and those for machine B are

$$
\begin{aligned}
J(0, 10): &\quad \sum_{j \in \{1,2,3,4,5\}} p_{Bj} x_{Bj} \leq 10 \\
J(2, 7): &\quad \sum_{j \in \{3,5\}} p_{Bj} x_{Bj} \leq 5 \\
J(2, 10): &\quad \sum_{j \in \{3,4,5\}} p_{Bj} x_{Bj} \leq 8 \\
J(4, 7) &\quad \sum_{j \in \{5\}} p_{Bj} x_{Bj} \leq 3
\end{aligned}
\tag{2.28}
$$

Section 7.13.3 shows how to identify nonredundant inequalities of this sort in a systematic way.

Further inequalities can be added to the master problem to constrain the makespan M:

$$M \geq \sum_{j \in \{1,2,3,4,5\}} p_{ij} x_{ij}, \quad i = A, B$$

$$v \geq 2 + \sum_{j \in \{3,4,5\}} p_{ij} x_{ij}, \quad i = A, B \qquad (2.29)$$

$$v \geq 4 + \sum_{j \in \{5\}} p_{ij} x_{ij}, \quad i = A, B$$

The three sets of inequalities correspond to the release times 0, 2, and 4. The first includes the jobs in $J(0, \infty)$, the second the jobs in $J(2, \infty)$, and the third the jobs in $J(4, \infty)$. The subproblem relaxation in the master problem (2.25) therefore consists of inequalities (2.27)–(2.29).

2.8.3 Inference: Benders Cuts

The inference stage consists of inferring Benders cuts from the subproblem that results when the master problem variables are fixed to their current values.

The subproblem separates into an independent scheduling problem on each machine. Thus, if $\bar{x}$ is the solution of the previous master problem, the subproblem on each machine i is

$$\min \; M_i$$
$$s_j + p_{ij} \leq d_j, \quad \text{all } j \text{ with } \bar{x}_{ij} = 1$$
$$\text{noOverlap}((s_j \mid \bar{x}_{ij} = 1) \mid (p_{ij} \mid \bar{x}_{ij} = 1))$$
$$s_j \in [r_j, \infty), \quad \text{all } j$$

If M_i^* is the optimal makespan on machine i for each i, then the optimal makespan overall is $\max_i \{M_i^*\}$.

The subproblem does not separate in this way if there are precedence constraints between jobs, because the time at which a job can be scheduled may depend on the times at which jobs are scheduled on other machines. Yet even when there are precedence constraints, separability can be preserved if they involve only jobs that must be scheduled on the same machine. Thus, if jobs j and k must be scheduled on the same machine, and j must precede k, one can add constraint $x_j = x_k$ to the master problem and the constraint $s_j + d_{ij} \leq s_k$ to the scheduling problem on every machine.

As noted above, the initial solution of the master problem assigns job 4 to machine B and the other jobs to machine A. The minimum

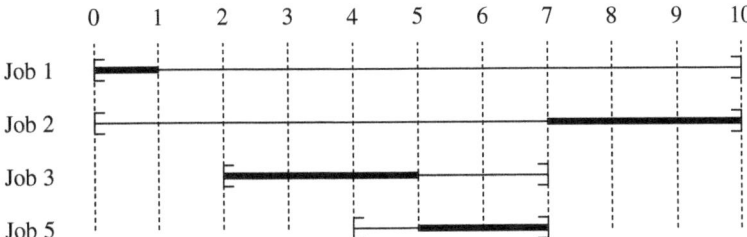

Fig. 2.5 A minimum makespan schedule on machine A for jobs 1, 2, 3, and 5 of Table 2.5. The horizontal lines represent time windows.

makespan schedule on machine B simply starts job 4 at its release time, resulting in a makespan of 8. Thus, whenever job 4 is assigned to machine B (i.e., whenever $x_{4B} = 1$), the makespan will be at least 8. This gives rise to the Benders cut

$$M \geq 8x_{B4}$$

Any solution that achieves a makespan better than 8 must avoid assigning job 4 to machine B.

The minimum makespan schedule for jobs 1, 2, 3, and 5 on machine A appears in Figure 2.5. The resulting makespan is 10, which produces the Benders cut

$$M \geq 10(x_{A1} + x_{A2} + x_{A3} + x_{A5} - 3) \tag{2.30}$$

Thus, any solution that obtains a makespan better than 10 must avoid assigning at least one of these jobs to machine A.

The master problem is now

$$\min M$$
$$\text{inequalities } (2.27)\text{--}(2.29)$$
$$M \geq 10$$
$$M \geq 10(x_{A1} + x_{A2} + x_{A3} + x_{A5} - 3)$$
$$x_{ij} \in \{0,1\}, \quad \text{all } i,j$$

Solution of this problem results in the same machine assignments as before, but the optimal value is now 10. Since this is a lower bound on the minimum makespan, and a feasible solution with makespan 10 was found by solving the subproblem, the algorithm terminates. The schedule found by solving the subproblem is optimal.

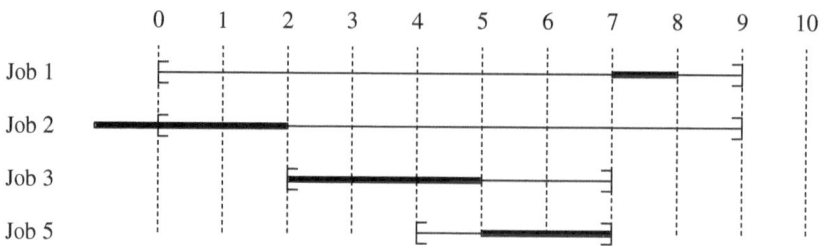

Fig. 2.6 Edge-finding proof of infeasibility of scheduling jobs 1, 2, 3, and 5 on machine A to achieve a makespan of 9.

In practice, the success of a Benders method often rests on finding strong Benders cuts. One way to deduce stronger cuts is to examine more closely the reasoning process that proves optimality in the subproblem. In the present case, a well-known *edge-finding* technique, deduces that the makespan on machine A is still 10 even if only jobs 2, 3, and 5 (and not job 1) are assigned to the machine. This is done by deducing that if the makespan is less than or equal to 9, job 2 must start before its release time (Fig. 2.6). One can therefore state a stronger Benders cut than (2.30):

$$M \geq 10(x_{A2} + x_{A3} + x_{A5} - 2) \tag{2.31}$$

Edge finding must often be combined with other procedures, such as branching, but one can nonetheless keep track of which jobs play a role in deducing the optimal makespan.

One difficulty with a cut of the form (2.31) is that it imposes no bound on M when a proper subset of jobs 2, 3, and 5 is assigned to machine A. One should be able to say something about the resulting minimum makespan if only jobs 2 and 3 are assigned to this machine, for example. In fact, one can often derive a cut that remains useful in such cases. Section 6.14.3 shows how to do this when all the release times are equal.

Exercises

2.17. Write (2.26) for each release time t_1 and each deadline t_2 ($t_1 < t_2$) for machine A in the problem of Table 2.5. Verify that (2.27) are the nonredundant inequalities.

2.18. Suppose that jobs 1, 2, 3, and 4 from Table 2.5 are assigned to machine A. What is the minimum makespan? Write the corresponding Benders cut containing the fewest variables.

2.19. A number of projects must be carried out in a shop, and each project j must start and finish within a time window $[r_j, d_j]$. Once started, the project must run p_j days without interruption. Only one project can be underway at any one time. Each month, the shop is shut down briefly to clean and maintain the equipment, and no project can be in process during this period. The goal is to find a feasible schedule. Formulate this problem and indicate how to solve it with logic-based Benders decomposition. *Hint:* Let each month's schedule be a subproblem.

2.20. A vehicle routing problem requires that a fleet of vehicles make deliveries to customers within specified time windows. Each customer j must take delivery between e_j and d_j, and vehicle i requires time p_{ijk} to travel from customer j to customer k. The objective is to minimize the number of vehicles. Describe a Benders-based solution method in which the master problem assigns customers to vehicles and the subproblem routes each vehicle. The subproblem for each vehicle is a traveling salesman problem with time windows. The subproblem can be written with variable indices and the circuit constraint (see Section 6.13).

2.9 Routing and Frequency Assignment

A final example illustrates decomposition in a more complex setting. The arcs of a directed network represent optical fibers with limited capacity (Fig. 2.7). There are requests for communication channels to be established between certain pairs of nodes. The channels must be routed over the network, and they must be assigned frequencies so that all the channels passing along any given arc have different frequencies. There are a limited number of frequencies available, and it may not be possible to establish all the channels requested. The objective is to maximize the number of channels established.

The problem can be decomposed into its routing and frequency-assignment elements. The routing problem is amenable to an MILP formulation, and the frequency assignment problem is conveniently written with alldiff constraints—provided that a subproblem constraint is used to fix the flows before the frequency assignment problem is stated.

The routing problem is similar to the well-known multicommodity network flow problem. This problem generalizes the capacitated

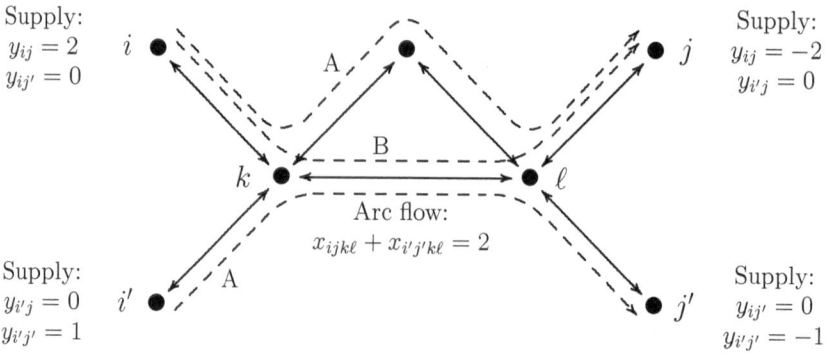

Fig. 2.7 A message routing and frequency assignment problem. Two message channels are requested from i to j and one from i' to j'. The arcs have capacity 2, and frequencies A, B are available. The dashed lines show an optimal solution.

min-cost network flow problem by distinguishing several commodities that must be transported over the network. There is a net supply of each commodity at each node, and the total flow on each arc must be within the arc capacity. In the message routing problem, each origin-destination pair represents a different commodity.

The message routing problem is not identical to the multicommodity flow problem because the net supplies are not fixed, due to the fact that some requests may not be satisfied. As a result, one would not be able to use a global constraint designed for multicommodity flow problems, even if one existed. Nonetheless, it is fairly easy to write the MILP constraints directly.

For each pair of nodes (i, j), let D_{ij} be the number of i-to-j channels requested (possibly zero). A key decision is which requests to honor, and one can therefore let integer variable y_{ij} be the number of channels from i to j that are actually established. (It is assumed here that different channels from i to j can be routed differently.) The net supply of commodity (i, j) is y_{ij} at node i, $-y_{ij}$ at node j, and zero at other nodes. Let $x_{ijk\ell}$ be the flow of commodity (i, j) on arc (k, ℓ), and $C_{k\ell}$ the capacity of the arc. To simplify notation, arcs missing from the network can be viewed as arcs with a capacity of zero. The flow model is

$$\text{MILP:} \begin{cases} \max \displaystyle\sum_{ij} y_{ij} \\[2ex] \displaystyle\sum_{\ell \neq i} x_{iji\ell} - \sum_{k \neq i} x_{ijki} = y_{ij}, \quad \text{all } i,j \\[2ex] \displaystyle\sum_{\ell \neq j} x_{ijj\ell} - \sum_{k \neq i} x_{ijkj} = -y_{ij}, \quad \text{all } i,j \\[2ex] \displaystyle\sum_{\ell \neq i,j,k} x_{ijk\ell} - \sum_{\ell \neq i,j,k} x_{ij\ell k} = 0, \quad \text{all } i,j,k \text{ with } k \neq i,j \\[2ex] \displaystyle\sum_{ij} x_{ijk\ell} \leq C_{k\ell}, \quad \text{all } k,\ell \\[2ex] x_{ijk\ell} \geq 0, \quad \text{all } i,j,k,\ell \\[1ex] 0 \leq y_{ij} \leq D_{ij}, \quad \text{all } i,j \end{cases}$$

$$\text{domains: } x_{ijk\ell}, y_{ij} \in \mathbb{Z}, \quad \text{all } i,j,k,\ell$$

After the communications channels are routed, a frequency f_{ij} can be assigned to each pair i,j so that the frequencies assigned to channels passing through any given arc are all different. The model is therefore completed by writing

$$\text{subproblem:} \begin{cases} \{x_{ijk\ell}, \quad \text{all } i,j,k,\ell\} \\[1ex] \text{alldiff: } \{f_{ij} \mid x_{ijk\ell} > 0\}, \quad \text{all } k,\ell \end{cases}$$

$$f_{ij} \in F, \quad \text{all } i,j \text{ with } i \neq j$$

where F is the set of available frequencies.

The problem can be solved by a branch-infer-and-relax method that propagates the alldiff constraint at every node of the search tree. The variables f_{ij} that appear in the propagated alldiff constraint for a given k,ℓ are those for which $x_{ijk\ell}$ is fixed to a positive number by branching. After the relaxation is solved, the search can branch on an alldiff that is violated by the solution of the relaxation. For this purpose, the variables f_{ij} in the alldiff are those for which $x_{ijk\ell}$ has a positive integer value in the solution of the current relaxation. One scheme for branching on a violated alldiff is described in Section 5.1.3.

2.10 Bibliographic Notes

Section 2.2. Various elements of the search-infer-and-relax framework presented here were proposed in [23, 101, 276, 278, 279, 282, 295, 297]. An extension to dynamic backtracking and heuristic methods is given in [286, 287].

Section 2.3. Cuts for general integer knapsack constraints are discussed in [124, 350, 397, 398]. A comprehensive treatment can be found in [26], which strengthens the inequalities discussed here. Nearly all development of knapsack cuts, however, has been concerned with 0-1 knapsack constraints, beginning with [30, 254, 384, 506].

Section 2.4. Conditional modeling is advocated in [295]. The idea of disjunctive modeling goes back at least to [228, 243] and is developed in [31, 32, 50, 104, 245, 295, 335, 409, 437, 483] and elsewhere.

Section 2.5. The employee scheduling model is similar to one presented in [422]. An overview of employee timetabling appears in [354].

Section 2.6. Continuous global optimization is extensively surveyed in [372] and a handbook [388]. The integrated approach taken here is roughly similar to that described in [468, 469, 470] and implemented in the solver BARON. Factorization of functions for purposes of relaxation was introduced by [352]. Global optimization problems can have discrete as well as continuous variables; recent solution technology is discussed in [70].

Section 2.7. The product configuration model is based on [474, 517]. The generic element constraint was introduced by [266]. The form of the constraint used here appears in [474].

Section 2.8. Classical Benders decomposition is due to [73] and was generalized to nonlinear programming in [224]. Logic-based Benders decomposition was introduced in [298] and developed in [279, 296]. Its application to planning and scheduling, with a CP subproblem, was proposed in [279], first implemented in [300], and extended in [257]. The machine scheduling example described here is adapted from [283, 517]. Edge finding originates in [117, 118]. The subproblem constraint is proposed in [293].

Section 2.9. The example is adapted from [459]. The subproblem constraint is unnecessary if the alldiff is an "open" constraint [490], meaning that the argument list depends on the value of another variable. Open constraints are not standard in CP solvers, however.

Chapter 3
Optimization Basics

Much of the material in this book presupposes a basic understanding of classical optimization methods, particularly linear programming, network flows, convex nonlinear programming, and dynamic programming. Linear programming is a ubiquitous relaxation tool and finds application in domain reduction, branch-and-price methods, and other techniques. Network flows play a central role in filtering methods for several global constraints. Nonlinear programming is used to solve convex relaxations and reduce variable domains in global optimization problems. Dynamic programming can be useful for domain reduction in sequencing problems and elsewhere.

This chapter presents the basic concepts related to these areas, with an emphasis on optimality conditions. No attempt is made to describe the solution technology in detail. The integrated optimization strategies in this book can be understood without familiarity with the numerical methods, data structures, and other specialized techniques used in practical implementations of the classical methods.

3.1 Linear Programming

A *linear programming* (LP) problem is a continuous optimization problem with a linear objective function and linear inequality constraints. It can be written

$$\text{linear:} \begin{cases} \min \ (\max) \ cx \\ Ax \geq b, \ x \geq 0 \end{cases} \tag{3.1}$$

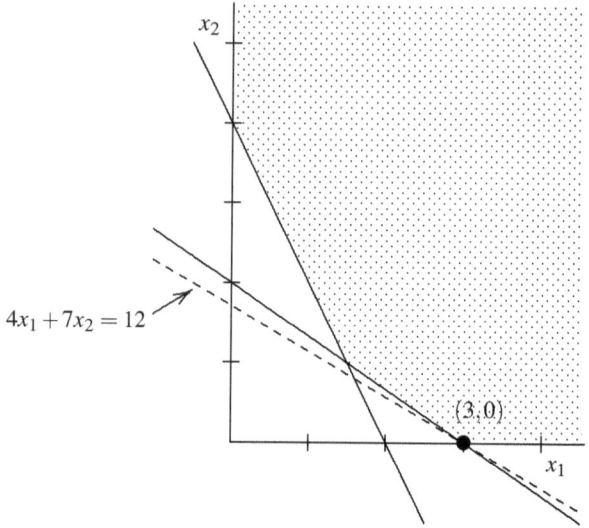

Fig. 3.1 Graph of a simple linear programming problem. The shaded area is the feasible set. The optimal solution (solid black circle) is $(x_1, x_2) = (3, 0)$, and the optimal value is 12.

where A is an $m \times n$ matrix. There is no loss of generality in assuming $x \geq 0$. If a variable x_j is unrestricted in sign, it can be replaced in each occurrence by $x_j^+ - x_j^-$, where the new variables x_j^+, $x_j^- \geq 0$.

As an example, consider the following small problem, illustrated in Fig. 3.1.

$$\text{linear:} \begin{cases} \min 4x_1 + 7x_2 \\ 2x_1 + 3x_2 \geq 6 \\ 2x_1 + x_2 \geq 4 \\ x_1, x_2 \geq 0 \end{cases} \tag{3.2}$$

The feasible set is a polyhedron. It is intuitively clear that if there is an optimal solution, then some vertex of the polyhedron is optimal. This idea is developed algebraically in the next two sections. In the present case, the optimal solution is the vertex $(x_1, x_2) = (3, 0)$.

3.1.1 Optimality Conditions

To state optimality conditions for an LP problem, it is convenient to write it in equality form:

$$\text{linear:} \begin{cases} \min \ cx \\ Ax = b, \quad x \geq 0 \end{cases} \tag{3.3}$$

An inequality constraint $ax \geq a_0$ can always be converted to an equality constraint by introducing a surplus variable $x_j \geq 0$ and writing $ax - x_j = a_0$. For example, problem (3.2) can be written in equality form as follows:

$$\text{linear:} \begin{cases} \min \ 4x_1 + 7x_2 \\ 2x_1 + 3x_2 - x_3 = 6 \\ 2x_1 + x_2 - x_4 = 4 \\ x_j \geq 0, \ j = 1, 2, 3, 4 \end{cases} \tag{3.4}$$

where x_3, x_4 are surplus variables. In matrix format, the problem is

$$\min \begin{bmatrix} 4 & 7 & 0 & 0 \end{bmatrix} x$$
$$\begin{bmatrix} 2 & 3 & -1 & 0 \\ 2 & 1 & 0 & -1 \end{bmatrix} x = \begin{bmatrix} 6 \\ 4 \end{bmatrix}, \quad x \geq 0 \tag{3.5}$$

If matrix A has rank m, any set of m linearly independent columns of A forms a basis for the space spanned by the columns of the matrix. A *basic solution* x of $Ax = b$ is one that uses only basic columns in the linear combination Ax. That is, $x_j = 0$ when column j of A is not part of the basis. This means that at most m variables are positive in a basic solution, and the remaining $n - m$ variables are equal to zero.

To identify the basic solution corresponding to a given basis, remove dependent rows from A, and partition A into a square submatrix B consisting of the basis and a submatrix N consisting of the remaining $n - m$ columns. Then, (3.3) may be written

$$\min \ c_B x_B + c_N x_N$$
$$B x_B + N x_N = b, \quad x_B, x_N \geq 0 \tag{3.6}$$

where x and c are similarly partitioned. The variables in x_B are the *basic variables*, and those in x_N are the *nonbasic variables*. Because B is a nonsingular $m \times m$ matrix, the equations may be solved for x_B in terms of x_N:

$$x_B = B^{-1}b - B^{-1}N x_N \tag{3.7}$$

If x_N is set to zero, the resulting solution $(x_B, x_N) = (B^{-1}b, 0)$ is a basic solution, because at most m variables are positive. It is a basic feasible solution if $B^{-1}b \geq 0$.

Basic feasible solutions are vertices of the feasible set, because they are points at which $n - m$ of the inequalities $x_j \geq 0$ are satisfied at equality. The basic solutions of the example problem are illustrated in Figure 3.2. The optimal solution is obtained when x_1, x_4 are basic and

$$B = \begin{bmatrix} 2 & 0 \\ 2 & -1 \end{bmatrix}$$

The optimal solution is

$$x_B = \begin{bmatrix} x_1 \\ x_4 \end{bmatrix} = B^{-1}b = \begin{bmatrix} \frac{1}{2} & 0 \\ 1 & -1 \end{bmatrix} \begin{bmatrix} 6 \\ 4 \end{bmatrix} = \begin{bmatrix} 3 \\ 2 \end{bmatrix}$$

$$x_N = \begin{bmatrix} x_2 \\ x_3 \end{bmatrix} = \begin{bmatrix} 0 \\ 0 \end{bmatrix}$$

It will be shown in the next section that if there is an optimal solution, then some basic feasible solution is optimal. To solve the problem, it therefore suffices to examine basic feasible solutions and check them for optimality. A test for optimality is easily derived. Using (3.7), the objective function $c_B x_B + c_N x_N$ may be written in terms of the nonbasic variables only. After collecting terms, this yields

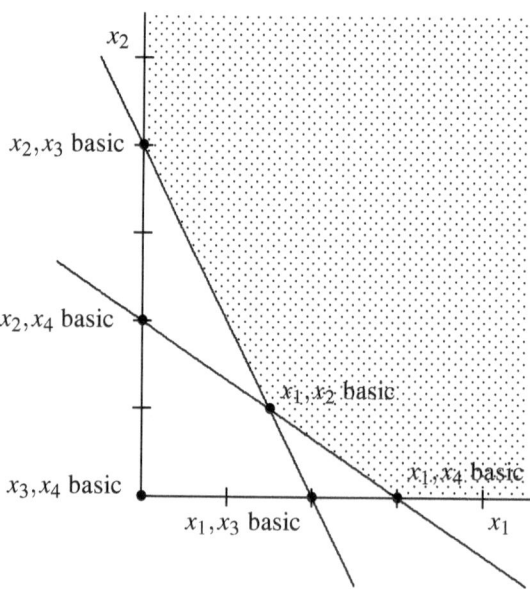

Fig. 3.2 Basic solutions (solid black circles) of a linear programming problem, showing the basic variables in each case. Three of the basic solutions are feasible.

$$c_B B^{-1} b + (c_N - c_B B^{-1} N) x_N = c_B B^{-1} b + r x_N$$

The quantity $c_B B^{-1} b$ is the objective function value of the basic solution $(B^{-1} b, 0)$, and $r = c_N - c_B B^{-1} N$ is the vector of *reduced costs*. If $r \geq 0$, no feasible solution can yield a smaller objective function value, because any feasible solution can be obtained by setting x_N to some nonnegative value. A basic feasible solution is therefore optimal if the corresponding reduced costs are nonnegative.

Theorem 3.1. *A basic feasible solution* $(x_B, x_N) = (B^{-1} b, 0)$ *of the LP problem (3.3) is optimal if the reduced cost vector* $r \geq 0$, *where* $r = c_N - c_B B^{-1} N$.

In the example, the reduced cost vector for basic variables x_1, x_4 is

$$r = c_N - c_B B^{-1} N = \begin{bmatrix} 7 & 0 \end{bmatrix} - \begin{bmatrix} 4 & 0 \end{bmatrix} \begin{bmatrix} \frac{1}{2} & 0 \\ 1 & -1 \end{bmatrix} \begin{bmatrix} 3 & -1 \\ 1 & 0 \end{bmatrix} = \begin{bmatrix} 1 & 2 \end{bmatrix}$$

Because $r \geq 0$, the solution $(x_1, x_2, x_3, x_4) = (3, 0, 0, 2)$ is in fact optimal.

An important variant of the LP problem places upper bounds on the variables.

$$\text{linear:} \begin{cases} \min \ cx \\ Ax = b, \quad 0 \leq x \leq q \end{cases} \tag{3.8}$$

A solution can be regarded as basic in this problem when $n - m$ variables are at one of their bounds (0 or q_j). As before, increasing the value of a nonbasic variable x_j at its lower bound improves the solution if $r_j < 0$. In addition, *reducing* the value of a nonbasic variable x_j at its upper bound improves the solution if $r_j > 0$.

Corollary 3.2 *A basic solution* (x_B, x_N) *of the upper bounded LP (3.8) is optimal if* $r_j \geq 0$ *for each nonbasic* x_j *with value 0, and* $r_j \leq 0$ *for each nonbasic* x_j *with value* q_j, *where* $r = c_N - c_B B^{-1} N$.

3.1.2 Simplex Method

The simplex method examines a sequence of basic feasible solutions. Each solution is obtained by increasing the value of a nonbasic variable with negative reduced cost in the previous solution. The process terminates when all reduced costs are nonnegative. If proper care is taken, termination is assured after finitely many iterations. It follows that some basic feasible solution is optimal if an optimal solution exists.

The simplex method receives its name from the fact that it examines a sequence of simplices in the "row space" in which its inventor George Dantzig originally conceived the method. An iteration is often called a "pivot" because each simplex is obtained by pivoting the previous one about one of its faces.

The modern simplex method begins with a basic feasible solution $(x_B, x_N) = (B^{-1}b, 0)$, which is obtained from the *phase I* procedure discussed below. If $r \geq 0$, the solution is optimal and the procedure terminates. Otherwise, suppose $r_j < 0$ for nonbasic x_j and increase x_j as much as possible without driving a basic variable negative. To write this algebraically, it is convenient to let $\tilde{b} = B^{-1}b$ and $\tilde{N} = B^{-1}N$. Then the expression (3.7) for the basic variables can be written $x_B = \tilde{b} - \tilde{N}x_N$. Because x_B must remain nonnegative, the new value of x_j is

$$x_j^{\text{new}} = \max\left\{ x_j \,\middle|\, \tilde{b} - \tilde{N}x_N \geq 0 \right\} = \max\left\{ x_j \,\middle|\, \tilde{b} - \tilde{N}_j\, x_j \geq 0 \right\}$$

where $\tilde{N}_j$ is the column of $\tilde{N}$ that corresponds to x_j. If $\hat{N}_j \leq 0$, then x_j can increase indefinitely, which means the problem is unbounded and the procedure terminates. Otherwise, x_j^{new} is a minimum ratio:

$$x_j^{\text{new}} = \min_i\left\{ \frac{\tilde{b}_i}{\tilde{N}_{ij}} \,\middle|\, \tilde{N}_{ij} > 0 \right\}$$

The basic variable x_i that hits zero first is the variable for which $\tilde{b}_i - \tilde{N}_{ij}\, x_j^{\text{new}} = 0$. It corresponds to the i that achieves the minimum ratio above. This is known as the *ratio test*.

At this point, x_j becomes a basic variable, and x_i becomes nonbasic. Column j of A replaces column i in the basis B, and B^{-1} is computed with a *rank 1 update*, for which there are a number of efficient numerical techniques. The procedure repeats until termination.

In the example (3.5), one might start with basic variables x_1, x_2, for which

$$B = \begin{bmatrix} 2 & 3 \\ 2 & 1 \end{bmatrix}, \quad B^{-1} = \begin{bmatrix} -\frac{1}{4} & \frac{3}{4} \\ \frac{1}{2} & -\frac{1}{2} \end{bmatrix}$$

$$\tilde{b} = \begin{bmatrix} 1 \\ \frac{3}{2} \end{bmatrix}, \quad r = [r_3 \ \ r_4] = [\tfrac{5}{2} \ \ -\tfrac{1}{2}] \tag{3.9}$$

Because x_4 has negative reduced cost, let

$$x_4^{\text{new}} = \max \left\{ x_4 \; \middle| \; \begin{bmatrix} \frac{3}{2} \\ 1 \end{bmatrix} - \begin{bmatrix} -\frac{3}{4} \\ \frac{1}{2} \end{bmatrix} x_4 \geq \begin{bmatrix} 0 \\ 0 \end{bmatrix} \right\} = \min \left\{ \begin{array}{c} 1 \\ \frac{1}{2} \end{array} \right\} = 2$$

Variable x_4 enters the basis and x_2 leaves the basis, so that x_1 and x_4 are now basic. Beginning the next iteration, one obtains

$$B = \begin{bmatrix} 2 & 0 \\ 2 & -1 \end{bmatrix}, \quad B^{-1} = \begin{bmatrix} \frac{1}{2} & 0 \\ 1 & -1 \end{bmatrix}, \quad \tilde{b} = \begin{bmatrix} 3 \\ 2 \end{bmatrix}, \quad r = [r_2 \; r_3] = [1 \; 2]$$

Because $r \geq 0$, the solution $(x_1, \ldots, x_4) = (3, 0, 0, 2)$ is optimal.

An initial basic feasible solution is obtained for (3.3) as follows. First, multiply rows of the given problem by -1 as necessary to make the right-hand sides nonnegative, and let (3.3) be the problem that results. Then solve the *phase I* problem

$$\min \sum_i y_i \tag{3.10}$$
$$Ax + y = b, \quad x, y \geq 0$$

where $b \geq 0$ and y is a vector of m "artificial" variables. An initial feasible basis is obvious, namely $B = I$, with basic variables $y_1, \ldots, y_m$. If the optimal solution of (3.10) has a positive value, then (3.3) is infeasible. Otherwise, pivot until all variables y_i vanish and become nonbasic, and let B be the resulting basis. Then B is a starting feasible basis for solving the original problem (*phase II*).

The phase I problem for the above example is

$$\min \begin{bmatrix} 0 & 0 & 0 & 0 & 1 & 1 \end{bmatrix} \begin{bmatrix} x \\ y \end{bmatrix}$$

$$\begin{bmatrix} 2 & 3 & -1 & 0 & 1 & 0 \\ 2 & 1 & 0 & -1 & 0 & 1 \end{bmatrix} \begin{bmatrix} x \\ y \end{bmatrix} = \begin{bmatrix} 6 \\ 4 \end{bmatrix}, \quad x, y \geq 0$$

where $y = (y_1, y_2)$ are the initial basic variables. After two pivots, y_1, y_2 become nonbasic, and the starting feasible basis (3.9) is obtained.

The simplex method as stated above need not converge, because it is possible that $x_j^{\text{new}} = 0$, which means there is no change in the solution. This can occur when one or more basic variables are equal to zero—a condition known as *degeneracy*. In principle, one can cycle through the same sequence of identical basic solutions indefinitely.

Degeneracy can be avoided by replacing each b_i in the problem with $b_i + \epsilon^i$, where $\epsilon > 0$ is a very small number. This ensures that cost increases by some minimum amount in each iteration, and the algorithm arrives at the optimal cost in finitely many steps. The effect of perturbing the problem in this way can be mimicked by *lexicographic pivoting*, which uses the original right-hand side but selects the same sequence of bases as in the perturbed problem. Thus the simplex method with lexicographic pivoting converges finitely to an optimal basic solution, if the problem has an optimal solution.

The *upper bounded simplex method* is designed for LP problems (3.8) with upper bounds on the variables. It is used by almost all simplex-based solvers, because it deals with upper bounds very efficiently by incorporating them implicitly into the pivoting logic, rather than treating them explicitly as constraints.

As noted earlier, the optimality conditions account for nonbasic variables x_j at either of their bounds, 0 or q_j. The upper bounded simplex method checks whether there is a nonbasic $x_j = 0$ with $r_j < 0$, or a nonbasic $x_j = q_j$ with $r_j > 0$. If there are no such variables, the solution is optimal. If a variable $x_j = 0$ with $r_j < 0$ is selected, x_j is increased until a basic variable hits one of its bounds, or x_j hits its upper bound. The basic variables x_B currently have value $\bar{x}_B = \tilde{b} - \tilde{N} x_N$, where x_N is the current value of the nonbasic variables. Therefore,

$$x_j^{\text{new}} = \max \left\{ x_j \ \middle| \ x_j \le q_j, \ 0 \le \bar{x}_B - \tilde{N}_j x_j \le p_B \right\}$$

If x_j hits its own bound first, it remains nonbasic but with value q_j, and the next iteration starts. Otherwise x_j enters the basis, a basic variable that hits its bound leaves the basis, B^{-1} is computed, and the next iteration starts. If a variable $x_j = q_j$ with $r_j > 0$ is selected,

$$x_j^{\text{new}} = \min \left\{ x_j \ \middle| \ x_j \ge 0, \ 0 \le \bar{x}_B + \tilde{N}_j (q_j - x_j) \le q_B \right\}$$

If x_j hits zero, it remains nonbasic but with value zero, and the next iteration starts. Otherwise x_j enters the basis, a basic variable that hits its bound leaves the basis, B^{-1} is computed, and the next iteration starts.

Even 60 years after its invention, the simplex method remains the algorithm of choice for most LP problems. State-of-the-art implementations are extremely fast and can solve almost any problem unless it is truly huge or very ill-conditioned numerically. *Interior point* methods can be superior on some large problems. These methods process

toward the optimal solution through the interior of the feasible polyhedron rather than by moving from one vertex to another.

3.1.3 Sensitivity Analysis

Sensitivity analysis investigates the sensitivity of the optimal solution,
and particularly the optimal cost, to perturbations in the problem
data. It is important in practice for identifying which problem data
have an important bearing on the solution, and it has applications to
bounds propagation. Sensitivity analysis is straightforward for linear
programming.

One can first observe that because the optimal value is $z^* = c_B B^{-1} b$,
the effect of a change Δb on the right-hand side of the problem is
immediately evident. If the constraint equations $Ax = b$ are modified
to $Ax = b + \Delta b$, and B remains the optimal basis, then the optimal
cost becomes

$$z^{\text{new}} = c_B B^{-1}(b + \Delta b) = z^* + u\Delta b$$

where $u = c_B B^{-1}$. If the optimal basis changes, then $z^* + u\Delta b$ remains
an upper bound on the new optimal cost. Thus a perturbation Δb in
the right-hand side increases the optimal cost by at most $u\Delta b$.

The vector of multipliers u plays a central role in LP duality theory,
as will be seen in Chapter 6. These multipliers are sometimes referred
to as *shadow prices*, because they indicate the marginal cost/benefit
of changing output requirements or resource limits. For example, if
b_i is the output requirement corresponding to constraint equation i,
then u_i indicates the marginal cost of increasing the requirement. It
is profitable at the margin to produce more of output i if its unit
revenue is greater than the shadow price u_i. If b_i is a resource limit,
then purchasing more of resource i is profitable at the margin if its
unit price is less than u_i.

It is easy to derive conditions under which a perturbation Δb does
not change the optimal basis, and therefore under which the optimal
cost changes exactly $u\Delta b$. A change in the right-hand side does not
affect the reduced costs $r = c_N - uN$, but the updated values $x_B^{\text{new}} = B^{-1}(b + \Delta b) = \tilde{b} + B^{-1}\Delta b$ of the basic variables could go negative. The
analysis is therefore exact ($z^{\text{new}} = z^* + u\Delta b$) when $x_B^{\text{new}} \geq 0$, which is
to say $B^{-1}\Delta b \geq -\tilde{b}$.

A similar analysis can be carried out for perturbations in the objective function coefficients (see exercises).

3.1.4 Feasibility Conditions

The classical Farkas lemma provides necessary and sufficient conditions for the feasibility of an LP problem (3.3). It is a fundamental result that permeates much of optimization theory. It will resurface, for example, when optimality conditions are derived for nonlinear programming later in this chapter, and when duality theory is developed in Chapter 4.

Theorem 3.3 (Farkas lemma). *The linear system $Ax = b$, $x \geq 0$ is infeasible if and only if $uA \leq 0$ and $ub > 0$ for some real vector u.*

The system $Ax = b$, $x \geq 0$ is clearly insoluble if such a u exists, because $Ax = b$ implies $uAx = ub$, which is nonpositive on the left-hand side and positive on the right-hand side.

The converse can be understood geometrically. If $Ax = b$, $x \geq 0$ is insoluble, b lies outside the polyhedral cone $C = \{Ax \mid x \geq 0\}$ spanned by the columns of A. The classical Weierstrass theorem implies that C contains a point $b - u$ that is the closest point in C to b (Figure 3.3). Furthermore, the hyperplane $\{y \mid uy = 0\}$ *separates* b from C, in the sense that $ub > 0$ and $uy \leq 0$ for all $y \in C$. But because any point of the form Ax for $x \geq 0$ lies in C, $uAx \leq 0$ for all $x \geq 0$. This could not be the case if any component of uA were positive, and it follows that $uA \leq 0$, as claimed.

Two elementary corollaries will prove useful. First, the system $Ax \geq b$, $x \geq 0$ is infeasible if and only if

$$\begin{bmatrix} A & -I \end{bmatrix} \begin{bmatrix} x \\ s \end{bmatrix} = b, \quad x, s \geq 0$$

is infeasible, where s is a vector of surplus variables and I is an identity matrix. Applying the Farkas lemma to this yields:

Corollary 3.4 *The system $Ax \geq b$, $x \geq 0$ is infeasible if and only if $uA \leq 0$ and $ub > 0$ for some $u \geq 0$.*

A second corollary is also easily shown.

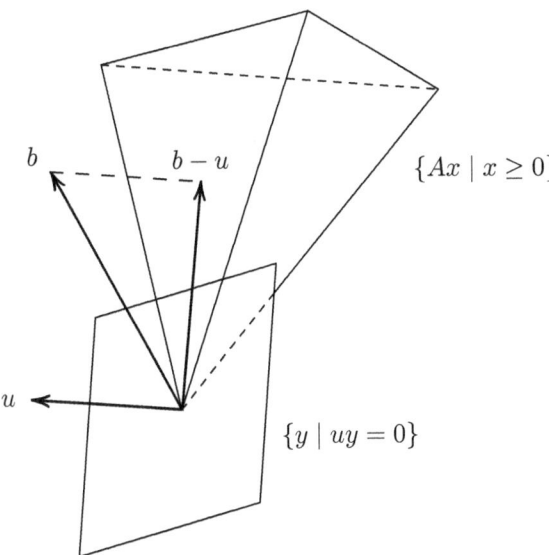

Fig. 3.3 *Illustration of the proof of the Farkas lemma.*

Corollary 3.5 *The system $Ax \geq b$ is infeasible if and only if $uA = 0$ and $ub > 0$ for some $u \geq 0$.*

Exercises

3.1. Let

$$A = \begin{bmatrix} 1 & 1 & 1 & 0 \\ 3 & 1 & 0 & 1 \end{bmatrix}, \quad b = \begin{bmatrix} 2 \\ 3 \end{bmatrix}, \quad c = [-5 \ -4 \ 0 \ 0]$$

Compute the basic solution (x_B, x_N) when $x_B = (x_1, x_2)$. Compute the reduced costs $r = c_N - c_B B^{-1} N$. Show that this solution is optimal. *Hint:* To obtain B^{-1}, solve $Ax = b$ for (x_1, x_2) by Gauss–Jordan elimination, which multiplies A and b by B^{-1}. Because the last two columns of A are I, they become B^{-1}. For example, if $x_B = (x_1, x_4)$, perform a row operation on $Ax = b$ to obtain

$$\begin{bmatrix} 1 & 1 & 1 & 0 \\ 0 & -2 & -3 & 1 \end{bmatrix} x = \begin{bmatrix} 2 \\ -3 \end{bmatrix}, \quad B^{-1} = \begin{bmatrix} 1 & 0 \\ -3 & 1 \end{bmatrix}$$

3.2. Solve $\min_{x \geq 0} \{cx \mid Ax = b\}$ with

$$c = [-3 \ 2 \ 0 \ 0 \ 0], \quad A = \begin{bmatrix} 1 & -2 & 1 & 0 & 0 \\ 1 & -1 & 0 & 1 & 0 \\ 2 & -1 & 0 & 0 & 1 \end{bmatrix}, \quad b = \begin{bmatrix} 1 \\ 1 \\ 3 \end{bmatrix}$$

For each iteration compute $x_B = B^{-1}b$, $r = c_N - c_B B^{-1} N$ as in the previous exercise. Let x_3 be the first variable to leave the basis. Two of the basic solutions are *degenerate* (at least one basic variable vanishes).

3.3. Attempt to find a feasible solution of

$$\max\ 3x_1 + 4x_2$$
$$x_1 + 2x_2 = 1$$
$$2x_1 + x_2 \geq 3$$
$$x \geq 0$$

using phase I of the simplex method. What is your conclusion?

3.4. Consider the problem of manufacturing widgets to maximize revenue, subject to labor and material availability:

$$\max\ 4x_1 + x_2$$
$$x_1 + x_2 \leq 2 \quad \text{(labor)}$$
$$3x_1 + x_2 \leq 3 \quad \text{(materials)}$$
$$x_1, x_2 \geq 0$$

where x_1, x_2 represent the production levels of deluxe and regular widgets, respectively.

(a) Solve the problem by inspecting the graph.

(b) Compute B^{-1} and the shadow prices of labor and materials. Note that x_B is (x_3, x_1), not (x_1, x_3), so that $B = \begin{bmatrix} 1 & 1 \\ 0 & 3 \end{bmatrix}$.

(c) Suppose an additional six cases of materials are made available. Use the shadow price of materials to derive an upper bound on the new objective function. What is the actual value of the new objective function?

3.5. Consider the problem

$$\min\ x_1 + x_2$$
$$x_1 \geq 1$$
$$2x_1 + x_2 \geq 4$$
$$2x_1 + 5x_2 \geq 10$$
$$x \geq 0$$

What are the shadow prices associated with the right-hand sides? Within what ranges are these prices valid if the right-hand sides are perturbed individually? *Hint:* The optimal solution is $x = (\frac{5}{4}, \frac{3}{2})$.

3.6. What happens to the optimal solution and cost of (3.3) when (c_B, c_N) is perturbed by $(\Delta c_B, \Delta c_N)$? Compute ranges for individual Δc_js within which this analysis is valid.

3.7. Perform a sensitivity analysis on the example in Exercise 3.5 for the objective function coefficients rather than the right-hand side. Compute ranges.

3.8. Prove Corollary 3.4.

3.9. Prove Corollary 3.5.

3.10. The Farkas lemma implies that some linear combination of the constraints in Exercise 3.3 proves their infeasibility. Exhibit such a linear combination. *Hint:* Consider the shadow prices associated with the two constraints.

3.11. Consider the linear system $Ax = b, x \geq 0$ given by

$$-x_1 + 2x_2 = -1$$
$$2x_1 - x_2 = 0$$
$$zx_1, x_2, \geq 0$$

Draw the polyhedral cone $C = \{Ax \mid x \geq 0\}$. What is the point $u - b$ in C closest to b? What is u? Prove that the linear system is infeasible.

3.2 Network Flows

The minimum-cost network flow problem is a highly versatile model with countless practical applications. Because it is a special case of the LP problem, it can be analyzed in much the same way. In fact, there is a network form of the simplex method that consumes much less time and space than the general method. This is due partly to the fact that basic solutions take a very simple form: they correspond to spanning trees on the network.

The network flow problem further specializes to the transportation problem, assignment problem, maximum-flow problem, and bipartite matching problem, all of which are useful in practice and can be solved with specialized forms of the simplex method.

Figure 3.4 shows a small instance of the minimum-cost network flow problem with $m = 5$ nodes. Each node i is associated with a net supply b_i at that node. Each arc (i, j) has a unit cost c_{ij}. A flow of x_{ij} over arc (i, j) incurs a cost of $c_{ij}x_{ij}$. The problem is to route flow over the arcs so as to minimize cost, while ensuring that the net flow out of each node i is equal to b_i.

In general, the minimum-cost network flow problem can be written,

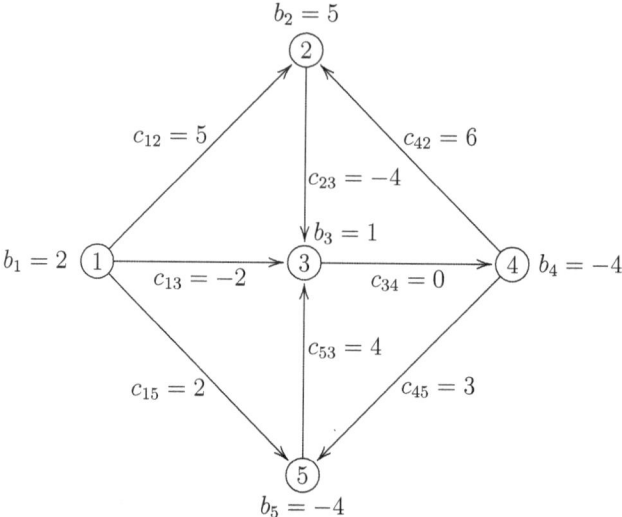

Fig. 3.4 A minimum-cost network flow problem.

$$\min \sum_{ij} c_{ij} x_{ij}$$

$$\sum_j x_{ij} - \sum_j x_{ji} = b_i, \text{ all } i \tag{3.11}$$

$$x_{ij} \geq 0, \text{ all } i, j$$

This is obviously a linear programming problem. If the instance of Fig. 3.4 is written with the equality constraints in matrix form $Ax = b$, the problem becomes

$$\min \sum_{ij} c_{ij} x_{ij}$$

$$\begin{bmatrix} 1 & 1 & 1 & & & & & \\ -1 & & & 1 & -1 & & & \\ & -1 & & -1 & 1 & & & -1 \\ & & & & -1 & 1 & 1 & \\ & & -1 & & & & -1 & 1 \end{bmatrix} \begin{bmatrix} x_{12} \\ x_{13} \\ x_{15} \\ x_{23} \\ x_{34} \\ x_{42} \\ x_{45} \\ x_{53} \end{bmatrix} = \begin{bmatrix} 2 \\ 5 \\ 1 \\ -4 \\ -4 \end{bmatrix} \tag{3.12}$$

$$x_{ij} \geq 0, \text{ all } i, j$$

where zeros are omitted for readability.

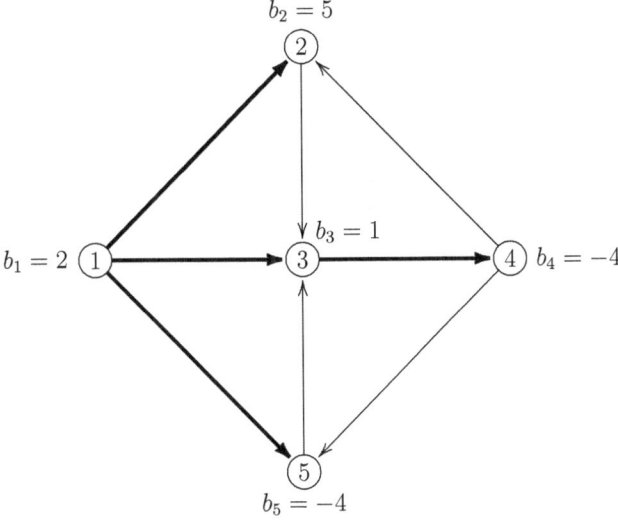

Fig. 3.5 A spanning tree (heavy arcs).

Note that the rows of A sum to a row of zeros. This is true in general, because each column contains a 1 and a -1, corresponding to the two endpoints of the arc. The rank of A must therefore be less than m. It is shown below that the rank is $m - 1$, which means that any basis consists of $m - 1$ columns.

3.2.1 Basis Tree Theorem

Any basis for the minimum-cost network flow problem corresponds to a spanning tree on the network (i.e., a tree that touches all the nodes). Consider, for example, the spanning tree shown by heavy arcs in Fig. 3.5. The corresponding columns of the constraint matrix A are

$$
\begin{array}{cccc}
x_{12} & x_{13} & x_{15} & x_{34}
\end{array}
$$

$$
\begin{bmatrix}
1 & 1 & 1 & \\
-1 & & & \\
& -1 & & 1 \\
& & & -1 \\
& & -1 &
\end{bmatrix}
\begin{array}{c}
1 \\
2 \\
3 \\
4 \\
5
\end{array}
$$

The rows and columns can be arranged so that the first four rows form a lower triangular matrix with nonzero diagonal elements. This shows that the columns are linearly independent and therefore form a basis.

To see how to triangularize the matrix, first pick any leaf node of the tree (say, node 4), and put the corresponding row first. This results in matrix (a) below.

$$
\begin{array}{cc}
\text{(a)} & \text{(b)} \\
x_{12}\ x_{13}\ x_{15}\ x_{34} & x_{34}\ x_{12}\ x_{13}\ x_{15} \\
\begin{bmatrix} & & & -1 \\ 1 & 1 & 1 & \\ -1 & & & \\ & -1 & & 1 \\ & & -1 & \end{bmatrix}\begin{matrix} 4 \\ 1 \\ 2 \\ 3 \\ 5 \end{matrix}
&
\begin{bmatrix} -1 & & & \\ & 1 & 1 & 1 \\ & -1 & & \\ 1 & & -1 & \\ & & & -1 \end{bmatrix}\begin{matrix} 4 \\ 1 \\ 2 \\ 3 \\ 5 \end{matrix}
\end{array}
$$

$$
\begin{array}{cc}
\text{(c)} & \text{(d)} \\
x_{34}\ x_{13}\ x_{12}\ x_{15} & x_{34}\ x_{13}\ x_{12}\ x_{15} \\
\begin{bmatrix} -1 & & & \\ 1 & -1 & & \\ & 1 & 1 & 1 \\ & & -1 & \\ & & & -1 \end{bmatrix}\begin{matrix} 4 \\ 3 \\ 1 \\ 2 \\ 5 \end{matrix}
&
\begin{bmatrix} -1 & & & \\ & -1 & & \\ & & -1 & \\ 1 & 1 & 1 & \\ & & & -1 \end{bmatrix}\begin{matrix} 4 \\ 3 \\ 2 \\ 1 \\ 5 \end{matrix}
\end{array}
$$

Because node 4 is a leaf node, there is only one nonzero entry in row 4. Move the column containing this entry to the first position to obtain matrix (b). Now remove node 4 from the tree, which leaves a smaller tree. Pick any leaf node (say, node 3), and move the corresponding row to the second position. Moving the column containing the nonzero entry to the second position results in matrix (c). After removing node 3, one more iteration of the process yields matrix (d), of which the first four rows are lower triangular with nonzero diagonal elements.

This shows that the four columns are linearly independent. Because the rank of A is less than 5, it must therefore be 4, and these columns form a basis. By applying this procedure in general, one can show that any spanning tree corresponds to $m - 1$ linearly independent columns and therefore to a basis.

Conversely, any basis corresponds to a spanning tree. A basis consists of $m - 1$ linearly independent columns. The corresponding edges cannot contain an (undirected) cycle, because otherwise the columns corresponding to the cycle would be linearly dependent. But any $m - 1$ edges that contain no cycle form a spanning tree.

To see why columns corresponding to a cycle must be linearly dependent, consider the undirected cycle 1–2–4–5 in Fig. 3.4. The corresponding columns are

$$
\begin{array}{cccc}
x_{12} & x_{42} & x_{45} & x_{15}
\end{array}
$$

$$
\begin{bmatrix}
1 & & & 1 \\
-1 & -1 & & \\
& & & \\
& & 1 & 1 \\
& & -1 & -1
\end{bmatrix}
\begin{array}{c}
1 \\
2 \\
3 \\
4 \\
5
\end{array}
$$

$$
\begin{array}{cccc}
1 & -1 & 1 & -1
\end{array}
$$

If one assigns multiplier 1 to forward arcs on the cycle and -1 to backward arcs (as shown beneath the columns), the resulting linear combination of columns is zero. This argument clearly applies to any cycle, and the Basis Tree theorem follows.

Theorem 3.6. *A set of columns of a network flow problem form a basis if and only if they correspond to a spanning tree of the network.*

3.2.2 Optimality Conditions

A basic feasible solution $x_B = B^{-1}b$ of an LP problem is optimal if the reduced cost vector $r = c_N - uN$ is nonnegative, where $u = c_B B^{-1}$. In the network flow problem, the x_{ij} column of N contains only two nonzeros, namely 1 in row i and -1 in row j. One can therefore write the column as $N_{ij} = e_i - e_j$, where e_i is the ith unit vector (a vector of all zeros except a 1 in the ith position). The reduced cost of x_{ij} is therefore

$$
r_{ij} = c_{ij} - u(e_i - e_j) = c_{ij} - u_i + u_j \tag{3.13}
$$

To evaluate the reduced costs r_{ij}, one need only compute the vector $u = c_B B^{-1}$. However, the $m - 1$ basis columns form an $m \times (m - 1)$ matrix B, which cannot be inverted. One can nonetheless solve the system $uB = c_B$, which consists of the equations

$$
u_i - u_j = c_{ij}, \quad \text{all basic arcs } (i, j) \tag{3.14}
$$

Because there are $m - 1$ equations and m variables, one of the variables u_i can be set to an arbitrary value, such as zero. This makes no difference because the reduced costs depend only on differences $u_j - u_i$. The system $uB = c_B$ is easy to solve because the first $m - 1$ rows of

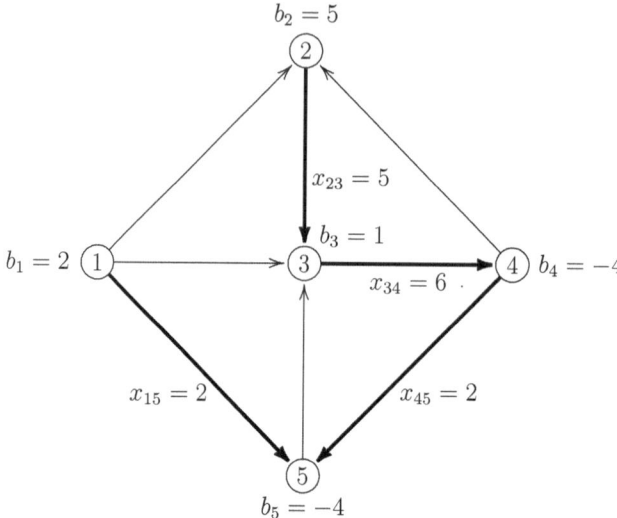

Fig. 3.6 A spanning tree corresponding to a basic feasible solution.

B are lower triangular (possibly after rearranging rows and columns), and u can be obtained by back substitution.

To illustrate this, consider again the example of Fig. 3.4. The basis tree of Fig. 3.6 corresponds to a basic feasible solution, which is shown in the figure. The system $uB = c_B$ for this basis is

$$\begin{bmatrix} u_1 & u_2 & u_3 & u_4 & u_5 \end{bmatrix} \begin{bmatrix} 1 & & & & \\ & 1 & & & \\ -1 & 1 & & & \\ & & -1 & 1 & \\ -1 & & & -1 & \end{bmatrix} = \begin{bmatrix} c_{15} & c_{23} & c_{34} & c_{45} \end{bmatrix} = \begin{bmatrix} 2 & -4 & 0 & 3 \end{bmatrix}$$

One can arbitrarily set $u_1 = 0$ and back out the other u_is as shown in Fig. 3.7, yielding $(u_1, \ldots, u_5) = (0, -3, 1, 1, -2)$. These values are sometimes called *potentials*, in analogy with an electrical circuit. They lead to the reduced costs shown in the figure. Only one reduced cost $r_{13} = -1$ is negative, which indicates that the solution is not optimal. It can be improved by bringing x_{13} into the basis.

The edges (i, j) in a network flow problem typically have capacities q_{ij}, which means that $x_{ij} \leq q_{ij}$. The optimality conditions for a *capacitated network flow problem* parallel that for an upper-bounded LP problem (3.8). A solution is optimal if the reduced costs are nonnegative on empty arcs and nonpositive on saturated arcs. That is, for each nonbasic x_{ij}, $r_{ij} \geq 0$ if $x_{ij} = 0$ and $r_{ij} \leq 0$ if $x_{ij} = q_{ij}$.

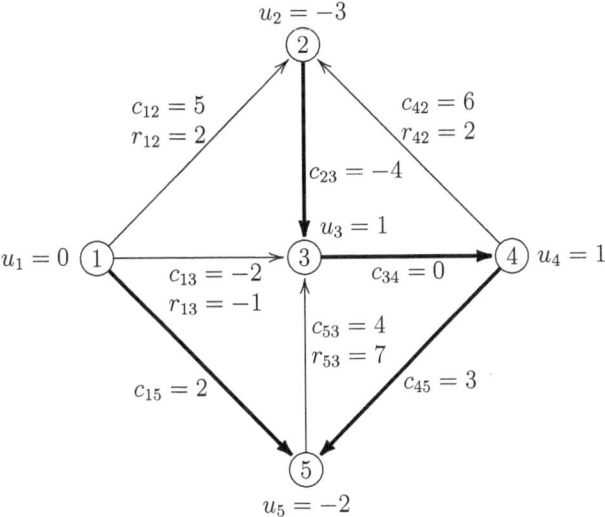

Fig. 3.7 Network potentials u_i and the resulting reduced costs r_{ij}.

3.2.3 Network Simplex Method

The general simplex method simplifies considerably for network flows, not only because of the Basis Tree Theorem, but also because the ratio test becomes a matter of identifying a cycle in the network.

The example of the previous section illustrates the idea. The non-basic variable x_{13} enters the basis because it has negative reduced cost $r_{13} = -1$. Adding edge $(1, 3)$ to the basis tree creates a cycle, as shown in Fig. 3.8. Only the flows on the cycle arcs can change. If the flow on $(1, 3)$ increases, then as one goes round the cycle, the flow on forward arcs must increase an equal amount, and the flow on the backward arcs must decrease an equal amount. At some point the flow on one of the backward arcs hits zero, and this arc will leave the basis.

In the present case, $(1, 5)$ is the only backward arc, and its flow of 2 hits zero when the flow on $(1, 3)$ rises to 2. This creates the new basis tree and corresponding flows shown in Fig. 3.9. At this point the new basic solution is already computed, and the next iteration begins by solving for potentials and reduced costs. As it happens, the new solution is optimal, and no further iterations are necessary.

A starting basic feasible solution can be found by solving a phase I problem, which is itself a network flow problem. The artificial variables correspond to artificial arcs to or from a dummy node 0 with supply

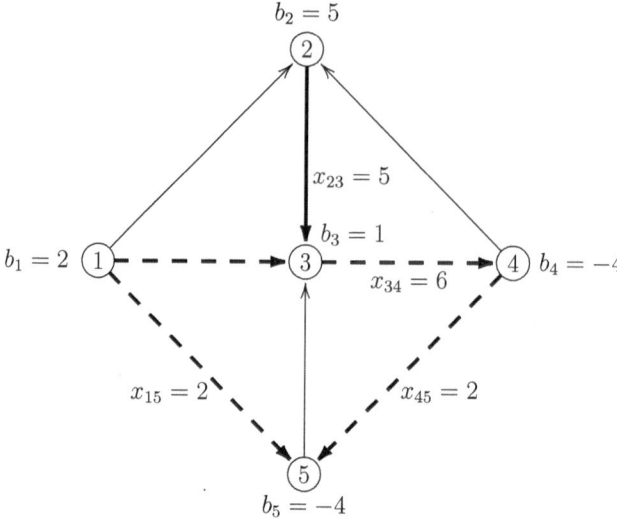

Fig. 3.8 Cycle created by adding arc (1,3) to the basis tree (dashed arcs).

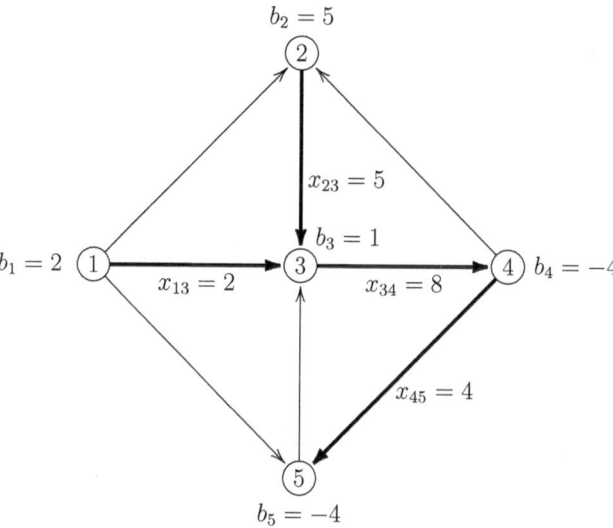

Fig. 3.9 An optimal basic solution.

zero. In particular, there are artificial arcs $(i, 0)$ for each node i with $b_i \geq 0$, and $(0, i)$ for each node i with $b_i < 0$. The artificial arcs obviously form a spanning tree, and the initial flow x_{i0} or x_{0i} on each artificial arc can be set to $|b_i|$. This provides a starting feasible basis, because $\sum_i b_i = 0$ in a feasible problem. Each artificial arc has cost 1,

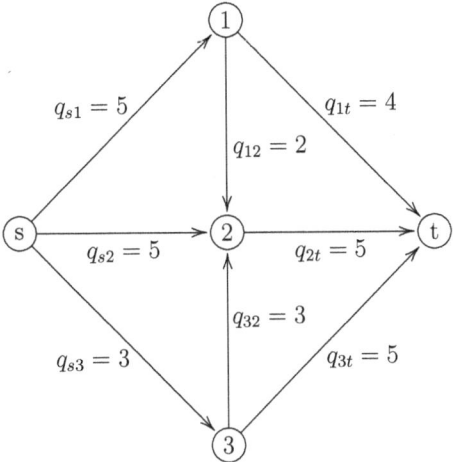

Fig. 3.10 A maximum-flow problem. Arc capacities are shown.

and the original arcs have cost zero. By minimizing cost in this phase I problem, one obtains a starting feasible basis for the original problem.

The capacitated network flow problem can be solved by extending the network simplex method to an upper-bounded method in the obvious way. This will become important in the solution of maximum-flow problems, discussed next.

3.2.4 Maximum Flow

The *maximum-flow problem* asks how to route a maximum amount of flow from a source node s to a sink node t, subject to arc capacities q_{ij}. A small instance of the problem appears in Fig. 3.10.

One can formulate the problem as a capacitated maximum-cost network flow problem by adding an arc (t, s) with cost 1 and placing a cost of zero on all other arcs (Fig. 3.11). The net supply at each node is zero, resulting in a *circulation*. The maximum-cost circulation in this network maximizes the flow from s to t. In general, the problem is

$$\text{max } x_{ts}$$

$$\sum_j x_{ij} - \sum_j x_{ji} = 0, \quad \text{all } i \tag{3.15}$$

$$0 \le x_{ij} \le q_{ij}, \quad \text{all } i, j$$

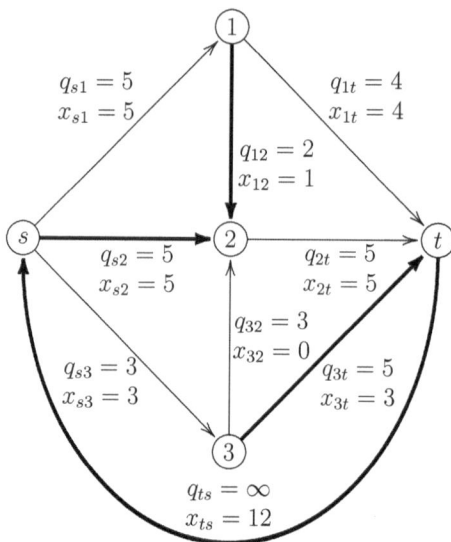

Fig. 3.11 Maximum-cost network flow formulation of the maximum-flow problem of Fig. 3.10. The cost is 1 on arc (t, s) and zero on the other arcs. A basis tree is shown (heavy arcs).

Optimality conditions follow from those for a general network flow problem. Because the objective is to maximize (rather than to minimize), a basic solution is optimal if the reduced costs are nonpositive on empty arcs and nonnegative on saturated arcs. That is, the solution is optimal if $r_{ij} = c_{ij} - u_i + u_j \leq 0$ when $x_{ij} = 0$ and $r_{ij} \geq 0$ when $r_{ij} = q_{ij}$ for all nonbasic arcs (i, j).

The potentials u_{ij} are particularly easy to calculate because only the return arc (t, s) has nonzero cost $c_{ts} = 1$. Because the potentials u_i must satisfy (3.14), u_s and u_t must satisfy $u_t - u_s = 1$. One can therefore set $u_s = 0$ and $u_t = 1$. To find the other potentials, let set S contain s and the nodes connected to s in the basis tree (nodes s, 1 and 2 in the example of Fig. 3.11), and let T contain t and the nodes connected to t (nodes t, 3). The pair (S, T) is called a *cut* because removing arcs from S to T cuts off all flow from s to t. Now because all remaining arcs have zero cost, $u_i = 0$ for $i \in S$ and $u_i = 1$ for $i \in T$. This simplifies the calculation of reduced costs using (3.13):

$$r_{ij} = \begin{cases} 0, \text{ if } i, j \in S \text{ or } i, j \in T \\ 1, \text{ if } i \in S \text{ and } j \in T \\ -1, \text{ if } i \in T \text{ and } j \in S \end{cases}$$

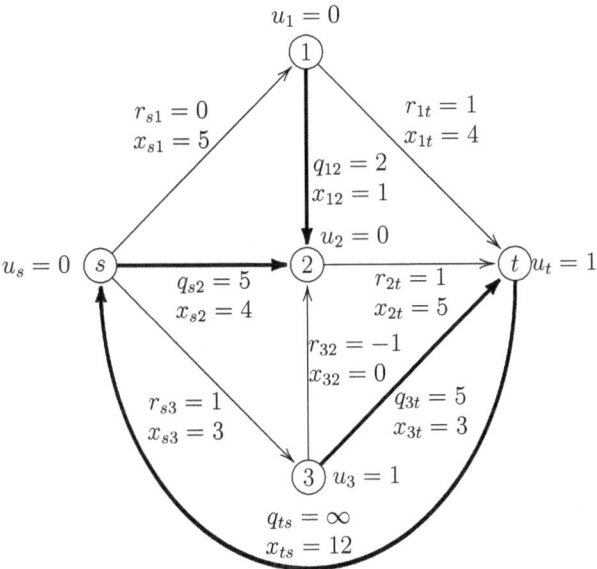

Fig. 3.12 Potentials and reduced costs for the basic solution of Fig. 3.11. The corresponding cut (S, T) is $S = \{s, 1, 2\}$ and $T = \{3, t\}$.

This is illustrated in Fig. 3.12. The optimality condition reduces to the following:

Corollary 3.7 *A basic solution of a maximum-flow problem that defines cut S, T is optimal if all arcs from S to T are saturated and all arcs from T to S are empty.*

It is easily checked that the basic solution of Fig. 3.12 is optimal.

A second optimality condition can be derived by examining how arcs enter the basis in the network simplex method. Consider the basis tree of Fig. 3.13, which is suboptimal. The flow can be increased by bringing saturated arc $(1, 2)$ from T to S into the basis. This creates the cycle consisting of the dashed arcs and the return arc (t, s) in Fig. 3.14. The flow on the cycle can be increased one unit, at which point the flow on $(1, t)$ hits its upper bound and becomes nonbasic.

The larger flow is obtained by increasing flow on forward arcs and reducing flow on backward arcs along the dashed path from s to t in Fig. 3.14. This can be viewed as augmenting the flow along a path in the *residual graph*, whose arcs represent residual capacity in each direction. The residual graph corresponding to the flow in Fig. 3.14

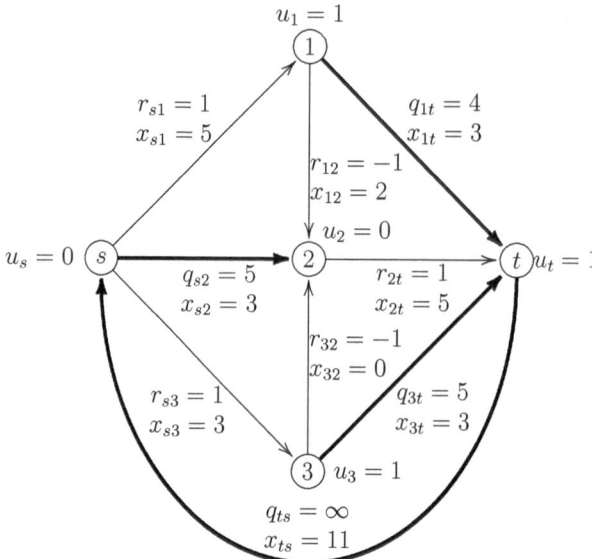

Fig. 3.13 A suboptimal basic solution of the problem in Fig. 3.10.

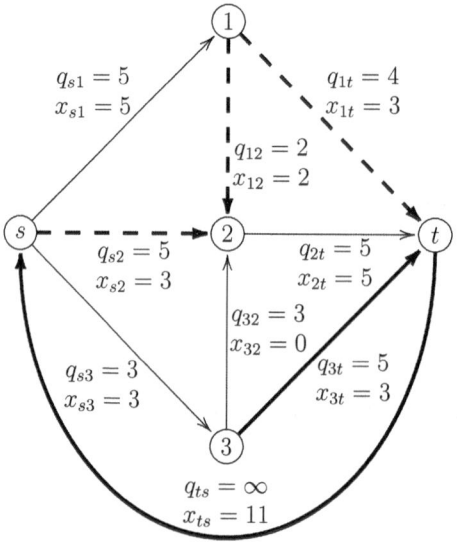

Fig. 3.14 Cycle created by adding arc $(1, 2)$ to the basic solution of Fig. 3.13 (dashed lines plus the return arc (t, s)).

appears in Fig. 3.15. Flow can be increased by one unit along the *augmenting path* shown in heavy arcs.

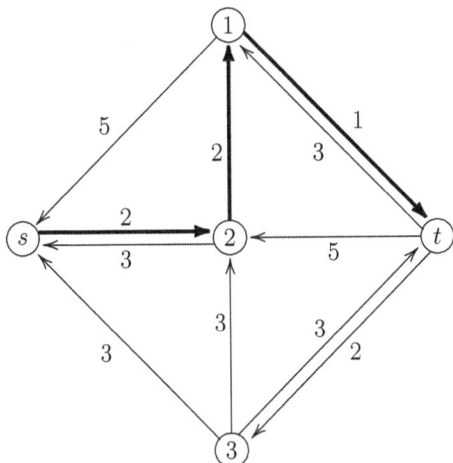

Fig. 3.15 Residual graph for the flow illustrated in Fig. 3.14. The residual capacity on each arc is shown.

In general, a residual graph is defined as follows. Suppose that each arc (i, j) of a graph G has capacity $[p_{ij}, q_{ij}]$; that is, the flow is bounded below by p_{ij} and above by p_{ij} (in the present context, $p_{ij} = 0$, but it will be convenient to consider positive lower bounds in subsequent sections). Let f_{ij} be the current flow on (i, j). The *residual graph* $R(f)$ of G for flow f is a graph on the same vertices as G that contains an edge (i, j) with capacity $[0, q_{ij} - f_{ij}]$ whenever $f_{ij} < q_{ij}$, and an edge (j, i) with capacity $[0, f_{ij} - p_{ij}]$ whenever $f_{ij} > p_{ij}$. The flow from t to s can be increased in G if there is an augmenting path from s to t; that is, if there is a path from s to t in $R(f)$.

The cycle created by adding an arc to the basis does not necessarily define an augmenting path from s to t, because it may be impossible to increase flow on the cycle due to degeneracy. Nonetheless, a solution is clearly suboptimal if an augmenting path exists, and the converse is true as well: a solution is optimal if there is no augmenting path from s to t. To show this, let S consist of s and all nodes to which there is an augmenting path from s. Thus $t \notin S$, and one can let T contain all nodes not in S. Then the total flow f_{ts} from s to t is at least the total capacity C of all the arcs from S to T, because these arcs are saturated. But (S, T) is a cut, which means that the flow from s to t can never be greater than C and therefore never greater than f_{ts}. This implies that f_{ts} is the maximum flow.

Corollary 3.8 *A feasible flow* f *in the maximum-flow problem is optimal if and only if there is no augmenting path from* s *to* t.

Because $f_{ts} \geq C$ and the the flow can never be greater than C, it follows that $f_{ts} = C$. Furthermore, no cut can have capacity less than C, because otherwise a flow of f_{ts} would be impossible. The maximum flow is therefore equal to the minimum cut capacity. This is an instance of *strong duality*—a concept introduced in the next chapter.

3.2.5 Bipartite Matching

Given a bipartite graph, a *matching* pairs some vertices on one side of the graph with vertices on the other side. More precisely, it is a set of edges of which at most one is incident to any given vertex. Figure 3.16(a) illustrates a matching in a small graph. It is in fact a *maximum cardinality* matching, because no matching on this graph has more than three edges.

A matching problem can be formulated as a maximum-flow problem as shown in Fig. 3.16(b). The arcs incident to s and t have unit capacity, and the remaining arcs have infinite capacity. All positive flows in a basic feasible solution must be unit flows, because otherwise the basic arcs would contain a cycle (possibly involving the return arc from t to s). All basic feasible solutions therefore define matchings.

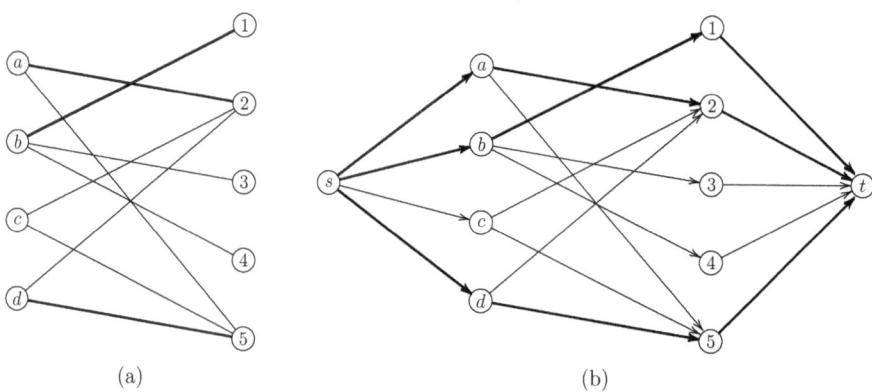

Fig. 3.16 (a) A bipartite matching problem, with a maximum cardinality matching shown as heavy edges. (b) Maximum-flow formulation of the matching problem, with a maximum flow shown as heavy arcs.

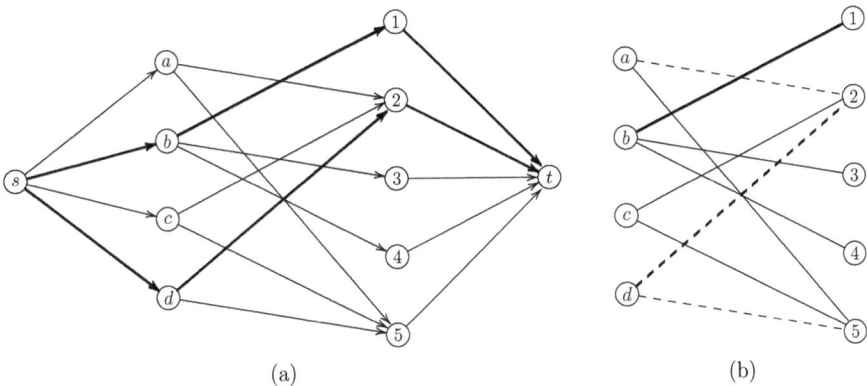

(a)   (b)

Fig. 3.17 (a) A flow that is not maximum (heavy arcs). (b) An alternating path (dashed arcs).

Conversely, every matching is clearly part of a basic solution of the flow problem. Maximum cardinality matchings therefore correspond to optimal basic solutions of the maximum-flow problem.

The optimality condition of Corollary 3.8 can be applied as follows. Because all positive flows in a basic feasible solution are unit flows, an augmenting path from s to t is one on which forward arcs are empty and backward arcs have unit flow. This is illustrated by Fig. 3.17(a), where there are three augmenting paths: $s \to a \to 5 \to t$, $s \to c \to 5 \to t$, and $s \to a \to 2 \to d \to 5 \to t$. By removing s and t from each of these, one obtains three *alternating paths* in the original graph; that is, paths on which edges in the matching alternate with edges not in the matching. For example, the third augmenting path defines the alternating path shown by dashed lines in Fig. 3.17(b).

In general, an alternating path for a given matching is a path of odd length in which every other edge is part of the matching, and the vertices at either end are not covered by the matching. Alternating paths correspond to augmenting paths in a basic feasible solution of the flow problem. An alternating path allows one to increase the size of the matching by reversing the out-in-out pattern to obtain an in-out-in pattern (which increases the flow on the augmenting path by one). Conversely, Corollary 3.8 implies that a basic feasible solution is optimal only if there is no augmenting path from s to t, which means there is no alternating path in the matching.

Corollary 3.9 *A bipartite matching has maximum cardinality if and only if there is no alternating path.*

Alternating paths also provide an algorithm that finds a maximum cardinality matching. Start with an arbitrary matching. If there is an alternating path P, let P' be the edges in P that belong to the matching. Remove the edges in P' from the matching and add the edges in $P \setminus P'$ to the matching. This creates a new matching that contains one more edge than the original matching. Continue until no alternating path exists.

3.2.6 Exercises

3.12. Verify that the basic solution of Fig. 3.9 is optimal by computing the potentials and reduced costs.

3.13. Construct the phase I network for the problem of Fig. 3.4. Compute the potentials and reduced costs for the starting basic solution. What is the next basic solution?

3.14. Apply the upper-bounded network simplex algorithm to the problem in Fig. 3.18, using the starting basic solution shown. Bring into the basis the variable whose reduced cost has the largest absolute value (and the correct sign).

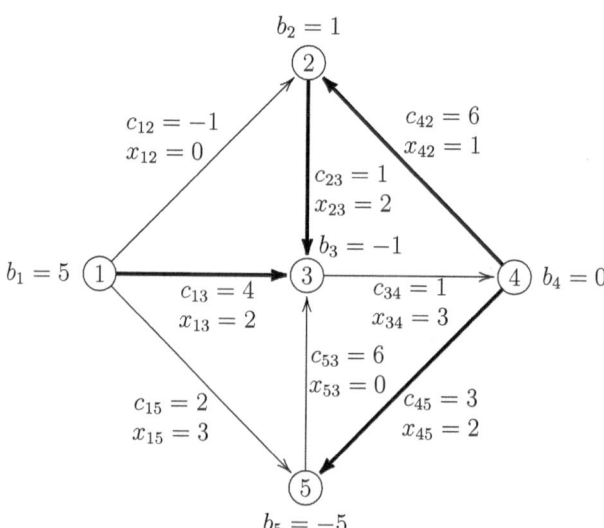

Fig. 3.18 Capacitated flow problem for Exercise 3.14. Each arc (i, j) has a capacity of $q_{ij} = 3$. A starting basic feasible solution is shown (bold arcs).

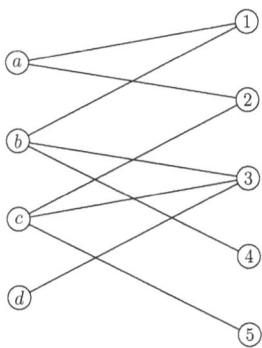

Fig. 3.19 Maximum cardinality matching problem for Exercise 3.16.

3.15. Use the network simplex algorithm to find a max flow for the network of Fig. 3.10. The starting basic solution can have a flow of zero on every arc. Arbitrarily let S contain all the nodes except t, and let the starting basis be an arbitrarily chosen spanning tree that includes arc (t, s). Thus, t is a leaf node of the basis tree. The solution may not have the same flow pattern as Fig. 3.11, but it should have the same maximum flow.

3.16. Solve the maximum cardinality matching problem in Fig. 3.19 as a maximum-flow problem by identifying augmenting paths. Then solve it by identifying alternating paths.

3.3 Nonlinear Programming

A *nonlinear programming* (NLP) problem is a continuous optimization problem with equality and/or inequality constraints. A problem with inequality constraints can be written

$$\text{nonlinear:} \begin{cases} \min \ f(x) \\ \mathbf{g}(x) \leq 0 \\ x \in S \subset \mathbb{R}^n \end{cases} \tag{3.16}$$

Here $\mathbf{g}$ is a vector of functions. An LP problem $\min\{cx \mid Ax \geq b\}$ is a special case of an NLP problem in which $f(x) = cx$ and $\mathbf{g}(x) = Ax - b$.

The problem (3.16) is *convex* when S is a convex set and the functions f and $\mathbf{g}$ are convex. A set $S \subset \mathbb{R}^n$ is convex if for all $x, x' \in S$ and all $\alpha \in [0, 1]$,

$$(1 - \alpha)x + \alpha x' \in S$$

A function $f : S \to \mathbb{R}$ is convex when for all $x, x' \in S$ and all $\alpha \in [0, 1]$,

$$f\left((1 - \alpha)x + \alpha x'\right) \leq (1 - \alpha)f(x) + \alpha f(x')$$

A function f is *concave* if $-f$ is convex. A convex problem (3.16) has a convex feasible set, because the constraints $\mathbf{g}(x) \leq 0$ for $x \in S$ define a convex set when $\mathbf{g}$ and S are convex.

A *local optimum* of (3.16) is one that is optimal in some neighborhood. That is, x^* is a local optimum if for sufficiently small δ, $f(x^*) \leq f(x)$ for all feasible x satisfying $||x^* - x|| \leq \delta$. Here $||v||$ denotes a norm of vector v, such as the euclidean norm $(\sum_j v_j^2)^{1/2}$. Convex problems have the advantage that any local optimum is a global optimum. Some necessary conditions for local optimality and sufficient conditions for global optimality are developed below.

3.3.1 Local Optimality Conditions

A local optimum can be characterized as a point at which one cannot improve the objective function by moving in a feasible direction. A vector d is an *improving direction* for the objective function f at x^* if one can move at least a short distance in direction d while reducing the value of f. That is, there is a $\delta > 0$ such that $f(x^* + \alpha d) < f(x^*)$ for all $\alpha \in (0, \delta]$. Vector d is a *feasible direction* for constraint $g_i(x) \leq 0$ at x^* if $g_i(x^*) \leq 0$, and one can move at least a short distance in direction d while continuing to satisfy the constraint. That is, there is a $\delta > 0$ such that $g_i(x^* + \alpha d) \leq 0$ for all $\alpha \in [0, \delta]$. A feasible point x^* for (3.16) is a local optimum if and only if no improving direction for the objective function is a feasible direction for all the constraints.

This fact leads to first-order necessary conditions for local optimality. If $f(x)$ is differentiable at x^*, let its *gradient* at $x = x^0$ be the row vector of partial derivatives evaluated at $x = x^0$:

$$\nabla f(x^0) = \left[\frac{\partial}{\partial x_1} f(x) \quad \cdots \quad \frac{\partial}{\partial x_n} f(x) \right]$$

Direction d is an improving direction for f at x^* if and only if $\nabla f(x^*)d < 0$. Let I index the binding constraints at x^*, so that $I = \{i \mid g_i(x^*) = 0\}$. Then if $i \notin I$, every direction is feasible for $g_i(x) \leq 0$, provided g_i is continuous. If $i \in I$ and g_i is differentiable, then any d for which $\nabla g_i(x^*)d < 0$ is a feasible direction. So, if no improving direction for f at x^* is a feasible direction for the constraints, the system

$$\nabla f(x^*)d < 0$$
$$\nabla g_i(x^*)d < 0, \quad i \in I$$

has no feasible solution. This means that for any given $\epsilon > 0$, the system

$$-\nabla f(x^*)d \geq \epsilon$$
$$-\nabla g_i(x^*)d \geq \epsilon, \quad i \in I$$

has no feasible solution. Corollary 3.5 of the Farkas lemma now implies that

$$u_0 \nabla f(x^*) + \sum_{i \in I} u_i \nabla g_i(x^*) = 0, \quad \epsilon u_0 + \epsilon \sum_{i \in I} u_i > 0 \qquad (3.17)$$

for some set of nonnegative multipliers u_0, u_i for $i \in I$. The inequality in (3.17) says that the multipliers are not all zero. Furthermore, if the gradients $\nabla g_i(x^*)$ for $i \in I$ are linearly independent, then one cannot have $u_0 = 0$. This means that $\nabla f(x^*)$ can be given a multiplier of 1 if the other gradients $\nabla g_i(x^*)$ have multiplier $\mu_i = u_i/u_0 \geq 0$.

Theorem 3.10. *Let x^* be a feasible solution of (3.16), and let $I = \{i \mid g_i(x^*) = 0\}$. Suppose that $\nabla f(x^*)$ exists, the gradients $\nabla g_i(x^*)$ for $i \in I$ are linearly independent, and g_i is continuous at x^* for $i \notin I$. Then x^* is a local optimum of (3.16) only if*

$$\nabla f(x^*) + \sum_{i \in I} \mu_i \nabla g_i(x^*) = 0 \qquad (3.18)$$

for some set of Lagrange multipliers $\mu_i \geq 0$, $i \in I$.

These necessary conditions for local optimality are known as the *Karush–Kuhn–Tucker (KKT) conditions*.

As an example, consider the problem of Fig. 3.20(a):

$$\begin{aligned} \min \ & x_1^2 + x_2 \\ & -x_1^3 + x_2 \leq 0 \\ & -x_2 + 1 \leq 0 \end{aligned} \qquad (3.19)$$

In this problem,

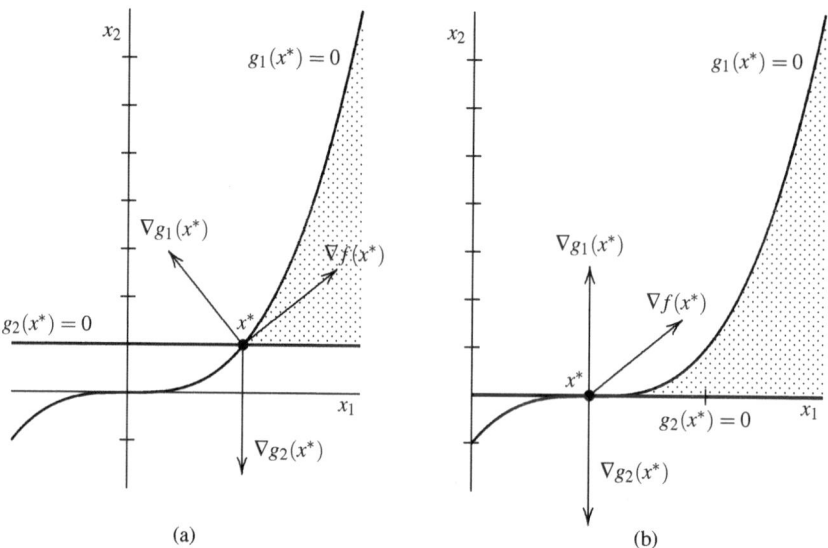

Fig. 3.20 (a) Illustration of the KKT conditions, where the shaded area is the feasible set. Here $\nabla f(x^*) + \mu_1 \nabla g_1(x^*) + \mu_2 \nabla g_2(x^*) = 0$ for some $\mu_1, \mu_2 \geq 0$. (b) A problem in which the KKT conditions do not apply because $\nabla g_1(x^*)$ and $\nabla g_2(x^*)$ are linearly dependent.

$$f(x) = x_1^2 + x_2, \quad \mathbf{g}(x) = \begin{bmatrix} -x_1^3 + x_2 \\ -x_2 + 1 \end{bmatrix}$$

$$\nabla f(x) = [2x_1 \ \ 1], \quad \begin{bmatrix} \nabla g_1(x) \\ \nabla g_2(x) \end{bmatrix} = \begin{bmatrix} -3x_1^2 & 1 \\ 0 & -1 \end{bmatrix}$$

A locally optimal (in fact, globally optimal) point is $x^* = (1, 1)$, at which both constraints are tight and the gradients $\nabla g_1(x^*), \nabla g_2(x^*)$ are linearly independent. The KKT conditions therefore imply that

$$[2 \ \ 1] + [\mu_1 \ \ \mu_2] \begin{bmatrix} -3 & 1 \\ 0 & -1 \end{bmatrix} = [0 \ \ 0]$$

for some $\mu_1, \mu_2 \geq 0$, namely $\frac{2}{3}$ and $\frac{5}{3}$. On the other hand, the problem of Fig. 3.20(b)

$$\min x_1^2 + x_2$$
$$-(x_1 - 1)^3 + x_2 \leq 0 \tag{3.20}$$
$$-x_2 \leq 0$$

has a locally (and globally) optimal solution $x^* = (1, 0)$, at which the gradients

$$\begin{bmatrix} \nabla g_1(x) \\ \nabla g_2(x) \end{bmatrix} = \begin{bmatrix} -3(x_1 - 1)^2 & 1 \\ 0 & -1 \end{bmatrix}$$

are linearly dependent. The KKT condition

$$[2 \ \ 1] + \begin{bmatrix} \mu_1 & \mu_2 \end{bmatrix} \begin{bmatrix} 0 & 1 \\ 0 & -1 \end{bmatrix} = [0 \ \ 0]$$

is not satisfied for any μ_1, μ_2.

The KKT conditions are easily extended to NLP problems with both equality and inequality constraints:

$$\text{nonlinear:} \begin{cases} \min(\max) \ f(x) \\ \mathbf{g}(x) \geq 0 \\ \mathbf{h}(x) = 0 \\ x \in \mathbb{R}^n \end{cases} \tag{3.21}$$

Let $\mathbf{h}(x)$ consist of functions $h_i(x)$ for $i \in I'$.

Corollary 3.11 *Let x^* be a feasible solution of (3.21), and let $I = \{i \mid g_i(x^*) = 0\}$. Suppose that $\nabla f(x^*)$ exists, the gradients $\nabla g_i(x^*)$ for $i \in I$, $\nabla h_i(x^*)$ for $i \in I'$ are linearly independent, and g_i is continuous at x^* for $i \notin I$. Then x^* is a local optimum of (3.16) only if*

$$\nabla f(x^*) + \sum_{i \in I} \mu_i \nabla g_i(x^*) + \sum_{i \in I'} \lambda_i \nabla h_i(x^*) = 0 \tag{3.22}$$

for some set of Lagrange multipliers $\mu_i \geq 0$ for $i \in I$ and λ_i for $i \in I'$.

3.3.2 Global Optimality Conditions

An NLP problem (3.16) tends to be much easier to solve when it is convex, because any local optimum is a global optimum. It suffices to find a local optimum, perhaps by finding a solution that satisfies the KKT conditions.

The underlying fact is that any local minimum of a convex function over a convex set is a global minimum.

Theorem 3.12. *If S is a convex subset of $\mathbb{R}^n$ and f a convex function on S, any local optimum of $\min\{f(x) \mid x \in S\}$ is a global optimum.*

Proof. If local optimum x^* is not a global optimum over S, then $f(\bar{x}) < f(x^*)$ for some $\bar{x} \in S$. Because S is convex, the line segment from x^* to $\bar{x}$ lies wholly within S. Given any neighborhood of x^*, there is a point $x = (1 - \alpha)x^* + \alpha\bar{x}$ in the interior of this line segment $(0 < \alpha < 1)$ within the neighborhood. Due to the convexity of f,

$$f(x) \leq (1 - \alpha)f(x^*) + \alpha f(\bar{x}) < f(x^*)$$

This violates the hypothesis that x^* is a local minimum. $\square$

The theorem applies to a convex NLP because it has a convex objective function and a convex feasible set. One can now state sufficient conditions for optimality: any solution of a convex NLP (3.16) that satisfies the KKT conditions is a global optimum. This can actually be strengthened somewhat, because only the functions g_i in binding constraints need be convex.

Corollary 3.13 *Let x^* be a feasible solution of (3.16) that satisfies the KKT conditions in Theorem 3.10. If S is a convex set, and f and g_i for $i \in I$ are convex, then x^* is a global optimum.*

Proof. Consider the relaxed problem R that results from dropping the nonbinding constraints from (3.16). The point x^* satisfies the KKT conditions for R as well as for (3.16). It is therefore a local optimum for R by Theorem 3.10, and a global optimum by Theorem 3.12. Because x^* is feasible for (3.16), it is a global optimum for (3.16). $\square$

A similar result holds for NLP problems (3.21) with equality constraints. Clearly, (3.21) has a convex feasible set if S is convex, the inequality constraint functions g_i are convex, and the equality constraint functions h_i are affine (i.e., have the form $A^i x - b_i$). Theorem 3.12 then implies that any local optimum is a global optimum. This can again be strengthened:

Corollary 3.14 *Let x^* be a feasible solution of (3.21) that satisfies the KKT conditions in Corollary 3.14. If S is a convex set, f and g_i for $i \in I$ are convex, h_i for all $i \in I'$ with $\lambda_i > 0$ are convex, and h_i for all $i \in I'$ with $\lambda_i < 0$ are concave, then x^* is a global optimum of (3.21).*

As an example, consider

$$\min \ x_1 + x_2$$
$$4x_1^2 - x_2 \le 0$$
$$2x_1^2 + x_2^2 = 1 \qquad\qquad (3.23)$$
$$x_1, x_2 \in \mathbb{R}$$

Supposing for the moment that the inequality constraint is not binding at the optimum, the the KKT equations (3.22) become

$$[1 \ \ 1] + \lambda[4x_1 \ \ 2x_2] = [0 \ \ 0]$$

This and the equality constraint imply that $(x_1, x_2) = (\pm 1/\sqrt{2}, \ 2/\sqrt{2})$, solutions that violate the inequality constraint. One must therefore suppose that the inequality constraint is binding at the optimum (if an optimum exists), and the KKT equations become

$$[1 \ \ 1] + \mu[8x_1 \ \ -1] + \lambda[4x_1 \ \ 2x_2] = [0 \ \ 0]$$

which imply
$$\lambda = -\frac{2x_1 + 1}{4x_1(x_2 + 1)}, \quad \mu = 2\lambda x_2 + 1$$

Only two points satisfy both constraints of (3.23) at equality:

$$(x_1, x_2) = \left(\pm \tfrac{1}{4}(\sqrt{17} - 1)^{1/2}, \ \tfrac{1}{16}(\sqrt{17} - 1) \right) \approx (\pm 0.261, 0.068)$$

The solution $(x_1, x_2) = (0.261, 0.068)$ yields $\mu \approx 0.814 \ge 0$ and $\lambda \approx -1.363$. Because $\lambda < 0$, this solution satisfies the sufficient conditions of Corollary 3.14 only if $h(x_1, x_2) = 2x_1^2 + x_2^2 - 1$ is concave. However, the function is convex and not concave. On the other hand, the solution $(x_1, x_2) \approx (-0.261, 0.068)$ yields $\mu \approx 1.058 \ge 0$ and $\lambda \approx 0.427 \ge 0$. Because $f(x) = x_1 + x_2$ and $g(x) = 4x_1^2 - x_2$ are also convex, this solution satisfies the conditions of Corollary 3.14 and is globally optimal.

Corollary 3.13 can be generalized to the case in which f is *pseudo-convex* and the functions $g_i(x)$ for $i \in I$ are *quasi-convex*. A differentiable function f is pseudo-convex if $f(x') \ge f(x)$ whenever $\nabla f(x)(x' - x) \ge 0$. A function $g_i(x)$ is quasi-convex if

$$g_i((1 - \alpha)x + \alpha x') \le \max\{g_i(x), g_i(x')\}$$

for all $x, x' \in S$ and all $\alpha \in [0, 1]$. Corollary 3.14 can be similarly generalized.

Exercises

3.17. Consider the problem

$$\min\ e^{x_1+x_2}$$
$$x_2 \geq e^{-x_1}$$
$$x_1 \geq 1$$

Find all local minima by considering the four cases defined by setting subsets of $\{\mu_1, \mu_2\}$ to zero. Be sure to check that the regularity condition is satisfied (linear indepedence of the constraint function gradients). Do any of the minima satisfy the sufficient conditions for a global optimum? *Hint:* $g_1(x_1, x_2) = e^{-x_1} - x_2$ is convex.

3.18. Exhibit a constrained optimization problem that has only one local minimum, such that (a) at least one Lagrange multipler is nonzero, and (b) the local minimum is not a global minimum but satisfies the KKT conditions. Show that the local minimum fails the sufficient conditions for a global minimum.

3.19. Write the KKT conditions for the unconstrained problem of minimizing $||Ax - b||^2$, which can also be written $(Ax - b)^T(Ax - b)$. Write a closed-form solution that is valid under appropriate assumptions. This is the *linear least squares* problem. *Hints:* The gradient of $x^T M x$ is $2x^T M$, and the gradient of cx is c.

3.20. Consider the linear programming problem (3.6) for a given basis B. Negate the equality constraint, so that is written $-Bx_B - Nx_n = -b$. Associate Lagrange multipliers u with the equality constraint and r with $x_N \geq 0$. State the KKT conditions for an optimal basic solution $(x_B, 0)$. Why is the regularity condition (i.e., the independence of constraint function gradients) satisfied? Show that any solution satisfying the KKT conditions satisfies sufficient conditions for a global optimum. Verify that the KKT conditions are identical with the LP optimality conditions in Theorem 3.1.

3.4 Dynamic Programming

Dynamic programming is another name for recursive optimization. It views the problem variables $x = (x_1, \ldots, x_n)$ as "controls" that govern transitions from one "state" to the next. Rather than enumerate exponentially many values of x, it solves the problem by evaluating states, which may be fewer in number. Dynamic programming can be deterministic or stochastic, but only the deterministic form is considered here.

3.4.1 State Variables

The idea of dynamic programming is best explained by example. Consider a knapsack problem:

$$\min \sum_{i=1}^{n} g_i(x_i)$$

$$\ell \le \sum_{i=1}^{n} a_i x_i \le u \tag{3.24}$$

$$x_i \in D_{x_i}$$

The problem is reformulated for dynamic programming by introducing *state variables* $s_1, \ldots, s_n$ that allow it to be solved in stages. In this case, the state s_i in stage i can be defined to be the partial sum $\sum_{j=1}^{i} a_i x_i$. The state in stage $i+1$ is obtained by adding $a_i x_i$ to s_i. The variable x_i can therefore be regarded as a *control variable* that governs the transition from state s_i to state s_{i+1}. The cost of this transition is $g_i(x_i)$.

The problem can now be stated in terms of states and controls:

$$\min \sum_{i=1}^{n} g_i(x_i)$$

$$s_{i+1} = s_i + a_i x_i, \quad i = 1, \ldots, n$$

$$x_i \in D_{x_i}, \quad i = 1, \ldots, n$$

$$s_1 = 0, \ \ell \le s_{n+1} \le u$$

Note that the initial state is $s_1 = 0$, and the final state must satisfy $\ell \le s_{n+1} \le u$.

A key property of this model is that it is *Markovian*. This means that the effect of applying a control and the resulting cost depend only on the current state s_i and the control x_i. It does not matter how the current state was reached.

Because the control variables are discrete in this problem, the states and transitions can be represented by a *state transition graph*. Suppose, for example, that the problem is to minimize $\sum_i g_i(x_i)$ subject to

$$7 \le 3x_1 + 2x_2 + x_3 + 4x_4 \le 9$$

Domains and costs are given in Table 3.1, where $g_i(D_{x_i})$ refers to a tuple of objective function values. For example, $g_1(D_{x_1})$ denotes

Table 3.1 Variable domains and costs for a small knapsack problem.

i	D_{x_i}	$g_i(D_{x_i})$
1	$\{0,1\}$	$(0,25)$
2	$\{0,1,2\}$	$(0,10,20)$
3	$\{1,3\}$	$(5,25)$
4	$\{0,1\}$	$(0,30)$

$(g_1(0), g_1(1)) = (0, 25)$. The state transition graph for this problem appears in Fig. 3.21. Each vertex corresponds to a state, and each edge to a state transition. The graph shows only the states that are *forward reachable*; that is, states to which there is a path from the initial state. Because each $a_i \geq 0$, transitions to $s_{i+1} > 9$ need not be considered. The terminal states $s_5 < 7$ are infeasible and are eliminated in Fig. 3.22. Traversing the graph backward from the three feasible states reveals which states are *backward reachable*. States that are backward unreachable can be dropped, leaving the simplified graph in Fig. 3.22. The feasible solutions of the problem correspond precisely to paths from stage 1 to stage 5 in this graph.

Each edge (s_i, s_{i+1}) of the state transition graph is assigned a length equal to the cost $g_i(x_i)$ of the transition it represents. The cost of

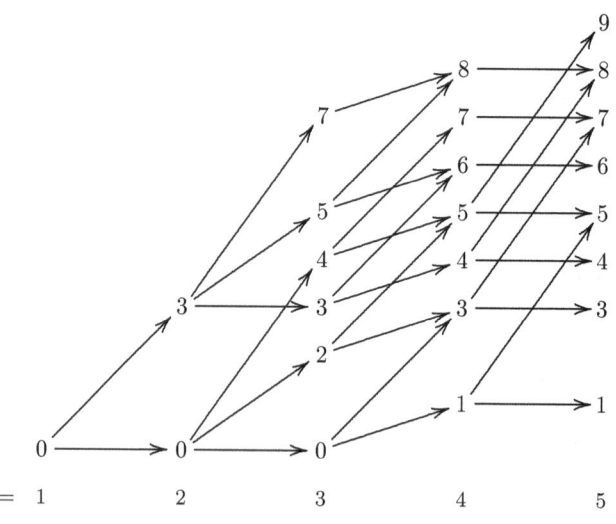

Fig. 3.21 State transition graph for a knapsack problem. The numbers at vertices represent the states s_i.

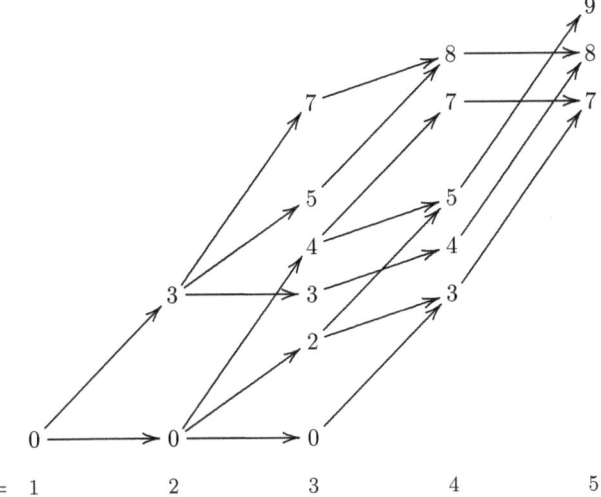

Fig. 3.22 State transition graph with unreachable states removed.

a solution is therefore the length of the corresponding path, and an
optimal solution corresponds to a shortest path.

In general, a dynamic programming model consists of control vari-
ables $x_1, \ldots, x_n$, state variables $s_1, \ldots, s_n$, and *transition functions*
$t_i(s_i, x_i)$ that determine the state s_{i+1} that results from applying control x_i in state s_i. In addition, each transition from s_i to $t_i(s_i, x_i)$
incurs a cost $c_i(s_i, x_i)$, and the cost of a solution x is the total cost
of the transitions it specifies. The problem of finding a minimum-cost
solution can be written

$$\min \sum_{i=1}^{n} c_i(s_i, x_i)$$
$$s_{i+1} = t_i(s_i, x_i), \quad i = 1, \ldots, n \qquad (3.25)$$
$$x_i \in X_i(s_i), \quad i = 1, \ldots, n$$
$$s_1 \in S_1, \quad s_{n+1} \in S_{n+1}$$

Here $X_i(s_i)$ is the set of available controls in state s_i, S_1 is the set of
feasible starting states, and S_{n+1} the set of feasible terminal states.

The state transition functions, along with the sets $X_i(s_i)$, S_1, and
S_{n+1}, encode the problem constraints. The constraints therefore need
have no particular form, so long as states can be defined to satisfy the
Markovian condition. It suffices to have oracles that deliver the value

of $t_i(s_i, x_i)$ and indicate whether a value or state belongs to $X_i(s_i)$, S_1, or S_{n+1}.

Only certain problems can be usefully written in the form (3.25). The art of dynamic programming consists in identifying state variables that encode enough information to satisfy the Markovian condition, but not so much as to result in an impracticably large number of states.

3.4.2 Recursive Solution

Due to the Markovian property, a dynamic programming problem (3.25) can be solved recursively. Let the *cost-to-go* $f(i, s_i)$ of state s_i be the length of a shortest path from s_i to a terminal state. The costs-to-go in stage i can be easily computed if the costs-to-go in stage $i+1$ are known. For the knapsack problem, one can use the recurrence relation

$$f(i, s_i) = \min_{x_i \in D_{x_i}} \{g_i(x_i) + f(i+1, s_i + a_i x_i)\} \qquad (3.26)$$

The costs-to-go are therefore computed in a backward pass through the graph.

The process starts with the boundary condition $f(5, s_5) = 0$ for all feasible terminal states s_5. The resulting costs-to-go are shown in Fig. 3.23, and the controls x_i that achieve the minimum in (3.26) are shown by heavy arc(s). The optimal value of the problem is the cost-to-go $f(1, s_1) = 45$ of the initial state $s_1 = 0$. The two paths from stage 1 to 5 consisting of heavy arcs represent the two optimal solutions, $x = (0, 1, 1, 1)$ and $(0, 2, 3, 0)$. Both have length 45.

In general, the cost-to-go is defined as

$$f(i, x_i) = \min \left\{ \sum_{j=i}^{n} c_j(s_j, x_j) \; \middle| \; \begin{array}{l} s_{j+1} = t_j(s_j, x_j), \; j = i, \ldots, n \\ x_j \in X_j(s_j), \; j = i, \ldots, n \\ s_{n+1} \in S_{n+1} \end{array} \right\}$$

The control variables x_i can be discrete or continuous, and the state variable s_i may be a tuple of variables. The cost-to-go can be computed recursively as follows:

$$f(i, s_i) = \min_{x_i \in X_i(s_i)} \{c_i(s_i) + f(i+1, t_i(s_i, x_i))\}, \quad i = 1, \ldots, n \quad (3.27)$$

given boundary conditions

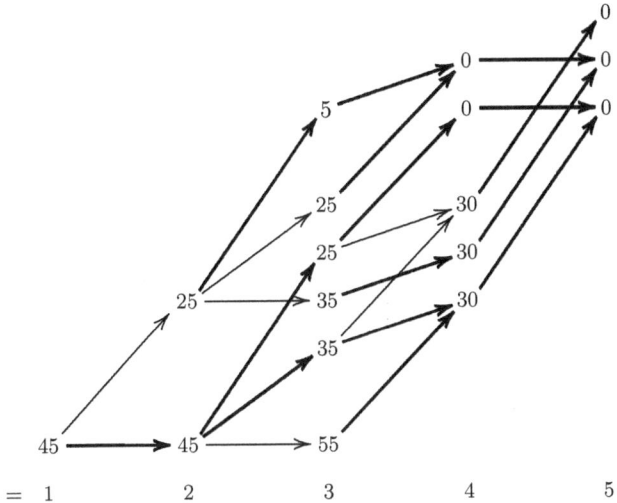

Fig. 3.23 State transition graph with costs-to-go. The cost-to-go at each vertex is shown and is obtained by following the heavy outgoing arc(s). The two heavy paths from stage 1 to stage 5 represent the optimal solutions.

$$f(n+1, s_{n+1}) = \begin{cases} 0 & \text{if } s_{n+1} \in S_{n+1} \\ \infty & \text{otherwise} \end{cases}$$

The optimal value is

$$\min_{x_1 \in S_1} \{f(1, s_i)\}$$

The recursion (3.27) is sometimes referred to as *Bellman's equations*.

A feasibility problem can be expressed in the form (3.25) by setting $c_i(s_i) = 0$ for all states s_i. Then all and only feasible solutions are optimal, with cost zero. In practice, however, it is common to define a cost-to-go that captures more information about the problem. In the knapsack problem, for example, $f(i, s_i)$ might represent the number of paths from s_i to a feasible terminal state. Then $f(1, 0)$ is the number of feasible solutions, zero if the problem is infeasible. The recursion is

$$f(i, s_i) = \sum_{x_i \in D_{x_i}} f(i+1, s_i + a_i x_i)$$

with boundary conditions

$$f(n+1, s_{n+1}) = \begin{cases} 1 & \text{if } \ell \leq s_{n+1} \leq u \\ 0 & \text{otherwise} \end{cases}$$

3.4.3 Complexity

Under the right conditions, dynamic programming can reduce the search time by an exponential factor. If n is the number of variables and m the size of the largest variable domain, then there are potentially m^n solutions to enumerate, each requiring $\mathcal{O}(n)$ function evaluations. Enumeration methods therefore have a worst-case complexity of $\mathcal{O}(nm^n)$. However, dynamic programming requires at most $\mathcal{O}(mN)$ function evaluations in each stage, where N is the maximum number of states in a stage. This results in worst-case complexity $\mathcal{O}(mnN)$, which may be polynomial in n, depending on how rapidly N grows.

For example, if the above knapsack problem has nonnegative integer coefficients a_i, N is at most the largest achievable sum $\sum_i a_i x_i$. If the magnitude of the coefficients a_i is bounded above by a constant, then $N = \mathcal{O}(n)$, and the complexity of dynamic programming is $\mathcal{O}(mn^2)$. The algorithm is said to be *pseudo-polynomial* because it is polynomial when the precision of data representation is bounded.

Unfortunately, many applications are afflicted with the *curse of dimensionality*, to use Bellman's term, which means that N grows exponentially with the number of variables. However, a careful choice of state variable may avoid the curse and keep N within bounds.

Exercises

3.21. A factory must meet a demand of d_t units at the end of each period t. The cost of producing x_t units is $c_t(x_t)$ in period t, where $x_t \in Q_t$ and Q_t is a finite set. The unit cost of holding inventory s_t during period t is $h_t(s_t)$, where s_t is the stock level at the beginning of the period. Any leftover stock after n periods has a unit salvage value of v. Write a dynamic programming recursion to find a production schedule that minimizes net cost while meeting demand and maintaining $s_t \geq 0$. Specify the boundary conditions, assuming $s_1 = 0$. *Hint:* Let x_t be the control and s_t the state variable.

3.22. The problem is the same as in the previous exercise, except that the factory must produce at one of two levels in each period t: 0 or q_t. Once the factory starts up, it must run at least k periods before shutting down, and there is a startup cost of f_t. Write a recursion to minimize cost. *Hint:* Use a second state variable to represent how long the factory has been running since the last startup.

3.23. A knapsack with capacity b must be filled, with no spare capacity, so as to maximize the value of the contents. There are n types of item to place in the knapsack, where each item of type i consumes space a_i and adds value c_i. Any (integral) number of items of each type may be selected. First write a dynamic programming recursion that defines $f(i, s_i)$ for stages $i = 1, \ldots, n + 1$. Then write a different recursion that dispenses with stages and defines $f(s)$ for each state s. The control decision in each state (except $s = b$) is which type of item to add to the knapsack (only one type, and only one item of that type). The state transitions must end up in state b, which means that $f(b) = 0$, and there are no available controls in state b. This formulates the problem as a deterministic finite automaton (Section 6.12.1).

3.24. A traveling salesman must visit each of the cities $1, \ldots, n$ exactly once, beginning at city 1. The distance from city i to j is c_{ij}. Write a dynamic programming recursion to minimize the distance traveled. *Hint:* To preserve the Markovian property, let the state be the set of cities visited so far. The curse of dimensionality afflicts this problem.

3.25. A ship carries cargo back and forth between two ports. In each period t, a quantity q_t of cargo arrives at port 1 to be shipped to port 2, and quantity q_i' of cargo arrives at port 2 to be shipped in the opposite direction. The unit cost of holding cargo at either port is c_t in period t. At the beginning of each period, the ship captain decides whether to sail to the other port at a cost of C, or to wait for more cargo to accumulate before sailing. The capacity of the ship is infinite for practical purposes. Write a dynamic programming recursion to find a minimum-cost sailing schedule.

3.5 Bibliographic Notes

Section 3.1. The simplex algorithm for linear programming is due to George Dantzig [155, 156]. Good expositions may be found in [136] and [491], the latter of which also presents interior point methods for linear programming. A theoretical treatment is [442]. The Farkas lemma was first stated in [192] and first correctly proved, using Fourier–Motzkin elimination, in [193].

Section 3.2. The basis tree theorem (Theorem 3.6) was stated in 1949 by Koopmans [323] for the transportation problem, a special case of the minimum-cost network flow problem. Early work on the primal simplex method for minimum-cost network flows includes [156, 232, 305, 463]. A good introduction to network flows is [46].

The augmenting-path algorithm for maximum-flow problems, also known as the the Ford–Fulkerson algorithm, appears in [207]. Other early results are developed in [208]. A survey of maximum-flow algorithms is provided by [7].

The alternating path condition for maximum cardinality bipartite matching (Corollary 3.9) is due to [79], but the basic idea is found in Egerváry's 1931 "Hungarian" algorithm for the assignment problem, as interpreted in [325]. A number of fast algorithms for maximum cardinality bipartite matching have been proposed, including one with $O(n^{1/2}m)$ complexity [299], where n is the number of vertices and m the number of edges, and an algorithm with $O(n^{1.5}(m/\log n)^{0.5})$ complexity [11].

Section 3.3. The Karush–Kuhn–Tucker conditions are developed in [303, 309, 326]. Good introductions to nonlinear programming include [47, 342].

Section 3.4. Dynamic programming is credited to Bellman [68, 69]. A good introductory text is [172], and a more advanced treatment [84].

Chapter 4
Duality

Duality provides a key to unifying optimization methods, because it connects search with the two primary mechanisms for exploiting problem structure: inference and relaxation. One can solve an optimization problem by searching for the best solution, but one can simultaneously search for a solution of the *inference dual* and/or the *relaxation dual*. The inference dual seeks to prove optimality by deducing from the constraints the tightest possible bound on the optimal value. The relaxation dual seeks the tightest bound that can be obtained by solving a relaxation of the problem.

Successful combinatorial optimization methods are almost always *primal–dual methods*: they combine search over solutions with inference and/or relaxation, perhaps to the point of solving the inference dual or the relaxation dual as they solve the *primal* (original) problem.

The best-known optimization duals—including the linear programming, Lagrangean, surrogate, and subadditive duals—can be viewed as inference duals or as relaxation duals, even if they seem historically to have been conceived as relaxation duals. When interpreted as inference duals, they provide a theoretical framework for constraint-directed search, which includes branching methods and Benders decomposition as special cases, as well as providing a basis for sensitivity analysis and some types of bounds propagation. When seen as relaxation duals, they allow one to obtain good bounds by removing or modifying the most troublesome constraints, a strategy that has made important contributions to optimization, particularly through Lagrangean duality.

4.1 Inference Duality

An optimization problem

$$\min\ f(x)$$
$$\mathcal{C}(x) \tag{4.1}$$
$$x \in D$$

can be viewed as posing the task of searching for a feasible solution x that minimizes $f(x)$ subject to $\mathcal{C}(x)$. The inference dual of (4.1) poses a complementary task: maximizing the lower bound on $f(x)$ that can be *inferred* from the constraints. This is likewise a search problem, the problem of searching for a *proof* of the optimal bound.

The dual problem can be written

$$\max\ v$$
$$\mathcal{C}(x) \overset{P}{\vdash} (f(x) \geq v) \tag{4.2}$$
$$v \in \mathbb{R},\ P \in \mathcal{P}$$

where $\mathcal{C}(x) \overset{P}{\vdash} (f(x) \geq v)$ indicates that proof P deduces $f(x) \geq v$ from $\mathcal{C}(x)$. The domain of variable P is a family $\mathcal{P}$ of proofs, and the dual solution is a pair (v, P). When the primal problem (4.1) is infeasible and the dual (4.2) therefore unbounded, the dual solution can be understood to contain, for any given x, a proof schema P that derives $f(x) \geq v$ from $\mathcal{C}(x)$ for any given v.

When discussing inference duals, it is important to distinguish inference from implication. The constraint set $\mathcal{C}$ *implies* $f(x) \geq v$ if any $x \in S$ satisfying $\mathcal{C}$ also satisfies $f(x) \geq v$. However, $f(x) \geq v$ can be inferred from $\mathcal{C}$ only when some proof in $\mathcal{P}$ derives $f(x) \geq v$ from $\mathcal{C}(x)$. A constraint inferred from $\mathcal{C}$ is always implied by $\mathcal{C}$, but the reverse need not be true.

The inference dual is meaningful when the proof family can derive such inequalities as $f(x) \geq v$. When (4.1) is an LP problem, for example, the family $\mathcal{P}$ of proofs can be taken to be nonnegative linear combinations of constraint inequalities that derive bounds $f(x) \geq v$. In this case, the dual solution can be identified with the multipliers in the linear combination that derive the tightest bound.

4.1.1 Weak and Strong Duality

A feasible value v of the inference dual can never be greater than a feasible value of the primal problem. For if v is feasible in the dual, then C implies $f(x) \geq v$, which means that $f(x) \geq v$ for all x that are feasible in the primal. This principle is known as *weak duality*.

Lemma 4.1 (Weak Inference Duality) *The optimal value of the primal problem is bounded below by the optimal value of any inference dual.*

The gap between the optimal value z^* of the primal and the optimal value v^* of the dual is the *duality gap*. A gap can exist because there may no proof in $\mathcal{P}$ that deduces $f(x) \geq z^*$ from C, even though C implies $f(x) \geq z^*$. This can occur when the proof family $\mathcal{P}$ is incomplete, meaning that it fails to deduce all implied constraints.

To make this more precise, let a proof family $\mathcal{P}$ be *complete* for C with respect to a family $\mathcal{F}$ of constraints if, given any $F \in \mathcal{F}$ implied by C, some proof $P \in \mathcal{P}$ deduces F from C. There is no duality gap if the family $\mathcal{P}$ of proofs is complete with respect to constraints of the form $f(x) \geq v$. The absence of a duality gap is known as *strong duality*.

Theorem 4.2. *Suppose the family $\mathcal{P}$ of proofs is complete with respect to constraints of the form $f(x) \geq v$. Then, the primal problem (4.1) and dual problem (4.2) have the same optimal value.*

Proof. Assume first that the primal problem has finite minimum value v^*. Then C implies $f(x) \geq v$ for $v = v^*$ but not for any $v > v^*$. Thus by completeness, some $P \in \mathcal{P}$ deduces $f(x) \geq v$ from C for $v = v^*$ but not for any $v > v^*$, which means that v^* is the optimal dual value. If the primal problem is infeasible, then C implies $f(x) \geq v$ for any v, because an infeasible constraint set implies everything. Thus by completeness, the dual is unbounded, and by convention both the primal and dual have optimal value ∞. If the primal is unbounded, then no v is a valid lower bound on $f(x)$, and both the primal and dual have optimal value $-\infty$. $\square$

Optimization methods frequently prove optimality by solving an inference dual, at least implicitly. It will be seen below that the simplex method, for example, proves optimality by solving the LP dual, which is an inference dual. When optimality is established by exhaustive search, the search generally provides a proof of optimality within some deductive framework. This is illustrated in Section 4.7 below. The inference dual is also related to complexity theory, because an optimization problem that belongs to co-NP has an inference dual that belongs to NP.

4.1.2 Certificates and Problem Complexity

A *certificate* of feasibility for a problem instance is a piece of information that allows one to verify that the instance is feasible. For example, a certificate might be a set of variable values that satisfy the constraints. A certificate of infeasibility is a proof that a problem instance has no feasible solution. A certificate of optimality is a special case: if v^* (possibly infinite) is the optimal value of the objective function $f(x)$, a certificate of optimality is a proof that the problem instance is infeasible when $f(x) < v^*$ is added to the constraints.

An optimization problem *belongs to NP* if there is a polynomial certificate of feasibility for any given feasible instance of the problem. This means that the amount of computation required to verify feasibility, using the certificate, is bounded by a polynomial function of the size of the problem instance (i.e., the number of binary digits required to encode the instance). An optimization problem *belongs to co-NP* if there is a polynomial certificate of optimality for any given instance. "NP" abbreviates *nondeterministic polynomial*, a concept from complexity theory that need not be further developed here.

Theorem 4.3. *If an optimization problem (4.1) belongs to co-NP, its inference dual (4.2) belongs to NP for some proof family $\mathcal{P}$.*

Proof. Suppose the problem belongs to co-NP. Then for any instance i of the problem, there is a polynomial proof P_i of the optimal value v_i^* of that instance. Let the proof family $\mathcal{P}$ consist of P_i for all instances i. Then given any instance i, there is a polynomial certificate (v_i^*, P_i) of feasibility for the dual. The dual therefore belongs to NP. $\square$

It is an interesting fact that most of the better known combinatorial problems belong to NP but not to co-NP. In other words, a combinatorial problem normally belongs to NP, and its inference dual typically does not. The primal problem might be expected to belong to NP, because a set of feasible variable values is normally a polynomial-size certificate. It is less predictable that a proof of optimality is typically much longer.

There are some notable exceptions, such as linear programming, which belongs to both NP and co-NP. Such problems are sometimes said to have *good characterizations*, in the sense that feasible solutions and proofs of optimality are easily encoded. In fact, one can plausibly conjecture that any problem belonging to both NP and co-NP is easy to solve, perhaps in the sense that it can be solved in polynomial time, as can an LP problem. No such conjecture has been proved, however.

4.1.3 Sensitivity Analysis

Sensitivity analysis is not only useful in practice but provides an important tool for domain reduction. It practical role stems from the fact that optimization models often require more data than can be accurately determined. While most of the numbers in a model typically have little bearing on the solution, the data that really matter generally cannot be identified in advance. In such cases, sensitivity analysis can be performed after the problem is solved to find the data to which the solution is sensitive. The key numbers are then adjusted or corrected and the problem re-solved. The cycle can be repeated until modelers are confident of the model's realism.

Sensitivity analysis is useful for domain reduction when there is a known upper bound $\bar{v}$ on the optimal value, perhaps the value of the best feasible solution discovered so far. The analysis is normally applied to a relaxation of the problem. If the solution of the relaxation is perturbed enough to lie outside of a certain range, sensitivity analysis may deduce that the optimal value of the relaxation will rise above $\bar{v}$. This shows that the optimal solution of the original problem must lie within this range, which provides a basis for reducing the variable domains.

The inference dual plays a central role in sensitivity analysis because it can bound the effect of changes in the problem data. By weak duality, the optimal solution (v^*, P^*) of the inference dual (4.2) provides a lower bound v^* on the optimal value of the primal. The proof P^* derives $f(x) \geq v^*$ using the constraint set $\mathcal{C}(x)$ as premises. Now if the problem data are changed, one can investigate whether *this same proof*

P^* still derives the bound v^*. Perhaps the premises that are essential to the proof are still intact. More generally, one can investigate what kind of alterations in the data can be made without invalidating P^* as a proof of the bound $v^* + \Delta v$. If there is no duality gap, this analysis identifies alterations in the problem that do not change the optimal value more than Δv.

In some cases, sensitivity analysis can identify a problem parameter b and a *value-bounding function* $v(b + \Delta b)$ that provides a valid lower bound on the optimal value for any perturbation Δb of b. For example, b might be the vector of right-hand sides of inequality constraints. A function $v(\cdot)$ is a value-bounding function when $(v(b + \Delta b), P^*)$ is dual feasible for any perturbation Δb. That is, the optimal proof P^* remains a valid proof as b is changed, but it proves a different lower bound $v(b + \Delta b)$. When $v(b + \Delta b)$ is exactly the optimal value of the perturbed problem, $v(\cdot)$ is known simply as a *value function*.

4.1.4 Constraint-Directed Search

The inference dual provides a general framework for constraint-directed search, of which branching search, Benders methods, and local search are special cases. Constraint-directed search was introduced in the context of the planning and schedule example of Section 2.8. It normally proceeds by assigning values to some of the variables, solving the subproblem that remains after fixing these variables, and generating a *nogood constraint* to direct the search toward possibly better assignments. The nogood constraint excludes the partial assignment just made and perhaps other assignments that are no better. The next partial assignment must satisfy the nogood constraints generated so far.

Constraint-directed search fits into the search-infer-and-relax framework because it enumerates a series of problem restrictions, namely subproblems in which some of the variables are fixed. The nogood constraints generated so far comprise a relaxation of the problem, which is solved to obtain the next partial assignment. A nogood constraint is obtained from each subproblem by an inference method, and in particular by solving an inference dual of the subproblem.

The inference dual plays a role similar to the one it plays in sensitivity analysis. Suppose the original problem minimizes $f(x, y)$ subject to

$\mathcal{C}$, and that variables x have been tentatively fixed to $\bar{x}$. The subproblem minimizes $f(x, y)$ subject to $\mathcal{C}$ and $x = \bar{x}$ to obtain an optimal value v^*. Solution of its inference dual obtains a proof of $f(x, y) \geq v^*$ based on the premises $\mathcal{C}$ and $x = \bar{x}$. One can now investigate what lower bound $v(x)$ can be obtained from *this same proof* when x takes other values. This gives rise to a nogood constraint $f(x, y) \geq v(x)$. The idea is developed formally in Chapter 5.

In the planning and scheduling example of Section 2.8, the nogood constraints are Benders cuts. Here the same variables are fixed in each iteration, namely variables x_j that indicate the machine to which job j is assigned. The next assignment to the x_js is obtained by solving the master problem, which contains all the Benders cuts generated so far. The subproblem is solved to find the minimum makespan M_i^* on each machine i when the jobs are so assigned. If M is the overall makespan, the solution of the inference dual is a proof that $M \geq M_i^*$ for each machine i, based perhaps on edge-finding, branching, or other inference methods. The proof for machine i is based on the premise that a certain set J_i of jobs are assigned to that machine. Examination of the proof, however, may reveal that a proper subset J_i' of these jobs actually play a role in the proof. This produces a Benders cut $M \geq v_i(x)$ where

$$v_i(x) = \begin{cases} M_i^*, & \text{if } x_j = i \text{ for all } j \in J_i' \\ 0, & \text{otherwise} \end{cases}$$

The cut can be formulated as a 0-1 inequality as in Section 2.8. It says that makespan can be reduced below M_i^* only by assigning at least one of the jobs in J_i' to another machine. The cut is added to the master problem, so that all future machine assignments must satisfy it. More sophisticated cuts are described in Section 6.14.3.

Section 5.2.2 presents a general Benders algorithm (under the rubric of logic-based Benders decomposition). To be effective, Benders cuts should be designed to exploit problem structure. Chapter 6 shows how to do this for a variety of problem types.

Exercises

4.1. Consider the optimization problem

$$\min\ 2x_1 + x_2$$
$$x_1 + x_2 \geq 1$$
$$x_1 - x_2 \geq 0 \tag{4.3}$$
$$x_1, x_2 \geq 0$$

where each x_j is a real number. Suppose that an inequality can be inferred from a constraint set if and only if it a sum of one or more constraints in the set. Solve the inference dual of (4.3) using this family of proofs. Exhibit two proofs, either of which solves the dual. What is the duality gap?

4.2. Consider the general LP problem $\min\{cx \mid Ax \geq b\}$, and let $A^i x \geq b_i$ be constraint i of $Ax \geq b$. Define an inference dual as in the previous exercise, and suppose that summing the constraints $A^i \geq b_i$ for $i \in I$ is a proof that solves the dual. Based on this solution, state a value-bounding function $v(b + \Delta b)$. What is this function for each of the two dual solutions identified in the previous exercise?

4.2 Relaxation Duality

In many cases, relaxations of a problem can be parameterized by a set of dual variables, thus providing a choice from an entire family of relaxations. The problem of finding the relaxation that provides the tightest bound can be posed as an optimization problem over the dual variables. This problem is the *relaxation dual.*

It is useful to define the concept of relaxation in terms of a problem's epigraph (Fig. 4.1). The *graph* of an optimization problem (4.1) is a plot of the objective function. Formally, it is the set $\{(f(x), x) \mid C(x),\ x \in D\}$. The *epigraph* of (4.1) is the set of points on or above the graph; namely,

$$\{(z, x) \mid z \geq f(x),\ C(x),\ x \in D\}$$

A *relaxation* of (4.1) is a problem whose epigraph contains the epigraph of (4.1). Clearly, the optimal value of a relaxation is a lower bound on the optimal value of the original problem.

A parameterized relaxation can be written

$$\min\ f(x, u)$$
$$C(x, u) \tag{4.4}$$
$$x \in D$$

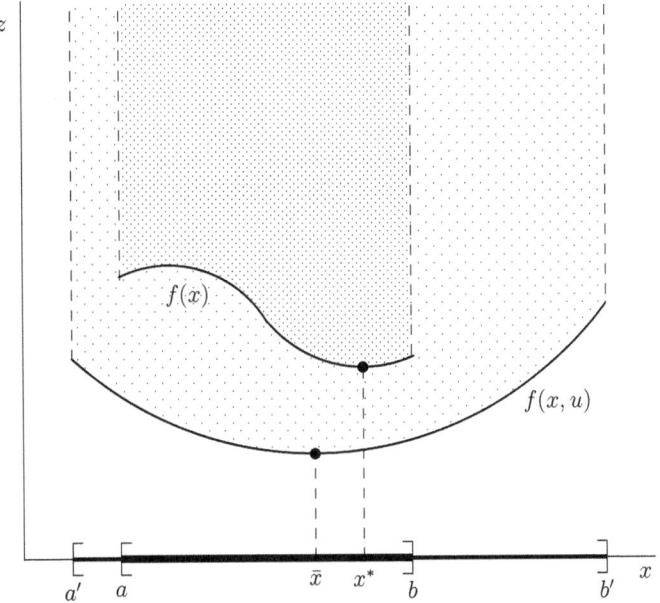

Fig. 4.1 Epigraph of an optimization problem (dark shading) with feasible set $[a, b]$, and epigraph of a relaxation of the problem (light shading) with feasible set $[a', b']$. The optimal solution $\bar{x}$ of the relaxation provides a lower bound $f(\bar{x}, u)$ on the optimal value $f(x^*)$ of the original problem. In this case, $\bar{x}$ is feasible for the original problem, but this is not true in general.

where $u \in U$ is a vector of *dual variables*. Each $u \in U$ defines a relaxation of (4.1). Let $\theta(u)$ be the optimal value of (4.4) for a given u, where $\theta(u)$ is ∞ if (4.4) is infeasible, and $-\infty$ if it is unbounded. Then the *relaxation dual* is the problem of finding the tightest relaxation:

$$\max_{u \in U} \theta(u) \tag{4.5}$$

Because $\theta(u)$ is a lower bound on the optimal value of (4.1) for each u, the same is true of the optimal dual value. The weak duality principle therefore holds.

Lemma 4.4 (Weak Relaxation Duality) *The optimal value of the primal problem is bounded below by the optimal value of any relaxation dual.*

There can be a duality gap, as in the case of inference duality. The relaxation dual is a strong dual when there is no duality gap.

The main utility of a relaxation dual is that it can provide a tight bound on the optimal value of the original problem. This may accelerate the search for an optimal solution or even prove optimality of a solution found during search. An obvious difficulty is that merely evaluating $\theta(u)$ requires the solution of an *inner* optimization problem. Nonetheless, if the dual is cleverly constructed, it may be practical to find a good bound. An optimal dual solution is not essential, because any feasible solution of the dual provides a valid lower bound. A local search algorithm, for example, may be perfectly satisfactory for solving the dual, as in the case of subgradient optimization (Section 4.5.6).

Exercises

4.3. Consider the problem of Exercise 4.1, and let $I = \{1, 2, 3, 4\}$ index the four constraints. Define a parameterized relaxation (4.4) by letting $f(x, u) = 2x_1 + x_2$ and letting $\mathcal{C}(x, u)$ be the set of constraints indexed by $u \subset I$, where $|u| \leq 2$. What is $\theta(u)$ for each u? What is an optimal solution of the relaxation dual? What is the duality gap?

4.4. In the previous exercise, let $\mathcal{C}(x, u)$ be the *sum* of constraints indexed by u. What is an optimal solution of the relaxation dual, and what is the duality gap?

4.3 Linear Programming Duality

The linear programming dual is one of the most elegant and useful concepts of classical optimization theory. It can be viewed as either an inference dual or a relaxation dual.

4.3.1 Inference Dual

The inference dual of a linear programming (LP) problem

$$\begin{aligned} \min \ & cx \\ & Ax \geq b \\ & x \geq 0 \end{aligned} \qquad (4.6)$$

can be written

$$\max v$$
$$(Ax \geq b) \overset{P}{\vdash} (cx \geq v) \tag{4.7}$$
$$v \in \mathbb{R}, \ P \in \mathcal{P}$$

Note that the domain of x is $\{x \in \mathbb{R}^n \mid x \geq 0\}$. To obtain the classical LP dual, one can let the proof family $\mathcal{P}$ consist of nonnegative linear combination and domination. That is, a bound $cx \geq v$ can be inferred when some nonnegative linear combination $uAx \geq ub$ of the system $Ax \geq b$ dominates $cx \geq v$. The inequality $uAx \geq ub$ dominates $cx \geq v$ when $uA \leq c$ and $ub \geq v$, or when no $x \geq 0$ satisfies $uAx \geq ub$.[1] Each proof P is therefore encoded as a vector $u \geq 0$ of multipliers that define the linear combination $uAx \geq ub$, which is sometimes referred to as a *surrogate* of $Ax \geq b$.

This inference dual is a strong dual because the inference method is complete with respect to linear inequalities, due to the Farkas lemma.

Corollary 4.5 *A linear system $Ax \geq b$ with $x \geq 0$ implies all and only linear inequalities dominated by its surrogates.*

Proof. Clearly, the system $Ax \geq b$ implies any surrogate, and therefore any inequality dominated by a surrogate. For the converse, it suffices to consider any inequality $cx \geq v$ that is implied by $Ax \geq b$, and show that some surrogate of $Ax \geq b$ dominates $cx \geq v$. Suppose first that $Ax \geq b, x \geq 0$ is feasible. This system together with $cx \leq v - \epsilon$ is infeasible for any $\epsilon > 0$, because $Ax \geq b$ implies $cx \geq v$. That is, the following is infeasible for any $\epsilon > 0$:

$$\begin{bmatrix} A \\ -c \end{bmatrix} x \geq \begin{bmatrix} b \\ \epsilon - v \end{bmatrix}, \quad x \geq 0$$

Applying Corollary 3.4 of the Farkas lemma, there is a vector $[\bar{u} \ u_0] \geq [0 \ 0]$ such that $\bar{u}A - u_0 c \leq 0$ and $\bar{u}b + u_0 \epsilon - u_0 v > 0$ for any $\epsilon > 0$. One may suppose $u_0 > 0$, because otherwise the system $Ax \geq b, x \geq 0$ is infeasible by Corollary 3.4. Dividing by u_0 yields $uA \leq c$ and $ub \geq v$, where $u = \bar{u}/u_0 \geq 0$. Thus, $cx \geq v$ is dominated by the surrogate $uAx \geq ub$ of $Ax \geq b$.

---

[1] One could perhaps more reasonably say that $uAx \geq ub$ dominates $cx \geq v$ when $\alpha uA \leq c$ and $\alpha ub \geq v$ for some $\alpha > 0$, or no $x \geq 0$ satisfies $uAx \geq ub$ (see Exercise 4.5). However, this broader definition complicates exposition and results in the same LP inference dual.

If the system $Ax \geq b, x \geq 0$ is infeasible, then by Corollary 3.4, $Ax \geq b$ has an infeasible surrogate, which dominates all inequalities, including $cx \geq v$. $\square$

Corollary 4.6 *The LP inference dual is a strong dual.*

The classical LP dual can now be derived. When the primal problem (4.6) is feasible, the inference dual (4.7) seeks the tightest bounding inequality $cx \geq v$ that is dominated by a surrogate $uAx \geq ub$. This can be written

$$\max \{v \mid uA \leq c, \ ub \geq v, \ u \geq 0\}$$

which can in turn be written

$$\max \ ub$$
$$uA \leq c \qquad\qquad (4.8)$$
$$u \geq 0$$

This is the classical LP dual, which is itself an LP problem with a polynomial certificate u of feasibility. The classical dual is therefore equivalent to the inference dual when the primal is feasible.

Corollary 4.7 *The inference dual (4.7) and classical dual (4.8) of a feasible LP problem (4.6) have the same optimal value, and any finite optimal solution of one is optimal for the other.*

The classical LP dual is also symmetric, in the sense that the dual of the dual is the primal.

As an example, consider the LP problem

$$\min \ 4x_1 + 7x_2$$
$$2x_1 + 3x_2 \geq 6$$
$$2x_1 + x_2 \geq 4 \qquad\qquad (4.9)$$
$$x_1, x_2 \geq 0$$

which was presented in Section 3.1. The optimal solution is $(x_1, x_2) = (3, 0)$, with optimal value 12. The classical dual is

$$\max \ 6u_1 + 4u_2$$
$$2u_1 + 2u_2 \leq 4$$
$$3u_1 + u_2 \leq 7$$
$$u_1, u_2 \geq 0$$

The dual problem can be solved as an LP problem to obtain the multipliers $(u_1, u_2) = (2, 0)$. This also solves the inference dual, because the surrogate $uAx \geq ub$ is $4x_1 + 6x_2 \geq 12$, which dominates the bound $4x_1 + 7x_2 \geq 12$.

Although strong duality holds without exception for the LP inference dual, the classical strong duality theorem is subject to a constraint qualification.

Theorem 4.8. *An LP problem (4.6) and its classical dual (4.8) have the same optimal value, unless both problems are infeasible.*

Proof. If the primal (4.6) is feasible, then the classical dual and inference dual are equivalent by Corollary 4.7, so that (4.6) and (4.8) have the same optimal value. If the classical dual (4.8) is feasible, then one can take advantage of the fact that, for classical duality, the dual of the dual is the primal. Thus, (4.8) can be regarded as the feasible primal and (4.6) the dual. These have the same optimal value, again by Corollary 4.7. $\square$

4.3.2 Dual Simplex Method

The simplex method maintains primal feasibility while moving toward a dual feasible solution. To see this, recall that the reduced cost vector of a basic solution $(x_B, 0) = (B^{-1}b, 0)$ is $r = c_N - uN \geq 0$, where $u = c_B B^{-1}$. If the solution is optimal in the primal (3.6), then $r \geq 0$, and it is easy to check that u is feasible in the dual:

$$\begin{aligned} \max \; & ub \\ & uB \leq c_B \\ & uN \leq c_N \end{aligned} \qquad (4.10)$$

Thus, achieving optimality is the same as achieving dual feasibility.

This suggests that one might use the opposite method of maintaining dual feasibility while moving toward primal feasibility. This is particularly useful when reoptimizing after the addition of one or more constraints, a situation that occurs repeatedly in branch-and-bound algorithms. Adding a constraint generally makes the optimal primal solution infeasible, but the optimal dual solution remains feasible if the new dual variable is set to zero. The current basis can therefore be used as a starting point to reoptimize the dual. This is accomplished by the *dual simplex method.*

The dual simplex method is not merely the result of applying the simplex method to the dual. Rather, it uses primal data structures to reoptimize the dual solution. It begins with a basic solution $x = (x_B, 0) = (B^{-1}b, 0) = (\tilde{b}, 0)$ and dual feasible solution $u = c_B B^{-1}$. If $\tilde{b} \geq 0$, the solution is already feasible in the primal. Otherwise, $\tilde{b}_i < 0$ for some i. As in the standard simplex method, the next step is to improve the dual objective function by moving some nonbasic dual variable as far from zero as possible. A suitable variable becomes evident after the change of variable $\tilde{u} = uB - c_B$, or equivalently $u = \tilde{u}B^{-1} + c_B B^{-1}$. The dual (4.10) becomes:

$$
\begin{aligned}
\max \ & \tilde{u}\tilde{b} + c_B\tilde{b} \\
& \tilde{u} \leq 0 \\
& \tilde{u}\tilde{N} \leq r
\end{aligned}
\tag{4.11}
$$

where $\tilde{N} = B^{-1}N$ and $r = c_N - c_B\tilde{N}$. Because $\tilde{b}_i < 0$, the goal is to make $\tilde{u}_i$ as negative as possible while maintaining dual feasibility:

$$
\tilde{u}_i^{\text{new}} = \min\left\{ \tilde{u}_i \ \middle| \ \tilde{u}\tilde{N} \leq r \right\} = \min\left\{ \tilde{u}_i \ \middle| \ \tilde{u}_i\tilde{N}^i \leq r \right\}
$$

where $\tilde{N}^i$ is row i of $\tilde{N}$. If $\tilde{N}^i \geq 0$, then the dual is unbounded, which means the primal is infeasible. Otherwise a ratio test is used:

$$
\tilde{u}_i^{\text{new}} = \max_j \left\{ \frac{r_j}{\tilde{N}_{ij}} \ \middle| \ \tilde{N}_{ij} < 0 \right\}
$$

However, u_i is not adjusted directly. Instead, the variable x_j for which j achieves the maximum ratio enters the basis, because the corresponding dual constraint $\tilde{u}\tilde{N}_j \leq r_j$ becomes tight. The variable that leaves the basis is the one that corresponds to the dual constraint $\tilde{u}_i \leq 0$ (i.e., the variable with value $\tilde{b}_i$), because this constraint is becoming slack.

The dual simplex method therefore proceeds as follows. If $\tilde{b} \geq 0$, the primal is already feasible. Otherwise select a $\tilde{b}_i < 0$, and let the corresponding primal variable with value $\tilde{b}_i$ leave the basis. If $\tilde{N}^i \geq 0$, the dual is unbounded and the primal infeasible. Otherwise let x_j enter the basis, where j is selected to minimize $r_j/(-\tilde{N}_{ij})$ subject to $\tilde{N}_{ij} < 0$. Update B^{-1} and repeat.

As an example, suppose the constraint $-x_1 + 2x_2 \geq -1$ is added to the problem (3.2), for which the optimal solution is $x_B = (x_1, x_4) = (3, 2)$ with

$$B = \begin{bmatrix} 2 & 0 \\ 2 & -1 \end{bmatrix}, \quad B^{-1} = \begin{bmatrix} \frac{1}{2} & 0 \\ 1 & -1 \end{bmatrix}, \quad \tilde{b} = \begin{bmatrix} 3 \\ 2 \end{bmatrix}, \quad u = \begin{bmatrix} 2 & 0 \end{bmatrix}$$

The altered problem is

$$\min \begin{bmatrix} 4 & 7 & 0 & 0 & 0 \end{bmatrix} x$$
$$\begin{bmatrix} 2 & 3 & -1 & 0 & 0 \\ 2 & 1 & 0 & -1 & 0 \\ -1 & 2 & 0 & 0 & -1 \end{bmatrix} x = \begin{bmatrix} 6 \\ 4 \\ -1 \end{bmatrix}, \quad x \geq 0 \qquad (4.12)$$

There is a new surplus variable x_5 in the primal and a new dual variable u_3. Keeping $(x_1, x_4) = (3, 2)$, x_5 becomes basic and takes the infeasible value -2. A row and column are added to B, and B^{-1} is updated with row operations. This yields

$$B = \begin{bmatrix} 2 & 0 & 0 \\ 2 & -1 & 0 \\ -1 & 0 & -1 \end{bmatrix}, \quad B^{-1} = \begin{bmatrix} \frac{1}{2} & 0 & 0 \\ 1 & -1 & 0 \\ -\frac{1}{2} & 0 & -1 \end{bmatrix}, \quad \tilde{b} = \begin{bmatrix} 3 \\ 2 \\ -2 \end{bmatrix}, \quad u = \begin{bmatrix} 2 & 0 & 0 \end{bmatrix}$$

with $r = c_N - uN = \begin{bmatrix} 1 & 2 \end{bmatrix}$. Note that the new dual variable is set to zero, and the dual solution u remains feasible because $r \geq 0$. The dual simplex method begins at this point. Because $x_5 = \tilde{b}_3 < 0$, x_5 leaves the basis. To find which variable enters the basis, select the smallest ratio $r_j/(-\tilde{N}_{3j})$ over all j for which $\tilde{N}_{3j} < 0$, where

$$r = \begin{bmatrix} r_2 & r_3 \end{bmatrix}, \quad \tilde{N}^3 = \begin{bmatrix} \tilde{N}_{32} & \tilde{N}_{33} \end{bmatrix} = \begin{bmatrix} -\frac{7}{2} & \frac{1}{2} \end{bmatrix}$$

Only $\tilde{N}_{32} = -\frac{7}{2}$ is negative, and x_2 therefore enters the basis. The new basic solution is $x_B = (x_1, x_4, x_2) = \tilde{b} = (\frac{15}{7}, \frac{4}{7}, \frac{6}{7})$. Because $\tilde{b} \geq 0$, the method terminates with a reoptimized solution.

4.3.3 Sensitivity Analysis

The LP inference dual leads immediately to sensitivity analysis for linear programming. Suppose that the LP problem (4.6) is feasible and is altered as follows:

$$\min (c + \Delta c)x$$
$$(A + \Delta A)x \geq b + \Delta b, \quad x \geq 0 \tag{4.13}$$

Let u^* be an optimal dual solution for the original problem. It encodes a proof of the surrogate inequality $u^*Ax \geq u^*b$ and therefore any dominated bound $cx \geq v$; that is, any bound for which $u^*A \leq c$ and $u^*b \geq v$. The *same proof* u^* derives the following surrogate inequality for the altered problem (4.13):

$$u^*(A + \Delta A)x \geq u^*(b + \Delta b) \tag{4.14}$$

If $u^*(A + \Delta A) \leq c + \Delta c$, this surrogate dominates and therefore proves the bound

$$(c + \Delta c)x \geq u^*b + u^*\Delta b \tag{4.15}$$

Because $u^*A \leq c$ by dual feasibility, inequality (4.14) and therefore the bound (4.15) hold whenever $u^*\Delta A \leq \Delta c$. Noting that u^*b is the optimal value of the original problem, one can state the following.

Theorem 4.9. *Let z^* be the optimal value of an LP problem (4.6) and u^* an optimal solution of its dual. The optimal value of the altered problem (4.13) is bounded below by $z^* + u^*\Delta b$, provided $u^*\Delta A \leq \Delta c$.*

If only the right-hand side b is perturbed, u^* remains dual feasible for any Δb, because b occurs only in the objective function of the dual (4.8). So by weak duality, $u^*(b + \Delta b) = z^* + u^*\Delta b$ is a lower bound on the optimal value of the perturbed problem. The components of u^* are therefore shadow prices, because they indicate the marginal sensitivity of the optimal cost to perturbations in the right-hand side.

Formally, $v(b + \Delta b) = z^* + u^*\Delta b$ is a value-bounding function for the LP problem, because $(z^* + u^*\Delta b, u^*)$ is feasible in the inference dual for any Δb.

Returning to the example (4.9), the optimal value is 12 and the optimal dual solution is $(u_1^*, u_2^*) = (2, 0)$. Applying Theorem 4.9, the perturbed problem (4.13) has an optimal value of at least $12 + 2\Delta b_1$, provided $2\Delta A_{11} \leq \Delta c_1$ and $2\Delta A_{12} \leq \Delta c_2$. For instance, if the right-hand side of the first inequality is reduced from 6 to 3, and the problem is otherwise unchanged, the minimum cost is bounded below by $12 + 2(-3) = 6$ (it is actually 8). Thus the solution is rather sensitive to the first constraint's right-hand side. Small perturbations of the second constraint (which is slack in the optimal solution) have no effect on the bound, because its shadow price $u_2^* = 0$.

In general, one would expect a shadow price u_i^* to be zero when the ith constraint $A^i x \geq b_i$ is slack, because small perturbations in the right-hand side have no effect on the solution. This is the *complementary slackness principle*, which states that if x^* is optimal in the primal and u^* is optimal in the dual, then $u_i^*(A^i x^* - b_i) = 0$ for all i. This is implied by the following, which has an easy algebraic proof.

Corollary 4.10 *If x^* is optimal in the primal problem (4.6), and u^* is optimal in the dual problem (4.8), then $u^*(Ax^* - b) = 0$.*

4.3.4 Relaxation Dual

A linear programming problem (4.6) can be relaxed by replacing its constraint set $Ax \geq b$ with a surrogate $uAx \geq ub$—whence the term "surrogate." The relaxation dual is the problem of finding a surrogate that dominates the tightest possible bound on the objective function. Formally, the dual maximizes $\theta(u)$ subject to $u \geq 0$, where

$$\theta(u) = \min_{x \geq 0} \{cx \mid uAx \geq ub\}$$

If v^* is the (finite) optimal value of the primal problem (4.6), Corollary 4.5 implies that there is a surrogate $u^* Ax \geq u^* b$ that dominates the bound $cx \geq v^*$. The LP relaxation dual is therefore solved by the same multipliers u^* that solve the LP inference dual, and there is no duality gap.

Corollary 4.11 *The LP relaxation dual has the same optimal value as the LP inference dual, and any finite optimal solution of one is an optimal solution of the other.*

Proof. If the primal has a finite optimal value v^*, the proof is as described above. If the primal is infeasible, the inference dual has optimal value ∞. By Corollary 3.4, $uA \leq 0$ and $ub > 0$ for some $u \geq 0$, which implies that the relaxation dual also has value ∞. If the primal is unbounded, the inference dual has value $-\infty$. $\square$

Exercises

4.5. Under what conditions does $ax \geq a_0$ imply $cx \geq c_0$ if $x \geq 0$? How do the conditions change if $\ell \leq x \leq u$?

4.6. Consider the linear programming problem

$$\min 4x_1 + 4x_2 + 3x_3$$
$$x_1 + 2x_2 + x_3 \geq 2$$
$$2x_1 + x_2 + x_3 \geq 3$$
$$x_1, x_2, x_3 \geq 0$$

Solve the classical dual by hand and use the solution to obtain the surrogate that provides the tightest bound on the optimal value of the primal. What is the optimal value? (There is no need to solve the primal directly.) Now use complementary slackness to find an optimal solution of the primal by solving two simultaneous equations.

4.7. Exhibit a linear programming problem with two variables for which both the primal and the dual are infeasible.

4.8. Show that the classical dual of $\min\{cx \mid Ax = b, \ x \geq 0\}$ is the problem $\max\{ub \mid uA \leq c\}$, where u is not restricted to be nonnegative.

4.9. Prove Corollary 4.10 (complementary slackness) by first showing $u^*b \leq u^*Ax^* \leq cx^*$.

4.10. Suppose that x and u are primal and dual feasible, respectively, and satisfy complementary slackness with respect to each other. Show that both are optimal. *Hint:* Use complementary slackness for both the primal and the dual; i.e., $u(Ax - b) = (uA - c)x = 0$.

4.11. The optimal solution of Exercise 3.2 has $x_B = (x_1, x_2, x_3) = (2, 1, 1)$ with

$$B^{-1} = \begin{bmatrix} 0 & -1 & 1 \\ 0 & -2 & 1 \\ 1 & -3 & 1 \end{bmatrix}$$

Suppose the constraint $x_1 + x_2 \leq 1$ is added to the problem, which makes the current optimum infeasible. Reoptimize using the dual simplex method.

4.12. Verify that (4.11) is the dual problem after the change of variable $\tilde{u} = uB - c_B$.

4.13. Write the classical dual of the minimum cost network flow problem (3.11). Show that if x and u are primal and dual feasible, respectively, then x satisfies the optimality conditions for the network flow problem.

4.14. Use LP duality to prove the max-flow/min-cut theorem. That is, prove that the maximum s-t flow in a network is equal the total capacity of a minimum capacity cut. *Hints:* Write the classical dual of the maximum-flow problem (3.15). Let cut (S,T) correspond to a maximum flow, and define a dual feasible solution with respect to (S,T) that satisfies complementary slackness.

4.4 Surrogate Duality

Like the linear programming dual, the surrogate dual is based on non-negative linear combinations of inequality constraints. The constraints, however, need not be linear, and the variables need not be continuous. When viewed as an inference dual, the surrogate dual uses nonnegative linear combination (and implication between inequalities) as an inference method. When viewed a a relaxation dual, it uses nonnegative linear combination as a relaxation.

The surrogate dual can be defined for any problem that contains inequality constraints:

$$\begin{aligned} &\min \ f(x) \\ &\mathbf{g}(x) \geq 0 \\ &x \in S \end{aligned} \qquad (4.16)$$

Here $\mathbf{g}(x)$ is a tuple of functions $g_1(x), \ldots, g_m(x)$, which may be linear or nonlinear. The domain constraint $x \in S$ represents an arbitrary constraint set that does not necessarily consist of inequalities. The variables themselves can be discrete or continuous. Linear programming is a special case of (4.16) in which $f(x) = cx$, $\mathbf{g}(x) = Ax - b$, and $S = \{x \in \mathbb{R}^n \mid x \geq 0\}$.

When defining the surrogate and Lagrangean duals, it is important to distinguish domination from implication. An inequality $g(x) \geq 0$ *dominates* inequality $h(x) \geq 0$ when $g(x) \leq h(x)$ for all $x \in S$, or when no $x \in S$ satisfies $g(x) \geq 0$.[2] Inequality $g(x) \geq 0$ *implies* $h(x) \geq 0$ when all $x \in S$ satisfying the former also satisfy the latter. An inequality $g(x) \geq 0$ may imply $h(x) \geq 0$ when there is no domination.

---

[2] Again, one could say that $g(x) \geq 0$ dominates $h(x) \geq 0$ when, for some $\alpha > 0$, $\alpha g(x) \leq h(x)$ for all $x \in S$, or when no $x \in S$ satisfies $g(x) \geq 0$. However, this complicates exposition and yields the same Lagrangean dual.

4.4.1 Inference Dual

The surrogate inference dual of (4.16) results when the inference method is nonnegative linear combination and implication (as opposed to domination):

$$\max v$$
$$(\mathbf{g}(x) \geq 0) \overset{P}{\vdash} (f(x) \geq v) \tag{4.17}$$
$$P \in \mathcal{P}, \quad v \in \mathbb{R}$$

Each proof $P \in \mathcal{P}$ corresponds to a vector $u \geq 0$ of multipliers. Proof P deduces $f(x) \geq v$ from $\mathbf{g}(x) \geq 0$ when the inequality $u\mathbf{g}(x) \geq 0$ implies $f(x) \geq v$, which is to say that the minimum of $f(x)$ subject to $u\mathbf{g}(x) \geq 0$ and $x \in S$ is at least v. So, (4.17) is equivalent to

$$\max_{u \geq 0} \left\{ v \;\middle|\; \min_{x \in S}\{f(x) \mid u\mathbf{g}(x) \geq 0\} \geq v \right\}$$

This can be written

$$\max_{u \geq 0}\{\sigma(u)\}, \quad \text{where } \sigma(u) = \min_{x \in S}\{f(x) \mid u\mathbf{g}(x) \geq 0\} \tag{4.18}$$

The surrogate dual therefore seeks a surrogate of $g(x) \geq 0$ for which $\min f(x)$ subject to the surrogate is as large as possible.

Theorem 4.12. *The surrogate inference dual is equivalent to (4.18).*

The surrogate dual is not a strong dual, because the inference method is not complete with respect to inequalities. Consider, for example, the integer programming problem

$$\min 4x_1 + 3x_2$$
$$6x_1 + x_2 \geq 9,$$
$$-x_1 + 3x_2 \geq 0 \tag{4.19}$$
$$x_1, x_2 \in \{0, 1, 2, 3\}$$

Here S consists of pairs of integers $0, 1, 2, 3$. The optimal solution of the problem is $(x_1, x_2) = (2, 1)$, with optimal value 11 (Fig. 4.2). The surrogate dual maximizes $\sigma(u_1, u_2)$ subject to $u_1, u_2 \geq 0$, where

$$\sigma(u_1, u_2) = \min_{x_j \in \{0, 1, 2, 3\}} \{4x_1 + 3x_2 \mid (6u_1 - u_2)x_1 + (u_1 + 3u_2)x_2 - 9u_1 \geq 0\}$$

For example, $\sigma(2, 1) = 8$, because the minimum is obtained for $u = (2, 1)$ when $(x_1, x_2) = (2, 0)$. In fact, this is the tightest bound that can

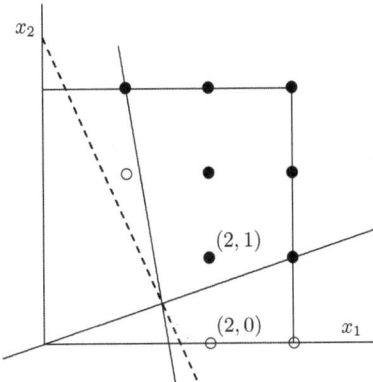

Fig. 4.2 Feasible set (solid black circles) of an integer programming problem, and feasible set (solid and open circles) of the surrogate $11x_1 + 5x_2 \geq 18$ (dashed line). The point $(2, 1)$ is optimal for the original problem, and $(2, 0)$ is optimal for the surrogate.

be inferred, because surrogates that cut off $(x_1, x_2) = (2, 0)$ allow the point $(x_1, x_2) = (1, 1)$, which has value 7. Therefore, $(u_1, u_2) = (2, 1)$ is an optimal dual solution. There is a duality gap because $\sigma(2, 1) < 11$. Yet $\sigma(2, 1) = 8$ is a tighter bound than that provided by the solution $(x_1, x_2) = (\frac{27}{19}, \frac{9}{19})$ of the LP relaxation, namely $7\frac{2}{19} \approx 7.105$.

4.4.2 Sensitivity Analysis

The surrogate inference dual provides a rudimentary form of sensitivity analysis. If u^* solves the surrogate dual, then $u^* \mathbf{g}(x) \geq 0$ implies $f(x) \geq \sigma(u^*)$. When $\mathbf{g}$ is replaced by another vector of functions $\mathbf{g}'$, one can still infer $f(x) \geq \sigma(u^*)$ if $u^* \mathbf{g}'(x) \leq u^* \mathbf{g}(x)$ for all $x \in S$. This is most useful when there is no duality gap, because it provides conditions under which a perturbation of the problem does not reduce the optimal value $\sigma(u^*)$.

When the constraints $\mathbf{g}(x) \geq 0$ are linear, as in the integer programming example above, one can write $\mathbf{g}(x) = Ax - b$. If A is perturbed by ΔA and b by Δb, the lower bound $\sigma(u^*)$ is valid as long as $u^* \Delta Ax \leq u^* \Delta b$ for all $x \in S$. If only the right-hand sides $(b_1, b_2) = (9, 0)$ in the example are perturbed, the lower bound 8 is valid as long as $2\Delta b_1 + \Delta b_2 \geq 0$.

4.4.3 Relaxation Dual

The surrogate relaxation dual is formed in the same way as the LP relaxation dual: by replacing the inequality constraints with a surrogate. This means that the parameterized relaxation is

$$\theta(u) = \min_{x \in S} \{f(x) \mid u\mathbf{g}(x) \geq 0\} \qquad (4.20)$$

for $u \geq 0$. Because the surrogate relaxation dual of (4.16) minimizes $\theta(u)$ subject to $u \geq 0$, it is identical to the surrogate inference dual (4.19).

The surrogate relaxation (4.20) may be easier to solve than the original problem because it contains only one constraint other than $x \in S$. For example, if the original problem is an integer programming problem, then (4.20) is an integer knapsack problem, which is normally much easier to solve than a general integer programming problem. Yet it may be difficult to find a value of u for which the solution of the knapsack problem yields a tight bound.

Exercise

4.15. Show that surrogate duality reduces to LP duality when the problem is linear.

4.5 Lagrangean Duality

Lagrangean duality is one of the most successful relaxation methods used in optimization. It can provide tighter bounds than LP relaxations while at the same time exploiting structure in the problem. In particular, it can "dualize" troublesome constraints by moving them into the objective function, leaving a simpler relaxation to solve.

Despite its historic association with relaxation, the Lagrangean dual can also be viewed as an inference dual. This perspective highlights its role in sensitivity analysis and reveals an unexpectedly close relationship with the surrogate dual.

4.5.1 Inference Dual

When viewed as an inference dual, the Lagrangean dual differs only slightly from the surrogate dual. The proof method consists of non-negative linear combination plus domination, rather than nonlinear combination plus implication as in the case of the surrogate dual. The Lagrangean dual nonetheless enjoys useful concavity and complementary slackness properties properties that the surrogate dual lacks, in addition to its ability to dualize constraints.

The Lagrangean inference dual of problem (4.16) is the problem (4.17), where $\mathcal{P}$ consists of proofs that combine linear combination with domination. That is, $f(x) \geq v$ is inferred from $\mathbf{g}(x) \geq 0$ when for some $u \geq 0$, $u\mathbf{g}(x) \geq 0$ dominates $f(x) \geq v$.

This results in a stronger form of inference than in the case of surrogate duality, because there can be implication between inequalities when there is no domination. This means that it is harder, in general, to infer good bounds. Weaker bounds are the cost of the Lagrangean dual's more desirable properties. Yet the popularity of the dual suggests that the benefit outweighs the cost.

Theorem 4.13. *The surrogate dual of (4.16) provides a lower bound that is at least as tight as that provided by the Lagrangean dual.*

This theorem illustrates an advantage of an inference-based interpretation of duality. It traditionally requires a nontrivial proof but follows immediately when the two duals are viewed as inference duals.

The Lagrangean inference dual can be analyzed in a manner similar to the LP inference dual. The surrogate $u\mathbf{g}(x) \geq 0$ dominates $f(x) \geq v$ when $u\mathbf{g}(x) \geq f(x) - v$ for all $x \in S$, or no $x \in S$ satisfies $u\mathbf{g}(x) \geq 0$ (i.e., the inequality is infeasible). Suppose first that $u\mathbf{g}(x) \geq 0$ is feasible for all $u \geq 0$. Then the Lagrangean dual maximizes v subject to $u \geq 0$, and subject to $u\mathbf{g}(x) \leq f(x) - v$ for all $x \in S$. Because the constraint $u g(x) \leq f(x) - v$ can be written $v \leq f(x) - u g(x)$, one can let

$$\theta(u, x) = f(x) - u g(x)$$

and write the Lagrangean dual as

$$\max_{u \geq 0}\{\theta(u)\}, \quad \text{where } \theta(u) = \min_{x \in S}\{\theta(u, x)\} \qquad (4.21)$$

The problem $\min_{x \in S}\{\theta(u, x)\}$ can be referred to as the *inner problem*. The constraints $g(x) \geq 0$ are said to be *dualized* because they are moved to the objective function. The components of u are *Lagrange multipliers*.

If $ug(x) \geq 0$ is an infeasible inequality for some $u \geq 0$, then the Lagrangean dual is unbounded. But in this case, (4.21) is likewise unbounded. The following theorem has been shown.

Theorem 4.14. *The Lagrangean inference dual is equivalent to (4.21).*

Like surrogate duality, Lagrangean duality is identical to LP duality when the problem is linear. This is because the inference method is the same as in the LP inference dual.

The Lagrangean dual for the problem instance (4.19) finds the maximum of $\theta(u_1, u_2)$ subject to $u_1, u_2 \geq 0$, where

$$\theta(u_1, u_2) = \min_{x_j \in \{0,1,2,3\}} \{(4 - 6u_1 + u_2)x_1 + (3 - u_1 - 3u_2)x_2 + 9u_1\}$$

(4.22)

Note that the value $\theta(u_1, u_2)$ of the inner problem is easy to compute, because values of x_1, x_2 that achieve the minimum are

$$x_1 = \begin{cases} 0, \text{ if } 4 - 6u_1 + u_2 \geq 0 \\ 3, \text{ otherwise} \end{cases} \qquad x_2 = \begin{cases} 0, \text{ if } 3 - u_1 - 3u_2 \geq 0 \\ 3, \text{ otherwise} \end{cases}$$

(4.23)

as illustrated in Figure 4.3. It will be seen in Section 4.5.6 that the optimal solution of the dual is $(u_1, u_2) = (\frac{15}{19}, \frac{14}{19})$, with optimal value

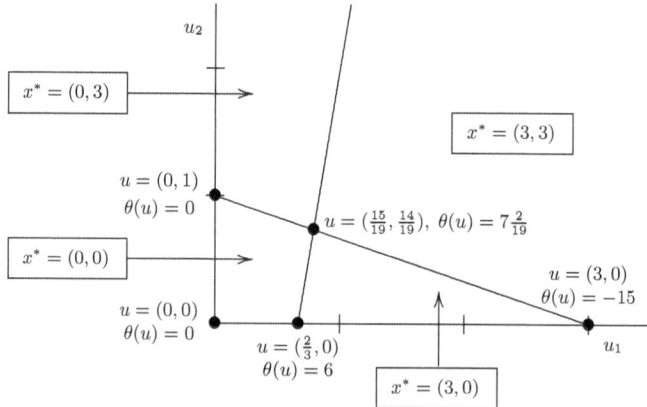

Fig. 4.3 Some values of the Lagrangean function $\theta(u)$ for the integer programming problem of Fig. 4.2. Regions in which x^* takes four possible values are shown, where $x = x^*$ maximizes $\theta(u, x)$ and thus $\theta(u) = \theta(u, x^*)$. The function $\theta(u)$ is affine within each of the four regions and is concave overall.

$7\frac{2}{19}$. This establishes a lower bound of $7\frac{2}{19}$ on the optimal value of (4.22). Because the optimal value of (4.22) is 11, there is a duality gap, and the Lagrangean bound is weaker than the bound of 8 obtained from the surrogate dual in the previous section.

In this example, the Lagrangean dual provides the same bound $7\frac{2}{19}$ as the LP relaxation of (4.19), obtained by replacing $x_1, x_2 \in \{0, 1, 2, 3\}$ with $x_1, x_2 \in [0, 3]$. This is because the inner problem has the same solution as its continuous relaxation and can therefore be solved as an LP problem. In general, the Lagrangean dual of an integer programming problem gives the same bound as the continuous relaxation when the inner problem can be solved as an LP problem. The Lagrangean dual is therefore useful only when the inner problem does not reduce to linear programming.

4.5.2 Sensitivity Analysis

The Lagrangean inference dual provides sensitivity analysis in a manner parallel to the LP inference dual. Suppose that the original problem (4.16) is perturbed to:

$$\min \ f(x)$$
$$\mathbf{g}(x) \geq \Delta b \qquad\qquad (4.24)$$
$$x \in S$$

Let u^* solve the Lagrangean dual, and let $v^* = \theta(u^*)$. The solution u^* encodes a proof of the surrogate $u^*\mathbf{g}(x) \geq 0$ and the dominated bound $f(x) \geq v^*$. This bound is dominated because

$$u^*\mathbf{g}(x) \leq f(x) - v^*, \quad \text{all } x \in S \qquad\qquad (4.25)$$

The solution u^* also encodes a proof of the surrogate $u^*\mathbf{g}(x) \geq u^*\Delta b$ for the perturbed problem. This surrogate dominates and therefore proves the bound $f(x) \geq v^* + \Delta v$ if

$$u^*(\mathbf{g}(x) - \Delta b) \leq f(x) - (v^* + \Delta v), \quad \text{all } x \in S$$

But due to (4.25), this inequality holds when $\Delta v = u^*\Delta b$. Thus $(v^* + u^*\Delta b, u^*)$ is dual feasible, which means that

$$v(b + \Delta b) = v^* + u^*\Delta b$$

is a value-bounding function.

Theorem 4.15. *Let* (v^*, u^*) *be an optimal solution of the Lagrangean dual of (4.16). The optimal value of the altered problem (4.24) is bounded below by* $v^* + u^* \Delta b$.

Lagrange multipliers can therefore be viewed as shadow prices when there is no duality gap (i.e., when v^* is the optimal value of the primal problem). There is also a complementary slackness property.

Corollary 4.16 *Let* x^* *be an optimal solution of (4.16) and* u^* *an optimal solution of the Lagrangean dual. If there is no duality gap, then* $u^* \mathbf{g}(x^*) = 0$.

This follows from

$$f(x^*) = \theta(u^*) = \min_{x \geq 0}\{\theta(u^*, x)\} \leq \theta(u^*, x^*) = f(x^*) - u^* \mathbf{g}(x^*) \leq f(x^*)$$

where the first equation is due to the lack of a duality gap, and the last inequality due to $u^* \geq 0$ and $\mathbf{g}(x^*) \geq 0$.

If the right-hand sides $b = (9, 0)$ in example (4.19) are perturbed, the lower bound is $v^* + u^* \Delta b = 7\frac{2}{19} + \frac{15}{19}\Delta b_1 + \frac{14}{19}\Delta b_2$.

4.5.3 Relaxation Dual

The Lagrangean relaxation dual is based on a very different kind of relaxation than the surrogate relaxation dual. Whereas the surrogate dual creates a relaxation by replacing the inequality constraints $\mathbf{g}(x) \geq 0$ with a surrogate and leaving the objective function $f(x)$ unaltered, the Lagrangean dual drops $\mathbf{g}(x) \geq 0$ altogether and replaces $f(x)$ with a lower bounding function. The result is a relaxation because its epigraph contains the epigraph of the original problem.

The lower bounding function is created by adding weighted penalties for violation of the inequality constraints. The weights become the dual variables u that parameterize the relaxation. Thus the Lagrangean relaxation dual of (4.16) maximizes $\theta(u)$, where

$$\theta(u) = \max_{x \in S}\{f(x) - u\mathbf{g}(x)\} \tag{4.26}$$

and $u \geq 0$. The inner problem (4.26) is known as a *Lagrangean relaxation*. One can check that it is in fact a relaxation of the primal problem

(4.16). The feasible set is larger, because the constraints $\mathbf{g}(x) \geq 0$ are dropped. Also the objective function bounds $f(x)$ below:

$$f(x) \geq f(x) - u\mathbf{g}(x), \quad \text{all } x \in S$$

because $u \geq 0$ and $\mathbf{g}(x) \geq 0$ for all x that are feasible in the primal. This dual is precisely the dual (4.21) obtained from the Lagrangean inference dual.

Theorem 4.17. *The Lagrangean relaxation dual is identical to the Lagrangean inference dual.*

Interpreting the Lagrangean dual as a relaxation dual draws attention to how one can create a relaxation that is easily solved but yields a reasonably tight bound. A common strategy is to dualize constraints that prevent the problem from decoupling into smaller problems. This occurs, for example, when the constraints are linear and the coefficient matrix has a block diagonal structure with connecting rows. That is, the problem has the form

$$\min \sum_{i=1}^{k} f_i(x^i)$$

$$\begin{bmatrix} A_1 \cdots A_k \\ D_1 \\ \ddots \\   D_k \end{bmatrix} \begin{bmatrix} x^1 \\ \vdots \\ x^k \end{bmatrix} \geq \begin{bmatrix} b \\ d^1 \\ \vdots \\ d^k \end{bmatrix}$$

The connecting rows are dualized, so that

$$\mathbf{g}(x) = \sum_{i=1}^{k} A_i x^i - b$$

and the Lagrangean relaxation decouples into k smaller problems.

A remarkable property of the Lagrangean function $\theta(u)$ that it is concave. This enables one to solve the dual by finding a local maximum, which can in turn be sought by a hill-climbing algorithm (Section 4.5.6).

Corollary 4.18 *The Lagrangean function $\theta(u)$ is concave.*

The concavity of $\theta(u)$ derives from the fact that $\theta(u, x)$, for fixed x, is an affine function of u. This makes $\theta(u)$ a minimum over affine functions and therefore a concave function. If S is finite, $\theta(u)$ is a minimum of finitely many affine functions and is therefore concave and piecewise linear.

In the example (4.19), the function $\theta(u_1, u_2)$ is linear except where $4 - 6u_1 + u_2 = 0$ or $3 - u_1 - 3u_2 = 0$. So, in this small example, the maximum can be found by examining points in the nonnegative quadrant at which the lines described by these two equations intersect with each other or with the axes (Figure 4.3). The maximum occurs at their intersection with each other, namely at $(u_1, u_2) = (\frac{15}{19}, \frac{14}{19})$.

4.5.4 Lagrangean Relaxation for LP

One application of Lagrangean relaxation is to LP problems in which some constraints are hard and some are easy. The hard constraints are dualized, and the resulting relaxation is an LP problem with easy constraints. This can be useful when one wishes to solve the LP relaxation at each node of a very large branch-and-bound tree, but solving the full LP relaxation is too slow. One therefore solves a Lagrangean relaxation of the LP problem and makes do with a somewhat weaker bound.

The first step is to obtain the optimal Lagrange multipliers at the root node by solving the full LP relaxation. At subsequent nodes, these same multipliers continue to define Lagrangean relaxations that provide valid bounds, albeit not the tightest possible bounds, because the branching process adds constraints to the problem.

Suppose, then, that the LP relaxation of a problem to be solved by branching is written

$$\min cx$$
$$Ax \geq b, \quad Dx \geq d, \quad x \geq 0 \tag{4.27}$$

where the linear system $Dx \geq d$ has some kind of special structure. If $Ax \geq b$ is dualized, then $\theta(u)$ is the optimal value of

$$\min_{\substack{Dx \geq d \\ x \geq 0}} \{\theta(x, u)\} \tag{4.28}$$

where $\theta(x, u) = cx - u(Ax - b) = (c - uA)x + ub$.

Theorem 4.19. *If (u^*, w^*) is an optimal dual solution of (4.27) in which u^* corresponds to $Ax \geq b$, then u^* is an optimal solution of the Lagrangean dual problem (4.28).*

Proof. Let x^* be an optimal solution of (4.27). By strong linear programming duality,

$$cx^* = u^*b + w^*d \qquad (4.29)$$

It will be shown below that $\theta(u^*) = cx^*$. This implies that u^* is an optimal dual solution of the Lagrangean dual problem, because $\theta(u)$ is a lower bound on cx^* for any $u \geq 0$.

To see that $\theta(u^*) = cx^*$, note first that x^* is feasible in

$$\min_{x \geq 0} \{(c - u^*A)x \mid Dx \geq d\} \qquad (4.30)$$

and w^* is feasible in its LP dual

$$\max_{w \geq 0} \{ud \mid wD \leq c - u^*A\} \qquad (4.31)$$

where the latter is true because (u^*, w^*) is dual feasible for (4.27). But the corresponding objective function value of (4.30) is

$$(c - u^*A)x^* = cx^* + u^*(b - Ax^*) - u^*b = cx^* - u^*b$$

where the second equation is due to complementary slackness. This is equal to the value w^*d of (4.31), due to (4.29). So, $cx^* - u^*b$ is the optimal value of (4.30), which means that cx^* is the optimal value $\theta(u^*)$ of (4.28) when $u = u^*$. $\square$

Due to Theorem 4.19, one can solve the Lagrangean dual of (4.27) at the root node by solving its LP dual. Let u^* be the Lagrangean dual solution obtained in this way. At subsequent nodes, one solves the specially structured LP problem (4.28), with u set to the value u^* obtained at the root node. Because the linear relaxation (4.27) at non-root nodes contains additional branching constraints, u^* is no longer optimal for the Lagrangean dual of (4.27). Yet due to weak duality, the optimal value of the specially structured LP problem (4.28) is a lower bound on the optimal value of (4.27) at that node. This bound can be used to prune the search tree.

4.5.5 Example: Generalized Assignment Problem

The generalized assignment problem is an assignment problem with additional knapsack constraints. By dualizing the knapsack constraints, one can solve a linear relaxation of the generalized assignment problem very rapidly using the Lagrangean technique described in the previous section.

The *generalized assignment problem* has the form

$$\min \sum_{ij} c_{ij}x_{ij} \qquad (a)$$

$$\sum_{j} x_{ij} = \sum_{j} x_{ji} = 1, \quad \text{all } i \qquad (b)$$

$$\sum_{j} a_{ij}x_{ij} \geq \alpha_i, \quad \text{all } i \qquad (c)$$

$$x_{ij} \in \{0,1\}, \quad \text{all } i,j$$

(4.32)

where constraints (b) are the assignment constraints and (c) the knapsack constraints.

Suppose that a linear programming relaxation of the problem is to be solved at each node of the search tree to obtain bounds and perhaps to fix variables using reduced costs. If solving this problem with a general-purpose solver is too slow, the complicating constraints (c) can be dualized, resulting in a pure assignment problem

$$\min \sum_{ij} (c_{ij} - u_i a_{ij})x_{ij} + \sum_{i} u_i \alpha_i$$

$$\sum_{j} x_{ij} = \sum_{j} x_{ji} = 1, \quad \text{all } i \qquad (4.33)$$

$$x_{ij} \geq 0, \quad \text{all } i,j$$

whose optimal value is $\theta(u)$ for $u \geq 0$.

The optimal dual solution u^* at the root node can be obtained by solving the linear relaxation of (4.32) and letting u^* be the optimal dual multipliers associated with constraints (c). At subsequent nodes of the search tree, one can solve the Lagrangean relaxation (4.33) very quickly using, for example, the Hungarian algorithm. The relaxation provides a weak bound, but the dual variables allow useful domain filtering as described in Section 6.2.3. The search branches by setting some x_{ij} to 0 or 1, which in turn can be achieved by giving x_{ij} a very large or very small cost in the objective function of the Lagrangean relaxation, thus preserving the problem structure.

4.5.6 Solving the Lagrangean Dual

If the Lagrangean dual is to provide a bound on the optimal value, it must somehow be solved. Fortunately, the concavity of $\theta(u)$ (Corollary 4.18) allows $\theta(u)$ to be maximized by a hill-climbing algorithm that finds a local maximum.

A popular hill-climbing procedure is *subgradient optimization*, which moves from one iterate to another in directions of steepest ascent until a local maximum is found. This approach has the advantage that the relevant subgradients are trivial to calculate, but the disadvantage that the step size is difficult to calibrate. In practice, the problem need not be solved to optimality, because only a valid lower bound is required.

Subgradient optimization begins with a starting value $u^0 \geq 0$ and sets each $u^{k+1} = u^k + \alpha_k \xi^k$, where ξ^k is a subgradient of $\theta(u)$ at $u = u^k$. Conveniently, if $\theta(u^k) = \theta(x^k, u^k)$, then $-g(x^k)$ is a subgradient. The stepsize α_k should decrease as k increases, but not so quickly as to cause the iterates to converge before a solution is reached.

A simple option is to set $\alpha_k = \alpha_0 / k$, because the resulting step sizes $\alpha_k \to 0$ as $k \to \infty$, while the series $\sum_{i=1}^k \alpha_i$ diverges. This can cause the algorithm to stall, however, because the stepsize soon becomes very small. Another approach is to set

$$\alpha_k = \gamma \frac{\bar{\theta} - \theta(u^k)}{\|\xi^k\|} \tag{4.34}$$

where $\bar{\theta}$ is a dynamically adjusted upper bound on the maximum value of $\theta(u)$. The step size calibration is highly problem-dependent.

In the example (4.19), one can easily compute $\theta(u)$ by setting $\theta(u) = \theta(u, x)$, where x is given by (4.23). Suppose the first iterate is $u^0 = (0, 0)$. Then $x^0 = (0, 0)$ and $\theta(u^0) = 0$. A direction of steepest ascent is given by the subgradient $\xi^0 = -g(x^0) = (9, 0)$. Using the L_∞ norm, $\|\xi^0\| = \max_j \{|\xi_j^0|\} = 9$. If the upper bound $\bar{\theta}$ is set to 11 and $\gamma = 0.1$, then $\alpha_0 = 0.1222$ and $u^1 = u^0 + \alpha_0 \xi^0 = (1.1, 0)$. As the algorithm continues, there is no convergence, but the value of $\theta(u^k)$ never exceeds 11. One option at this point is to re-run the algorithm with smaller values of $\bar{\theta}$ until convergence is achieved at a value very close to $\bar{\theta}$. This occurs when $\bar{\theta} \approx 7\frac{2}{19}$, which is the optimal value of the Lagrangean dual. The iterates u^k converge to $u = (\frac{15}{19}, \frac{14}{19})$, which is the optimal solution.

Another option is to reduce $\bar{\theta}$ gradually during the algorithm but increase it somewhat when $\theta(u^k) > \bar{\theta}$. There are also *block halving* schemes that divide the iterations into contiguous blocks that use successively smaller values of γ. The stepsize α_k is computed with (4.34) at the beginning of each block and is constant thereafter, except that it is halved when the incumbent solution (the best solution so far) fails to improve over a certain number of iterations. The current iterate u^k is set to the incumbent solution at the beginning of each block and when the stepsize is halved.

Exercises

4.16. State the Lagrangean dual of the problem

$$\min 15x_1 + 15x_2$$
$$3x_1 + 5x_2 \geq 9, \quad x_1, x_2 \in \{0, 1, 2\}$$

in which $3x_1 + 5x_2 \geq 9$ is dualized. Solve it by plotting $\theta(u)$ against u. What is the duality gap?

4.17. Consider a proof system in which $f(x) \geq v$ is inferred from the system $\mathbf{g}(x) \geq 0$ when there is a $u \geq 0$ such that $u\mathbf{g}(x) \leq f(x) - v - \alpha \min\{0, u\mathbf{g}(x)\}$ for all $x \in S$, where $\alpha \geq 0$ is fixed. Write the corresponding inference dual of (4.16). Show that its solution provides a bound that is no better than that of the surrogate dual and no worse than that of the Lagrangean dual.

4.18. Show that Lagrangean duality reduces to linear programming duality when $g(x) = Ax - b$ and $S = \{x \mid x \geq 0\}$. *Hint:* It suffices to show that the form of domination used in the Lagrangean dual reduces to the form of domination used to define the linear programming dual.

4.19. In an integer programming problem $\min_{x \geq 0}\{cx \mid Ax \geq b, \ x \text{ integral}\}$, the Lagrangean relaxation is

$$\theta(u) = \min_{x \geq 0}\{(c - uA)x + ub \mid x \text{ integral}\}$$

Show that when the Lagrangean relaxation can be solved as a linear programming problem, the Lagrangean dual gives the same bound as the linear programming relaxation. That is, if

$$\theta(u) = \min_{x \geq 0}\{(c - uA)x + ub\}$$

then $\max_{u \geq 0}\{\theta(u)\} = \min_{x \geq 0}\{cx \mid Ax \geq b\}$.

4.20. Consider the problem of maximizing cx subject to $\sum_j a_j x_j^2 \leq b$, where each $a_j > 0$ and $b > 0$, and x_j is any real number. (a) Solve the problem by applying KKT conditions. (b) Solve the Lagrangean dual. Show that there is no duality gap, and that the optimal dual solution u is identical to the vector μ of Lagrange multipliers that satisfies the KKT conditions. *Hint:* For part (a), note that $g(x)$ is convex.

4.21. Given an integer programming problem

$$\begin{aligned} \min \ & cx \\ & Ax \geq a \\ & Bx \geq b \\ & x \geq 0 \text{ and integral} \end{aligned} \tag{4.35}$$

suppose that $Bx \geq b, x \geq 0$ describes a polyhedron with all integral extreme points. Show that a Lagrangean dual in which $Ax \geq a$ is dualized provides the same bound as the continuous relaxation of (4.35).

4.22. An optimization problem minimizes $\sum_i f_i(z_j)$ subject to constraint sets $\mathcal{C}_1, \ldots, \mathcal{C}_m$. Each constraint set uses its own variables, except for a few variables that occur in more than one constraint set. Specifically, for each pair k, ℓ $(k < \ell)$, constraint sets $\mathcal{C}_k$ and $\mathcal{C}_\ell$ have the variables x_j for $j \in J_{k\ell}$ in common. Also each z_i occurs only in $\mathcal{C}_i$. Show how Lagrangean relaxation might take advantage of this structure to obtain a lower bound on the optimal value.

4.23. A subgradient of a concave function $\theta(u)$ at $u = u^*$ (where u is a row vector) is a vector ξ such that $\theta(u) - \theta(u^*) \leq (u - u^*)\xi$ for all u. Show that if $\theta(u^*) = f(x^*) - u^*\mathbf{g}(x^*)$ for the Lagrangean function $\theta(u)$, then $\xi = -\mathbf{g}(x^*)$ is a subgradient of $\theta(u)$ at $u = u^*$.

4.6 Subadditive Duality

Lagrangean and surrogate duality are not strong duals and therefore cannot in general prove optimality. In addition, they are not ideal for sensitivity analysis, because they predict the effect of problem perturbations only on the optimal relaxation value, not necessarily on the optimal value of the original problem. It would therefore be useful to identify strong duals for problem classes larger than LP problems.

One such dual is the *subadditive dual*, which has been defined for integer linear programming. It is a generalization of an inference dual in which the inference method is a cutting-plane algorithm. One can solve the primal and subadditive dual problems simultaneously if one solves the primal by a pure cutting-plane algorithm (Gomory's method). Other strong duals for combinatorial problems will be discussed in Section 4.7.

An integer linear programming problem may be written

$$\min \ cx$$
$$Ax \geq b, \quad x \geq 0 \text{ and integer} \tag{4.36}$$

For technical reasons, it is assumed that $Ax \geq b$ is a *bounded* system, meaning that the maximum and minimum of x_j subject to the constraints of (4.36) are finite for every j.

4.6.1 Inference Dual

An inference dual of (4.36) has the form

$$\max \ v$$
$$(Ax \geq b) \overset{P}{\vdash} (cx \geq v) \tag{4.37}$$
$$v \in \mathbb{R}, \ P \in \mathcal{P}$$

Nonnegative linear combination yields incomplete conference methods for this problem, and the resulting surrogate and Lagrangean duals are not strong duals. However, linear combination combined with rounding gives rise to a complete inference method.

Consider, for example, the linear integer programming problem

$$\min \ 2x_1 + 3x_2$$
$$x_1 + 3x_2 \geq 3 \qquad (a)$$
$$4x_1 + 3x_2 \geq 6 \qquad (b) \tag{4.38}$$
$$x_1, x_2 \geq 0 \text{ and integral}$$

The optimal solution is $(x_1, x_2) = (1, 1)$ with value 5. The following sequence of nonnegative linear combinations and roundings yields the optimal bound of 5. First, take a linear combination of constraints (a) and (b) using multipliers $\frac{5}{9}$ and $\frac{1}{9}$. This yields $x_1 + 2x_2 \geq \frac{21}{9}$, which after rounding is

$$x_1 + 2x_2 \geq 3 \qquad (c)$$

Rounding up the right-hand side preserves validity because the left-hand side is integral. If any coefficients on the left are nonintegral, they can be rounded up to ensure that the left-hand side is integral. Now, take a linear combination of (b) and (c) using multipliers $\frac{3}{5}$ and $\frac{3}{5}$ to obtain the inequality $3x_1 + 3x_2 \geq \frac{27}{5}$, which rounds to

$$3x_1 + 3x_2 \geq 6 \qquad (d)$$

Finally, take a linear combination of (c) and (d) using multipliers 1 and $\frac{1}{3}$, which yields the desired bound $2x_1 + 3x_2 \geq 5$.

Any inequality that can be obtained by such a process of repeated nonnegative linear combination and rounding is a *Chvátal–Gomory cut*. The process is equivalent to applying an appropriate function $h(\cdot)$ to the constraint set $Ax \geq b$ to obtain $h(A)x \geq h(b)$, where $h(A)$ denotes the row vector of function values $[h(A_1) \cdots h(A_n)]$, and where A_j is column j of A. In the above example $h(\cdot)$ is given by

$$h(d) = u \lceil M_2 \lceil M_1 d \rceil \rceil$$

and

$$u = [1 \ \tfrac{1}{3}], \quad M_2 = \begin{bmatrix} 0 & 1 \\ \frac{3}{5} & \frac{3}{5} \end{bmatrix}, \quad M_1 = \begin{bmatrix} 0 & 1 \\ \frac{5}{9} & \frac{1}{9} \end{bmatrix}$$

At this point, it can be checked that $h(A)x \geq h(b)$ is the desired inequality $2x_1 + 3x_2 \geq 5$. In particular, $h(b) = h(3,6) = 5$ is the optimal value.

A function $h(\cdot)$ of this kind is a *Chvátal function*, which in general is a function $h : \mathbb{R}^m \to \mathbb{R}$ having the form

$$h(d) = \lceil u \lceil M_{k-1} \lceil M_{k-2} \cdots \lceil M_2 \lceil M_1 d \rceil \rceil \cdots \rceil \rceil \rceil$$

for some sequence of matrices $M_1, \ldots, M_{k-1}$ and some row vector u in which all components are nonnegative.[3] Thus a Chvátal function operates on a tuple d by taking several nonnegative linear combinations of components of d, rounding up each result to the nearest integer, and creating a tuple of the resulting numbers. The process repeats a finite number of times. A scalar value is obtained at the end by taking a single linear combination.

---

[3] The components are generally defined to be rational as well, but this condition is not required here.

The inference dual can now be formally defined as one in which $cx \geq v$ is inferred from $Ax \geq b$ when $cx \geq v$ has the form $h(A)x \geq h(b)$ for some Chvátal function h.

A fundamental result of cutting-plane theory, to be proved in Section 7.3.1, is that the Chvátal–Gomory procedure generates all valid cutting planes for an integer linear system. It follows that the procedure is a complete inference method, and that the corresponding inference dual is a strong dual.

Corollary 4.20 *Any inequality implied by a bounded integer system $Ax \geq b$ has the form $h(A)x \geq h(b)$ for some Chvátal function h.*

4.6.2 Subadditive Dual

The inference dual based on the Chvátal–Gomory procedure can be generalized to obtain a *subadditive dual*. This is an inference dual in which each proof $P \in \mathcal{P}$ corresponds to a subadditive, homogeneous, nondecreasing function $h : \mathbb{R}^m \to \mathbb{R}$. Function h is *subadditive* if

$$h(d^1 + d^2) \leq h(d^1) + h(d^2)$$

for all d^1, d^2, and it is *homogeneous* if $h(0) = 0$. The bound $cx \geq v$ can be deduced from $Ax \geq b$ when $h(Ax) \geq h(b)$ implies, or dominates, $cx \geq v$ for some h satisfying these properties. Implication and domination give rise to different duals.

The mere fact that h is nondecreasing makes this a valid inference method. It will be seen shortly that it is a complete inference method as well, whether implication or domination is used. This is because Chvátal functions are a special case of subadditive functions, and they alone provide a complete inference method. The subadditive dual is therefore a strong dual. Because a domination-based dual has desirable properties that make it easier to analyze, as in the case of Lagrangean duality, the domination-based dual is studied here.

The *superadditive dual*, also a strong dual, is defined in an analogous way for maximization problems., A function h is superadditive if $h(d^1 + d^2) \geq h(d^1) + h(d^2)$ for all d^1, d^2.

The inequality $h(Ax) \geq h(b)$ *dominates* $cx \geq v$ when

$$h(Ax) - h(b) \leq cx - v, \text{ all integral } x \geq 0 \qquad (4.39)$$

The subadditive dual therefore maximizes v over all subadditive, homogeneous, and nondecreasing functions h subject to (4.39). But because h is homogeneous, (4.39) is equivalent to saying that $h(b) \geq v$ and $h(Ax) \leq cx$ for all integral $x \geq 0$. The subadditive dual (4.37) can therefore be written

$$\max_{h \in H} \{h(b) \mid h(Ax) \leq cx \text{ for all integral } x \geq 0\} \qquad (4.40)$$

where H is the set of subadditive, homogeneous, nondecreasing functions. Subadditivity and homogeneity are useful properties because they allow one to simplify the dual, due to the following fact.

Lemma 4.21 *If h is nondecreasing, subadditive and homogeneous, then $h(Ax) \leq cx$ for all integral $x \geq 0$ if and only if $h(A) \leq c$.*

Proof. First, if $h(Ax) \leq cx$ for all integral $x \geq 0$, then in particular $h(Ae^j) \leq ce^j$ for each unit vector e^j. This says $h(A_j) \leq c_j$ for each j, or $h(A) \leq c$. For the converse, suppose $h(A_j) \leq c_j$ for each j. Then

$$h(Ax) = h\left(\sum_j A_j x_j\right) \leq \sum_j h(A_j) x_j \leq \sum_j c_j x_j = cx$$

If $x \neq 0$, the first inequality is due to subadditivity and the fact that each x_j is a nonnegative integer. If $x = 0$, it is due to homogeneity. The second inequality follows from the hypothesis $h(A_j) \leq c_j$. $\square$

Due to Lemma 4.21, the subadditive dual can be written very simply.

Theorem 4.22. *The subadditive inference dual can be written*

$$\max_{h \in H} \{h(b) \mid h(A) \leq c\} \qquad (4.41)$$

where H is the set of subadditive, homogeneous, and nondecreasing functions.

This is closely analogous to the classical linear programming dual of (4.36) without the integrality constraint:

$$\max_{u \geq 0} \{ub \mid uA \leq c\}$$

The subadditive dual is a strong dual because the associated inference method is complete. The inference method is complete *a fortiori* when domination is replaced by implication.

Theorem 4.23. *A bounded integer system $Ax \geq b$ with $x \geq 0$ implies $cx \geq v$ if and only if $h(A)x \geq h(b)$ dominates $cx \geq v$ for some subadditive, homogeneous, nondecreasing function h.*

Proof. By Corollary 4.20, any inequality $cx \geq v$ implied by a bounded integer system $Ax \geq b$ has the form $h(A)x \geq h(b)$ for some Chvátal function h. It can be shown that Chvátal functions are subadditive, homogeneous, and nondecreasing, and the theorem follows. $\square$

Corollary 4.24 *The subadditive inference dual is a strong dual.*

Because there is no duality gap, $h(b)$ is the optimal value of the primal problem (4.36) when h is an optimal solution of the dual (4.41). The dual can therefore be solved by identifying a Chvátal function h for which $h(b)$ is the optimal value of the primal. Section 7.3.2 will show that this can be accomplished by solving (4.36) with the Gomory cutting-plane algorithm.

4.6.3 Sensitivity Analysis

Let the integer programming problem (4.36) be perturbed as follows:

$$\min (c + \Delta c)x$$
$$(A + \Delta A)x \geq (b + \Delta b), \quad x \geq 0 \text{ and integer} \tag{4.42}$$

If h^* solves the subadditive dual, the inferred inequality $h^*(A)x \geq h^*(b)$ dominates the bound $cx \geq h^*(b)$, which means $h^*(A) \leq c$. Thus the inequality

$$h^*(A + \Delta A)x \geq h^*(b + \Delta b)$$

inferred from (4.42) dominates $(c + \Delta c)x \geq h^*(b + \Delta b)$ if

$$h^*(A + \Delta A) \leq c + \Delta c$$

Because h^* is subadditive, it suffices that

$$h^*(A) + h^*(\Delta A) \leq c + \Delta c$$

and therefore that $h^*(\Delta A) \leq \Delta c$.

Theorem 4.25. *Let (z^*, h^*) be an optimal solution of the subadditive dual of (4.36). The optimal value of the altered problem (4.42) is bounded below by $z^* + h^*(b + \Delta b)$ if $h^*(\Delta A) \leq \Delta c$.*

If only the right-hand side b is perturbed, h^* is a value-bounding function because $(h^*(b + \Delta b), h^*)$ is dual feasible. The optimal value is therefore bounded below by $h^*(b + \Delta b)$ for all perturbations Δb.

Suppose the right-hand side of example (4.38) is perturbed to $(3 + \Delta_1, 6 + \Delta_2)$. The resulting optimal value is at least

$$h^*(3 + \Delta_1, 6 + \Delta_2) =$$
$$\left\lceil \tfrac{7}{3} + \tfrac{5}{9}\Delta_1 + \tfrac{1}{9}\Delta_2 \right\rceil + \tfrac{1}{3} \left\lceil \tfrac{3}{5} \lceil 6 + \Delta_2 \rceil + \tfrac{3}{5} \left\lceil \tfrac{7}{3} + \tfrac{5}{9}\Delta_1 + \tfrac{1}{9}\Delta_2 \right\rceil \right\rceil$$

For instance, if the right-hand side is changed to $(5, 8)$, the optimal value is at least $h^*(5, 8) = 6\tfrac{1}{3}$. It is actually 7.

A practical difficulty with the subadditive dual is that the only reasonably practical method for solving it is to solve the original problem with the Gomory cutting-plane algorithm. Although Gomory cuts are very useful in a branch-and-cut algorithm, problems are rarely solved with Gomory's method alone. Section 4.7 presents a dual whose solution can be read directly from a branching tree that is created to solve the problem.

4.6.4 Relaxation Dual

A subadditive relaxation dual can be defined in a manner parallel to Lagrangean duality. The problem (4.36) can be relaxed by dualizing the constraints $Ax \geq b$:

$$\theta(h) = \min_{x \in \mathbb{Z}^n} \left\{ cx - (h(Ax) - h(b)) \mid x \geq 0 \right\}$$

The subadditive relaxation dual maximizes $\theta(h)$ over all subadditive, homogeneous, and nondecreasing functions h. Note that $\theta(h)$ is in fact a relaxation because its objective function $cx - h(Ax) + h(b)$ is a lower bound on cx for all feasible x.

The subadditive relaxation dual is equivalent to the subadditive inference dual. Recall that the latter maximizes v over all subadditive, homogeneous, and nondecreasing functions h subject to (4.39), or equivalently, subject to

$$v \leq cx - h(Ax) + h(b), \text{ all integral } x \geq 0$$

It therefore maximizes $\theta(h)$ over all subadditive, homogeneous, and nondecreasing functions h.

Theorem 4.26. *The subadditive relaxation dual is equivalent to the subadditive inference dual and can be written as in (4.41).*

Exercises

4.24. Show that all Chvátal functions are subadditive, homogeneous, and nondecreasing.

4.25.
Consider the integer programming problem

$$\min\ 3x_1 + 4x_2$$
$$-x_1 + 3x_2 \geq 0$$
$$2x_1 + x_2 - 5 \geq 0$$
$$x_1, x_2 \geq \{0, 1, 2, 3\}$$

which has optimal value 10. The bound $3x_1 + 4x_2 \geq 10$ can be obtained by first taking a linear combination of the constraints with multipliers $\frac{2}{7}$ and $\frac{1}{7}$ and rounding up to obtain a new inequality, then taking a linear combination of the second constraint and the new inequality with multipliers $\frac{3}{2}$ and $\frac{5}{2}$. Write the corresponding Chvátal function in the form $h(d) = u\lceil Md \rceil$. Write an expression for a lower bound on the optimal value if the right-hand side is perturbed to $(0 + \Delta_1, 5 + \Delta_2)$.

4.7 Branching Duality

Combinatorial optimization problems are very frequently solved by some kind of branching method. The branching process solves an inference dual while it solves the primal problem, much as the simplex method solves the primal and dual simultaneously. The branching tree represents a search procedure when viewed from top down and a proof of optimality when viewed from bottom up. This proof can be viewed as a solution of a suitably defined inference dual. It is obviously a strong dual, because there is no duality gap. When viewed from this perspective, the branching tree can provide the kind of sensitivity analysis that one obtains from inference duals in general (Section 4.1.3).

4.7.1 Inference Dual

Branching methods typically obtain bounds by solving a relaxation of the problem restriction at each node of the branching tree. In the freight transfer problem of Section 2.3, for example, a continuous relaxation is solved at each node of the branching tree in Fig. 2.1 after domain filtering. Solution of the relaxation proves a lower bound on the optimal value at that node, and the branching tree as a whole proves a lower bound of 530 for the original problem (the optimal value).

The proof of optimality can be obtained by reasoning from the bottom up in the branching tree, as shown in Fig. 4.4. First, at each leaf node i, a lower bound LB_i is obtained for the optimal value cx of the restricted problem at node i. The bound is inferred in two stages. For example, at the leftmost leaf node in Fig. 4.4, a reduced domain D_i' is

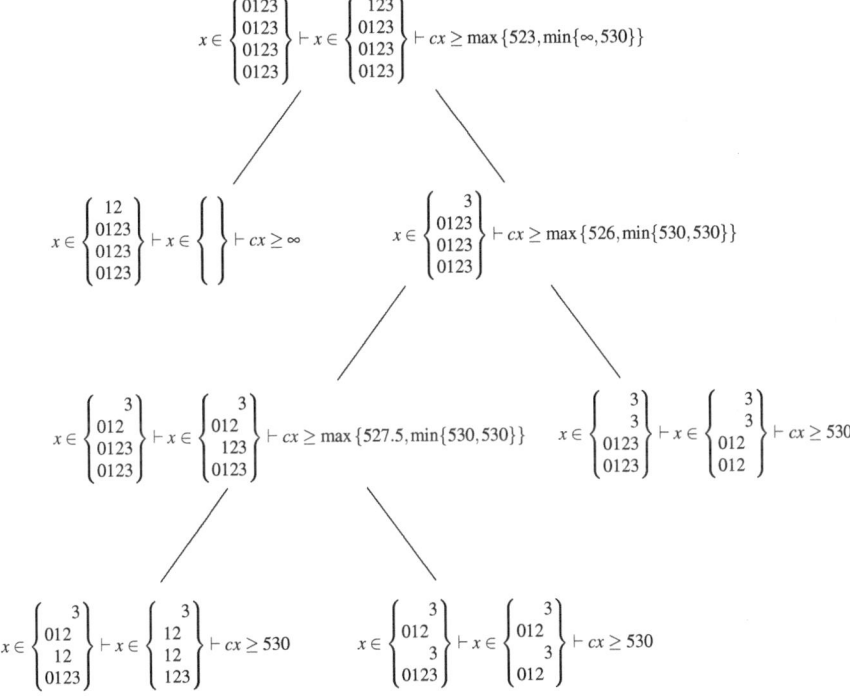

Fig. 4.4 Inference dual solution based on the branching tree of Fig. 2.1, where cx is the objective function. A filtered domain D_i' and a lower bound $cx \geq \mathrm{LB}_i$ are inferred at each node i.

inferred from the initial domain D_i by filtering out two zeros. Then the continuous relaxation is solved with domain D_i' to obtain its optimal value $v_i^* = 530$. One can then set the lower bound LB_i to $v_i^* = 530$.

The procedure is slightly different at a nonleaf node. Consider, for example, the parent of the leftmost leaf node in Fig. 4.4. The initial domain D_i is again filtered to obtain D_i', and the continuous relaxation solved with domain D_i' to obtain lower bound $v^* = 527.5$. But additional lower bounds $\mathrm{LB}_j, \mathrm{LB}_k$ were obtained from the two child nodes j and k, and these are valid for the smaller domains D_j and D_k, respectively. Because $D_i' = D_j \cup D_k$, it follows that $\min\{\mathrm{LB}_j, \mathrm{LB}_k\} = \min\{530, 530\}$ is a valid lower bound for the domain D_i' at the parent node i, and therefore for the domain D_i deduced from D_i'. One can therefore use the tighter of this bound and v^*:

$$\mathrm{LB}_i = \max\left\{v_i^*, \min\left\{\mathrm{LB}_j, \mathrm{LB}_k\right\}\right\} = \max\left\{527.5, \min\{530, 530\}\right\} \tag{4.43}$$

The process continues to the root node, at which the optimal value 530 is established as a valid lower bound for the original domain. The proof depicted in Fig. 4.4 is therefore an optimal solution of an inference dual in which inference takes the form just described.

The term v_i^* in (4.43) may seem redundant, because v_i^* can never be larger than bounds obtained at child nodes. However, this term can be helpful in sensitivity analysis, where the same proof schema is applied to a perturbed problem. It is possible that the proof schema yields a tighter bound at node i than at child nodes when the problem is perturbed. This will be illustrated in the discussion of integer programming below.

In general, a branching inference dual can be defined for any branching algorithm that solves $\min\{f(x) \mid \mathcal{C}(x),\, x \in D\}$. It is assumed that the algorithm solves a relaxation

$$v_i^* = \min\left\{f(x) \mid \mathcal{C}'(x),\, x \in D_i'\right\} \tag{4.44}$$

at each node i. A feasible solution of the dual is a tree structure in which each node i is associated with the proposition

$$(x \in D_i) \vdash (f(x) \geq \mathrm{LB}_i) \tag{4.45}$$

Each leaf node i infers $x \in D_i'$ from $x \in D_i$ and $\mathcal{C}(x)$, perhaps by means of a filtering algorithm. It then infers $f(x) \geq v_i^*$ by solving (4.44), thus establishing (4.45) with $\mathrm{LB}_i = v_i^*$. Each nonleaf node i infers $x \in D_i'$ from $x \in D_i$ and $\mathcal{C}(x)$. It also infers

$$\left(x \in D_i'\right) \vdash \left(f(x) \geq \min_{j \in J_i}\{\mathrm{LB}_j\}\right) \tag{4.46}$$

from

$$(x \in D_j) \vdash (f(x) \geq \mathrm{LB}_j), \ j \in J_i$$

and $D_i' = \bigcup_{j \in J} D_j$, where J_i indexes the children of node i. This establishes (4.45) with

$$\mathrm{LB}_i = \max\left\{v_i^*, \min_{j \in J_i}\{\mathrm{LB}_j\}\right\}$$

At the root node, D_i must be the original domain.

4.7.2 A Value-Bounding Function

A branching dual can provide the basis for detailed sensitivity analysis. The analysis can take the form of a value-bounding function or a more general approach that accounts for perturbations in all the problem data. The former is developed in the present section, and the latter in the next section.

The ideas are best illustrated by example, such as the following 0-1 linear programming problem.

$$\begin{aligned}
\min \ & 5x_1 + 6x_2 + 7x_3 \\
& 4x_1 + 3x_2 - x_3 \geq 2 \quad (a) \\
& -x_1 + x_2 + 4x_3 \geq 3 \quad (b) \\
& x_1, x_2, x_3 \in \{0, 1\}
\end{aligned} \tag{4.47}$$

A search tree for the problem appears in Fig. 4.5. To simplify discussion, no domain filtering or cutting planes are used. The optimal solution $(x_1, x_2, x_3) = (1, 0, 1)$ has value 12 and is found at node 3.

At each node i of the search tree, the continuous relaxation of the problem has the following form:

$$\min 5x_1 + 6x_2 + 7x_3$$

$$
\begin{array}{llll}
4x_1 + 3x_2 - \ x_3 \geq 2 & (u_1^i) \\
-x_1 + \ x_2 + 4x_3 \geq 3 & (u_2^i) \\
x_1 \qquad\qquad\qquad \geq p_1^i & (\sigma_1^i) \\
\quad\ x_2 \qquad\qquad\ \geq p_2^i & (\sigma_2^i) \\
\qquad\qquad x_3 \geq p_3^i & (\sigma_3^i) \\
-x_1 \qquad\qquad\quad\ \geq -q_1^i & (\tau_1^i) \\
\quad -x_2 \qquad\qquad \geq -q_2^i & (\tau_2^i) \\
\qquad\qquad -x_3 \geq -q_3^i & (\tau_3^i)
\end{array}
\qquad (4.48)
$$

At the root node (node 0), $p_j^0 = 0$ and $q_j^0 = 1$ for all j. At subsequent nodes i, the bounds p_j^i, q_j^i are tightened to reflect how the node was reached by branching. For example, branching to node 1 (where x_2 is fixed to 0) corresponds to setting $q_2^1 = 0$. Dual variables appear on the right, and dual solutions at each node are shown in Fig. 4.5. A proof of optimality, following the pattern of the previous section, is indicated by the derivation of lower bounds LB_i in the figure.

For sensitivity analysis, suppose first that only the right-hand side $b = (b_1, b_2)$ of (4.47) is perturbed, namely by $\Delta b = (\Delta b_1, \Delta b_2)$. A value-bounding function v can be developed as follows by defining a value-bounding function v_i at each node. At a feasible leaf node ($i = 3$ or 6 in the example), v_i is given by LP sensitivity analysis:

$$v_i(b + \Delta b) = v_i^* + u^i \Delta b$$

where v_i^* is the optimal value of the relaxation at node i. Thus

$$v_3(b + \Delta b) = 12 + \tfrac{7}{4}\Delta b_2, \qquad v_6(b + \Delta b) = 13 + \tfrac{5}{4}\Delta b_1$$

At an infeasible leaf node i, the lower bound of ∞ applies as long as the dual solution $[u^i \ \sigma^i \ \tau^i]$ proves infeasibility; that is, as long as

$$u^i(b + \Delta b) + \sigma^i p^i - \tau^i q^i > 0$$

The value-bounding function at an infeasible node is therefore

$$v_i(b + \Delta b) = \begin{cases} \infty & \text{if } u^i \Delta b > \tau^i q^i - \sigma^i p^i - u^i b \\ -\infty & \text{otherwise} \end{cases}$$

Thus

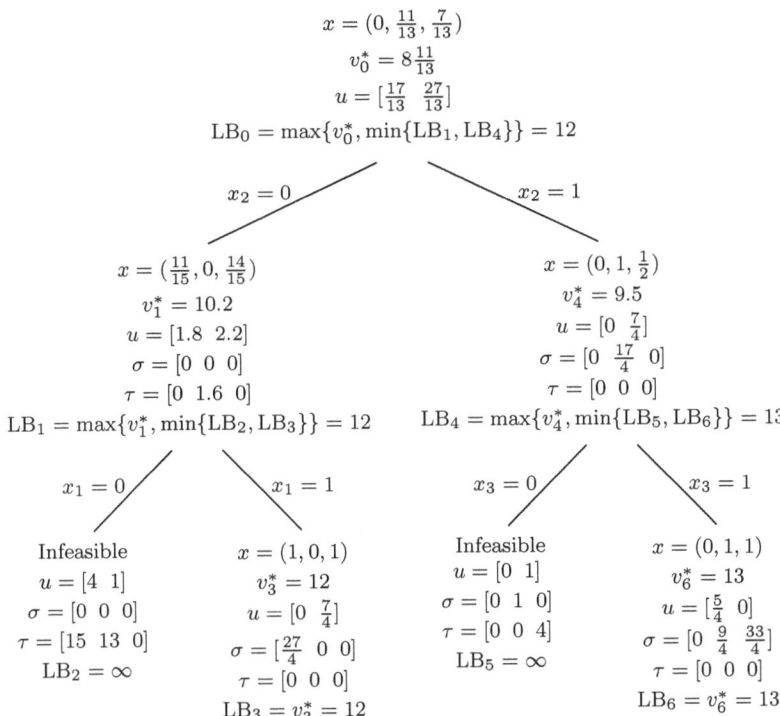

Fig. 4.5 Branching solution of a 0-1 linear integer programming problem. Each node i shows the solution x and the value v_i^* of the continuous relaxation at that node, as well as the corresponding dual solution u, σ, τ. The bounds LB_i indicate how the optimality of the solution is proved.

$$v_3(b + \Delta b) = \begin{cases} \infty & \text{if } 4\Delta b_1 + \Delta b_2 > -11 \\ -\infty & \text{otherwise} \end{cases}$$

$$v_6(b + \Delta b) = \begin{cases} \infty & \text{if } \Delta b_2 > -2 \\ -\infty & \text{otherwise} \end{cases}$$

The value-bounding function at a nonleaf node i combines the bounds at child nodes with the bound obtained from the relaxation at node i. The latter is again $v_i^* + u^i \Delta b$. The value-bounding function is therefore

$$v_i(b + \Delta b) = \max \left\{ v_i^* + u^i \Delta b, \min_{j \in J_i} \{v_j(b + \Delta b)\} \right\} \qquad (4.49)$$

The value-bounding function $v_0(b + \Delta b)$ at the root node becomes the overall value-bounding function $v(b + \Delta b)$. Thus

$$v_1(b + \Delta b) = \max\left\{10.2 + 1.8\Delta b_1 + 2.2\Delta b_2,\ \min\left\{v_2(b + \Delta b), v_3(b + \Delta b)\right\}\right\}$$

$$v_4(b + \Delta b) = \max\left\{9.5 + \tfrac{7}{4}\Delta b_2,\ \min\left\{v_5(b + \Delta b), v_6(b + \Delta b)\right\}\right\}$$

$$v(b + \Delta b) = \max\left\{8\tfrac{11}{13} + \tfrac{17}{13}\Delta b_1 + \tfrac{21}{13}\Delta b_2,\ \min\left\{v_1(b + \Delta b), v_4(b + \Delta b)\right\}\right\}$$

The term $v_i^* + u^i\Delta b$ of (4.49) is redundant when $\Delta b = 0$, because in this case $v_i^* + u^i\Delta b$ cannot be larger than $v_j(b + \Delta b)$ at child nodes j. However, the situation changes when $\Delta b > 0$. If an infeasible child node j becomes feasible after perturbation, for example, then $v_j(b + \Delta b)$ drops to $-\infty$, in which case $v_i^* + u^i\Delta b$ dominates.

The value-bounding function $v(b + \Delta b)$ is therefore recursively defined and consists of nested min and max functions. It remains valid if c and A are perturbed, provided all of the dual solutions $[u^i\ \sigma^i\ \tau^i]$ remain feasible after the perturbation. However, this condition is rather complicated, particularly if the search tree is large.

4.7.3 General Sensitivity Analysis

The value-bounding function derived in the previous section can be complex and difficult to interpret for large search trees. Also it is valid only for perturbations of the right-hand side, unless other perturbations satisfy a complicated condition involving all of the dual values u^i, σ^i, τ^i.

A more practical sensitivity analysis, and one that applies to perturbations in all the data, can be obtained by viewing the inference dual in a slightly different way. The resulting analysis derives a set of linear conditions under which perturbations reduce the optimal value no more than a stated amount Δz.

The analysis examines only the leaf nodes of the search tree. Let a *feasible node* of the tree be a leaf node at which a feasible solution is found, an *infeasible node* be a leaf node at which the relaxation is infeasible, and a *fathomed* node be a leaf node at which the search backtracks because the value of the relaxation is no better than the value of the incumbent solution. If z^* is the optimal value of the original problem, the search tree remains a proof that the new optimal value is at least $z^* - \Delta z$ if two conditions are satisfied: (a) The perturbed relaxation at every infeasible node is infeasible, and (b) the optimal value of the perturbed relaxation at every feasible node and every fathomed node is at least $z^* - \Delta z$.

This is again illustrated by the search tree of Fig. 4.5. At the infeasible node 2, the dual multipliers $[u_1^2 \ u_2^2] = [4 \ 1]$ define a surrogate $15x_1 + 13x_2 \geq 11$ that is violated by the branching constraints $(x_1, x_2) = (0, 0)$. Now suppose the problem data A, b, c are changed to $\tilde{A}, \tilde{b}, \tilde{c}$. This same dual solution remains a proof that the relaxation is infeasible if $(x_1, x_2) = (0, 0)$ violates the surrogate

$$(4\tilde{A}_{11} + \tilde{A}_{21})x_1 + (4\tilde{A}_{12} + \tilde{A}_{22})x_2 + (4\tilde{A}_{13} + \tilde{A}_{23})x_3 \geq 4\tilde{b}_1 + \tilde{b}_2 \quad (4.50)$$

It is easy to write a necessary and sufficient condition for when $(x_1, x_2) = (0, 0)$ violates (4.50). In general, a partial assignment $x_j = t_j$ for $j \in J$ violates a 0-1 inequality $dx \geq \delta$ if and only if

$$\sum_{j \in J} d_j t_j + \sum_{j \notin J} \max\{0, d_j\} < \delta \quad (4.51)$$

Thus, $(x_1, x_2) = (0, 0)$ violates (4.50) as long as the perturbed data satisfy

$$\max\{0, 4\tilde{A}_{13} + \tilde{A}_{23}\} < 4\tilde{b}_1 + \tilde{b}_2 \quad (4.52)$$

Similarly, node 5 remains infeasible if

$$\max\{0, \tilde{A}_{21}\} + \tilde{A}_{22} < \tilde{b}_2 \quad (4.53)$$

The feasible leaf nodes require a slightly different treatment. The optimal value of the relaxation at a feasible node 3 is at least $12 - \Delta$ if the relaxation is infeasible when $5x_1 + 5x_2 + 7x_3 < 12 - \Delta$ is added to its constraint set. This inequality can be written

$$-5x_1 - 6x_2 - 7x_3 > -12 + \Delta \quad (4.54)$$

By assigning dual multipliers $u^4 = [0 \ \frac{7}{4}]$ to $Ax \geq b$ and multiplier 1 to (4.54), one obtains a surrogate

$$-\tfrac{27}{4}x_1 - \tfrac{17}{4}x_2 > \Delta - \tfrac{27}{4}$$

that is violated by the branching constraints $(x_1, x_2) = (1, 0)$ when $\Delta z = 0$. If the problem data are changed to $\tilde{A}, \tilde{b}, \tilde{c}$, the surrogate that results from this same dual solution is

$$(\tfrac{7}{4}\tilde{A}_{21} - \tilde{c}_1)x_1 + (\tfrac{7}{4}\tilde{A}_{22} - \tilde{c}_2)x_2 + (\tfrac{7}{4}\tilde{A}_{23} - \tilde{c}_3)x_3 > \tfrac{7}{4}\tilde{b}_2 - 12 + \Delta z$$

The partial assignment $(x_1, x_2) = (1, 0)$ violates this surrogate when

$$\tfrac{7}{4}\tilde{A}_{21} - \tilde{c}_1 + \max\{0, \tfrac{7}{4}\tilde{A}_{23} - \tilde{c}_3\} \le \tfrac{7}{4}\tilde{b}_2 - 12 + \Delta z \qquad (4.55)$$

The parallel inequality for node 6 is

$$\tfrac{5}{4}\tilde{A}_{12} - \tilde{c}_2 + \tfrac{5}{4}\tilde{A}_{13} - \tilde{c}_3 + \max\{0, \tfrac{5}{4}\tilde{A}_{11} - \tilde{c}_1\} \le \tfrac{5}{4}\tilde{b}_1 - 12 + \Delta z \quad (4.56)$$

The conclusion of this analysis is that for a given Δz, altered problem data $\tilde{A}, \tilde{b}, \tilde{c}$ will reduce the optimal value no more than Δz if the data satisfy (4.52), (4.53), (4.55), and (4.56). Conversely, if the perturbations are fixed, one can easy read from these inequalities a lower bound on Δz.

In general, sensitivity analysis proceeds as follows. Let J_i be the set of indices j for which branching has fixed x_j to some value $t_{ij} \in \{0,1\}$ at the current node i. At an infeasible leaf node i, the partial assignment $x_{J_i} = t_{iJ_i}$ (i.e., $x_j = t_{ij}$ for $j \in J_i$) violates the surrogate $u^i A x \ge u^i b$. For alternate data $\tilde{A}, \tilde{b}$, this partial assignment violates $u\tilde{A}x \ge u\tilde{b}$ if

$$\sum_{j \in J_i} u^i \tilde{A}_j t_{ij} + \sum_{j \notin J_i} \max\{0, u^i \tilde{A}_j\} < u^i \tilde{b} \qquad (4.57)$$

At each feasible node and each fathoming node i, the partial assignment $x_{J_i} = t_{iJ_i}$ violates the surrogate $(u^i A - c)x > u^i b - z^*$. For any $\Delta z \ge 0$ and alternate data $\tilde{A}, \tilde{b}, \tilde{c}$, this partial assignment violates the surrogate $(u^i \tilde{A} - \tilde{c})x > u^i \tilde{b} - z^* + \Delta z$ if

$$\sum_{j \in J_i} (u^i \tilde{A}_j t_{ij} - \tilde{c}_j) + \sum_{j \notin J_i} \max\{0, u^i \tilde{A}_j - \tilde{c}_j\} \le u^i \tilde{b} - z^* + \Delta z \quad (4.58)$$

Thus the new data $\tilde{A}, \tilde{b}, \tilde{c}$ reduce the optimal value no more than Δz if they satisfy (4.57) at every infeasible node i and (4.58) at every feasible node and fathoming node i.

Also, for a given Δz, one can use linear programming to calculate a range of values for any data item A_{kj}, b_i, or c_i within which the optimal value does not drop below $z^* - \Delta z$. This is because condition (4.51) can be linearized by writing it

$$\sum_{j \in J} d_j t_j + \sum_{j \notin J} w_j < \delta$$

$$w_j \ge d_j, \; w_j \ge 0, \text{ all } j \notin J$$

Now one can solve two LP problems: minimize and maximize $\tilde{A}_{kj}, \tilde{b}_i,$ or $\tilde{c}_i$ subject to the linearized form of (4.57) and (4.58).

If cutting planes are used in the tree search, then one must write conditions under which the cutting planes remain valid and impose these conditions as well on the problem data.

4.7.4 Relaxation Dual

Given a seach tree for the problem $\min\{f(x) \mid C(x),\ x \in D\}$, any subtree of it with the same root can be regarded as defining a relaxation of the problem. Again, it is supposed that a relaxation (4.44) is solved at each node i of the search tree. A bound LB_i is derived at each node i as described in Section 4.7.1, including an overall bound LB_0 on the optimal value at the root node. The search tree will in general be incomplete, because it can be any subtree that contains the root of the original search tree.

If $\theta(T) = \mathrm{LB}_0$ is the bound derived for subtree T, then the relaxation dual is $\max\{\theta(T) \mid T \in \mathcal{T}\}$, where $\mathcal{T}$ is the family of subtrees over which one wishes to search. There is no duality gap if $\mathcal{T}$ contains at least one exhaustive search tree.

In an exhaustive search tree, $\theta(T)$ is the minimum of v_i^* over all leaf nodes i (where v_i^* is the optimal value of the relaxation as before). However, if T is inexhaustive, at least one node i is *incomplete*, meaning that one or more of its children are missing. The tightest bound that one can establish at node i is v_i^*. Thus the overall bound $\theta(T)$ is the minimum of v_i^* over all leaf nodes and incomplete nodes i.

Exploring a node in a branching algorithm corresponds to moving from the current search tree T to a tree T' with one additional node. The relaxation function θ is monotone nondecreasing in the sense that $\theta(T') \geq \theta(T)$ whenever $T' \supset T$. It can therefore be maximized by a hill-climbing search, much like the Lagrangean function. Each step of the algorithm moves from T to a neighboring solution obtained by adding one node to T.

All branching strategies can be viewed as hill-climbing algorithms for solving the branching relaxation dual. These include depth-first search, breadth-first search, limited discrepancy search, and strong branching, which are discussed in Chapter 5. Strong branching makes use of *pseudo-costs*, which estimate the effect on the objective function of branching to a given node. A pseudo-cost can be conceived in general as the rate of change in $\theta(T)$ as one moves from T to a neighboring solution T' of the dual. A search strategy that maximizes pseudo-cost at each iteration is analogous to a subgradient optimization method for the Lagrangean dual.

Exercises

4.26. Construct a branching tree that solves the 0-1 programming problem $\min\{cx \mid Ax \geq b\}$ given by

$$\min 3x_1 + 4x_2$$
$$6x_1 + 2x_2 \geq 3$$
$$3x_1 + 6x_2 \geq 4$$
$$x_1, x_2 \in \{0, 1\}$$

Branch on x_1 first. Recursively define a value-bounding function $v(b + \Delta b)$. Then write a closed-form expression for $v(b + \Delta b)$ in terms of Δb that distinguishes three cases ($\Delta b_1, \Delta b_2 \geq -1$; $\Delta b_1 \geq -1$ and $\Delta b_2 < -1$; and $\Delta b_1 < -1$).

4.27. Write a set of inequalities (one for each leaf node) whose satisfaction by a perturbation of the problem in Exercise 4.26 is sufficient to ensure that the resulting optimal value is reduced no more than Δz.

4.28. Consider the problem

$$\min x_1 + x_2 + x_3$$
$$6x_1 + 7x_2 + 8x_3 \geq 51$$
$$x_i \in \{0, 1, 2, 3\}, \text{ all } i$$

Solve the problem by branching. At each node, first propagate the inequality constraint to reduce domains (if possible). Then solve the LP relaxation, unless the domains are already singletons or empty. Branch on fractional variables. Construct a proof tree analogous to Fig. 4.4. Now derive conditions under which perturbation of the inequality constraint (coefficients and RHS) does not reduce the optimal value. *Hints:* In the analysis of Secction 4.7.3, there is no propagation, and sensitivity analysis depends only on LP relaxations at the leaf nodes. Here, the situation is reversed. LP relaxations affect only the branching, and the current optimal value remains optimal as long as the domain reduction inferences remain valid. Domain reductions at all nodes (not just leaf nodes) must be considered.

4.29. Indicate how the sensitivity analysis of Secction 4.7.3 can be extended to general integer programming. Assume that branching occurs in the usual fashion by splitting the domain of a variable.

4.8 Bibliographic Notes

Section 4.1. Inference duality is developed in [277, 279, 288, 298]. The connection between inference duality and sensitivity analysis appears in [277] and is applied to mixed-integer programming in [164]. The value-bounding function discussed here generalizes the value function for integer programming introduced in [95, 96] and further studied in [461, 476, 507, 508]. The connection between inference duality and constraint-directed search is described in [279, 286, 287, 298]. Introductions to computational complexity theory include [24, 233, 386]

Section 4.2. Relaxation duals, particularly the Lagrangean dual, have long been studied. The abstract concept of a relaxation dual is discussed in [279].

Section 4.3. Linear programming duality is based on the Farkas lemma and but is usually credited [157] to John von Neumann, who formulated it in connection with the theory of noncooperative two-person games. The dual simplex method is due to [48, 338].

Section 4.4. Surrogate duality was introduced by [230].

Section 4.5. Lagrangean duality is an old idea that is first applied to integer programming in [262, 263], where the dual is solved by subgradient optimization. A good exposition of the application of Lagrangean duality to integer programming can be found in [200] and to nonlinear programming in [47]. General discussions of integer programming duality appear in [280, 476, 499].

Section 4.6. The subadditive (superadditive) dual is stated in [306]. Its connection with integer programming is explored in [95], and in particular with Chvátal–Gomory cuts (Corollary 4.20) in [96, 301, 507]. These ideas are presented in Section II.3 of [371].

Section 4.7. A branching dual is used in [441] for integer programming sensitivity analysis (with respect to the right-hand side only) by computing a piecewise linear value function; see also [461]. A more general branching dual is used in [277] for sensitivity analysis and applied to mixed-integer programming for the same purpose in [164].

Chapter 5
Search

The main issue in search is deciding where to look next. There are many ways to do this, but exhaustive search methods have generally taken one of two forms—branching and constraint-directed search.

Branching is a divide-and-conquer strategy that is essentially guided by problem difficulty. If a problem is too hard to solve, it is broken into two or more subproblems that are more highly restricted and perhaps easier to solve. The process generates a search tree whose leaf nodes correspond to subproblems that can be solved or shown infeasible. The best solution found is optimal for the original problem.

Constraint-directed search is guided by past experience. Whenever a solution is examined, a nogood constraint is generated that excludes it and perhaps other solutions that can be no better. The next solution examined must satisfy the nogood constraints generated so far, to avoid covering the same ground. The search is over when there is no ground left to cover, and the best solution found is optimal. Branching is actually a special case of constraint-directed search, but a very important special case that deserves separate discussion.

The chapter is divided into three sections, corresponding to branching, constraint-directed search, and incomplete search. The first section examines node and variable selection schemes, primal heuristics, and branch-and-price methods. Airline crew scheduling is presented as an example. The second section develops constraint-directed search in the abstract and then describes three varieties of it: logic-based Benders decomposition, constraint-directed branching, and partial-order dynamic backtracking. Propositional satisfiability algorithms provide examples. The chapter concludes by showing how branching and constraint-directed search schemes can be seen at work in local search as well as exact methods. This allows one to understand exhaustive and local search as belonging to a single algorithmic framework.

5.1 Branching Search

Branching search uses a recursive divide-and-conquer strategy. If the original problem P_0 is too hard to solve as given, the branching algorithm creates a series of restrictions $P_1, \ldots, P_m$ of P_0 and tries to solve them. In other words, it *branches* at P_0. Each P_i is solved, if possible. If a solution is obtained, it becomes a candidate solution of P_0. The best candidate solution found so far is the *incumbent solution*. If P_i is too hard to solve, the search procedure attacks P_i in a similar manner by branching at P_i, and so on, recursively.

A problem restriction is *processed* when an attempt is made to solve it. The search begins with a single unprocessed problem (the original problem) and terminates when no unprocessed problems remain. At this point, the incumbent solution is optimal.

Figure 5.1 summarizes the basic algorithm, which can be viewed as generating a search tree with the original problem P_0 at the root. Each restriction P is identified with a node of the tree, and the restrictions $P_1, \ldots, P_m$ of P with children of this node. The set S contains the currently unprocessed nodes. Node P is *complete* when P has been solved, or the restrictions $P_1, \ldots, P_m$ have been processed. The algorithm

---

Let $S = \{P_0\}$ and $v_{\mathrm{UB}} = \infty$.
While S is nonempty repeat:
    Select a problem restriction $P \in S$ and remove P from S.
    If P is too hard to solve then
        Define restrictions $P_1, \ldots, P_m$ of P and add them to S.
    Else
        Let v be the optimal value of P and let $v_{\mathrm{UB}} = \min\{v, v_{\mathrm{UB}}\}$.
The optimal value of P_0 is v_{UB}.

Fig. 5.1 Generic branching algorithm for solving a minimization problem P_0. Set S contains the problem restrictions so far generated but not yet attempted, and v_{UB} is the best solution value obtained so far.

terminates when S is empty, which is to say that all nodes in the current tree are complete.

The most popular branching mechanism is to branch on a variable x_j. The domain of x_j is divided into two or more subsets, and restrictions are created by successively restricting x_j to each of these subsets. This is a special case of branching on constraints, in which restrictions are created by imposing additional constraints in each branch.

To ensure an exhaustive search, the restrictions $P_1, \ldots, P_m$ created at each branch should be exhaustive. Normally, their feasible sets also partition the feasible set of P (i.e., they are pairwise disjoint), which is unnecessary but more efficient, as it avoids covering the same ground more than once.

To ensure that the search terminates, the branching mechanism must be designed so that problems become easy enough to solve as they are increasingly restricted. For instance, if the variable domains are finite, then branching on variables will eventually reduce the domains to singletons, thus fixing the value of each x_j and making the restriction trivial to solve.

5.1.1 Branch-Infer-and-Relax Methods

Inference and/or relaxation may be combined with branching to reduce the size of the search tree.

A branching method incorporates inference by inferring new constraints for a problem restriction before it is solved. This results in a *branch-and-infer* method. When inference takes the form of domain filtering, for example, some of the variable domains are reduced in size. When one branches on variables, this tends to reduce the size of the branching tree because the domains are more rapidly reduced to singletons or the empty set. The employee scheduling example of Section 2.5 is solved by a branch-and-infer method.

Branching can also be combined with relaxation to obtain a *branch-and-relax* method. If bounds obtained from the relaxations are used to prune the search tree, the method becomes *branch and bound*.

A branch-and-relax method processes a node by solving a relaxation R of the restriction P, rather than P itself. If the solution of R happens to be feasible in P, it is optimal for P and becomes a candidate solution. If R is infeasible, then so is P. Otherwise, the algorithm

Let $S = \{P_0\}$ and $v_{\mathrm{UB}} = \infty$.
While S is nonempty repeat:
    Node selection. Select a problem restriction $P \in S$ and
        remove P from S.
    Inference. Repeat as desired:
        Add inferred constraints to P.
        Let v_{R} be the optimal value of a relaxation R of P.
    Bounding. If $v_{\mathrm{R}} < v_{\mathrm{UB}}$ then
        If R's optimal solution is feasible for P let $v_{\mathrm{UB}} = \min\{v_{\mathrm{R}}, v_{\mathrm{UB}}\}$.
        Branching.
        Else define restrictions $P_1, \ldots, P_m$ of P and add them to S.
The optimal value of P_0 is v_{UB}.

Fig. 5.2 Generic branching algorithm, with inference and relaxation, for solving a minimization problem P_0. The inner repeat loop is typically executed only once, but it may be executed several times, perhaps until no more constraints can be inferred or R becomes infeasible. The inference of constraints can be guided by the solution of previous relaxations.

branches at P. To ensure termination, the branching mechanism must be designed so that solving R becomes sufficient to solve P if one descends deeply enough into the search tree.

Branch and bound uses the optimal value v^* of R to prune the search. If v^* is greater than or equal to the value of the incumbent solution, then there is no point in solving P and no need to branch on P. In this case, the search tree is said to be "pruned" at P.

Inference and relaxation can work together effectively in branch-infer-and-relax methods. The addition of inferred constraints (e.g., cutting planes) to P can result in a tighter bound when one solves its relaxation R. Conversely, the solution of R can provide guidance for generating further constraints, as for instance when separating cuts are generated to exclude a solution of R that is infeasible in P.

A generic branch-infer-and-relax algorithm appears in Fig. 5.2. The freight transfer, production planning, continuous global optimization, and product configuration examples of Chapter 2 illustrate specific realizations of the algorithm. The most widely used solvers in constraint programming, integer programming, and continuous global optimization combine inference and relaxation with branching. Table 5.1 shows how they fit into the general search-infer-and-relax framework.

Table 5.1 How some selected branching methods fit into the search-infer-and-relax framework.

Solution method	Restriction P	Relaxation R	Inference
Constraint programming	Created by splitting domain, etc.	Domain store	Domain filtering, propagation
Mixed-integer programming (branch and cut)	Created by branching on fractional variables	LP relaxation + cutting planes	Cutting planes, preprocessing, reduced-cost variable fixing
Continuous global optimization	Created by splitting intervals	LP or convex NLP* relaxation	Interval propagation, Lagrangean bounding

*Nonlinear programming.

5.1.2 Node Selection

A perennial issue in branching search is *node selection*, or the choice of which unprocessed node to consider next. In the generic algorithm of Fig. 5.2, it is the choice of which problem restriction P to remove from S.

Node selection is complicated by the fact that most branching methods for optimization are primal/dual methods. They obviously seek good feasible solutions (the primal problem), but they also seek a proof of optimality (the dual problem). The difficulty is that the primal and dual problems call for different search strategies that must somehow be combined in a single branching algorithm.

As noted in Section 4.7.4, the dual problem maximizes $\theta(T)$ over partial and complete search trees, where $\theta(T)$ is the bound established by tree T. At any point in a branching search, the current set of processed nodes forms a tree T. Selecting an unprocessed node corresponds to adding a new leaf node to T, which creates a new tree $T' \supset T$. The branching process can be regarded as a hill climbing algorithm for maximizing $\theta(T)$, because $T' \supset T$ implies $\theta(T') \geq \theta(T)$.

A natural strategy for the primal problem is *depth-first search*, which results from a node selection rule that chooses the most recently created node. In the algorithm of Fig. 5.2, it selects the restriction P that was most recently added to S. Depth-first search immediately probes to the bottom of the tree, which opens the possibility of discovering feasible solutions early in the search. It also has the advantage of requiring little space. If d is the current tree depth and b the *branching factor* (the maximum number of children of a node), the number of unprocessed nodes at any one time is at worst proportional to bd.

However, depth-first search may make very slow progress toward solving the dual; that is, toward obtaining a tight lower bound on the optimal value. The lower bound at any given point in the search is the minimum of the relaxation bounds obtained at the leaf nodes, and shallow nodes tend to provide weaker bounds. Because shallow leaf nodes remain in the tree until the very end of the search, the bound remains weak until the end.

A natural approach to solving the relaxation dual is the opposite strategy of *breadth-first search*, because it explores all the nodes on one level before moving to the next. It maximizes the depth of the shallowest node and may therefore converge to a tight bound early in the search.

Unfortunately, breadth-first search can be very slow to find feasible solutions, because it does not fix many variables until it has generated a great many nodes. To fix d variables, it must explore about b^d nodes. Its space requirement is also impractically large, because it is also proportional to b^d.

Primal/dual strategies attempt to obtain some of the advantages of both depth-first and breadth-first search. One simple approach is *iterative deepening*, which conducts a depth-first search to maximum depth d, then conducts a depth-first search (from scratch) to maximum depth $d+1$, and so forth until no unprocessed nodes remain at the end of the depth-first phase. This may seem wasteful because it re-creates a tree of depth d when generating a tree to depth $d + 1$, but the former requires only about $1/b$ the work of the latter. Also, the space requirement is at worst proportional to bd.

Iterative deepening inherits the advantage of breadth-first search while avoiding the exponential space requirement: it obtains the best lower bound available at depth d before moving to depth $d + 1$. However, it lacks the ability of pure depth-first search to probe deeply after generating only a few nodes. It can reach depth d only after an amount of work that grows exponentially with d. This is not a serious handicap in many contexts, such as game trees, but it is a shortcoming when searching for feasible solutions of an optimization problem.

Another primal–dual strategy is *limited discrepancy search*, which explores nodes in a band of gradually increasing width. Iteration 0 consists of a probe directly to the bottom of the tree, following a heuristic that is designed to find a feasible solution. Iteration k consists of a depth-first search in which at most k variables are set to values different from those in iteration 0. The space requirement is proportional to bd as in ordinary depth-first search.

This strategy is clearly suitable for the primal problem, because it has basically the same advantages as depth-first search. One might also argue that it inherits the bound-proving capability of breadth-first search without paying an inordinate cost. Iteration k of limited discrepancy search is exhaustive down to level k of the tree. It therefore delivers a bound at least as good as breadth-first search to level k, which creates a tree whose size grows exponentially with k. The limited discrepancy search tree contains many nodes below level k, but its size still grows exponentially with k.

A related strategy is *local branching*, which was originally developed for integer programming. It begins with a feasible solution x^0, perhaps obtained by a heuristic algorithm. In the left branch it considers all solutions that are within distance k of x^0, as in step k of limited discrepancy search, and in the right branch it considers solutions beyond distance k of x^0. Let x^1 be the best feasible solution found in the left branch. In the right branch it immediately branches again in the same manner, except that the left branch considers solutions that are close to x^1 rather than x^0. The process continues until no new solution is found in the left branch, in which case the right branch completes the branching search in the usual manner.

Supposing for simplicity that the variables are 0-1, the left branch at level i is formed by adding the constraint on the left below, and the right branch by adding the constraint on the right:

$$\sum_{\substack{j \\ x_j^i = 0}} x_j + \sum_{\substack{j \\ x_j^i = 1}} (1 - x_j) \le k \qquad \sum_{\substack{j \\ x_j^i = 0}} x_j + \sum_{\substack{j \\ x_j^i = 1}} (1 - x_j) \ge k + 1$$

The motivation for local branching is that the left branch can perhaps find feasible solutions quickly because it addresses a relatively

easy subproblem. Local branching is actually more than a node selection strategy, because it specifies how to partition the subproblems
as well. Furthermore, the subproblem in the left branch need not be
solved by branching, but can be solved by any exact algorithm.

5.1.3 Variable and Branch Point Selection

Most branching algorithms branch on variables. They execute the
branching step in the algorithm of Fig. 5.2 by splitting the current
domain of some variable x_j into subsets $D_1, \ldots, D_m$. The restrictions P_i are formed by augmenting the current subproblem P with
$x_j \in D_i$. Branching therefore requires two decisions: *variable selection*,
which determines the variable x_j on which to branch, and *branch point
selection*, which determines how to partition the domain of x_j.

Unfortunately, existing theory does not provide a reliable guide to
variable and branch point selection, and one typically relies on heuristic principles and/or trial and error. The literature contains a wide
variety of specialized branching schemes that seem to work for particular problems, but no attempt is made here to survey these in detail.

A natural way to select branching variables and branch points is to
observe which constraints are violated by the solution $\hat{x}$ of the current relaxation R_i. The branches can then be designed to repair the
violation.

The simplest sort of violation is a domain violation, which occurs
when some $\hat{x}_j \notin D_{x_j}$. If variables are required to be integer, for
example, one can branch on a variable with a fractional value $\hat{x}_j$. The
branch point is defined by creating a left branch with $x_j \leq \lfloor \hat{x}_j \rfloor$ and a
right branch with $x_j \geq \lceil \hat{x}_j \rceil$. Similarly, if $\hat{x}_j = 5$ and $D_{x_j} = \{1, 3, 6, 8\}$,
then one can branch by adding constraints $x_j \leq 3$ and $x_j \geq 6$. The
branching constraints must become part of the relaxations at the child
nodes, or else the infeasible value will not be repaired.

Branching can be designed for other types of constraint violations as
well. For example, if $\hat{x}$ violates alldiff$(x_1, \ldots, x_n)$ because $\hat{x}_1 = \hat{x}_2 = a$,
one can set $x_1 = a$ and delete a from D_{x_j} for $j \neq 1$ in the left branch,
and set $x_2 = a$ and delete a from D_{x_j} for $j \neq 2$ in the right branch.

Variable selection need not be based on constraint violations, and
even when it is, one may be obliged to select from several variables that
are involved in violations. In such cases, three related principles are

commonly used: first-fail branching, constrainedness branching, and hierarchical branching. One can also use cost-based branching (Section 5.1.4), which applies to variable selection as well as node selection.

First-fail branching selects a variable that is more likely to result in infeasibility when fixed. In CP solvers, this is normally implemented by branching first on variables with small domains. Constrainedness branching selects variables that are more highly constrained, on the theory that fixing them will have more consequences. This might be measured by the number of constraints containing the variable, or perhaps the number of constraints containing it that are violated by $\hat{x}$. Hierarchical branching branches first on variables that are associated with major decisions, such as what kind of factory to build, and last on variables that specify details, such as what kind of forklift to use in the factory. All three principles are based on the idea that unimportant decisions have little effect on how many decisions must be made subsequently, and one should therefore begin with the important decisions.

Variables that represent important decisions may have the property that, when fixed, the problem simplifies radically. A set of variables with this property has been called a *backdoor* in recent literature. Backdoors are obviously key to Benders decomposition, because fixing the master problem variables should result in a much simpler subproblem. They are also important for branching, because branching first on backdoor variables can reduce the search tree by several orders of magnitude.

In most cases it is hard to recognize a backdoor in advance, but *random restarts* may discover one. A random restart strategy incorporates some randomness into the variable selection. If the search tree starts to grow large without solving the problem, the search is restarted, perhaps repeatedly. This simple idea has proved remarkably successful in a variety of solvers that seek feasible solutions, although less is known about its value for optimization.

The success of random restarts may be related to the *heavy-tailed* behavior of randomized branching algorithms. The distribution of running times in a random sample of searches is heavy-tailed at the upper end, so that the sample mean tends to grow without bound as the sample size increases. The existence of backdoors may help explain this behavior, because the search tree size may depend critically on whether one happens to start searching with backdoor variables.

A few additional observations should be made about branch point selection. One rule of thumb is that the resulting subtrees should be roughly balanced in size. A well-known example is the set-partitioning constraint $x_1 + \cdots + x_n = 1$, where each $x_j \in \{0, 1\}$. If $\hat{x}_1$ is fractional and one simply creates two branches in which $x_1 = 0$ and $x_1 = 1$, then the former branch may result in a much larger subtree than the latter. A more balanced scheme creates n branches in which each branch i sets $x_j = 0$ for $j \neq i$. The same scheme can be used when at most one of a set of nonnegative continuous variables $x_1, \ldots, x_n$ is allowed to be positive. In traditional mixed-integer modeling, $\{x_1, \ldots, x_n\}$ is known as a *special ordered set of type 1* (SOS1), and the balanced branching scheme is *SOS1 branching*.

Branch point selection is especially critical in nonlinear global optimization, which frequently branches on continuous variables. Typically, $\hat{x}$ is obtained by solving a linear relaxation of the problem at the current node. The object of branching is to partition the problem space near the optimal solution into smaller and smaller boxes, so that the linear relaxations within the boxes become progressively more accurate and $\hat{x}$ eventually satisfies the original constraints within a suitable tolerance. The search normally branches on a variable x_j that appears in a nonlinear constraint C violated by $\hat{x}$. The left branch imposes $x_j \leq \bar{x}_j$ and the right branch imposes $x_j \geq \bar{x}_j$. The branch point $\bar{x}_j$ is not necessarily the same as $\hat{x}_j$ but can be chosen to result in tight linear relaxations of C in the two branches. A compromise between possibly conflicting criteria might be used: the tightness of the relaxations should be roughly equal, and their average tightness should in some sense be maximized. One way to measure tightness is to estimate the volumetric difference between the feasible set of C's linear relaxation and the feasible set of C.

5.1.4 Cost-Based Branching

Cost-based branching estimates the effect of branching decisions on the relaxation bounds that result. It is normally associated with mixed-integer programming but can be applied generally. One popular cost-based node selection strategy is *strong branching*, which chooses the unprocessed node with the largest relaxation value. The same idea can be applied to variable selection by choosing a variable that results in the largest relaxation values after branching on it.

Strong branching maximizes the relaxation value by actually solving the relaxations at unprocessed nodes. In mixed-integer programming, this is accomplished by the dual simplex method. Recall that one branches to a node i by rounding some $\hat{x}_j$ down or up. The branching constraint $x_j \leq \lfloor \hat{x}_j \rfloor$ or $x_j \geq \lceil \hat{x}_j \rceil$ is added to the LP relaxation, which is then reoptimized using the dual simplex method to obtain the relaxation value v_i^*. Because full reoptimization is time consuming, a variant of strong branching executes only a few iterations of the dual simplex method, perhaps even a single iteration. This results in an approximation of v_i^* that bounds it below.

The relaxation values can also be approximated by using *pseudo-costs*, which measure the rate of increase of the relaxation value when the value of x_j is forced to change. In mixed-integer programming, a pseudo-cost is the rate of increase when $\hat{x}_j$ is rounded up or down. Let $f = \hat{x}_j - \lfloor \hat{x}_j \rfloor$ be the fractional part of $\hat{x}_j$, and let v^* be the optimal value of relaxation at the current node. The pseudo-cost at child node i is $\Delta_j^- = (v_i^* - v^*)/f$ if $\hat{x}_j$ is rounded down to create node i, and $\Delta_j^+ = (v_i^* - v^*)/(1 - f)$ if it is rounded up. Thus if $\tilde{\Delta}_j^-$ and $\tilde{\Delta}_j^+$ are estimates of the pseudo-costs, then

$$v_i^* \approx \begin{cases} v^* + \tilde{\Delta}_j^- f, & \text{if } \hat{x}_j \text{ is rounded down} \\ v^* + \tilde{\Delta}_j^+ (1 - f), & \text{if } \hat{x}_j \text{ is rounded up} \end{cases}$$

In practice, various methods are used to estimate pseudo-costs. If the search has branched on x_j at previous nodes, the pseudo-cost can be estimated to be the average of Δ_j^- or Δ_j^+ (whichever applies) at all such nodes. If no branching on x_j has occurred, the pseudo-cost can be computed or estimated by applying the dual simplex method. A variant of this approach, known as *reliability branching*, uses the average of previous pseudo-costs only if the search has branched on x_j a certain minimum number of times.

Strong branching can be interpreted as an approach to solving the relaxation dual. As noted above, the branching process is in effect a hill-climbing algorithm for maximizing $\theta(T)$ in the relaxation dual. Strong branching proposes a particular criterion for selecting the next search tree T'. The dual perspective can suggest alternative criteria.

Recall from Section 4.7.4 that $\theta(T)$ is the minimum of the relaxation values v_i^* over all leaf nodes and incomplete nodes in T. An non-leaf node i is incomplete when at at least one currently unprocessed node has node i as its parent. If T' is formed from T by attaching unprocessed node k to its parent node i, then $\theta(T')$ can be larger than $\theta(T)$ only if adding node k completes node i. Even in this case, there can be improvement only if v_i^* is smaller than all the other relaxation values. Specifically,

$$\theta(T') = \begin{cases} \min_{j \in I}\{v_j^*\}, & \text{if node } i \text{ is complete in } T' \text{ and } v_i^* < \min_{j \in J}\{v_j^*\} \\ \theta(T), & \text{otherwise} \end{cases}$$

$$(5.1)$$

where I is the set of node i's children in T', and J is the set of leaf nodes and incomplete nodes in T'.

Strong branching selects T' by adding the node k with the largest relaxation value v_k^*. This rarely improves the dual solution, because generally the relaxation value v_i^* at the parent node is already rather large, and it is unlikely that $v_i^* < \min_{j \in J}\{v_j^*\}$. A more dual-oriented branching scheme might strive to select a T' for which $\theta(T') > \theta(T)$. For example, it could select an unprocessed node who parent node i has the minimum relaxation value v_i^*. Or it could simply select a node for which $\theta(T')$ as defined in (5.1) is largest.

Strong branching might be viewed as a primal-oriented strategy on the ground that it tends to find good feasible solutions. A primal–dual strategy might use strong branching initially to obtain good upper bounds, and then move to a dual-oriented scheme to tighten the lower bound.

5.1.5 Primal Heuristics

A theme that runs through the foregoing discussion is the challenge of designing a branching scheme to solve both the primal and dual problems simultaneously. The challenge is difficult because the primal and dual generally call for different search strategies. One solution to this problem is to avoid it—in particular, by conducting a primal search for good feasible solutions outside the branching tree, and allowing the branching scheme to focus more on proving optimality.

The primal search can be accomplished by a *primal heuristic*, which is an algorithm that seeks feasible solutions. It is normally executed at the root node, but potentially at any node of the search tree. Primal heuristics have become popular in the optimization community in the last decade or so and are now a standard feature of all major solvers.

A wide variety variety of primal heuristics are used, but they can be generally classified as local search and diving heuristics. *Local search* seeks feasible solutions by modifying the solution of the relaxation. Variables that contribute to infeasibility are reset to a value that may help to restore feasibility. Because a limited number of changes are considered, local search in effect examines a neighborhood of the relaxation solution. *Diving heuristics* repeatedly fix the value of a variable and then re-solve the relaxation that results. They in effect probe deeply into the tree, as would occur at the beginning of depth-first search, but the branches that result do not become part of the search tree. Rather, the feasible solutions found by either type of heuristic provide upper bounds that help prune the tree.

Local search is illustrated by rounding heuristics in integer programming, where the constraints are $Ax \geq b$, $x \geq 0$. To simplify discussion, assume that all variables are restricted to integer values. In this context, the solution $\hat{x}$ of the relaxation is infeasible when some $\hat{x}_j$ is fractional. A very simple heuristic rounds $\hat{x}_j$ down when none of its coefficients have the "wrong sign" for rounding down (i.e., $A_{ij} < 0$ for no i), and up when none of its coefficients have the wrong sign for rounding up ($A_{ij} > 0$ for no i). This sort of rounding obviously cannot violate $Ax \geq b$. The heuristic fails if some fractional variables remain.

A related heuristic begins by setting $\bar{x} = \hat{x}$, and each iteration proceeds as follows. If $\bar{x}$ satisfies $Ax \geq b$, then select a fractional $\bar{x}_j$ having the smallest number of coefficients with the wrong sign for rounding (up or down), and replace $\bar{x}_j$ with the result of rounding $\bar{x}_j$ up or down accordingly. If $\bar{x}$ violates some constraint in $Ax \geq b$, then select a fractional $\bar{x}_j$ than when rounded (up or down) most reduces the constraint violation, and replace $\bar{x}_j$ with the result of rounding $\bar{x}_j$ up or down accordingly. Continue until all $\bar{x}_j$ are integer. The heuristic fails if $\bar{x}$ violates $Ax \geq b$.

A simple diving heuristic for integer programming selects a fractional $\hat{x}_j$ and fixes x_j to the result of rounding $\hat{x}_j$ up or down. It then reoptimizes the LP relaxation using the dual simplex method and repeats. The process continues until the LP becomes infeasible, its optimal value exceeds the value of the incumbent solution, a time limit is exceeded, or all the variables are integer. The heuristic fails in the first three cases. There are a number of criteria for selecting the fractional $\hat{x}_j$ and the rounding direction. One is to select the $\hat{x}_j$ that is closest to an integer and round it to that integer. Another is to select an $\hat{x}_j$ having the smallest number of coefficients with the wrong sign.

Still others make use of pseudo-costs, the incumbent solution, and so forth.

The "feasibility pump" heuristic is similar to diving in that it re-peatedly re-solves an LP relaxation, but it does not dive to the bottom of the search tree. It begins by setting $\bar{x} = \hat{x}$. In each iteration it rounds $\bar{x}$ to the nearest integer y and terminates with success if y is feasible. Otherwise, it sets $\bar{x}$ to the point x that minimizes the distance $|y - x|$ to y subject to $Ax \geq b$, $x \geq 0$. The minimum distance can be found by solving an LP problem that minimizes $\sum_j d_j$ subject to $d_j \geq y_j - x_j$, $d_j \geq x_j - y_j$, and $Ax \geq b$, $x \geq 0$. The procedure can cycle through the same values y, in which case y is randomly perturbed to break the cycle. The heuristic fails if it reaches the time limit.

5.1.6 Branch and Price

Branch and price is a specialized branch-and-bound method designed for problems with an integer programming formulation. Despite this restriction, the method is of interest here because it has become one of the more popular settings for integrated problem solving. It has proved particularly successful for airline crew scheduling and other transport-related applications.

Branch and price can be attractive when the integer programming model has a huge number of variables. In such cases the variables are added to the model only as needed to solve the LP relaxation. A *pricing* subproblem is solved to deliver a variable with negative reduced cost. If there is no such variable, the algorithm terminates. Typically, only a small fraction of the total variable set is required to reach an optimal solution. When the continuous relaxation is solved, one can branch as in ordinary branch-and-bound methods and solve the contin-uous relaxations at the branches by adding further variables. Because adding a variable to a problem corresponds to adding a column to the constraint matrix, branch and price is a *column generation* method.

The pricing problem can be solved by any convenient method, and this is where other technologies may enter the picture. Constraint pro-gramming, in particular, can be useful for this purpose, because it can deal with the often complex constraints that must be observed when the variables are generated. Airline crew scheduling, for example, is normally constrained by a host of complicated work rules.

The problem at any given node of the branch-and-bound tree has the form $\min\{cx \mid Ax \geq b,\ x \geq 0,\ x \text{ integer}\}$. Its LP relaxation drops the integrality constraint. When A has a huge number of columns, the LP problem is initially formulated with only a small subset of the columns:

$$\min\ cx$$
$$\sum_{j \in J} A_j x_j \geq b \qquad\qquad (5.2)$$
$$x_j \geq 0,\ \text{all } j \in J$$

This is the *restricted master problem*. It is a restriction because any feasible solution of it is a feasible solution of the full problem with the remaining variables set to zero. The algorithm for solving the full LP begins by obtaining the optimal dual solution u of (5.2). The reduced cost of any column A_j that is missing from (5.2) is $r_j = c_j - uA_j$. To find a column with negative reduced cost, the pricing subproblem is solved:

$$\min\ y_0 - uy$$
$$\mathcal{C}(y_0, y)$$

where the constraints $\mathcal{C}(y_0, y)$ are designed to be satisfied by all and only the columns of (c, A). If the optimal value of the pricing subproblem is zero, the current solution of the restricted master is optimal, and the algorithm terminates. Otherwise, column (y_0, y) is added to (5.2), and the process repeats.

5.1.7 Example: Airline Crew Scheduling

A simplified airline crew scheduling problem shows how CP and integer programming can be combined in a branch-and-price framework. It also illustrates set-valued variables, which have become important in CP solvers.

The goal is to assign flight crew members to flights so as to minimize cost while covering all the flights and observing a number of work rules. Each flight j starts at time s_j, finishes at time f_j, and requires n_j crew members. A small example with six flights appears in Table 5.2. Whenever a crew member staffs two consecutive flights, the rest period between the flights must be at least $\Delta_{\min}$ and at most $\Delta_{\max}$. The total

Table 5.2 Start times s_j and finish times f_j for flights in a small crew-scheduling problem.

j	s_j	f_j
1	0	3
2	1	3
3	5	8
4	6	9
5	10	12
6	12	14

flight time assigned to a crew member must be between $T_{\min}$ and $T_{\max}$. There may be other restrictions on scheduling as well.

Integer Programming Formulation

The set of flights assigned to a crew member is known as a *roster*. The problem can in principle be formulated by generating all possible rosters and assigning one roster to each crew member in such a way as to cover each flight. The cost c_{ik} of assigning crew member i to roster k depends on a number of factors, such as seniority, the timing of flights and rest periods, and so forth.

Let δ_{jk} be 1 when flight j is part of roster k, and 0 otherwise. Also let 0-1 variable $x_{ik} = 1$ when crew member i is assigned roster k. Then, the problem can be formulated with an integer programming model in which dual multipliers u_i, v_j are associated with the constraints as shown:

$$
\text{0-1 linear:} \begin{cases}
\min \displaystyle\sum_{ik} c_{ik} x_{ik} \\[2ex]
\displaystyle\sum_k x_{ik} = 1, \quad \text{all } i \qquad (u_i) \\[2ex]
\displaystyle\sum_i \sum_k \delta_{jk} x_{ik} \geq n_j, \quad \text{all } j \ (v_j)
\end{cases}
\tag{5.3}
$$

domains: $x_{ik} \in \{0,1\}$, all i, k

Suppose for the example of Table 5.2 that the minimum and maximum gap between flights are $(\Delta_{\min}, \Delta_{\max}) = (2, 3)$, and the minimum and maximum flight times are $(T_{\min}, T_{\max}) = (6, 10)$. Thus, flight 6 cannot immediately follow flight 1, because the gap is too large (11), and flight 5 cannot immediately follow flight 4, because the gap is too small (1). The constraints permit four possible rosters: $\{1,3,5\}$, $\{1,4,6\}$, $\{2,3,5\}$, and $\{2,4,6\}$. If there are two crew members, and each flight requires one crew member, the continuous relaxation of problem (5.3) becomes:

$$\min z$$

$$
\begin{bmatrix}
10 & 12 & 7 & 13 & 9 & 11 & 7 & 12 \\
1 & 1 & 1 & 1 & 0 & 0 & 0 & 0 \\
0 & 0 & 0 & 0 & 1 & 1 & 1 & 1 \\
1 & 1 & 0 & 0 & 1 & 1 & 0 & 0 \\
0 & 0 & 1 & 1 & 0 & 0 & 1 & 1 \\
1 & 0 & 1 & 0 & 1 & 0 & 1 & 0 \\
0 & 1 & 0 & 1 & 0 & 1 & 0 & 1 \\
1 & 0 & 1 & 0 & 1 & 0 & 1 & 0 \\
0 & 1 & 0 & 1 & 0 & 1 & 0 & 1
\end{bmatrix}
\begin{bmatrix}
x_{11} \\ x_{12} \\ x_{13} \\ x_{14} \\ x_{21} \\ x_{22} \\ x_{23} \\ x_{24}
\end{bmatrix}
\begin{matrix}
= \\ = \\ = \\ \geq \\ \geq \\ \geq \\ \geq \\ \geq \\ \geq
\end{matrix}
\begin{bmatrix}
z \\ 1 \\ 1 \\ 1 \\ 1 \\ 1 \\ 1 \\ 1 \\ 1
\end{bmatrix}
\qquad (5.4)
$$

$$x_{ik} \geq 0, \text{ all } i,k$$

The problem is written in matrix form to show the eight columns, which correspond to the four possible rosters for each of the two crew members. The top row of the matrix contains the costs c_{ik}.

Rather than solve the complete problem (5.4), the problem is first solved with a subset of the columns, perhaps the following:

$$\min z$$

$$
\begin{bmatrix}
10 & 13 & 9 & 12 \\
1 & 1 & 0 & 0 \\
0 & 0 & 1 & 1 \\
1 & 0 & 1 & 0 \\
0 & 1 & 0 & 1 \\
1 & 0 & 1 & 0 \\
0 & 1 & 0 & 1 \\
1 & 0 & 1 & 0 \\
0 & 1 & 0 & 1
\end{bmatrix}
\begin{bmatrix}
x_{11} \\ x_{14} \\ x_{21} \\ x_{24}
\end{bmatrix}
\begin{matrix}
= \\ = \\ = \\ \geq \\ \geq \\ \geq \\ \geq \\ \geq \\ \geq
\end{matrix}
\begin{bmatrix}
z \\ 1 \\ 1 \\ 1 \\ 1 \\ 1 \\ 1 \\ 1 \\ 1
\end{bmatrix}
\begin{matrix}
\\ (10) \\ (9) \\ (0) \\ (0) \\ (0) \\ (0) \\ (0) \\ (3)
\end{matrix}
\qquad (5.5)
$$

$$x_{ik} \geq 0, \text{ all } i,k$$

The optimal solution is $(x_{11}, x_{14}, x_{21}, x_{24}) = (0,1,1,0)$. The corresponding dual multipliers $u = (10,9)$ and $v = (0,0,0,0,0,3)$ are shown on the right.

The Pricing Subproblem

The pricing subproblem attempts to identify a variable of (5.3) with a negative reduced cost. A column associated with x_{ik} contains a 1 in the row corresponding to u_i and the rows corresponding to v_j for each flight j in the roster. The reduced cost of x_{ik} is therefore

$$c_{ik} - u_i - \sum_{j \text{ in roster } k} v_j \tag{5.6}$$

One way to formulate the pricing subproblem is to model it with a directed graph. The graph contains a node j for each flight, plus a source node s and sink node t. The graph contains a directed arc (j, j') when flight j can immediately precede j' in a roster; that is, $\Delta_{\min} \leq s_{j'} - f_j \leq \Delta_{\max}$. There are also arcs (s, j) and (j, t) for every flight j. Every possible roster corresponds to a path from s to t, although not every path corresponds to a roster, because the total fight time may not be in the range $[T_{\min}, T_{\max}]$.

The graph corresponding to the example appears in Fig. 5.3. Only some of the arcs incident to s and t are shown, since the others can be removed by elementary preprocessing of the graph. For example, arc $(s, 3)$ is removed because every path from 3 to t results in a total flight

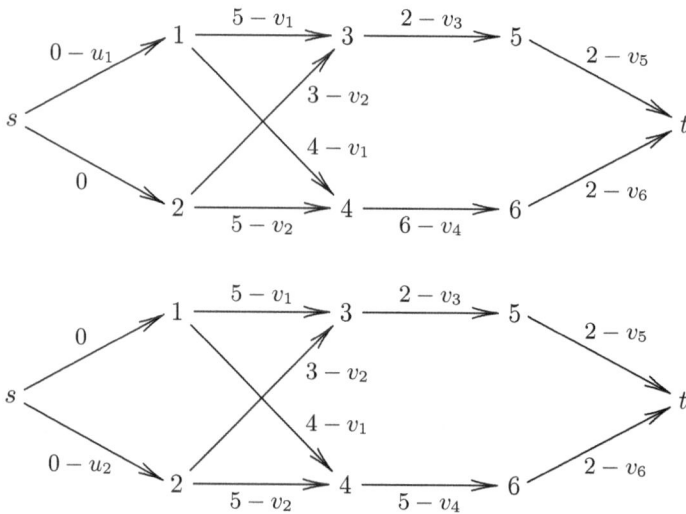

Fig. 5.3 Path model for crew rostering. The top graph pertains to crew member 1, and the bottom graph to crew member 2. The arc lengths reflect costs offset by dual multipliers u_i, v_j.

time less than the minimum of $T_{\min} = 6$. This can be ascertained by computing the longest path from 3 to t when the length of each arc (j, j') is set to the duration of flight j.

The graph-based model assumes that the cost c_{ik} associated with x_{ik} can be equated with the length of the corresponding path, if the arc lengths are suitably defined. This is possible if the cost depends only on the cost $\alpha_{ijj'}$ of staffing each flight j with crew member i and transferring him or her to the next flight j'. The length of each arc (j, j') is set equal to $\alpha_{ijj'}$, the length of each (s, j) to the cost α_{isj} of transferring from home to flight j, and the length of (j, t) to the cost α_{ijt} of operating flight j and returning home. The arc costs associated with each crew member in the example problem are shown in Fig. 5.3 (ignore the terms u_i and v_j for the moment).

The reduced cost (5.6) of a variable x_{ik} can also be equated with the length of the corresponding path, if the arc lengths are offset by the dual multipliers. The cost term c_{ik} in (5.6) is the path length using the arc lengths just defined. The path length becomes the reduced cost if the length of each (j, j') is set to $\alpha_{ijj'} - v_j$, the length of each (s, j) to $\alpha_{isj} - u_i$, and the length of each (j, t) to $\alpha_{ijt} - v_j$. These adjustments are shown in Fig. 5.3.

The pricing subproblem is now the problem of finding a path in the directed graph G with negative length, for which the total flight time lies in the interval $[T_{\min}, T_{\max}]$. Let X_i be the set of flights assigned to crew member i. Since each path is specified by some X_i, the problem can be written as follows:

$$\text{path: } (X_i, z_i \mid G, c, s, t), \text{ all } i$$

$$\text{setSum: } T_{\min} \leq \sum_{j \in X_i} p_j \leq T_{\max}, \text{ all } i \tag{5.7}$$

$$\text{domains: } X_i \subset \{\text{flights}\}, \ z_i < 0, \text{ all } i$$

where $p_j = f_j - s_j$ is the duration of flight j. The path metaconstraint requires that the nodes in X_i define a path from s to t in G with length z_i, where c contains the edge lengths c_{ij}. Since z_i's domain elements are negative, the constraint enforces a negative reduced cost. The set sum constraint simply requires that the total flight duration be within the prescribed bounds.

In the example of Fig. 5.3, the arc lengths are defined by the dual multipliers $u = (10, 9)$ and $v = (0, 0, 0, 0, 0, 3)$. The shortest path in the graph for crew member 1 is s-1-4-6-t, which has length -1. This defines the column corresponding to variable x_{12} in (5.4). The shortest path for crew member 2 is s-2-3-5-t with length -2, which defines the column corresponding to x_{23}. So variables x_{12} and x_{23} have negative reduced costs, and their columns are added to the restricted master problem (5.5). The new solution is $(x_{11}, x_{12}, x_{14}, x_{21}, x_{23}, x_{24}) = (0, 1, 0, 0, 1, 0)$ with dual multipliers $u = (10, 5)$ and $v = (0, 1, 0, 0, 0, 2)$. When the arc lengths are updated accordingly, the shortest path for both crew members has length zero. This means there is no improving column, and the solution of the restricted master is optimal. Because this solution is integral, there is no need to branch. The optimal rostering assigns flights 1, 4 and 6 to crew member 1, and the remaining flights to crew member 2.

Bounds Propagation for Set-Valued Variables

The pricing subproblem (5.7) can, in general, be solved by a combination of bounds propagation and branching. Bounds propagation must be reinterpreted for the variables X_i because they are set valued, but the idea is straightforward. The domain of each X_i is stored in the form of bounds $[L_{X_i}, U_{X_i}]$, where L_{X_i} is a set that X_i must contain, and U_{X_i} is a set that must contain X_i. Initially L_{X_i} is empty and U_{X_i} is the set of all flights, but branching and bounds propagation may tighten these bounds. For instance, if branching fixes $j \in X_j$, then $L_{X_i} = \{j\}$.

The path constraint in (5.7) tightens bounds for both z_i and X_i. It updates the lower bound L_{z_i} on z_i by finding a shortest path from s to t whose node set lies in the range $[L_{X_i}, U_{X_i}]$. If ℓ is the length of the shortest path, then L_{z_i} is updated to $\max\{L_{z_i}, \ell\}$. If $\ell \geq 0$, the domain of z_i becomes empty because U_{z_i} is initially negative, and there is no variable with negative reduced cost (the constraint also updates U_{z_i} by finding a longest path, but this is not relevant here).

Since G is acyclic, there are very fast algorithms for finding a shortest path, despite the possibility of negative arc lengths. These algorithms can be adapted to ensure that the node set lies in the range $[L_{X_i}, U_{X_i}]$ by temporarily modifying G. To make sure the path nodes belong to U_{X_i}, simply delete all nodes in G that are not in U_{X_i}. To make sure all nodes in L_{X_i} belong to the path, avoid routing around nodes in L_{X_i}. This is accomplished by first performing a topological sort on the graph. That is, index the nodes so that there is a directed path from node j to node j' if and only if $j < j'$. Then delete any arc (j, j'') such that $j < j' < j''$ for some $j' \in L_{X_i}$.

The modified graph G' can be used to tighten the bounds for X_i as well. If every path from s to t in G' contains node j, then add j to L_{X_i}. If no path contains j, then remove j from U_{X_i}. This can be accomplished by supposing that the arcs of G' carry flow in the direction of their orientation. It is assumed that a flow of 1 unit volume enters G' at node s and exits at node t. If the minimum flow through node j subject to these conditions is 1, then j contains all paths from s to t, and if the maximum is 0, it contains no paths. Minimum and maximum flow problems of this sort can be solved very quickly. Thus, if $L_{X_i} = \{3\}$ in the example, the minimum flow through node 5 in G' is 1 and the maximum through 4 is 0, which indicates that node 5 may be added to L_{X_i} and node 4 removed from U_{X_i}.

The domain $[L_{z_i}, U_{z_i}]$ of z_i may also permit filtering. A node can be removed from U_{x_i} is no path through j in G' has length at least L_{z_i}, or if no path through j has length less than or equal to U_{z_i}.

The set sum constraint can also be used for propagation in the obvious way. Again let $[L_{z_i}, U_{z_i}]$ be the current domain for z_i. One can update L_{z_i} to

$$\max\left\{ L_{z_i}, \sum_{j \in L_{X_i}} p_j + \sum_{j \in U_{X_i} \setminus L_{X_i}} \min\{0, p_j\} \right\} \tag{5.8}$$

and analogously for U_{z_i}. Also, if

$$p_k + \sum_{j \in L_{X_i}} p_j > U_{z_i} \quad \text{or} \quad p_k + \sum_{j \in L_{X_i}} p_j < L_{z_i}$$

for $k \notin L_{X_i}$, then k can be removed from U_{X_i}. One can also write a sufficient condition for adding k to L_{X_i}.

Branching as well as propagation may be required to solve the pricing problem. A standard way to branch on a set-valued variable X_i is to consider the options $j \in X_i$ and $j \notin X_i$, where $j \in U_{X_i} \setminus L_{X_i}$. The former branch is implemented by adding j to L_{X_i}, and the latter by removing j from U_{X_i}.

While the pricing problem discussed here contains only two meta-constraints, realistic problems must deal with many complicated work rules. Specialized constraints and associated filtering methods have been developed for some of these.

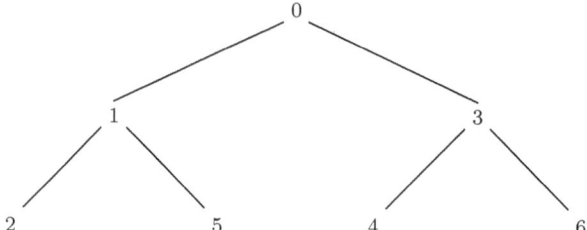

Fig. 5.4 Search tree for Exercises 5.1 and 5.2. The numbers at nodes are relaxation values.

Exercises

5.1. Consider the complete search tree Fig. 5.4 for a problem with two binary variables. The value of the relaxation is shown at each node. (a) Suppose the tree is traversed using strong branching. In what order are the nodes processed? What is the relaxation bound $\theta(T)$ for each partial tree T constructed during the search? (b) Now suppose the tree is traversed by maximizing $\theta(T)$ with a steepest ascent algorithm. If adding either of two or more nodes results in the same bound $\theta(T)$, break the tie by processing the leftmost node next. In what order are the nodes processed? What is the bound $\theta(T)$ for each T? Note that steepest ascent increases the bound more rapidly than strong branching.

5.2. Consider again the search tree of Fig. 5.4. (a) Suppose the tree is traversed using branch and bound in a depth-first fashion. At each nonleaf node, the left or right branch is taken first with equal probability. Compute the expected lower bound obtained, and the expected best incumbent solution value obtained, after k (nonroot) nodes are processed, for $k = 1, \ldots, 6$. (b) Now do the same for a breadth-first search. For $k \geq 4$, one of the level-one nodes is completed before any nodes under the other level-one node are processed. Note that in the early part of the search, depth-first obtains good incumbent solutions more quickly, and breadth-first obtains good bounds more quickly.

5.3. Recall the crew rostering problem illustrated by Fig. 5.3. Suppose the current domain of the set-valued variable X_1 is the interval $[L_{X_1}, U_{X_1}]$ where $L_{X_1} = \{2\}$ and $U_{X_1} = \{1, 2, 3, 4, 6\}$, and the current domain of z is $[L_{z_1}, U_{z_1}] = [0, 8]$. Use the min and max flow test to filter the domain.

5.4. If the third column is omitted from (5.5), the resulting dual multipliers are $(u_1, u_2) = (10, 9)$ and $(v_1, \ldots, v_6) = (0, 3, 0, 0, 0, 0)$. Use the graphs of Fig. 5.3 to identify one or more columns with negative reduced cost.

5.5. Write an expression analogous to (5.8) for updating U_{z_i}.

5.6. Consider the set sum constraint $\ell \leq \sum_{j \in X} p_j \leq u$, where X is a set-valued variable with domain $[L_X, U_X]$. Write a sufficient condition under which $k \in U_X$ can be added to L_X, and a sufficient condition under which $k \in U_X$ can be deleted from U_X.

5.7. In a two-dimensional cutting stock problem, customers have placed orders for q_i rectangular pieces of glass having size i (which has dimensions $a_i \times b_i$) for $i = 1, \ldots, m$. These orders must be filled by cutting them from standard $a \times b$ sheets. There are many patterns according to which a standard sheet can be cut into one or more of the pieces on order. Let y_j be the number of standard sheets that are cut according to pattern j, and let A_{ij} be the number of pieces of size i the pattern yields from one sheet. Write an integer programming problem that minimizes the number of standard sheets cut while meeting customer demand. Indicate how to solve the problem by column generation by formulating a subproblem to find a pattern with a negative reduced cost. The subproblem should use the diffn global constraint (see Chapter 8). *Hint:* The subproblem can contain the constraint

$$\text{linearDisjunction:} \quad \begin{bmatrix} \delta_k = 1 \\ \Delta x_k = (a_i, b_i) \end{bmatrix} \vee \begin{bmatrix} \delta_k = 0 \\ \Delta x_k = (0, 0) \end{bmatrix}$$

for each i.

5.2 Constraint-Directed Search

An ever-present issue when searching over problem restrictions is the choice of which restrictions to consider and in what order. Branching search addresses this issue in a general way by letting problem difficulty guide the search. If a given restriction is too hard to solve, it is split into problems that are more highly restricted, and otherwise one moves on to the next restriction, thus determining the sequence of restrictions in a recursive fashion.

Another general approach is to create the next restriction on the basis of lessons learned from solving past restrictions. At the very least, one would like to avoid solving restrictions that are no better than past ones, in the sense that they cannot produce solutions any better those already found.

This can be accomplished as follows. Whenever a restriction is found to be infeasible, one can generate a constraint that excludes that restriction, and perhaps other restrictions that are infeasible for the same reason. Or when an optimal solution is found for the restriction, one can write a constraint stating that no restriction similar to it in certain ways can have a better solution. Such a constraint is a *nogood*. When the next restriction is selected, it must satisfy the nogoods generated so far. The process continues until no restriction satisfying the nogoods can have a better solution than those already found.

Nogoods are obtained by solving an inference dual of the restricted problem. The dual solution is a proof of optimality, a proof whose premises include the fact that the problem was restricted in a certain way. The nogood states that if the problem is restricted in this way again, then the same optimal value will result. More generally, the same proof schema may yield the same bound (or perhaps a weaker bound) if the premises are altered by restricting the problem in some other way. The nogood then states that if the problem is restricted in certain ways, the optimal value can be no better than a certain bound.

Varieties of constraint-directed search include logic-based Benders decomposition, constraint-directed branching, and dynamic backtracking. The Davis–Putnam–Loveland–Logemann (DPLL) method for solving propositional satisfiability problems is an instance of constraint-directed branching. Table 5.3 indicates briefly how these methods exemplify the search-infer-and-relax scheme. They are discussed in separate sections below.

5.2.1 The Search Algorithm

Constraint-directed search creates a sequence of restrictions $P_1, P_2, \ldots$ of the original problem P. The optimal value v_i^* of each restriction P_i is computed and a nogood N_i derived. The nogood states that to obtain a solution value better than v_i^*, one must select a restriction that is unlike P_i. The next restriction P_{i+1} is selected by considering the the set $\mathcal{N}$ of nogoods $N_1, \ldots, N_i$ generated so far. The algorithm continues until no restriction satisfying $\mathcal{N}$ can have a solution value z better than those already found. At this point the optimal value of P is the minimum of $v_1^*, \ldots, v_i^*$.

The search necessarily terminates if there are a finite number of possible restrictions, because each iteration excludes at least one restriction. If there are infinitely many restrictions, some care must be taken in designing the nogoods to ensure that the search is finite.

Table 5.3 How some selected constraint-directed search methods fit into the search-infer-and-relax framework. The selection criterion is the criterion for selecting a feasible solution of the nogood set $\mathcal{N}$.

Solution method	Restriction $P(\bar{x}^i)$	Relaxation $R(\bar{x}^i)$	Selection criterion	Inference
Logic-based Benders decomposition	Subproblem defined by solution of master	Master problem (Benders cuts)	Optimal solution of master	Derivation of Benders cuts (nogoods)
Constraint-directed branching	Subproblem at leaf node	Processed nogoods	Conformity with $\mathcal{N}$, variable selection consistent with branching order	Nogood generation + parallel resolution
DPLL for SAT	Subproblem at leaf node	Processed conflict clauses	Conformity with $\mathcal{N}$, variable selection consistent with branching order	Implication graph + parallel resolution
Partial-order dynamic backtracking	Subproblem after fixing some variables	Processed nogoods	Conformity with $\mathcal{N}$, variable selection consistent with partial order	Nogood generation + parallel resolution

In practice, restrictions are normally defined by fixing the values of some of the variables $x_1, \ldots, x_n$. This can be captured in notation by letting x^i be the tuple of variables that are fixed in iteration i. Suppose that x^i is fixed to $\bar{x}^i$, and let $P(\bar{x}^i)$ be the subproblem that results. The inference dual of $P(\bar{x}^i)$ is solved, and a nogood N_i derived from its solution. The nogood in general has the form $z \geq v_i(x^i)$, where $v_i(\bar{x}^i) = v_i^*$. When the subproblem is infeasible, $v_i(\bar{x}^i) = v_i^* = \infty$, in which case one can drop z from the nogood and regard it as a constraint on x^i only. The nogood N_i is added to $\mathcal{N}$, and a solution $(\bar{z}, \bar{x})$ of the nogood set $\mathcal{N}$ is found. A tuple x^{i+1} is selected and set to $\bar{x}^{i+1}$ to define the next subproblem.

The procedure terminates when the minimum of z subject to $\mathcal{N}$ is equal to $\min_i \{v_i^*\}$, where i ranges over all previous iterations. In feasibility problems, the procedure terminates when $\mathcal{N}$ is infeasible. A generic constraint-directed search algorithm using this notation appears in Fig. 5.5. An analogous algorithm for feasibility problems is given in Fig. 5.6.

Let $z^* = -\infty$, $v_{\mathrm{UB}} = \infty$, $\mathcal{N} = \emptyset$, $i = 1$. Select $\bar{x}$.
While $z^* < v_{\mathrm{UB}}$ repeat:
    Select a tuple x^i of variables to fix to $\bar{x}^i$.
    Let v_i^* be the optimal value of $P(\bar{x}^i)$, and let $v_{\mathrm{UB}} = \min\{v_i^*, v_{\mathrm{UB}}\}$.
    Derive a nogood N_i from the solution of an inference dual of $P(\bar{x}^i)$.
    Add N_i to $\mathcal{N}$ and process $\mathcal{N}$.
    Let $i = i + 1$ and let $(z^*, \bar{x})$ be an optimal solution of $\min\{z \mid \mathcal{N}\}$.
The optimal value of P is v_{UB}.

Fig. 5.5 Generic constraint-directed search algorithm for solving a minimization problem P. $\mathcal{N}$ contains the nogoods generated so far.

Let $\mathcal{N} = \emptyset$, $i = 1$. Select $\bar{x}$.
While $\mathcal{N}$ is feasible repeat:
    Select a tuple x^i of variables to fix to $\bar{x}^i$.
    If $P(\bar{x}^i)$ is feasible, stop; P is solved.
    Else derive a nogood N_i from solution of an inference dual of $P(\bar{x}^i)$.
    Add N_i to $\mathcal{N}$ and process $\mathcal{N}$.
    Let $i = i + 1$ and select a solution $\bar{x}$ of $\mathcal{N}$.
P is infeasible.

Fig. 5.6 Generic constraint-directed search algorithm for solving a feasibility problem P.

There is normally a good deal of freedom in how to select a solution $(\bar{z}, \bar{x})$ of $\mathcal{N}$, and a constraint-directed search method is partly characterized by how the selection is made. Benders decomposition, for example, selects a solution that minimizes z. Certain selection criteria can make $\mathcal{N}$ easier to solve in subsequent iterations. In some constraint-directed methods, such as dynamic backtracking methods, it may be necessary to *process* the nogood set $\mathcal{N}$ to make it easier to solve. In such cases, the selection criterion is designed to make processing easier. In partial-order dynamic backtracking, for example, a solution of $\mathcal{N}$ is selected to *conform* to previous solutions, so that $\mathcal{N}$ can be processed by a fast version of resolution (parallel resolution).

Constraint-directed search is an instance of the search-infer-and-relax paradigm. It obviously enumerates problem restrictions, which take the form of subproblems. Furthermore, each nogood set $\mathcal{N}$ is a relaxation of the original problem.

Lemma 5.1 *A set $\mathcal{N}$ of nogoods for a minimization problem P is a relaxation of P.*

Proof. Suppose P is the problem $\min\{f(x) \mid \mathcal{C}\}$. It suffices to show that the nogoods exclude only points (z, x) that lie outside the epigraph of P; that is points for which x is infeasible or $z < f(x)$. Let $x = (x^i, y^i)$. A nogood is either a constraint $C(x^i)$ on x^i or a bound $z \geq v_i(x^i)$. In the former case, $C(x^i)$ excludes (x^i, y^i) only when $P(x^i)$ is infeasible, which means (x^i, y^i) is infeasible. In the latter case, $z \geq v_i(x^i)$ excludes (z, x^i, y^i) only when z is less than the optimal value of $P(x^i)$, which means $z < f(x^i, y^i)$. $\square$

The search process therefore solves a series of relaxations whose solutions guide the choice of the next restriction. Unlike branching methods, constraint-directed search requires the solution of every restriction, regardless of the outcome when the previous relaxation is solved.

A variant of constraint directed search generates nogoods to exclude solutions even when they are feasible. That is, whenever a subproblem $P(\bar{x}^i)$ is solved, the resulting nogood N_i excludes $\bar{x}^i$ (and perhaps other values of x^i) even when $P(\bar{x}^i)$ is feasible. Such nogoods might be called *enumerative nogoods*, because their purpose is to keep track of all solutions that have been examined. The procedure terminates when the nogood set $\mathcal{N}$ is infeasible, whereupon the best feasible solution found is optimal. A set $\mathcal{N}$ of enumerative nogoods is therefore not a relaxation of the original problem.

This variant is equivalent to a standard nogood-based search that always selects a solution (z, x) of $\mathcal{N}$ for which $z < \min_i\{v_i^*\}$. The search terminates when there is no such solution. When enumerative nogoods are used in subsequent sections, it is understood that the search can always be viewed as an instance of standard constraint-directed search.

5.2.2 Logic-Based Benders Decomposition

Benders decomposition is a constraint-directed search that defines all the problem restrictions by fixing the same tuple x^i of variables. They might be called the *search variables*, because the search procedure in effect enumerates values of these variables. In a Benders context, the nogoods are known as *Benders cuts*, the nogood set $\mathcal{N}$ as the *master problem*, and the restriction $P(\bar{x}^i)$ as the *subproblem* or *slave problem*. The master problem is solved only for the variables in x^i, because the remaining variables never appear in the master.

The primary rationale for Benders decomposition is that the problem may simplify considerably when certain variables are fixed to any value. The choice of search variables is therefore crucial to the success of the method, as is the ability to formulate strong Benders cuts.

Classical Benders decomposition is defined for the case in which the subproblem is an LP problem. The Benders cuts are obtained from the solution of the LP dual. The root idea of the Benders method, however, can be generalized to great advantage. The Benders cuts can be obtained from a logical analysis of the proof that solves the inference dual, resulting in a *logic-based* Benders method. This, in principle, allows the subproblem to take any form, but a separate analysis must be conducted for each class of subproblems.

It is convenient to let $x = x^i$ and let y represent the variables in the subproblem. The original problem P can be written

$$\min\ f(x, y)$$
$$\mathcal{C}(x, y)$$
$$x \in D_x,\ y \in D_y$$

where $\mathcal{C}(x, y)$ is a constraint set that contains variables x, y. The subproblem $P(\bar{x})$ is

$$\min\ f(\bar{x}, y)$$
$$\mathcal{C}(\bar{x}, y)$$
$$y \in D_y$$

The resulting Benders cut is $z \geq v_i(x)$, where $v_i(\bar{x})$ is the optimal value v_i^* of the subproblem ($v_i^* = \infty$ if the subproblem is infeasible). The cut is added to $\mathcal{N}$ to obtain the next master problem.

The kth master problem selects a (z, x) to minimize the objective function z while satisfying the Benders cuts so far generated:

$$\min\ z$$
$$z \geq v_i(x),\quad i = 1, \dots, k$$
$$x \in D_x$$

Thus, a feasible solution of $\mathcal{N}$ is selected so as to minimize z.

The algorithm terminates when the optimal value z_k^* of the master problem is equal to $\min_i\{v_i^*\}$. At any step k, z_k^* and $\min_i\{v_i^*\}$ provide lower and upper bounds, respectively, on the optimal value of the

Let $z^* = -\infty$, $v_{\mathrm{UB}} = \infty$, $\mathcal{N} = \emptyset$, and $i = 1$. Select $\bar{x}$.
While $z^* < v_{\mathrm{UB}}$ repeat:
    Let v_i^* be the optimal value of $P(\bar{x}) = \min\{f(\bar{x}, y) \mid \mathcal{C}(\bar{x}, y), y \in D_y\}$.
    Let $v_{\mathrm{UB}} = \min\{v_i^*, v_{\mathrm{UB}}\}$.
    Derive a Benders cut $z \geq v_i(x)$ from the solution of an inference
       dual of $P(\bar{x})$.
    Let $\mathcal{N} = \mathcal{N} \cup \{z \geq v_i(x)\}$.
    Let $i = i + 1$ and let $(z^*, \bar{x})$ be an optimal solution of $\min\{z \mid \mathcal{N}\}$.
The optimal value of P is v_{UB}.

Fig. 5.7 Generic logic-based Benders algorithm for minimizing $f(x, y)$ subject to $\mathcal{C}(x, y)$ and $(x, y) \in D_x \times D_y$. When solving a feasibility problem, the algorithm terminates immediately if $P(\bar{x})$ is feasible.

original problem. The algorithm appears in Fig. 5.7. The first master problem $\mathcal{N}$ may be augmented with precomputed cuts for a "warm start." This and subsequent master problems may also contain constraints from P that involve only x, as well as other constraints that involve only x and are valid for P.

Logic-based Benders decomposition is illustrated by the planning and scheduling problem of Section 2.8. In this example, the master problem assigns jobs to machines by fixing the assignment variables x_{ij}, and the subproblem schedules the jobs on each machine by solving for the start-time variables s_j. The solution of the subproblem dual consists of a proof of the minimum makespan, based perhaps on edge finding and branching. A simple Benders cut is based on which job assignments play a role in the proof, while more sophisticated cuts are derived in Section 6.14.3. The example also illustrates how a relaxation of the subproblem can be included in the master problem, an important maneuver in practice to accelerate convergence.

The classical Benders methods is presented in Section 6.2.4. Benders cuts for a variety of other problems are given in subsequent sections of Chapter 6, classified by the relevant constraint type.

5.2.3 Constraint-Directed Branching

Constraint-directed branching is a branching algorithm directed by no-goods. Each leaf node of the branching tree is viewed as corresponding to one iteration of constraint-directed search. Certain variables at a leaf node are fixed by branching, and the restricted problem that remains can be treated as a subproblem.

Unlike Benders decomposition, constraint-directed branching may fix a different tuple x^i of search variables in each iteration—namely, the variables that are fixed by branching down to a leaf node. The value $\bar{x}^i$ to which x^i is fixed at the ith leaf node defines a subproblem. Solving the inference dual of the subproblem (actually, a relaxation of the subproblem) yields a nogood that helps to direct the rest of the search. The nogood excludes $x^i = \bar{x}^i$ and perhaps other partial assignments. When feasible leaf nodes (as well as infeasible and pruned leaf nodes) generate nogoods, the nogoods should be viewed as enumerative because they exclude feasible solutions.

Conventional branching is a particular case of constraint-directed branching in which each nogood is simply $x^i \neq \bar{x}^i$. An example appears in Fig. 5.8. Suppose for simplicity that the tree proves infeasiblity, so that the subproblem relaxation at every leaf node is infeasible. The branching order x_1, x_2, x_3 to leaf node 2 is first determined as specified by the variable selection rule. A solution of $\mathcal{N} = \emptyset$ is $x^1 = (x_1, x_2, x_3) = (0, 0, 0)$, as shown in the second line of Table 5.4. Suppose that the subproblem relaxation becomes infeasible only after all three variables are fixed to these values. This yields the nogood $x^1 \neq (0, 0, 0)$ at leaf node 1. The nogood can be written as the logical clause $x_1 \vee x_2 \vee x_3$, which states that at least one of the literals x_1, x_2, x_3 must be true

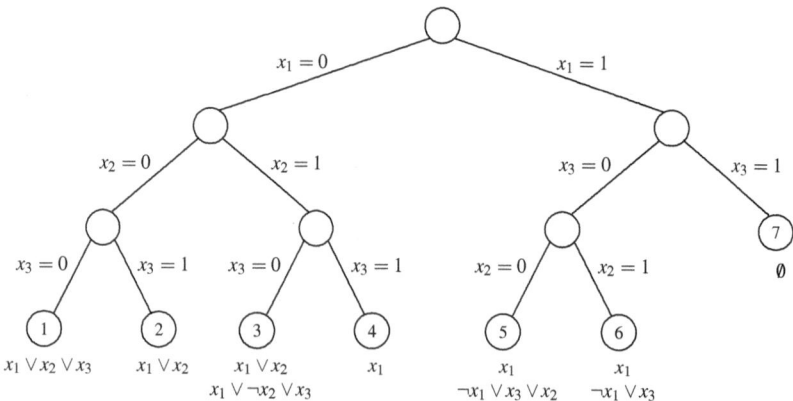

Fig. 5.8 Branching tree interpreted as constraint directed search. The processed nogoods are shown below each leaf node.

Table 5.4 Interpretation of branching as constraint-directed search.

i	Solution $\bar{x}^i$ of $\mathcal{N}$	Nogoods generated	Processed nogood set $\mathcal{N}$
0			$\emptyset$
1	$(x_1, x_2, x_3) = (0,0,0)$	$x_1 \vee x_2 \vee x_3$	$\{x_1 \vee x_2 \vee x_3\}$
2	$(x_1, x_2, x_3) = (0,0,1)$	$x_1 \vee x_2 \vee \neg x_3$	$\{x_1 \vee x_2\}$
3	$(x_1, x_2, x_3) = (0,1,0)$	$x_1 \vee \neg x_2 \vee x_3$	$\left\{ \begin{matrix} x_1 \vee x_2, \\ x_1 \vee \neg x_2 \vee x_3 \end{matrix} \right\}$
4	$(x_1, x_2, x_3) = (0,1,1)$	$x_1 \vee \neg x_2 \vee \neg x_3$	$\{x_1\}$
5	$(x_1, x_3, x_2) = (1,0,0)$	$\neg x_1 \vee x_3 \vee x_2$	$\left\{ \begin{matrix} x_1, \\ \neg x_1 \vee x_3 \vee x_2 \end{matrix} \right\}$
6	$(x_1, x_3, x_2) = (1,0,1)$	$\neg x_1 \vee x_3 \vee \neg x_2$	$\left\{ \begin{matrix} x_1, \\ \neg x_1 \vee x_3 \end{matrix} \right\}$
7	$(x_1, x_3) = (1,1)$	$\neg x_1 \vee \neg x_3$	$\{\emptyset\}$

(i.e., equal to 1). In general, a *literal* is a variable x_j or its negation, and a *clause* is a disjunction of literals.

At this point $\mathcal{N}$ contains one nogood, and its variables have a particular order (x_1, x_2, x_3). A feasible solution of $\mathcal{N}$ is chosen to *conform* to $\mathcal{N}$; that is, each x_j in $\mathcal{N}$ except the last receives a value (namely, 0) opposite its sign in the nogood. The remaining variables are set to values that satisfy $\mathcal{N}$. Infeasibility is again detected only after all three variables are fixed, so that $\bar{x}^2 = (0,0,1)$. This takes the search to leaf node 2, and the resulting nogood is $x_1 \vee x_2 \vee \neg x_3$.

The nogoods in $\mathcal{N} = \{x_1 \vee x_2 \vee x_3, \ x_1 \vee x_2 \vee \neg x_3\}$ are now *processed* to infer $x_1 \vee x_2$, which is the *parallel resolvent* of the two nogoods. Two clauses have a resolvent when exactly one variable x_j occurs positively in one and negatively in the other, and the resolvent contains the literals of both clauses except $x_j, \neg x_j$. Two clauses have a parallel resolvent when x_j is the last variable in both (the idea is developed further in Section 5.2.6). Once the nogood $x_1 \vee x_2$ is obtained by resolution, the two nogoods $x_1 \vee x_2 \vee x_3$ and $x_1 \vee x_2 \vee \neg x_3$ become redundant because they are *parallel absorbed* by $x_1 \vee x_2$, and they are dropped.

A *conforming* solution of the new nogood set $\mathcal{N} = \{x_1 \vee x_2\}$ must set $x_1 = 0$ because x_1 occurs positively. Then x_2 is set to 1 to satisfy the nogood, whereupon x_3 can be arbitrarily set to zero. Infeasibility is detected after fixing all these values, which generates leaf node 3 and the nogood $x_1 \vee \neg x_2 \vee x_3$. The nogood does not parallel resolve with $x_1 \vee x_2$ because $\neg x_2$ is not the last literal in the nogood.

At leaf node 4, parallel resolution simplifies $\mathcal{N}$ to $\{x_1\}$, and the next solution is $\bar{x} = (1, 0, 0)$. The ordering of variables x_2, x_3 can now be redefined, because they do not occur in $\mathcal{N}$ and their previous ordering is forgotten. The process continues through the remaining leaf nodes. At leaf node 7, infeasiblity is detected when only two variables are fixed, and $\bar{x}^7 = (\bar{x}_1, \bar{x}_3) = (1, 1)$. Parallel resolution simplifies the nogood set $\mathcal{N}$ to $\{\emptyset\}$, where $\emptyset$ is the empty clause (which is necessarily false). Because $\mathcal{N}$ is infeasible, the search is complete.

Because relaxations are normally solved at nodes of a branching tree, nogoods are derived from the inference dual of the relaxation $R(\bar{x}^i)$ rather than from the inference dual of $P(\bar{x}^i)$. As usual in branching, it is assumed that solving $R(\bar{x}^i)$ becomes tantamount to solving $P(\bar{x}^i)$ if one branches deeply enough into the tree. This ensures an exhaustive search even when nogoods are derived from the relaxation.

Nontrivial learning occurs when the nogoods exclude partial assignments other than $x^i = \bar{x}^i$. This may allow the search to backtrack to a higher level and bypass a part of the search tree, a maneuver known historically as *backjumping*. This is illustrated in the next section.

A generic algorithm appears in Fig. 5.9. In each iteration i, a conforming solution $\bar{x}$ of the nogood set $\mathcal{N}$ is first selected. Since parallel

---

Let $v_{\mathrm{UB}} = \infty$, $\mathcal{N} = \emptyset$, and $i = 1$. Select $\bar{x}$.
While $\mathcal{N}$ is feasible repeat:
    Let $x^i = (x_{j_1}, \ldots, x_{j_d})$ contain the variables in $\mathcal{N}$.
    Select an ordering $j_{d+1}, \ldots, j_n$ of the variables not in x^i.
    For $k = d + 1, \ldots, n$ until $R(\bar{x}^i)$ is terminal: let $x^i = (x^i, x_{j_k})$.
    If $R(\bar{x}^i)$ is infeasible or its solution is feasible in $P(\bar{x}^i)$ then
        Let v_i^* be the optimal value of $R(\bar{x}^i)$ and let $v_{\mathrm{UB}} = \min\{v_i^*, v_{\mathrm{UB}}\}$.
    Derive a nogood N_i from the solution of the inference dual of $R(\bar{x}^i)$.
    Add N_i to $\mathcal{N}$ and process $\mathcal{N}$ with parallel resolution,
        based on ordering $j_1, \ldots, j_n$.
    Let $i = i + 1$, and select a conforming solution $\bar{x}$ of $\mathcal{N}$.
The optimal value of P is v_{UB}.

Fig. 5.9 Constraint-directed branching. The subproblem relaxation $R(\bar{x}^i)$ is *terminal* when a leaf node is reached; that is, $R(\bar{x}^i)$ is infeasible, $v_i^* \geq v_{\mathrm{UB}}$, or the solution of $R(\bar{x}^i)$ is feasible in $P(\bar{x}^i)$. When solving a feasibility problem, the search terminates immediately if the solution of $R(\bar{x}^i)$ is feasible in $P(\bar{x}^i)$.

resolution was applied to $\mathcal{N}$, a comforming solution can always be found without backtracking (as proved in Section 6.4.6). The variables not in $\mathcal{N}$ are ordered $x_{j_{d+1}}, \ldots, x_{j_n}$ as specified by the variable selection rule (the variables in $\mathcal{N}$ are already ordered $x_{j_1}, \ldots, x_{j_d}$). The variables $x_{j_{d+1}}, \ldots, x_{j_n}$ are fixed to $\bar{x}_{j_{d+1}}, \ldots, \bar{x}_{j_n}$, one at a time, until a leaf node is reached; that is, until the relaxation of the resulting subproblem is infeasible, has an optimal value no better than the incumbent solution, or has a solution that is feasible in the original problem. At this point, the variables in x^i have been fixed to $\bar{x}^i$. A nogood is obtained by solving the inference dual of the relaxation $R(\bar{x}^i)$ of the subproblem $P(\bar{x}^i)$. The nogood is added to $\mathcal{N}$, which is processed with parallel resolution.

5.2.4 Example: Propositional Satisfiability

Propositional satisfiability is one of the fundamental problems of combinatorial optimization, partly because a wide range of combinatorial problems can be formulated in the language of propositional logic. The currently fastest algorithms for propositional satisfiability use a form of the Davis-Putnam–Logemann–Loveland (DPLL) algorithm with clause learning. This algorithm can be interpreted as constraint-directed branching, and a small example will illustrate the basic idea. The example also prepares the ground for the discussion of partial-order dynamic backtracking in the next section.

The example is artificial but is contrived to show how nogoods, known in this context as *conflict clauses*, help solve a satisfiability problem. Suppose a company must some hire some staff to complete a task and has workers 1, ..., 6 to choose from. Workers 3 and 4 are temporary workers. Due to the qualifications of the workers and nature of the task, the hiring must satisfy the following conditions:

(a) The company must hire at least one of the workers 1, 5 and 6.
(b) The company cannot hire 6 unless it hires 1 or 5.
(c) The company cannot hire 5 unless it hires 2 or 6.
(d) If the company hires 5 and 6, it must definitely must hire 2.
(e) If the company hires 1 or 2, then it must hire at least one temporary worker.
(f) The company can hire neither 1 nor 2 if it hires any temporary workers.

The company wishes to know whether it is possible to satisfy these conditions simultaneously.

Let $x_j = T$ (for *true*) if the company hires worker j, and $x_j = F$ otherwise. Conditions (a)–(f) can be written in logical form as follows:

$$
\begin{aligned}
x_1 \vee x_5 \vee x_6 &\qquad\qquad (a)\\
x_6 &\to (x_1 \vee x_5) &\qquad (b)\\
x_5 &\to (x_2 \vee x_6) &\qquad (c)\\
(x_5 \wedge x_6) &\to x_2 &\qquad (d)\\
(x_1 \vee x_2) &\to (x_3 \vee x_4) &\qquad (e)\\
(x_3 \vee x_4) &\to (\neg x_1 \wedge \neg x_2) &\qquad (f)
\end{aligned}
\qquad (5.9)
$$

The conjunction of these formulas is a proposition Q that must be true if the conditions are met. The question is whether Q is *satisfiable*; that is, whether some assignment of truth values to its variables makes it true.

To check the satisfiability of Q, it is convenient to write Q in *conjunctive normal form* (CNF), which is to say as a conjunction of logical clauses. An implication $x_1 \to x_2$ can be written $\neg x_1 \vee x_2$ because it is interpreted as a material conditional—that is, it states that either x_2 is true or x_1 is false. Thus, formula (b) in (5.9) can be written $x_1 \vee x_5 \vee \neg x_6$, and similarly for (c). Formula (d) can be written $\neg x_5 \vee \neg x_6 \vee x_2$. Formula (e) is equivalent to the conjunction of two conditionals, $x_1 \to (x_3 \vee x_4)$ and $x_2 \to (x_3 \vee x_4)$, and formula (f) is similarly equivalent to four conditionals. Proposition Q is therefore the conjunction of the clauses in the following feasibility problem:

$$
\text{logic:}\begin{cases}
x_1 & \vee\ x_5 \vee\ x_6 & (a)\\
x_1 & \vee\ x_5 \vee \neg x_6 & (b)\\
x_2 & \vee \neg x_5 \vee\ x_6 & (c)\\
x_2 & \vee \neg x_5 \vee \neg x_6 & (d)\\
\neg x_1 \vee\ x_3 \vee\ x_4 & & (e1)\\
\neg x_2 \vee\ x_3 \vee\ x_4 & & (e2)\\
\neg x_1 \vee \neg x_3 & & (f1)\\
\neg x_1 \vee \neg x_4 & & (f2)\\
\neg x_2 \vee \neg x_3 & & (f3)\\
\neg x_2 \vee \neg x_4 & & (f4)
\end{cases}
\qquad (5.10)
$$

domains: $x_j \in \{T, F\}$, $j = 1, \dots, 6$

where *logic* indicates that the constraint is a conjunction of the clauses listed.

The DPLL algorithm solves the problem by branching and applying the *unit clause rule* at each node. The rule says that when all but one of the literals in a clause have been determined to be false, the remaining literal must be true. The false literals can be deleted, leaving a *unit clause*. The unit clause rule is applied repeatedly until no further literals can be deleted. The resulting procedure is an incomplete inference method, *unit resolution*, which proves unsatisfiability when it generates the empty clause. Unit resolution is normally regarded as a propagation method, but it can be viewed as solving a relaxation of the problem (i.e., a relaxation whose feasible set consists of solutions not excluded by the unit clause rule).

Fig. 5.10 depicts a branching tree for problem (5.10) that is generated by constraint-directed branching. The nogood set $\mathcal{N}$ is solved by a simple greedy algorithm: for $j = 1, \ldots, 6$, let $\bar{x}_j = F$ if unit resolution detects no unsatisfiability when $(x_1, \ldots, x_j)$ is set to $(\bar{x}_1, \ldots, \bar{x}_{j-1}, F)$ in the original problem, and otherwise set $\bar{x}_d = T$.

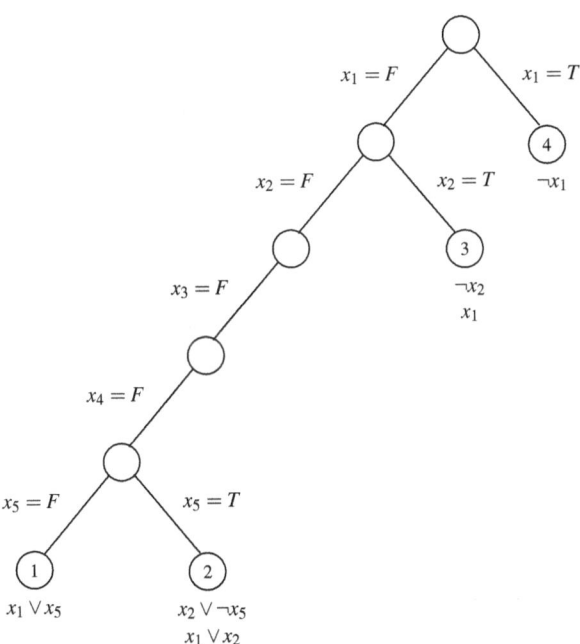

Fig. 5.10 Branching tree for a propositional satisfiability problem. Conflict clauses are shown below the nodes with which they are associated.

Table 5.5 Interpretation of the DPLL procedure with conflict clauses as constraint-directed branching.

i	Solution $\bar{x}^i$ of $\mathcal{N}$	Nogoods generated	Processed nogood set $\mathcal{N}$
0			$\emptyset$
1	$(x_1,\ldots,x_5) = (F,\ldots,F)$	$x_1 \vee x_5$	$\{x_1 \vee x_5\}$
2	$(x_1,\ldots,x_5) = (F,F,F,F,T)$	$x_2 \vee \neg x_5$	$\{x_1 \vee x_2\}$
3	$(x_1,x_2) = (F,T)$	$\neg x_2$	$\{x_1\}$
4	$(x_1) = (T)$	$\neg x_1$	$\{\emptyset\}$

Initially, $\mathcal{N}$ is empty and $\bar{x} = (F,\ldots,F)$. Unit resolution finds no infeasiblity in setting $(x_1,\ldots,x_j) = (F,\ldots,F)$ for $j = 1,\ldots,4$, but for $j = 5$ it uses clauses (a) and (b) to prove infeasibility. Thus $\bar{x}^1 = (\bar{x}_1,\ldots,\bar{x}_5) = (F,\ldots,F)$, and the relaxation $R(\bar{x}^1)$ is infeasible. Examination of the proof reveals that the settings $x_1 = x_5 = F$ are enough to cause infeasibility. The resulting nogood is $(x_1,x_5) \neq (F,F)$ rather than $(x_1,\ldots,x_5) = (F,\ldots,F)$. The nogood can be written as a *conflict clause* $x_1 \vee x_5$ (Table 5.5).

Leaf node 2 similarly yields the nogood $x_2 \vee \neg x_5$. The nogood set is processed with parallel resolution as in the previous section. A conforming solution of the nogood set $\mathcal{N} = \{x_1 \vee x_2\}$ is $\bar{x} = (F,T,F,F,F,F)$. Leaf node 3 is therefore enumerated next, bypassing much of the search tree. This is called *backjumping*. Unit resolution at leaf node 3 derives infeasibility from clauses $(f3)$, $(f4)$, and $(e2)$ and generates the nogood $\neg x_2$. Processing $\mathcal{N}$ obtains x_1. The branch to leaf node 4 generates the nogood $\neg x_1$, which parallel resolves with x_1 to yield the empty clause. The search is complete without finding a feasible solution.

Because nogoods are logical clauses, they can be added to the original problem as implied constraints. This may increase the effectiveness of unit resolution, as illustrated in the next section.

Unit resolution is frequently implemented with a data structure called *watched literals*, which can substantially improve performance. Initially, two literals are arbitrarily selected in each clause to be watched literals. The key idea is that if unit resolution reduces a clause to a single literal, it must at some point fix one of the two watched literals. It therefore suffices to examine a clause only when one of its watched literals is fixed. The data structure consists of two lists for each variable x_j: a list of clauses containing the watched literal x_j, and a list of clauses containing the watched literal $\neg x_j$. When x_j is fixed (say, to true), $\neg x_j$ is removed from each clause in the second list, and an unfixed literal in the clause is selected to replace it as a watched literal. If only one unfixed literal remains (the other watched literal), it is fixed. The scheme is also efficient for backtracking, because there is no need to retrace how watched literals were assigned during the branching process. This is an instance of a "lazy" data structure that updates only as much information as is really needed.

5.2.5 Implication Graph

In satisfiability solvers, the subproblem at a node is typically represented by an *implication graph*. Proofs of infeasibility (i.e., solutions of the inference dual) are subgraphs of the implication graph, known as *conflict graphs*. The structure of a conflict graph reveals how to derive conflict clauses.

The implication graph for leaf node 1 of Fig. 5.10 appears in part (a) of Fig. 5.11. The graph is built as follows. It initially contains a vertex for each branching literal (i.e., literal that is fixed to true by branching). It adds directed edges (unless already present) from literals $\ell_1, \ldots, \ell_m$ to literal ℓ when $\neg \ell_1 \vee \cdots \vee \neg \ell_m \vee \ell$ is a clause C in the constraint set, and labels these edges with C. Some edges may receive two or more labels. The process is repeated until no new edges can be added.

Unit resolution proves infeasibility if and only if there is a *conflict variable* x_j in the implication graph. This is a variable for which literals x_j and $\neg x_j$ appear in the graph. The literals x_j, $\neg x_j$ are then *conflict literals*. Edges are added from each conflict literal to a vertex representing the empty clause $\emptyset$. In Fig. 5.11(a), x_6 is the only conflict variable (there can be several).

The implication graph G may contain several proofs of infeasibility, each corresponding to a conflict graph. A conflict graph H is built by starting with any two conflict literals x_j, $\neg x_j$ in G and retracing the steps of a proof. To accomplish this: (a) select a nonbranching literal ℓ in H with no incoming edges, (b) select a clause C that labels at least one edge entering ℓ in G, and (c) add to H all of G's edges into ℓ

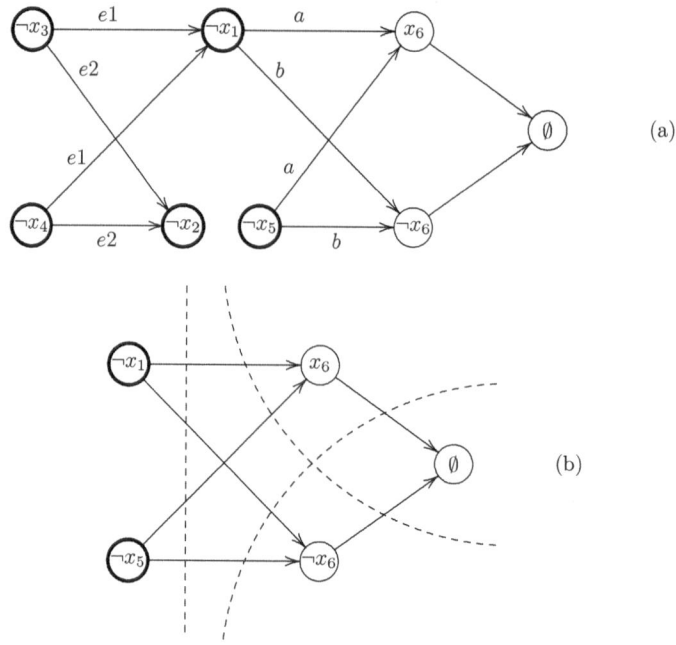

Fig. 5.11 (a) Implication graph for node 5 of the branching tree of Fig. 5.10. (b) Conflict graph, with three cuts indicated by dashed lines. The heavy circles indicate branching literals.

that are labeled by C. Repeat this action until no further edges can be added. The implication graph of Fig. 5.11(a) contains only one conflict graph, which appears in Fig. 5.11(b).

Conflict clauses are identified by finding certain cuts in the conflict graph H. A cut is a subset of edges whose removal separates the graph. To find a conflict clause, identify a cut that puts all branching literals on one side (the "reason side") and at least one conflict literal on the other side (the "conflict side"). Let $\ell_1, \ldots, \ell_k$ be the literals in the *frontier* of the cut; that is, the literals on the reason side having at least one outgoing edge in the cut. Then $\neg\ell_1 \vee \cdots \vee \neg\ell_k$ is a conflict clause. This is because premises in the frontier are sufficient to prove infeasibility. In Fig. 5.11(b), there are three such cuts (indicated by dashed lines), giving rise to conflict clauses $x_1 \vee x_5 \vee x_6$, $x_1 \vee x_5 \vee \neg x_6$, and $x_1 \vee x_5$. In this case, one conflict clause ($x_1 \vee x_5$) dominates the other two.

The simplest cut places all and only branching literals on the reason side, so that the conflict clause contains only branching literals. In the

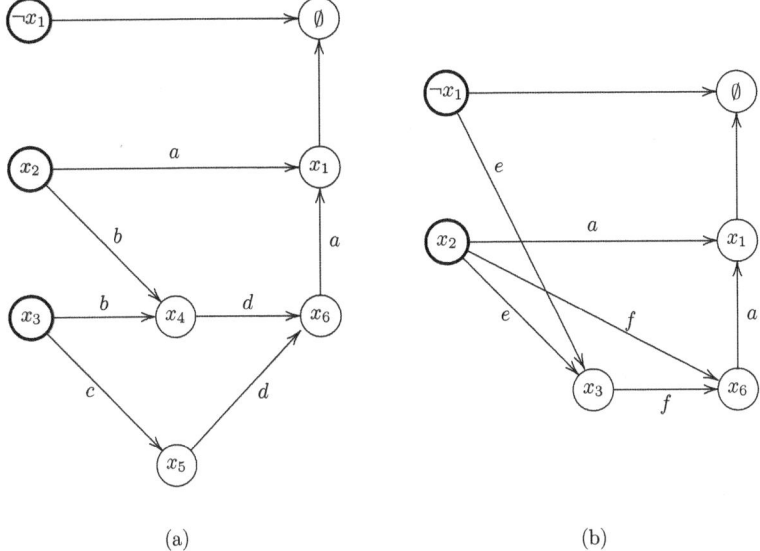

(a)   (b)

Fig. 5.12 (a) Conflict graph with branching literals $\neg x_1$, x_2, x_3. (b) Conflict graph after adding two inferred clauses.

example, this is the conflict clause $x_1 \vee x_5$. However, some solvers use other cuts.

One can also define cuts that separate vertices other than $\emptyset$ from the branching literals. The clauses inferred from such cuts can be added to the implication graph, perhaps resulting in stronger nogoods. This is best explained by example. Suppose that the clause set at a leaf node is the following, where $\neg x_1, x_2$, and x_3 are the branching literals:

$$
\begin{array}{ll}
x_1 \vee \neg x_2 \vee \neg x_5 & (a) \\
\neg x_2 \vee \neg x_3 \vee x_4 & (b) \\
\neg x_3 \vee x_5 & (c) \\
\neg x_4 \vee \neg x_5 \vee x_6 & (d)
\end{array}
\qquad (5.11)
$$

The implication graph contains several conflict graphs, one of which appears in Fig. 5.12(a). The cut that places only the branching variables on the reason side yields the conflict clause $x_1 \vee \neg x_2 \vee \neg x_3$.

Now consider cuts that separate branching literals from x_6. One such cut places x_2 and x_3 on the reason side, and x_4, x_5, x_6 on the other side. This cut reveals that x_2 and x_3, taken together, imply x_6. The clause $\neg x_2 \vee \neg x_3 \vee x_6$ is therefore valid. This is called a *reconvergence cut* because there are two or more paths to x_6 from some vertex in the frontier or on the nonreason side. In this case, there are two paths from x_3 to x_6. Suppose the original clause set (5.11) is augmented by the conflict clause inferred above and the reconvergence clause:

$$x_1 \lor \neg x_2 \lor \neg x_3 \qquad (e)$$
$$\neg x_2 \lor \neg x_3 \lor x_6 \qquad (f)$$

Then a stronger nogood can be deduced. Using only $\neg x_1$ and x_2 as branching literals, the resulting implication graph contains the conflict graph of Fig. 5.12(b). It establishes a nogood $x_1 \lor \neg x_2$ that could not be proved using the original conflict graph.

5.2.6 Partial-Order Dynamic Backtracking

A slight generalization of the selection criterion and nogood processing converts constraint-directed branching to a more flexible search procedure: partial-order dynamic backtracking. Ordinary constraint-directed branching is a special case.

In the constraint-directed branching algorithm of Section 5.2.4, the order in which variables in the nogood set are instantiated is fixed by previous branching decisions. In partial-order dynamic backtracking, the variables in the nogood set are only partially ordered, and they may be instantiated in any order that is consistent with this partial order. The result need not be a search tree in the usual sense. The algorithm allows more freedom in the search than branching does, without sacrificing completeness.

When a nogood (conflict clause) is generated, one of the variables in the clause is selected to be *last*. The remaining variables in the clause are *penultimate*. The last variables in the current nogood set define a partial order in which each penultimate variable in a nogood precedes the last variable in that nogood. The last variable in a new nogood must be selected so as to be consistent with the partial order defined by the existing nogoods. In constraint-directed branching, the last variable is always the one on which the search last branched.

As in constraint-directed branching, the nogood set is processed with parallel resolution, in which the eliminated variable must be the last variable in both clauses. A conflict clause is redundant and can be eliminated when it is *parallel absorbed* by another clause; that is, one of its penultimate literals is last in another clause. The solution of the nogood set must again conform to the nogood set, meaning that each variable that occurs penultimately in some nogood must take a value opposite its sign in the nogood. This criterion is well defined, because it can be shown (Section 6.4.6) that a variable will have the same sign in all its penultimate occurrences. If a variable does not occur penultimately in any nogood, it can be set to any value that violates none of the nogoods. Section 6.4.6 also shows that a conforming solution can always be found without backtracking, unless of course the nogoods are infeasible, in which case the search is over. In addition, processing the nogood set with parallel resolution is shown to be a polynomial-time algorithm.

A partial-order dynamic backtracking algorithm for feasibility problems is stated in Fig. 5.13. In this algorithm, $\leq_{\mathcal{N}}$ denotes the partial order that is defined by the penultimate and last variables in each clause.

Let $v_{\mathrm{UB}} = \infty$, $\mathcal{N} = \emptyset$, and $i = 1$. Select $\bar{x}$.
While $\mathcal{N}$ is feasible repeat:
    Select an ordering $j_1, \ldots, j_n$ that is consistent with $\leq_{\mathcal{N}}$, and
        let $x^i = ()$.
    For $k = 1, \ldots, n$ until $R(\bar{x}^i)$ is terminal: let $x^i = (x^i, x_{j_k})$.
    If $R(\bar{x}^i)$ is infeasible or its solution is feasible in $P(\bar{x}^i)$ then
        Let v_i^* be the optimal value of $R(\bar{x}^i)$; let $v_{\mathrm{UB}} = \min\{v_i^*, v_{\mathrm{UB}}\}$.
    Derive a nogood N from solution of the inference dual of $R(\bar{x}^i)$.
    Select the last variable x_N in N so that $x_N \not\leq_{\mathcal{N}} x_j$ for all x_j
        that are penultimate in N.
    Add N to $\mathcal{N}$ and process $\mathcal{N}$ with parallel resolution.
    Let $i = i + 1$ and select a conforming solution $\bar{x}$ of $\mathcal{N}$.
The optimal value of P is v_{UB}.

Fig. 5.13 Partial-order dynamic backtracking algorithm for a minimization problem P with Boolean variables. The subproblem relaxation $R(\bar{x}^i)$ is terminal when $R(\bar{x}^i)$ is infeasible, $v_i^* \geq v_{\mathrm{UB}}$, or the solution of $R(\bar{x}^i)$ is feasible in $P(\bar{x}^i)$. The partial order $\leq_{\mathcal{N}}$ is the transitive closure of the relations $x_j \leq_{\mathcal{N}} x_N$ for all penultimate variables x_j in N and every $N \in \mathcal{N}$. When solving a feasibility problem, the algorithm terminates immediately if the solution of $R(\bar{x}^i)$ is feasible in $P(\bar{x}^i)$.

5.2.7 Example: Propositional Satisfiability

The satisfiability instance (6.24) solved earlier is convenient for illustrating partial-order dynamic backtracking. Table 5.6 summarizes the procedure. Initially, the conformity principle imposes no restriction since the nogood set is empty. For purposes of this illustration, the nogood sets are solved by setting variables unaffected by the conformity principle to false, if possible, or to true, if necessary, to avoid violating a nogood. Unit resolution is applied after each setting.

For $i = 0$ in Table 5.6, the variables are assigned values in the order in which they are indexed, but any order would be acceptable. When x_5 is reached, a clause in the original constraint set is violated, and the nogood $x_1 \vee x_5$ is generated. The variable x_1 is arbitrarily selected as last, as is indicated by writing the nogood as $x_5 \vee x_1$.

At this point, x_5 occurs positively as a penultimate variable, and it must therefore be set to false in the solution of $\mathcal{N}_1$. Variable x_1 is arbitrarily assigned next, and it must be assigned true to avoid violating the nogood $x_5 \vee x_1$. At this point a clause in the original constraint set is already violated. The restriction P_2, which contains the original clauses and $(x_1, x_5) = (T, F)$, is infeasible, and the nogood $x_5 \vee \neg x_1$ is generated (one could generate the stronger nogood $\neg x_1$, since it alone creates an infeasibility). Variable x_1 must be selected as last, since it occurs after x_5 in the partial order defined by the one existing nogood. Now the two nogoods $x_5 \vee x_1$ and $x_5 \vee \neg x_1$ can be parallel-resolved, resulting in the new nogood x_5, whose only variable is necessarily chosen as last. The other two nogoods are now redundant and are dropped, since the last literal of the clause x_5 occurs penultimately in both.

Table 5.6 Partial-order dynamic backtracking solution of a propositional satisfiability problem. The "last" variable in each nogood is written last.

i	Solution $\bar{x}^i$ of $\mathcal{N}$	Nogoods generated	Processed nogood set $\mathcal{N}$
0			$\emptyset$
1	$(x_1, \ldots, x_5) = (F, \ldots, F)$	$x_5 \vee x_1$	$\{x_5 \vee x_1\}$
2	$(x_1, x_5) = (T, F)$	$x_5 \vee \neg x_1$	$\{x_5\}$
3	$(x_2, x_5) = (F, T)$	$\neg x_5 \vee x_2$	$\left\{ \begin{matrix} x_5, \\ \neg x_5 \vee x_2 \end{matrix} \right\}$
4	$(x_2, x_5) = (T, T)$	$\neg x_2$	$\{\emptyset\}$

The current nogood set $\mathcal{N}$ can be solved without regard to conformity, because no variables occur penultimately in it. Variable x_5 must be set to true, and variable x_2 is arbitrarily set to false next. This already violates a constraint and yields the nogood $\neg x_5 \vee x_2$, in which x_2 is arbitrarily chosen to be last. The current nogoods x_5 and $\neg x_5 \vee x_2$ have the resolvent x_2, but they do not have a parallel resolvent because x_5 does not occur last in both clauses. Both clauses are therefore retained in $\mathcal{N}_3$. When nogood x_2 is generated by the solution of $\mathcal{N}_3$, two steps of parallel resolution yield the empty clause, and the search terminates without finding a feasible solution.

Exercises

5.8. A group of medications are commonly used to treat a form of cancer, but they can be taken only in certain combinations. A patient who takes medications 1 and 2 must take medication 5 as well. At least one of medications 3, 4, and 5 must be taken. If 5 is taken, then 3 or 4 must be taken. If 4 is taken, then 3 or 5 must be taken. Medication 3 must be taken if both 4 and 5 are taken. Medication 3 cannot be taken without 1, and 1 can be taken if and only if 5 is not taken. Let x_j be true when medication j is taken, and write these conditions in propositional form. Convert them to CNF without adding variables.

5.9. Find a feasible solution of the CNF expression in Exercise 5.8 using a DPL algorithm with clause learning. Branch on variables in the order $x_1, \ldots, x_5$, and take the false branch first.

5.10. Interpret the branching search of Exercise 5.9 as constraint-directed search by writing a table similar to Table 5.5.

5.11. Find an optimal solution of Exercise 5.9 using constraint-directed search, where the objective is to minimize the number of medications taken. Solve the current nogood set by setting a variable to false whenever possible. When a feasible solution is found, generate a nogood that rules it out, and continue the search. Thus, if a solution $x = (T, F, T, T, F)$ is found, generate the nogood $\neg x_1 \vee x_2 \vee \neg x_3 \vee \neg x_4 \vee x_5$. The nogoods are therefore enumerative, and the optimal solution is the best feasible solution found. The search table should indicate the value of any feasible solutions as well as the information in Table 5.5. How can bounding reduce the search?

5.12. Find a feasible solution of the problem in Exercise 5.8, 5.9, and 5.10 by partial-order dynamic backtracking. Experiment with various choices of the last literal in a nogood, and with various heuristics for solving the problem restriction.

5.13. Find an optimal solution of Exercise 5.12 using partial-order dynamic backtracking, where the objective is to minimize the number of medications taken. Solve the current nogood set by setting a variable to false whenever possible. Use enumerative nogoods as in Exercise 5.11. The search can be substantially shorter than with constraint-directed branching. How can bounding reduce the search even further?

5.14. Consider the propositional satisfiability problem

$$
\begin{aligned}
& x_1 \lor \; x_2 \qquad \quad \lor \; x_4 \\
& \qquad \quad x_2 \quad \; \lor \neg x_4 \lor \neg x_5 \\
& x_1 \lor \; x_2 \qquad \lor \neg x_4 \lor \; x_5 \lor \; x_6 \\
& x_1 \lor \; x_2 \qquad \lor \neg x_4 \lor \; x_5 \lor \neg x_6 \\
& \neg x_1 \lor \; x_2 \lor x_3 \lor \neg x_4 \lor \; x_5 \\
& \neg x_1 \lor \; x_2 \lor x_3 \lor \neg x_4 \\
& x_1 \lor \neg x_2 \qquad \qquad \qquad \lor \; x_6 \\
& \qquad \neg x_2 \lor x_3 \qquad \qquad \lor \neg x_6
\end{aligned}
$$

The variables x_1, x_2, x_3 form a backdoor set, because if their values are fixed, the problem that remains simplifies to a *renamable Horn* subproblem (Section 6.4.3).[1] The main problem can be solved by logic-based Benders decomposition in which the master problem contains x_1, x_2, x_3 and is solved by any convenient method (such as DPLL). The subproblem is solved by unit resolution in linear time, and nogoods are generated in the same way as conflict clauses in constraint-directed search. Solve the above problem in this fashion. (One can also exploit the backdoor by solving the problem by a DPLL method that branches on x_1, x_2, x_3 first.)

5.15. The Benders approach of the previous exercise can be applied to logic circuit verification. The object is to test whether circuit A has the correct Boolean output for every possible Boolean input. This is done by comparing with circuit B, which is known to be correct. In Fig. 5.14(a), x_1, x_2, x_3 are the inputs, and there are two outputs. If circuits A and B agree on both outputs for all inputs, the combined circuit in Fig. 5.14(a) has output 1 for all inputs; that is, it represents a tautology. The Benders method can be used for tautology checking, as illustrated by the small example in Fig. 5.14(b). The question is whether $y_6 = 1$ for all possible inputs x_1, x_2, x_3. The results of input $x = (1, 0, 1)$ are shown in the figure. Let the master problem contain the input variables. The subproblem is renamable Horn as in the previous exercise, and Benders cuts can be obtained in the same fashion. But they can be generated more efficiently by direct examination of the circuit. If the

---

[1] The variable x_3 can be eliminated althogther, along with the clauses containing it, by setting it to true, because all of its occurences are positive. This is an instance of the *pure literal rule*, which is not often used in state-of-the-art solvers because there are too few pure literals to justify the time investment of finding them.

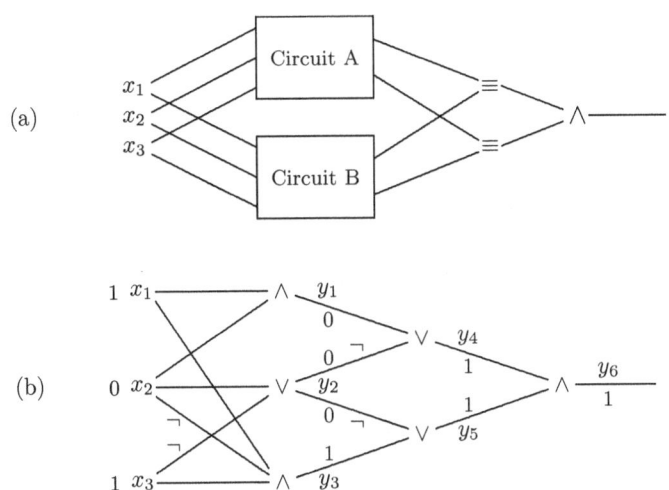

Fig. 5.14 (a) Comparing circuits A and B for equivalence. The output of logic gate $\equiv$ is 1 if and only if both inputs agree. The output of gate $\wedge$ is 1 if and only if both inputs are 1. (b) A tautology checking problem. The output of gate $\vee$ is 1 if at least one input is 1, and $\neg$ negates the signal.

input is $x = (1, 0, 0)$, for example, one can scan backward to determine which signals are necessary to result in an output of 1. Both $y_4 = 1$ and $y_5 = 1$ are necessary, but for both of these it suffices that $y_2 = 0$. For this it suffices that $(x_2, x_3) = (0, 1)$. This results in the Benders cut $x_2 \vee \neg x_3$. Solve this tautology checking problem by logic-based Benders decomposition.

5.3 Local Search

Local search methods attack a problem by solving it repeatedly over small subsets of the solution space, each of which is a *neighborhood* of the previous solution. The neighborhood consists of solutions obtained by making small changes in the previous solution, perhaps by changing the value of one variable or swapping the values of two variables.

The motivation for local search is that a neighborhood is more easily searched than the entire solution space. By moving from neighborhood to neighborhood, the search may happen upon a good solution. Well-designed local search methods can, in fact, deliver remarkably good solutions within a reasonable time, although tuning them to work efficiently is more an art than a science. Local search has become indispensable for attacking many practical problems that are too large to solve by exact methods.

In general, the neighborhoods examined during the search cover only a small portion of the solution space. Even the neighborhoods themselves may not be examined exhaustively. Local search therefore provides no guarantee that the solution is optimal or even lies within any given distance from the optimum.

Local search fits naturally into the solution scheme presented here. Because each neighborhood is the feasible set of a problem restriction, local search in effect solves a sequence of problem restrictions. Inference and relaxation can also play a role. In fact, many local search strategies can be viewed as analogs of branching or constraint-directed search, and these analogies suggest how techniques from exhaustive search can be transferred to heuristic methods. The role of relaxation in branching, for example, can be mirrored in such branching-related local search methods as *greedy randomized adaptive search procedures* (GRASPs). Inference is already a part of local search methods related to constraint-directed search, such as tabu search, where the tabu list can be viewed as consisting of enumerative nogoods. The analogy can be exploited further, because ideas from such techniques as partial-order dynamic backtracking can be imported into tabu search, resulting in a more sophisticated heuristic method.

5.3.1 Some Popular Metaheuristics

Such popular local search schemes or *metaheuristics* as simulated annealing, tabu search, and GRASP algorithms are easily seen to be searches over problem restrictions (Table 5.7). Simulated annealing randomly chooses a solution x' in the neighborhood of the current solution x. If x' is better than x, then x' is *accepted* and becomes the current solution, whereupon the process repeats. If x' is no better than x, x' is nonetheless accepted with a certain probability p. If x' is not accepted, another solution x' is chosen randomly from the neighborhood of x, and the process repeats. The algorithm mimics a cooling process in which molecules seek a minimum energy configuration. The probability p decreases with the *temperature* as the process continues. The search may be terminated at will, and it may be rerun with several

Table 5.7 How some selected heuristic methods fit into the search-infer-and-relax framework.

Solution method	Restriction P_i	Relaxation R_i	Solution of R_i	Inference
Simulated annealing	Neighborhood of current solution	P_i	Random solution in neighborhood	None
Tabu search	Neighborhood subject to tabu list	P_i	Best solution in neighborhood subject to tabu list	Addition of nogoods to tabu list
GRASP	Neighborhood of partial solution	Problem specific	Random or greedy selection of solution in neighborhood	None

different starting points. Clearly the neighborhoods are not examined exhaustively in this method. Each restriction is "solved" simply by selecting a solution, or at most a few solutions, randomly from the current neighborhood.

Tabu search differs in that it exhaustively searches each neighborhood. The best solution x' in the neighborhood of the current solution x becomes the current solution. To reduce the probability of cycling repeatedly through the same solutions, a *tabu list* of the last few solutions is maintained. Solutions on the tabu list are excluded from the neighborhood of x (the tabu list can also contain the types of alterations or *moves* performed on the last few solutions to obtain the next solution, rather than the solutions themselves). The items on the tabu list can be viewed as enumerative nogoods that rule out solutions or moves that have recently been examined.

Tabu search is therefore an inexhaustive form of constraint-directed search in which the tabu list contains enumerative nogoods. It is distinguished by its selection criterion: the solution of the nogood set is selected to be within the current neighborhood. It is in fact optimal within the neighborhood, subject to the nogoods.

Each iteration of a GRASP has two phases. The first constructs a solution in a greedy fashion, and the second uses this solution as a starting point for a local search. The greedy algorithm of the first phase assigns values to one variable at a time until all variables are fixed. The possible values that might be assigned to each variable x_i are ranked according to an easily computable criterion. The algorithm is adaptive in the sense that this ranking depends on what values were assigned to $x_1, \ldots, x_{i-1}$. One of the highly ranked values is then randomly selected as the value of x_i. This random component allows different iterations of the GRASP to construct different starting solutions.

The local search phase can be seen as a search over problem restrictions for reasons already discussed. The greedy phase is likewise a search over problem restrictions in a sense that is reminiscent of a branching search. Recall that a branching search typically branches on a problem P by assigning some variable its possible values. This creates a series of restrictions $P_1, \ldots, P_m$ whose feasible sets partition the feasible set of P. The search may then create restrictions of each P_i by branching on a second variable, and so on recursively.

The greedy algorithm is analogous, except that it generates only one restriction of P rather than an exhaustive list of restrictions $P_1, \ldots, P_m$. Specifically, it creates a restriction P_1 by setting x_1 to a value that is highly ranked. It then restricts P_1 by setting x_2 to a highly ranked value (given the value of x_1), and so forth, until all variables are assigned values.

5.3.2 Local Search Conceived as Branching

Simulated annealing and GRASPs can be seen as special cases of a generic local search procedure that is analogous to branching but does not explore all possible branches. This interpretation of local search also incorporates relaxation in a natural way.

The generic local search algorithm of Fig. 5.15 keeps "branching" until it arrives at a problem that is easy enough to solve, at which point it solves the problem (by searching a neighborhood) and backtracks. When branching on a given problem restriction P, however, the algorithm creates only one branch. The search may backtrack to P later and generate additional branches. The branches eventually created at P differ in two ways, however, from those in a normal branching search: (a) they need not be exhaustive, which is to say the union of their feasible sets need not be the feasible set F of P, and (b) their feasible sets need not partition F.

Let $v_{\mathrm{UB}} = \infty$ and $S = \{P_0\}$.
While S is nonempty repeat:
    Select a restriction $P \in S$ and remove P from S.
    If P is too hard to solve then
        Add a restriction of P to S.
    Else
        Let v be the value of P's solution and let $v_{\mathrm{UB}} = \min\{v, v_{\mathrm{UB}}\}$.
        Remove P from S.
The best solution found for P_0 has value v_{UB}.

Fig. 5.15 Generic algorithm for local search conceived as branching. The algorithm solves a minimization problem P_0. Set S contains the problem restrictions generated so far. v_{UB} is the value of the incumbent solution. Note that the algorithm is almost identical to the generic branching algorithm of Fig. 5.1.

Local search and GRASPs are special cases of this generic algorithm in which each restriction P is specified by setting one or more variables. If all the variables $x = (x_1, \ldots, x_n)$ are set to values $v = (v_1, \ldots, v_n)$, P's feasible set is a neighborhood of v. P is easily solved by searching the neighborhood. If only some of the variables $(x_1, \ldots, x_i)$ are set to $(v_1, \ldots, v_i)$, P is regarded as too hard to solve.

A pure local search algorithm, such as simulated annealing, branches on the original problem P_0 by setting all the variables at once to $v = (v_1, \ldots, v_n)$. The resulting restriction P is solved by searching a neighborhood of v. Supposing P's solution is v', the search backtracks to P_0 and branches again by setting $x = v'$. Thus, in pure local search, the search tree is never more than one level deep. The algorithm stops generating branches whenever the user terminates the search, generally long before the search is exhaustive.

In simulated annealing, P is "solved" by randomly selecting one or more elements of the neighborhood until one of them, say v', is accepted. The search backtracks to P_0 and branches by setting $x = v'$.

In a GRASP-like algorithm, the branching choices differ in the constructive and local search phases. In the constructive phase, the search branches by setting variables one at a time. At the original problem P_0, it branches by setting one variable, say x_1, to a value v_1 chosen in a randomized greedy fashion. It then branches again by setting x_2, and so forth. The resulting restrictions P are regarded as too hard to solve until all the variables x are set to some value v. When this occurs, a solution v' of P is found by searching a neighborhood of v, and the algorithm moves into the local search phase. It backtracks directly to P_0 and branches by setting $x = v'$ in one step. Local search continues as long as desired, whereupon the search returns to the constructive phase.

It was noted earlier that branching need not create a partition, and this is true in particular of a GRASP scheme. Fig. 5.16, for instance, illustrates a small GRASP search in which the initial constructive phase assigns variables x_1, x_2, and x_3 the values A, B, and C, respectively, thus arriving at restriction 3. At this point, the algorithm moves into the local search phase. It searches a neighborhood of $x = (A, B, C)$ by considering all interchanges of two components of x and selects $x = (B, A, C)$. It backtracks to the root and immediately generates a branch (restriction 4), at which the feasible set is a neighborhood of $x = (B, A, C)$. After searching this neighborhood, the local search is terminated and a new constructive phase assigns B, A, and C, respectively, to x_1, x_2, and x_3, thus arriving at restriction 7. The neighborhood here is the same as for restriction 4. Thus the branches at the root node do not create a partition: $x = (B, A, C)$ is consistent with two of the branches.

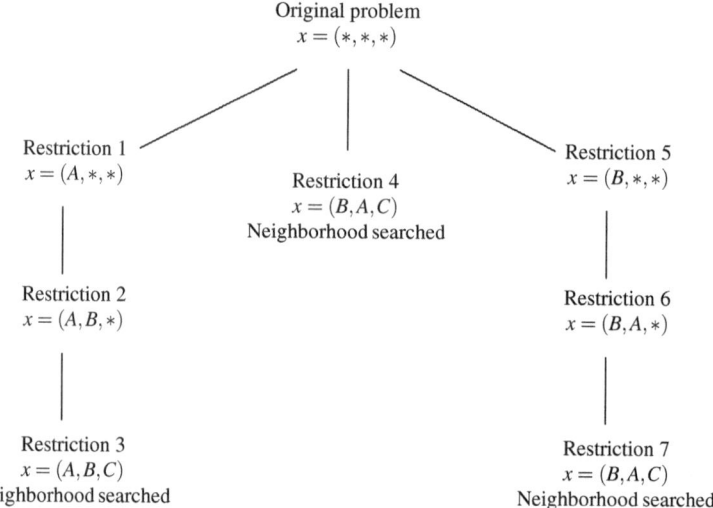

Fig. 5.16 Branching tree for a GRASP search. $x = (A, *, *)$ indicates that x_1 is set to A, but x_2 and x_3 are not set.

5.3.3 Relaxation

Conceiving local search as part of a quasi-branching scheme has the advantage of revealing an analogy with branch-and-relax algorithms, and thereby suggesting how relaxation can be used to accelerate the search.

The idea can be illustrated in the example of Fig. 5.16. Suppose that an objective function $f(x)$ is to be minimized. Thus the solution of restriction 3 has value $f(B, A, C)$. Suppose that $x = (B, A, C)$ is still the incumbent solution when restriction 5 is encountered. If a *relaxation* of restriction 5 is solved and its value is no less than $f(B, A, C)$, there is no need to branch further at restriction 5. Restrictions 6 and 7 are pruned from the tree. Fig. 5.17 contains a generic local search algorithm with relaxation.

In ordinary branch-and-bound algorithms, pruning the tree at some node ensures that no problem below the node will be solved. This is not true of local search. For example, restriction 7 in Fig. 5.16 is solved despite the pruning because it is identical to restriction 4. In general, a restriction may be reached via several paths in the tree, and pruning one path may leave other access routes open. Nonetheless, pruning by relaxation reduces the size of the search tree that would otherwise be traversed. This is illustrated by the example in Section 5.3.5.

Let $v_{UB} = \infty$ and $S = \{P_0\}$.
While S is nonempty repeat:
    Select a restriction $P \in S$ and remove P from S.
    If P is too hard to solve then
        Let v_R be the optimal value of a relaxation of P.
        If $v_R < v_{UB}$ then
            Add a restriction of P to S.
    Else
        Let v be the value of P's solution and $v_{UB} = \min\{v, v_{UB}\}$.
The best solution found for P_0 has value v_{UB}.

Fig. 5.17 Generic local-search-and-relax algorithm for solving a minimization problem P_0. The notation is the same as in Fig. 5.15.

5.3.4 Example: Single-Vehicle Routing

The idea of a local-search-and-relax algorithm can be further illustrated with a single-vehicle routing problem with time windows, also known as a traveling salesman problem with time windows. A vehicle must deliver packages to several customers and then return to its home base. Each package must be delivered within a certain time window. The truck may arrive early, but it must wait until the beginning of the time window before it can drop off the package and proceed to the next stop. The problem is to decide in what order to visit the customers so as to return home as soon as possible, while observing the time windows.

The data for a small problem appear in Table 5.8. The home base is at location A, and the four customers are located at B, C, D and E. The travel times are symmetric, and so the time from A to B and from B to A is 5, for instance. The time windows indicate the earliest and latest time at which the package may be dropped off. The vehicle leaves home base (location A) at time zero and returns when all packages have been delivered.

Exhaustive enumeration of the twenty-four possible routings would reveal six feasible ones: ACBDEA, ACDBEA, ACDEBA, ACEDBA, ADCBEA, ADCEBA. The last one is optimal and requires thirty-four time units to complete.

A simple heuristic algorithm adds one customer at a time to the route in a greedy fashion, by adding the customer that can be served the earliest. The search creates a branch whenever a customer is added. When all customers have been served, or when it is no longer possible to observe time windows, the search jumps to a random node N in the current search tree. It deletes from the tree all successors of N

Table 5.8 Travel times and delivery time windows for a small single-vehicle routing problem.

Origin	Travel time to: B	C	D	E		Customer	Time window
A	5	6	3	7		B	[20,35]
B		8	5	4		C	[15,25]
C			7	6		D	[10,30]
D				5		E	[25,35]

to keep memory requirements under control. It creates a branch at N by adding a random customer. At subsequent branches, customers are added according to the greedy criterion. The process can start over repeatedly as desired by returning to the root node.

This algorithm can be viewed as a generalized GRASP. It is a GRASP in the sense that it alternates between a greedy phase and a local search phase. The greedy phase constructs a solution as in an ordinary GRASP. The local search phase, however, does not necessarily select the next solution from a neighborhood of the current solution, as in a conventional GRASP. Rather, it randomly jumps to a previously enumerated partial solution and randomly instantiates one more variable. If the random jump is restricted to a jump to the immediate successor of the current leaf node, then the random instantiation is equivalent to randomly selecting a solution in a neighborhood of the current solution, where the neighborhood consists of solutions that differ in one variable. Thus, when the random jump is restricted in this way, a generalized GRASP becomes a conventional GRASP.

Figure 5.18 illustrates a possible search. Starting from the home base (node 0), the earliest possible delivery is to customer D at time 10. The travel time to D is only 3, but D's time window starts at 10. The search therefore branches to node 1. Departing customer D at time 10, the earliest possible delivery is to customer C at time 17, and so forth. The greedy procedure is fortunate enough to obtain a feasible solution at node 3 without backtracking. The search jumps randomly to node 1, whereupon nodes 2 and 3 are deleted. A randomly chosen customer, E, is added to the route, and the greedy criterion adds customer B at node 5. This violates the time windows, and the search randomly jumps to node 0, where it randomly adds customer B. Nodes 1–5 are deleted, and the greedy process obtains another infeasible routing at node 8. The search is arbitrarily terminated at this point.

This can be viewed as a local search algorithm in the sense that the greedy procedure searches a neighborhood in the space of problem restrictions. The neighborhood consists of all restrictions that can be formed from the current restriction by adding a customer to the end of the route. A deleted node can reappear due to subsequent branching.

A relaxation mechanism can help the search avoid unproductive areas of the search tree. One way to relax the problem is to replace the travel times for unscheduled trip segments with lower bounds on the travel times. A segment is the portion of the trip between two customers, i and j, that are adjacent on the route. If customer j has

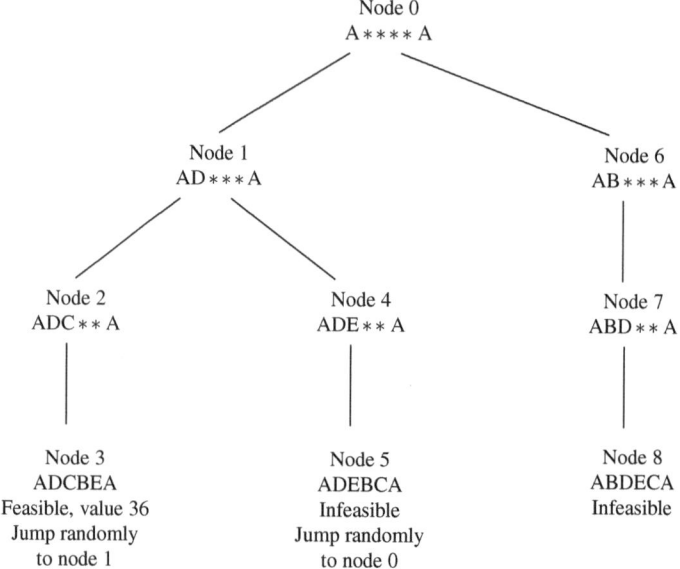

Fig. 5.18 Local-search tree for a single-vehicle routing problem with time windows. The notation AD ∗ ∗ ∗ A indicates a partial routing that runs from A to D, through three unspecified stops, and back to A.

not been scheduled, then the preceding customer i and the segment travel time are unknown. Yet a simple lower bound on this time is the travel time to j from the nearest customer that could precede j.

To make this more precise, let t_{ij} be the travel time between customers i and j, and let variable x_i be the ith customer visited (where x_0 is fixed to be the home base, Customer 0). Suppose a partial route consisting of the first k customers has been formed, so that $x_0, \ldots, x_k$ have been assigned distinct values. For $j \notin \{x_0, \ldots, x_k\}$, the travel time to customer j from the customer that precedes it in the route will be at least

$$L_j = \min_{i \notin \{j, x_0, \ldots, x_{k-1}\}} \{t_{ij}\}$$

and the travel time from the last customer served to the base will be at least

$$L_0 = \min_{j \notin \{x_0, \ldots, x_k\}} \{t_{j0}\}$$

Then, if T is the earliest time the vehicle can depart customer k,

$$T + L_0 + \sum_{j \notin \{x_0, \ldots, x_k\}} L_j$$

is a lower bound on the duration of any completion of the partial route. If this value is greater than or equal to the value of the incumbent solution, there is no need to branch further.

The search algorithm can be amended so that whenever the tree can be pruned at some node by bounding, that node is deleted from the tree. The search then proceeds exactly as it does when it constructs a feasible route or encounters infeasibility: it jumps to a randomly chosen node that remains in the tree and branches by adding a random customer to the end of the route. A node deleted by bounding may reappear due to subsequent branching, whereupon it will again be deleted. Unlike a conventional GRASP, this particular algorithm will never find an alternate route to solutions below a node that is pruned by bounding.

It is illustrative to rerun the search of Fig. 5.18 with bounding, and the result appears in Fig. 5.19. In the partial route ADE at node 4, the vehicle cannot depart E before time 25. Since B and C are unscheduled, the lower bound on the duration of the completed route is

$$25 + \min\{t_{CB}, t_{EB}\} + \min\{t_{BC}, t_{EC}\} + \min\{t_{BA}, t_{CA}\} = 40$$

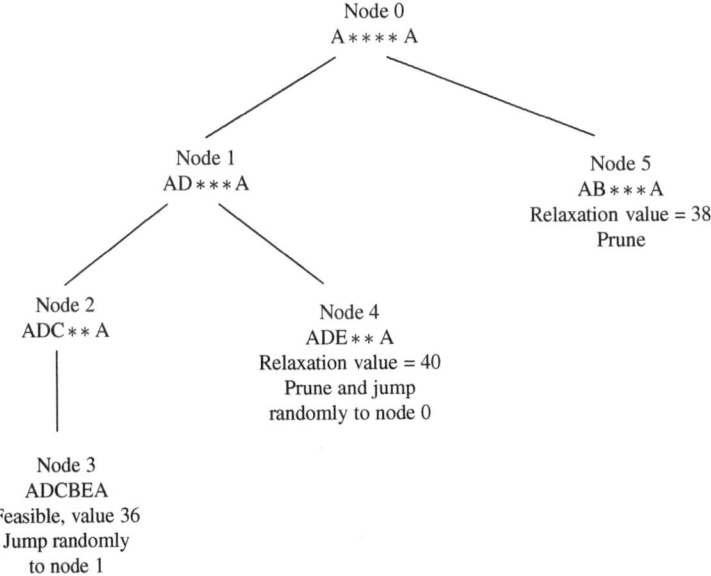

Fig. 5.19 *Local-search-and-relax tree for a single-vehicle routing problem with time windows. The notation is the same as in Fig. 5.18.*

Since this is larger than the incumbent value of 36, Node 4 is deleted. The search randomly jumps to node 0 and randomly adds customer B at node 5. Here the relaxation value is 38, which again allows the node to be pruned.

5.3.5 Constraint-Directed Local Search

Nothing in the generic local search algorithm of Fig. 5.15 or 5.17 prevents enumeration of the same solution several times. Repetition can be reduced or eliminated by maintaining a list of enumerative nogoods (e.g., a tabu list) and rejecting any solution or partial solution that violates one of the nogoods. The list can be of restricted length, as in tabu search, or it can remember all nogoods generated. In the latter case, the search would eventually become complete.

Constraint-directed local search examines only those solutions in the current neighborhood that satisfy the nogood set. It therefore solves the nogood set and the subproblem simultaneously. Formally, the neighborhood search can be regarded as a selection criterion: a method of selecting a solution of the nogood set. A generic constraint-directed local search algorithm appears in Fig. 5.20.

Any exhaustive constraint-directed search method can be converted to an inexhaustive search method by dropping constraints from the nogood set. This is a possibility, for example, in partial-order dynamic backtracking. Since the nogood set remains small, it may be practical to process the nogoods more intensely than with the parallel resolution.

Let $\mathcal{N} = \emptyset$, $v_{\mathrm{UB}} = \infty$. Select a starting neighborhood S.
Repeat as desired:
    Select a solution $\bar{x} \in S$ that satisfies $\mathcal{N}$.
    Let $\bar{v}$ be the value of $\bar{x}$, and let $v_{\mathrm{UB}} = \min\{v_{\mathrm{UB}}, \bar{v}\}$.
    Derive a nogood N that excludes $\bar{x}$ and possibly other solutions x'
        with $f(x') \geq f(x)$.
    Add N to $\mathcal{N}$ and process $\mathcal{N}$, removing nogoods as desired.
    Define a neighborhood S of $\bar{x}$.

Fig. 5.20 Generic constraint-directed local search algorithm for solving a minimization problem P with objective function $f(x)$. R is the relaxation of the current problem restriction.

This allows more freedom in the solution of the current nogood set, because the solution may be allowed to conform to previous solutions in a weaker sense. This is illustrated in the next section.

A Benders method can also be converted to a heuristic method by dropping older Benders cuts from the master problem, or perhaps generating cuts that are too weak to ensure termination. Such techniques may be used by practitioners when a Benders algorithm bogs down.

5.3.6 Example: Single-Vehicle Routing

The vehicle routing problem of Section 5.3.4 can solved in a manner similar to partial-order dynamic backtracking, as illustrated in Table 5.9. However, since the size of the nogood set will be limited, it is practical to process the nogood set more thoroughly to avoid backtracking while solving it.

Initially there are no nogoods, and a greedy algorithm selects the first solution ADCBEA by moving from each customer to the next cus-

Table 5.9 Solution of a single-vehicle problem with time windows by incomplete constraint-directed search.

i	Solution of $\mathcal{N}_i$	Value	Nogoods generated	Processed nogood set $\mathcal{N}_i$
0				$\emptyset$
1	ADCBEA	36	ADCB	{ADCB}
2	ADCEBA	34	ADCE	{ADC}
3	ADBECA	∞	EC	$\left\{ \begin{array}{l} \text{ABDE, ABEC,} \\ \text{ADBE, ADC,} \\ \text{ADEC, AEC} \end{array} \right\}$
4	ADBCEA	∞	BC	$\left\{ \begin{array}{l} \text{ABC, ABEC,} \\ \text{AD, AEC, AEDB} \end{array} \right\}$
5	ACDBEA	38	ACDB	$\left\{ \begin{array}{l} \text{ABC, ACDB,} \\ \text{AD, AEDB} \end{array} \right\}$
6	ACDEBA	36	ACDE	$\left\{ \begin{array}{l} \text{ABC, ACD,} \\ \text{AD, AEDB} \end{array} \right\}$
$\vdots$				

tomer that can be served most quickly. The greedy solution is feasible, and the nogood ADCB is generated to rule out this particular nogood. The meaning of the nogood ADCB is that no solution beginning ADCB can be considered. In iteration 1, the greedy algorithm is constrained by the nogood ADCB and selects ADCEBA, which generates nogood ADCE. These two nogoods obtained so far exclude all solutions beginning ADC, and so the nogood ADC comprises the nogood set $\mathcal{N}_2$. In effect ADCB and ADCE are resolved to yield ADC.

The greedy solution subject to ADC is the infeasible solution ADBECA. Some analysis reveals that the cause of the infeasibility is the subsequence EC, which is therefore generated as a nogood. To avoid backtracking in the solution of $\mathcal{N}_3$, all excluded subsequences beginning with A must be spelled out: AEC, ABEC, ADEC, ABDE, ADBE. These are added to $\mathcal{N}_3$, which has solution ADBCEA, again infeasible. Because subsequence BC is the cause of the infeasibility, all excluded subsequences beginning with A are added to the nogood set, and all possible resolutions performed to obtain $\mathcal{N}_4$. The resulting feasible solution generates nogood ACDB.

At this point, some of the older nogoods are dropped before adding ACDB to keep the nogood list short. Since the nogoods in $\mathcal{N}_1$ and $\mathcal{N}_2$ are no longer present in $\mathcal{N}_4$, the nogoods in $\mathcal{N}_3$ that are still present are dropped, leaving ABC, AD, and AEDB. Now, the new nogood ACDB is added to obtain $\mathcal{N}_5$. The process continues in this fashion until one wishes to terminate it. As it happens, the algorithm discovers the optimal solution ADCEBA in iteration 1.

Exercises

5.16. Consider a knapsack packing problem in which the objective is to maximize cx subject to $ax \leq 23$ and each $x_j \in \{0, 1\}$, where the data appear in Table 5.10. Generate part of a local-search-and-relax-tree similar to that of

Table 5.10 Data for a small knapsack packing problem.

i	1	2	3	4	5
c_i	24	14	15	9	14
a_i	11	7	8	5	9
c_i/a_i	2.182	2.000	1.875	1.800	1.556

Fig. 5.19. At each step of the greedy phase, fix to 1 the variable x_i that has not already been fixed and that has the largest ratio c_i/a_i. A leaf node is reached when no more variables can be fixed to 1. Thus, each leaf node will correspond to a feasible solution. After evaluating a leaf node, backtrack to a random node in the current tree, and randomly select the next variable to instantiate before resuming the greedy approach. As a relaxation, maximize cx subject to $ax \leq 23$, $x_j \in [0,1]$, and the currently fixed values (this is trivial to solve).

5.17. Consider again the problem of Exercise 5.16. (a) First apply an exhaustive constraint-directed search algorithm. Solve the current nogood set by setting the variables in the order $x_1, \ldots, x_5$. Set each x_i to 1 if this, in combination with the variables already set, does not violate a nogood, and otherwise set it to zero. Continue until the constraint $ax \leq 23$ is violated or all variables are set. Then generate an enumerative nogood. Since the order of instantiation is constant, parallel resolution of the nogoods is adequate. The first few steps of the search appear in Table 5.11. Add the next few steps to the table. (b) Now solve the problem with an incomplete constraint-directed search, as follows. Instantiate the variables in any order, but apply full resolution to the nogoods. If the nogood set grows large, drop the older nogoods. This might be described as a sophisticated form of tabu search.

5.18. A *genetic algorithm* mimics evolution by natural selection. It begins with a set of solutions (i.e., a population) and allows some pairs of solutions, perhaps the best ones, to mate. A crossover operation produces an offspring that inherits some characteristics of the parent solutions. At this point, the less desirable solutions are eliminated from the population so that only the fittest survive. The process repeats for several generations, and the best solution in the resulting population is selected. Indicate how this algorithm can

Table 5.11 A few iterations of constraint-directed search for Exercise 5.17.

i	Solution $(\bar{x}_1, \ldots, \bar{x}_5)$ of $\mathcal{N}$	Nogoods generated	Value	Processed nogood set $\mathcal{N}$
0				$\emptyset$
1	$(1,1,1,\cdot,\cdot)$	$\neg x_1 \vee \neg x_2 \vee \neg x_3$	∞	$\{\neg x_1 \vee \neg x_2 \vee \neg x_3\}$
2	$(1,1,0,1,1)$	$\neg x_1 \vee \neg x_2 \vee x_3 \vee \neg x_4 \vee \neg x_5$	∞	$\left\{\begin{array}{l}\neg x_1 \vee \neg x_2 \vee \neg x_3, \\ \neg x_1 \vee \neg x_2 \vee x_3 \vee \neg x_4 \vee \neg x_5\end{array}\right\}$
3	$(1,1,0,1,0)$	$\neg x_1 \vee \neg x_2 \vee x_3 \vee \neg x_4 \vee x_5$	47	$\left\{\begin{array}{l}\neg x_1 \vee \neg x_2 \vee \neg x_3, \\ \neg x_1 \vee \neg x_2 \vee x_3 \vee \neg x_4\end{array}\right\}$
4	$(1,1,0,0,1)$	$\neg x_1 \vee \neg x_2 \vee x_3 \vee x_4 \vee \neg x_5$	∞	$\left\{\begin{array}{l}\neg x_1 \vee \neg x_2 \vee \neg x_3, \\ \neg x_1 \vee \neg x_2 \vee x_3 \vee \neg x_4, \\ \neg x_1 \vee \neg x_2 \vee x_3 \vee x_4 \vee \neg x_5\end{array}\right\}$
$\vdots$				

be viewed as examining a sequence of problem restrictions. In what way does generation of offspring produce a relaxation of the current restriction? What is the role of the selection criterion? Why is relaxation bounding unhelpful in this algorithm? *Hint:* Relaxation bounding is helpful when it obviates the necessity of solving the current restriction. Think about how the current relaxation is obtained.

5.19. *Ant colony optimization* can be applied to the traveling salesman problem on n cities as follows. Initially all the ants of the colony are in city 1. In each iteration, each ant crawls from its current location i to city j with probability proportional to u_{ij}/d_{ij}, where u_{ij} is the density of accumulated pheromone deposit on the trail from i to j, and d_{ij} is the distance from i to j. Each ant deposits pheromone at a constant rate while crawling, and a certain fraction of the pheromone evaporates between each iteration and the next. Each ant remembers where it has been and does not visit the same city twice until all cities have been visited. After returning to city 1, the ants forget everything and start over again. When the process terminates, the shortest tour found by an ant is selected. Show how this algorithm can be understood as enumerating problem restrictions. How can relaxation bounding be introduced into the algorithm?

5.20. *Particle swarm optimization* can be applied to global optimization, as follows. The goal is to search a space of many dimensions for the best solution. A swarm of particles are initially distributed randomly through the space. Certain particles have two-way communication with certain others. In each iteration, each particle moves randomly to another position, but with higher probability of moving closer to a communicating particle that occupies a good solution. After many iterations, the best solution found is selected. How can this process be viewed as enumerating a sequence of problem restrictions? Why is there no role for relaxation bounding here?

5.4 Bibliographic Notes

Section 5.1. Depth-first and breadth-first search are standard techniques. Iterative deepening was proposed by a number of investigators in the 1970s. Limited discrepancy search is due to [259], and local branching to [197]. The first-fail principle is formulated in [256].

Backdoors, conceived as sets of variables that yield a polynomial-time problem when fixed, were first studied in [127] with respect to Horn clauses in propositional logic. The term *backdoor* is introduced in [504], and the idea is further analyzed in [175, 373]. The connection between backdoors and random restarts is explored in [505].

Pseudocosts were introduced in [74, 218] for mixed-integer programming, and their extension to CP is discussed in [415]. Strong branching is proposed in [93]. Uses of these and related techniques in mixed-integer programming are surveyed in [339]. The feasibility pump is introduced in [196].

Column generation methods have been used for decades. A unifying treatment of branch and price for mixed-integer programming can be found in [41]. Branch-and-price with CP-based column generation originated with [308, 518], and the area is surveyed in [185]. The airline crew rostering example described here is based on [190]. CP-based branch-and-price methods are surveyed in [185, 430].

Section 5.2. Constraint-directed search is discussed in connection with dynamic backtracking in [226, 227, 351], which also point out the connection between nogood-based search and branching. The ideas are further developed in [279]. Constraint-directed branching originated as conflict-directed backtracking [217, 464].

Classical Benders decomposition is due to [73] and was generalized to nonlinear programming in [224]. Logic-based Benders decomposition was introduced in [298] and developed in [279, 296].

The Davis–Putnam–Logemann–Loveland (DPLL) method for the propositional satisfiability problem was originally a resolution method proposed by Davis and Putnam [162]. Davis, Logemann, and Loveland [161] replaced the resolution step with branching. Clause learning grew out of research in the artificial intelligence community during the 1980s [163, 167, 168, 223]. The fastest satisfiability algorithms, such as CHAFF [361], combine DPLL with clause learning [49]. Learning techniques based on implication graphs were developed along with satisfiability solvers in the 1990s. Watched literals were introduced in [361]. A recent survey of these and other techniques is given in [234].

Partial-order dynamic backtracking was introduced in [351] and generalized in [99]. It is unified with other forms of dynamic backtracking and further generalized in [279], which also proves the completeness and polynomial complexity of parallel resolution for partial-order dynamic backtracking. The Benders approach to circuit verification in Exercise 5.15 is from [298].

Section 5.3. The integrated approach to heuristic methods presented here follows [287]. Tabu search is due to [231, 255]. GRASP originated with [458]. The idea of using relaxations in local search appears in [400].

There is a large literature on metaheuristics. Simulated annealing is described in [318] and has origins in the Metropolis algorithm [356]. Genetic algorithms, which have antecedents in the 1950s, were developed by [42, 412] among others and popularized by [272]. Ant colony optimization is introduced in [178, 179] and particle swarm optimization in [312, 456].

A survey on the integration of local search and CP appears in [454], and survey on hybrid metaheuristics in [100]. Recent work includes [72, 174, 251, 314, 396]

Chapter 6
Inference

Inference brings hidden information to light. When applied to a constraint set, it deduces valid constraints that were only implicit. These constraints can reveal that certain regions of the search space contain no solutions, or at least no optimal solutions, and one wastes less time in unproductive search.

Inference is most useful when applied to specially structured subsets of constraints, or metaconstraints. Most of this chapter is therefore organized around inference methods that are tailored to specific types of constraints.

The chapter begins by defining the fundamental concepts of completeness for inference methods and consistency for constraint sets. Various forms of consistency, particulary domain and bounds consistency, have played a major role in constraint programming solvers.

At this point, the chapter takes up inference methods for specific constraint classes, starting with linear inequalities. Linear programming and Lagrangean duality provide tools for domain reduction and classical Benders decomposition. Next comes propositional logic, for which inference methods are well developed. Integer and 0-1 inequalities can be regarded as logical propositions, and a theory of inference is developed here for them.

Following this, inference methods in the form of domain filters are presented for several global constraints, including the element, all-different, cardinality, nvalues, among, sequence, stretch, regular, and circuit constraints. The chapter concludes with domain reduction methods for disjunctive and cumulative scheduling, which are among the most successful application areas for constraint programming.

The inference of nogoods and Benders cuts for constraint-directed search, is discussed when it has been studied—for linear programming, integer linear inequalities, and disjunctive and cumulative scheduling. The use of nogoods for solving propositional satisfiability problems is described in Chapter 5, as are constraint-directed methods for local search. One should bear in mind, however, that whenever an inference dual can be defined, a constraint-directed search method can be developed.

6.1 Completeness and Consistency

If the purpose of inference is to make explicit what is implicit, one measure of an inference method is whether it derives all implied constraints of a given form—that is to say, whether it is *complete* with respect to constraints of a given form. The concept of completeness is particularly important in constraint programming, where it forms the basis for *consistency*. An inference method that is complete in a given sense, when applied to a constraint set, makes it consistent in a corresponding sense.

Following some basic definitions, several forms of consistency are reviewed below. Domain consistency and bounds consistency are the simplest and perhaps the most important for constraint solvers. Domain consistency can be generalized as k-consistency, which is related to the amount of backtracking necessary to solve a problem. Strong k-consistency can actually eliminate backtracking if the problem's dependency graph has width less than k.

6.1.1 Basic Definitions

It is necessary to begin by clarifying what a constraint is. Suppose there is a stock of variables $x_1, \ldots, x_n$, and each variable x_j takes values in its *domain* D_{x_j}. It is convenient to write $x \in D$, where $x = (x_1, \ldots, x_n)$ and $D = D_{x_1} \times \cdots \times D_{x_n}$. A *constraint* C is associated with a function $C(x_C)$ whose value is *true* or *false*. The variables $x_C = (x_{j_1}, \ldots, x_{j_d})$ are the variables that appear in C. A partial assignment $x_C = v_C$ to these particular variables *satisfies* C if $C(v_C) = true$, and it *violates* C otherwise.

A full assignment $(x_1, \ldots, x_n) = (v_1, \ldots, v_n)$ satisfies C if it assigns values to all the variables in C and makes C true. That is, $\{x_{j_1}, \ldots, x_{j_d}\} \subset \{x_1, \ldots, x_k\}$ and $C(v_{j_1}, \ldots, v_{j_d}) = true$. The assignment violates C if it assigns values to all the variables in x_C and makes C false. If the assignment does not assign values to all the variables in C, it neither satisfies nor violates C.

A fundamental concept of inference is *implication*. Constraint C_1 implies constraint C_2 (with respect to domain D) if any $x \in D$ that satisfies C_1 also satisfies C_2. A constraint set $\mathcal{C}$ implies constraint C if any $x \in D$ that is feasible for $\mathcal{C}$ (i.e., satisfies all the constraints in $\mathcal{C}$) also satisfies C. Two constraints are equivalent if they imply each other, and similarly for two constraint sets.

An *inference method* is a procedure that derives implied constraints, and a *complete* inference method derives all implied constraints in a given family $\mathcal{F}$. To make this more precise, suppose that an inference method is *applied* to constraint set $\mathcal{C}$ when one adds to $\mathcal{C}$ all constraints in $\mathcal{F}$ that the method can derive from $\mathcal{C}$, and that are not already implied by some constraint in $\mathcal{F} \cap \mathcal{C}$. The operation repeats until no further constraints can be added to $\mathcal{C}$ in this fashion. If the procedure does not terminate, then $\mathcal{C}$ is set equal to the infinite union of all constraint sets obtained by the procedure. This can occur, for instance, when applying interval propagation to linear inequalities.

An inference method is complete for $\mathcal{C}$ with respect to $\mathcal{F}$ when it yields a constraint set that contains all implications of $\mathcal{C}$ belonging to $\mathcal{F}$. So, a complete inference method in some sense brings out all the relevant information, where relevance is understood as expressibility by a constraint in $\mathcal{F}$.

Completeness can be similarly defined for constraint sets. A constraint set $\mathcal{C}$ is *complete* with respect to $\mathcal{F}$ if $\mathcal{C}$ contains all implications of $\mathcal{C}$ belonging to $\mathcal{F}$. Thus, a complete inference method creates complete constraint sets. That is, applying to a constraint set $\mathcal{C}$ an inference method that is complete with respect to $\mathcal{F}$ yields a constraint set $\mathcal{C}'$ that is complete with respect to $\mathcal{F}$.

6.1.2 Domain Consistency

In a search procedure, it is often useful to know whether an individual variable assignment $x_j = v$ is *consistent* with a constraint set, that is, whether x_j takes value v in at least one feasible solution of the constraint set. If not, then no time should be wasted enumerating solutions in which $x_j = v$.

The desired property is *domain consistency*. A constraint set $\mathcal{C}$ is domain consistent if it is complete with respect to *domain constraints*, which are constraints of the form $x_j \in D$. That is, each inconsistent value is excluded by a domain constraint in $\mathcal{C}$. For historical reasons, domain consistency is known as *hyperarc consistency* or *generalized arc consistency* in the CP community.

Domain consistency is generally achieved by removing all infeasible values from the domain D_{x_j} of each variable x_j and including the domain constraints $x_j \in D_{x_j}$ in $\mathcal{C}$. This ensures that any domain constraint $x_j \in D$ implied by $\mathcal{C}$ is implied by a domain constraint in $\mathcal{C}$, namely $x_j \in D_{x_j}$. This makes $\mathcal{C}$ complete with respect to domain constraints.

Domain consistency is closely related to projection. If S is a set of tuples $(x_1, \ldots, x_n)$, the *projection* of S onto variables $x_1, \ldots, x_k$ is the set of all k-tuples $(x_1, \ldots, x_k)$ that can be extended to a tuple in S; that is, the set of all $(x_1, \ldots, x_k)$ such that $(x_1, \ldots, x_k, x_{k+1}, \ldots, x_n) \in S$ for some $(x_{k+1}, \ldots, x_n)$. If S consists of the feasible solutions of a constraint set, domain consistency is achieved by ensuring that each domain D_{x_j} is equal to the projection of S onto x_j.

The process of removing infeasible values from domains is known as *domain reduction* or *domain filtering*. Domain filtering, combined with its weaker counterpart bounds propagation, is the workhorse of CP solvers and can play a key role in integrated solvers. Domain consistency is normally achieved, if at all, for individual constraints rather than the problem as a whole. Domains reduced by one constraint are passed on to the next constraint (*constraint propagation*), as illustrated by several examples in Chapter 2. The set of current variable domains is often referred to as the *domain store*.

Domain consistency can be illustrated with a small instance of the traveling salesman problem. A salesman must decide in which order to visit four cities so that the distance traveled is at most 28 km, including the distance from the last city back to the city of origin. Let x_j denote the city visited immediately after city j, and c_{ij} the distance between cities i and j in either direction (Table 6.1). The problem can be written

Table 6.1 Distances $c_{ij} = c_{ji}$ between cities i and j in a small instance of the traveling salesman problem.

		j		
		2	3	4
	1	5	8	7
i	2		6	9
	3			9

linear: $\displaystyle\sum_{j=1}^{4} c_{j}x_{j} \leq 28$

circuit: (x_1, x_2, x_3, x_4)

domains: $x_j \in \{1, 2, 3, 4\} \setminus \{j\},\ j = 1, \ldots, 4$

(6.1)

The circuit constraint requires that the sequence of cities visited form a single circuit that covers all cities. Only six solutions satisfy the circuit constraint: $(x_1, \ldots, x_4) = (2, 3, 4, 1),\ (2, 4, 1, 3),\ (3, 1, 4, 2),\ (3, 4, 2, 1),$ $(4, 1, 2, 3),\ (4, 3, 1, 2)$. Of these, only two satisfy the distance constraint:

$$(x_1, \ldots, x_4) = (2, 3, 4, 1), (4, 1, 2, 3) \tag{6.2}$$

These two solutions form the feasible set.

The constraint set (6.1) is not domain consistent. For instance, the assignment $x_1 = 3$ is inconsistent but $3 \in D_{x_1}$. Domain consistency can be achieved by removing the one inconsistent value in each domain, resulting in the domain constraints

$$x_1 \in \{2, 4\}, \quad x_2 \in \{1, 3\}, \quad x_3 \in \{2, 4\}, \quad x_4 \in \{1, 3\} \tag{6.3}$$

Each D_{x_j} is now the projection of the feasible set onto x_j.

6.1.3 Bounds Consistency

Bounds consistency is useful when domain elements have a natural ordering, as in the case of real numbers or integers. Bounds consistency is a relaxation of domain consistency and easier to achieve, but it can nonetheless accelerate solution.

Let I_{x_j} be the interval that spans the domain D_{x_j} of variable x_j. This means that $I_{x_j} = \{L_j, \ldots, U_j\}$ if x_j is integer valued, and $I_{x_j} = [L_j, U_j]$ if x_j is continuous, where $L_j = \min D_{x_j}$ and $U_j = \max D_{x_j}$.

A constraint set is *bounds consistent* if it is complete with respect to constraints of the form $L \leq x_j \leq U$ when the domains D_{x_j} are taken to be the intervals I_{x_j}. Bounds consistency therefore implies that the smallest and largest elements of each domain are consistent, when the other domains are replaced by intervals.

Like domain consistency, bounds consistency can be maintained by reducing domains. Let C' be the result of replacing the variable domains D_{x_j} by the interval domains I_{x_j} in constraint set C. Then C is bounds consistent when, for every j, $L_j = \min\{x_j \mid C'\}$ and $U_j = \max\{x_j \mid C'\}$. That is, L_j is the smallest value in the projection of C' onto x_j, and U_j is the largest.

In practice, it is common for solvers not to maintain the precise domain of an integer- or real-valued variable, but only lower and upper bounds. In this case bounds consistency is achieved by making the bounds as tight as possible. Domain consistency is a stronger property because it requires the solver to keep track of *holes* in the domain, or intermediate values that the variable cannot take in a feasible solution.

Sections 2.3 and 2.6 provide examples of bounds filtering for inequalities and equations.

6.1.4 k-Completeness

A constraint set is domain consistent when each infeasible assignment to some variable violates some constraint in the set. A natural extension of the concept ensures that each infeasible assignment to k variables violates some constraint. This property can be called *k-completeness*. It allows one to avoid unpromising assignments to subsets of variables. In particular, it can reduce backtracking in a branching algorithm.

The name k-completeness rather than k-consistency is used because a weaker property reduces backtracking to the same degree. This weaker property is known as k-consistency in the CP literature, and it is discussed in the next section.

A constraint set C is k-complete if every infeasible assignment $(x_{j_1}, \ldots, x_{j_k}) = (v_1, \ldots, v_k)$ to k variables (where each $v_i \in D_{x_{j_i}}$) violates some constraint in C. Thus, a domain consistent constraint set is 1-complete.

The traveling salesman problem of Section 6.1.2 is domain consistent after the addition of domain constraints (6.3). Yet it is not 2-complete. For example, the assignment $(x_1, x_2) = (2, 1)$ violates no constraint but is not part of a feasible solution.

One can achieve k-completeness by projecting the feasible set onto each subset of k variables. For each subset $\{x_{j_1}, \ldots, x_{j_k}\}$, one writes new constraints (containing only these variables) to describe the projection, and adds these constraints to the constraint set. The new constraints are violated by any assignment to $x_{j_1}, \ldots, x_{j_k}$ that cannot be extended to a feasible solution. The augmented constraint set is therefore k-complete.

k-completeness can reduce backtracking in the following way. Suppose that at level $k - 1$ of the branching tree, the branching process has assigned values to $k - 1$ variables

$$(x_{j_1}, \ldots, x_{j_{k-1}}) = (v_1, \ldots, v_{k-1}) \tag{6.4}$$

without violating any constraints in $\mathcal{C}$ so far. Suppose further that no value can be assigned to the next variable x_{j_k} without violating a constraint. That is, there is no $v_k \in D_{x_{j_k}}$ such that $(x_{j_1}, \ldots, x_{j_{k-1}}, x_{j_k}) = (v_1, \ldots, v_{k-1}, v_k)$ violates no constraint. It is therefore necessary to backtrack.

This backtrack could have been avoided if some constraint in $\mathcal{C}$ explicitly ruled out the assignment (6.4), so that this point in the branching tree was never reached. $\mathcal{C}$ contains such a constraint if it is $(k - 1)$-complete.

Let $\mathcal{C}$ be *strongly k-complete* if it is i-complete for $i \leq k$. If $\mathcal{C}$ is strongly $(k - 1)$-complete, one can branch at least down to level k without backtracking, if the branching algorithm uses *1-step lookahead*. That is, the algorithm checks whether assigning a given value violates some constraint before making the assignment. Thus, at the root node (level 1) of the search tree, it assigns x_1 some value that violates no constraint in $\mathcal{C}$; such a value exists unless $\mathcal{C}$ is infeasible. At level 2, 1-completeness implies that $x_1 = v_1$ can be extended to a feasible solution of $\mathcal{C}$. So, in particular, there is some value $v_2 \in D_{x_2}$ such that $(x_1, x_2) = (v_1, v_2)$ violates no constraint. The procedure continues until variables $x_{j_1}, \ldots, x_{j_k}$ are all assigned values.

6.1.5 k-Consistency

A property that slightly relaxes strong $(k - 1)$-completeness suffices to avoid backtracking down to level k if one uses 1-step lookahead. It is known as strong k-consistency, and one can see why it suffices by reexamining the branching procedure.

To avoid backtracking at level $k - 1$, it is enough that some constraint in $\mathcal{C}$ rule out assignment (6.4) when it cannot be extended to a kth variable; it is not necessary to rule out all assignments that cannot be extended to a feasible solution. The constraint set $\mathcal{C}$ has the required property if it is k-consistent. Thus, $\mathcal{C}$ is k-consistent if for every assignment $(x_{j_1}, \ldots, x_{j_{k-1}}) = (v_1, \ldots, v_{k-1})$ that violates no constraints in $\mathcal{C}$, and for every variable $x_{j_k} \notin \{x_{j_1}, \ldots, x_{j_{k-1}}\}$, there is a value $v_k \in D_{x_{j_k}}$ for which setting $(x_{j_1}, \ldots, x_{j_{k-1}}, x_{j_k}) = (v_1, \ldots, v_{k-1}, v_k)$ violates no constraints in $\mathcal{C}$.

$\mathcal{C}$ is *strongly k-consistent* if it is i-consistent for $i = 1, \ldots, k$. Thus, if $\mathcal{C}$ is feasible and strongly k-consistent, one reaches level k of the search tree without backtracking.

$(k-1)$-completeness obviously implies k-consistency, but the traveling salesman example shows that the reverse is not true. The constraint set (6.1) is 3-consistent but not 2-complete. It is 3-consistent simply because no constraint contains fewer than four variables. Because no assignment to three variables violates any constraints, any assignment to two variables can be extended to another variable by assigning it any value one wants. However, (6.1) is not 2-complete because solution $(x_1, x_2) = (2, 4)$ violates no constraint but is not part of a feasible solution.

6.1.6 Backtracking and Width

If the variables are loosely coupled, strong k-consistency can eliminate backtracking altogether, even when k is relatively small. To achieve this, however, it is necessary to branch on variables in a certain order.

Two variables are *coupled* in a constraint set when they occur in a common constraint. The pattern of variable coupling is indicated by its *dependency graph*, sometimes called its *primal graph*. The dependency graph contains a vertex for each variable and an edge connecting two variables when they occur in a common constraint.

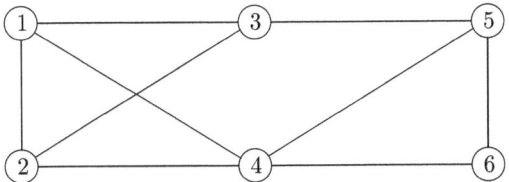

Fig. 6.1 Dependency graph for an integer knapsack problem.

For instance, the following constraint set has the dependency graph in Fig. 6.1, where each vertex j corresponds to variable x_j.

$$
\begin{aligned}
-3x_1 + 2x_2 - x_3 && \geq -8 \\
2x_1 - 3x_2 && + x_4 && \geq -4 \\
x_3 && + x_5 && \geq 4 \\
2x_4 - x_5 + x_6 && \geq 2
\end{aligned}
\qquad (6.5)
$$
$$
x_j \in \{1, 2, 3\}, \text{ all } j
$$

The amount of backtracking depends on the branching order, as well as on the nature of the variable coupling. To account for this, it is useful to define the *width* of the dependency graph with respect to a given ordering of the vertices. Let an edge connecting vertices i and j be directed from i to j when i occurs before j in the ordering. The *in-degree* of a vertex j is the number of edges incident to j that are directed toward j. The width of the graph, with respect to the ordering, is the maximum in-degree of its vertices. Thus, the graph in Fig. 6.1 has width 2 with respect to ordering $1, \ldots, 6$ and width 3 with respect to ordering $6, 5, \ldots, 1$.

A strongly k-consistent problem can be solved without backtracking if its dependency graph has width less than k with respect to some ordering.

Theorem 6.1. *If a feasible constraint set $\mathcal{C}$ is strongly k-consistent, and its dependency graph has width less than k with respect to some ordering of the variables, then branching in that order with 1-step lookahead obtains a feasible solution for $\mathcal{C}$ without backtracking.*

Proof. Suppose the variables are ordered $x_1, \ldots, x_n$. Branching with 1-step lookahead assigns x_1 some value that violates no constraint; such a value exists because $\mathcal{C}$ is feasible. Arguing by induction, suppose that the first $i - 1$ variables have been assigned values $(x_1, \ldots, x_{i-1}) =$

$(v_1, \ldots, v_{i-1})$ that violate no constraint. It suffices to show that x_i can be assigned a value that, together with the previous assignments, violates no constraint. It is enough to check the constraints that contain x_i, because by hypothesis none of the other constraints are violated so far. But since the dependency graph of S has width less than k with respect to the ordering $1, \ldots, n$, variable x_i occurs in a common constraint with fewer than k other variables $x_{j_1}, \ldots, x_{j_d}$ in the set $\{x_1, \ldots, x_{i-1}\}$. Because C is strongly k-consistent, the assignment $(x_{j_1}, \ldots, x_{j_d}) = (v_{j_1}, \ldots, v_{j_d})$ can be extended to $(x_{j_1}, \ldots, x_{j_d}, x_i) = (v_{j_1}, \ldots, v_{j_d}, v_i)$ without violating any constraint. Thus, $(x_1, \ldots, x_{i-1}) = (v_1, \ldots, v_{i-1})$ can be extended to $(x_1, \ldots, x_{i-1}, x_i) = (v_1, \ldots, v_{i-1}, v_i)$ without violating any constraint. $\square$

An immediate corollary is that a strongly $(k-1)$-complete problem can be solved without backtracking if its dependency graph has width less than k for some ordering of the variables.

The integer knapsack instance (6.5) is strongly 3-consistent (it is 2-complete as well). Thus, since its dependency graph has width 2 with respect to the ordering $1, \ldots, 6$, 1-step lookahead always finds a feasible solution. Suppose, for instance, that each x_i is assigned the smallest value that violates no constraint. This produces the feasible solution $(x_1, \ldots, x_6) = (1, 1, 1, 1, 3, 3)$ without backtracking. However, 1-step lookahead applied to the variables in reverse order results in $(x_6, \ldots, x_1) = (1, 1, 1, 3, 1, ?)$, with no feasible value for x_1, thus requiring a backtrack.

Exercises

6.1. Consider the constraint set C consisting of the equation $2x_1 + 3x_2 = 7x_3$ and domains $x_j \in \{0, 1, 2, 3\}$ for $j = 1, 2, 3$. Reduce the domains so as to achieve bounds consistency. Reduce them further to achieve domain consistency. Add to C one or more 2-variable equations that, together with the reduced domains, make the set 2-complete.

6.2. Show by counterexamples that a k-complete constraint set is not necessarily $(k-1)$-complete, and not necessarily $(k+1)$-complete.

6.3. Show that a $(k-1)$-complete constraint set is k-consistent.

6.4. Show by counterexample (other than the example in the text) that a k-consistent constraint set is not necessarily $(k-1)$-complete.

6.5. Show by counterexamples that a k-consistent constraint set is not necessarily $(k-1)$-consistent and not necessarily $(k+1)$-consistent.

6.6. Consider the constraint set $\mathcal{C}$, consisting of

$$x_1 + x_2 + x_4 \geq 1$$
$$x_1 + (1 - x_2) + x_3 \geq 1$$
$$x_1 + (1 - x_4) \geq 1$$

with domains $x_j \in \{0,1\}$ for $j = 1, 2, 3, 4$. Draw the dependency graph and note that it has width 2 with respect to the ordering 1, 2, 3, 4. Show that the constraint set is not 3-consistent, and show that a branching algorithm that follows the ordering 1, 2, 3, 4 may be required to backtrack. Recall that a constraint is not violated until all of its variables are fixed. Add the constraints $x_1 + x_2 \geq 1$ and $x_1 + x_3 \geq 1$ to $\mathcal{C}$ and verify that $\mathcal{C}$ is now strongly 3-consistent. Check that the sequence of branches that led to the backtrack is no longer possible.

6.2 Linear Inequalities

The theory of linear programming provides several inference-based tools for linear inequalities. These include methods for achieving domain consistency and k-completeness, and dual methods for reducing domains and deriving classical Benders cuts.

6.2.1 Domain and Bounds Consistency

Domain and bounds consistency are identical for linear inequalities, because the projection of a linear system

$$Ax \geq b, \quad x \geq 0 \tag{6.6}$$

onto a variable x_j is always an interval $[L_j, U_j]$ of real numbers. The bound L_j can be computed by minimizing x_j subject to (6.6), and U_j computed by maximizing x_j subject to (6.6). Because both are LP problems, domain consistency can be achieved by solving $2n$ LP problems over the same constraint set.

Bounds filtering may be faster than solving $2n$ LP problems, although in general it does not achieve domain consistency. Each round of filtering propagates the inequalities of (6.6) one at a time. Prior to propagating inequality i, let the current domain of each x_j be $[L_j, U_j]$. Inequality i is used to compute an updated lower bound for each x_j in the obvious way:

$$
L'_j = \begin{cases} \max \left\{ L_j, \dfrac{1}{A_{ij}} \left(b_i - \displaystyle\sum_{k \in J_i^+} A_{ik} U_k - \sum_{k \in J_i^-} A_{ik} L_k \right) \right\} & \text{if } A_{ij} > 0 \\[2em] L_j & \text{otherwise} \end{cases}
$$

and similarly for the upper bound:

$$
U'_j = \begin{cases} \min \left\{ U_j, \dfrac{1}{A_{ij}} \left(b_i - \displaystyle\sum_{k \in J_i^+} A_{ik} L_k - \sum_{k \in J_i^-} A_{ik} U_k \right) \right\} & \text{if } A_{ij} < 0 \\[2em] U_j & \text{otherwise} \end{cases}
$$

Here, $J_i^+ = \{k \neq j \mid A_{ik} > 0\}$, and analogously for J_i^-. An illustration is presented in Section 2.3.

An example in which bounds filtering does not achieve domain consistency is the system $x_1 + x_2 \geq 1$, $x_1 - x_2 \geq 0$. If the initial domain is $[0, 1]$ for each variable, bounds filtering has no effect, but the projection onto x_1 is $[\frac{1}{2}, 1]$.

Bounds filtering need not converge to a fixed point in finitely many iterations. Consider, for example, the system

$$
\begin{aligned} \alpha x_1 - x_2 &\geq 0 \\ -x_1 + x_2 &\geq 0 \end{aligned} \tag{6.7}
$$

with $0 < \alpha < 1$ and with initial domain $[0, 1]$ for each variable. In each round, the first inequality yields $U'_2 = \alpha U_1$, and the second inequality yields $U'_1 = U'_2 = \alpha U_1$. So the upper bounds U_1, U_2 converge asymptotically to zero.

6.2.2 k-Completeness

One can achieve k-completeness for a linear system (6.6) by projecting its feasible set onto each subset of k variables. This is the polyhedral projection problem, which is traditionally approached in two ways— Fourier–Motzkin elimination and generation of surrogates. Both tend to be computationally difficult.

The general polyhedral projection problem is to project a linear system of the form

$$Ax + By \geq b$$
$$x \in \mathbb{R}^p, \ y \in \mathbb{R}^q$$
(6.8)

onto x. The projection is itself a polyhedron and is therefore described by a finite set of linear inequalities.

Fourier–Motzkin elimination computes the projection by removing one variable at a time. It first projects (6.8) onto $(x, y_1, \ldots, y_{q-1})$. Then, it projects the system thereby obtained onto $(x, y_1, \ldots, y_{q-2})$, and so forth until $y_1, \ldots, y_q$ are eliminated.

The idea may be illustrated by projecting the following system onto $x = (x_1, x_2)$:

$$x_1 + 2x_2 + y_1 + 2y_2 \geq 10$$
$$2x_1 - x_2 - 2y_1 - y_2 \geq 4$$
$$3y_1 + 2y_2 \geq 2$$
$$3x_1 - x_2 + y_1 \geq 6$$
(6.9)

First, y_2 is eliminated by "solving" for y_2 the inequalities that contain y_2:

$$y_2 \geq -\tfrac{1}{2}x_1 - x_2 - \tfrac{1}{2}y_1 + 5$$
$$2x_1 - x_2 - 2y_1 - 4 \geq y_2$$
$$y_2 \geq -\tfrac{3}{2}y_1 + 1$$

The expression(s) on the left are paired with those on the right:

$$2x_1 - x_2 - 2y_1 - 4 \geq -\tfrac{1}{2}x_1 - x_2 - \tfrac{1}{2}y_1 + 5$$
$$2x_1 - x_2 - 2y_1 - 4 \geq -\tfrac{3}{2}y_1 + 1$$

The resulting inequalities are simplified and combined with the inequalities in (6.9) that do not contain y_2:

$$5x_1 - 3y_1 \geq 18$$
$$4x_1 - 2x_2 - y_1 \geq 10$$
$$3x_1 - x_2 + y_1 \geq 6$$
(6.10)

Next, y_1 is eliminated in the same manner, leaving the system

$$7x_1 - 3x_2 \geq 16$$
$$14x_1 - 3x_2 \geq 36$$
(6.11)

This is the projection of (6.8) onto x.

A second method of projection eliminates all of the y_j's at once. It takes advantage of the completeness of nonnegative linear combination as an inference method. The projection of (6.8) onto x is described by all the linear inequalities containing only x that can be inferred from (6.8). By Corollary 4.5, these are all dominated by the surrogates that contain only x. These surrogates, in turn, are all nonnegative linear combinations $uAx \geq ub$ for which $uB = 0$. Thus, the projection of (6.8) onto x is the set

$$P(x) = \{x \mid uAx \geq ub, \text{ all } u \in C\}$$

where C is the polyhedral cone $C = \{u \geq 0 \mid uB = 0\}$.

This defines $P(x)$ in terms of infinitely many surrogates, but fortunately only finitely many are necessary. The *extreme rays* of C are vectors that define the edges of the cone. A ray is extreme if it is a nonnegative linear combination of no two distinct rays. All vectors in C are nonnegative linear combinations of a finite set of extreme rays. Thus, the projection can be finitely computed

$$P(x) = \{x \mid uAx \geq ub, \text{all } u \in \text{extr}(C)\}$$

where $\text{extr}(C)$ is the set of extreme rays of the cone C.

In the example above, the cone is

$$C = \left\{ [u_1 \, u_2 \, u_3 \, u_4] \geq [0\,0\,0\,0] \; \middle| \; [u_1 \, u_2 \, u_3 \, u_4] \begin{bmatrix} 1 & 2 \\ -2 & -1 \\ 3 & 2 \\ 1 & 0 \end{bmatrix} = \begin{bmatrix} 0 \\ 0 \end{bmatrix} \right\}$$

C has two extreme rays, $u = [0\,2\,1\,1]$ and $u = [1\,2\,0\,3]$. The resulting surrogates $uA \geq ub$ are the two inequalities (6.11) already found to define the projection $P(x)$.

A step of Fourier–Motzkin elimination is actually a special case of this approach that eliminates only one variable. The first inequality in (6.10), for example, was obtained from the first two inequalities of (6.9). It is a linear combination of these inequalities with multipliers 1 and 2, respectively, which causes y_2 to vanish.

6.2.3 Domain Reduction with Dual Multipliers

Integer programming, constraint programming, and global optimization frequently use reduced costs to reduce variable domains. This technique requires knowledge of an upper bound U on the optimal value, perhaps the value of a feasible solution found earlier in the search process.

First, a linear programming relaxation of the problem is solved. If the reduced cost of a nonbasic variable x_j indicates that increasing its value from zero to v would raise the optimal value of the relaxation to U or, then only values of x_j less than or equal to v need be considered. If x_j is restricted to integers in the original problem, only integral values less than or equal to v need be considered, and in particular x_j can be fixed to zero if $v < 1$. This last inference is known as *reduced-cost variable fixing*.

This kind of domain reduction is actually a special case of a general method for deriving valid inequalities that is based on the dual solution. It is convenient to state the general method first and then specialize it to domain reduction.

Suppose that

$$\min \{cx \mid Ax \geq b, \ x \geq 0\} \qquad (6.12)$$

is an LP relaxation of the original problem with optimal solution x^*, optimal value v^*, and optimal dual solution u^*. Suppose further than $u_i^* > 0$, which means the ith constraint $A^i x \geq b_i$ is tight (i.e., $A^i x^* = b_i$), due to complementary slackness. The value v^* is a lower bound on the value of any solution that is feasible in the relaxation (6.12). Now suppose the right-hand side b_i of constraint i is increased to $b_i + \Delta b_i$. Because u_i^* is a shadow price, this increases the optimal value of (6.12) by at least $u_i^* \Delta b_i$. So $v^* + u_i^* \Delta b_i$ is a lower bound on the value of any solution that is feasible in the perturbed relaxation. If $v^* + u_i^* \Delta b_i \geq U$ then the optimal value of (6.12) rises at least to the value of a known feasible solution. This implies that a solution x that is feasible in the relaxation (6.12) and satisfies $A^i x \geq b_i + \Delta b_i$ can be optimal only if $v^* + u_i^* \Delta b_i \leq U$, or equivalently

$$\Delta b_i \leq \frac{U - v^*}{u_i^*}$$

Any optimal solution x must therefore satisfy

$$A^i x \leq b_i + \frac{U - v^*}{u_i^*} \qquad (6.13)$$

for each constraint i of (6.12) with $u_i^* > 0$. The inequality (6.13) can be propagated, which is particularly useful if some of the variables x_j have integer domains in the original problem.

One can reduce the domain of a particular nonbasic variable x_j by considering the nonnegativity constraint $x_j \geq 0$. Because the reduced cost r_j of x_j measures the effect on cost of increasing x_j, r_j can play the role of the dual multiplier in (6.13). So, (6.13) becomes $x_j \leq (U - v^*)/r_j$. If x_j has an integer domain in the original problem, one can say $x_j \leq \lfloor (U - v^*)/r_j \rfloor$. In particular, if $(U - v^*)/r_j < 1$, one can fix x_j to zero.

Suppose, for example, that (4.9) is an LP relaxation of a problem with integer-valued variables. The optimal dual solution of (4.9) is

$$u^* = c_B B^{-1} = \begin{bmatrix} 4 & 0 \end{bmatrix} \begin{bmatrix} \frac{1}{2} & 0 \\ 1 & -1 \end{bmatrix} = \begin{bmatrix} 2 & 0 \end{bmatrix}$$

Because the optimal value of (4.9) is $v^* = 12$ and the dual multiplier for the first constraint $2x_1 + 3x_2 \geq 6$ is $u_1^* = 2$, the inequality (6.13) becomes $2x_1 + 3x_2 \leq \frac{1}{2}U$. In addition, the nonbasic variable x_2 has reduced cost $r_2 = 2$, which yields the bound $x_2 \leq \lfloor \frac{1}{2}(U - 12) \rfloor$. Thus, if $U < 14$, x_2 can be fixed to zero.

6.2.4 Classical Benders Cuts

Logic-based Benders decomposition can be specilized to linear programming. In fact, if the Benders subproblem is an LP problem, the LP dual yields the linear Benders cuts that were originally developed for Benders decomposition.

Classical Benders decomposition applies to problems of the form

$$\min\ f(x) + cy$$
$$g(x) + Ay \geq b$$
$$x \in D_x,\ y \geq 0$$

which become linear when x is fixed to the solution $\bar{x}$ of the previous master problem. The subproblem is

$$\min\ f(\bar{x}) + cy$$
$$Ay \geq b - g(\bar{x}),\ y \geq 0 \tag{6.14}$$

Its classical LP dual, ignoring the constant $f(\bar{x})$, is

$$\max\ u\,(b - g(\bar{x}))$$
$$uA \leq c,\ u \geq 0 \tag{6.15}$$

Suppose first that the primal (6.14) as a finite optimal value $f(\bar{x}) + v^*$, which means the dual (6.15) has a finite optimal solution u^*. Then u^* encodes a proof of the lower bound

$$f(\bar{x}) + cy \geq f(\bar{x}) + u^* (b - g(\bar{x})) \tag{6.16}$$

because it defines a surrogate

$$u^* Ay \geq u^* (b - g(\bar{x}))$$

that dominates $cy \geq u^*(b - g(\bar{x}))$ and therefore (6.16). There is domination because u^* is dual feasible (i.e., $u^* A \leq c$). This is a proof of the optimal value $f(\bar{x}) + v^*$ because $v^* = u^*(b - g(\bar{x}))$ due to strong duality.

The key to generating the classical Benders cut is the fact that u^* remains feasible in the dual problem (6.15) when $\bar{x}$ is replaced by any x. It therefore defines a surrogate

$$u^* Ay \geq u^* (b - g(x))$$

that dominates $cy \geq u^*(b - g(x))$ and therefore proves the bound

$$f(x) + cy \geq f(x) + u^* (b - g(x))$$

for any x. If z represents total cost in the master problem, this yields the Benders cut

$$z \geq f(x) + u^* (b - g(x)) \tag{6.17}$$

for any x.

If the dual (6.15) is unbounded, there is a direction or ray u^* along which its solution value can increase indefinitely. In this case, the Benders cut is

$$u^* (b - g(x)) \leq 0 \tag{6.18}$$

rather than (6.17). The proof of this is left as an exercise.

Consider, for example, the problem

$$\min x_1 + 2x_2 + 3y_1 + 4y_2$$
$$4x_1 + x_2 - y_1 - 2y_2 \geq -2$$
$$-x_1 - x_2 - y_1 + y_2 \geq 2$$
$$x_j, y_j \geq 0, \quad x_j \text{ integral}, \quad j = 1, 2$$

Because the problem becomes an LP problem when $x = (x_1, x_2)$ is fixed, it decomposes into a master problem containing x and an LP subproblem containing y. If $\bar{x}$ is the solution of the master problem, the resulting subproblem is:

$$\min \bar{x}_1 + 2\bar{x}_2 + 3y_1 + 4y_2$$
$$-y_1 - 2y_2 \geq -2 - 4\bar{x}_1 - \bar{x}_2$$
$$-y_1 + y_2 \geq 2 + \bar{x}_1 + \bar{x}_2$$
$$y_j \geq 0, \ j = 1, 2$$

The initial master problem minimizes z subject to no constraints and is solved by setting $z = -\infty$ and x to, say, $\bar{x} = (0,0)$. The subproblem is infeasible, and its dual is unbounded with an extreme ray solution $u^* = [1 \ 2]$. The resulting Benders cut (6.18) is added to the master problem:

$$\min z$$
$$2x_1 - x_2 \geq 2$$
$$x_j \geq 0 \text{ and integer}$$

An optimal solution is $\bar{x} = (1,0)$ with $z = -\infty$. The next subproblem has optimal solution $y = (0,3)$ with dual solution $u^* = [0 \ 4]$ and value 13. The resulting Benders cut (6.17) is added to the master problem:

$$\min z$$
$$2x_1 - x_2 \geq 2$$
$$z \geq 8 + 5x_1 + 4x_2$$
$$x_j \geq 0 \text{ and integer}$$

The optimal solution of this problem is $\bar{x} = (1,0)$ with value 13. Since this is equal to the value of a previous subproblem, the algorithm terminates with optimal solution $x = (1,0)$ and $y = (0,3)$.

Exercises

6.7. Consider the system of inequalities $x_1 + x_2 \geq 1$, $x_1 - x_2 \geq 0$ with each $x_j \in [0, \alpha]$. For what values of $\alpha \geq 0$ does bounds propagation achieve bounds consistency?

6.8. Use Fourier–Motzkin elimination to project the feasible set of (6.7) and $x_1, x_2 \in [0, 1]$ onto x_1, assuming $0 < \alpha < 1$.

6.9. Suppose that every component of A is strictly positive. Use Fourier–Motzkin elimination to verify that bounds propagation is enough to achieve bounds consistency for $Ax \geq b, x \geq 0$.

6.10. Suppose that the system $Ax + By \geq b, x \geq 0$ is given by

$$x + 2y \geq 3$$
$$x - y \geq 1$$
$$x - 2y \geq 0$$
$$x, y \geq 0$$

and let $P(x)$ be the projection of $P = \{x \mid Ax + Bx \geq b, \; x \geq 0\}$ onto x. What are the two extreme rays of the polyhedral cone $C = \{u \geq 0 \mid uB = 0\}$? What are the two resulting projection cuts $uAx \geq ub$? What is $P(x)$?

6.11. Let $P(x) = \{x \mid uAx \geq ub, \; \text{all } u \in C\}$ be the projection of $P = \{x \mid Ax + By \geq b, \; x \geq 0\}$ onto x, where $C = \{u \geq 0 \mid uB = 0\}$. Show that

$$P(x) = \{x \mid uAx \geq ub, \; \text{all } u \in \text{extr}(C)\}$$

by showing that any inequality $uAx \geq ub$ for $u \in C$ is implied by inequalities $uAx \geq ub$ for $u \in \text{extr}(C)$.

6.12. Suppose that the problem in Exercise 4.6 is the continuous relaxation of an integer programming problem. Suppose further that the best known integral solution has value 8. Derive two inequalities from the dual solution that can be propagated to reduce domains. Also, derive an upper bound on the nonbasic variable x_3 from its reduced cost. (The reduced cost can be deduced from the slack in the corresponding dual constraint. Why?)

6.13. Solve the following problem by Benders decomposition:

$$\min 6x_1 + 5x_2 + 4y_1 + 3y_2$$
$$7x_1 + 5x_2 + 3y_1 + 2y_2 \geq 16$$
$$x_1, x_2 \geq 0 \text{ and integral}, \quad y_1, y_2 \geq 0$$

Hint: The master problem remaions unbounded after the first Benders cut is added. Solve it by setting $x_1 = x_2 = \infty$. The second Benders cut will appear twice before the algorithm terminates.

6.14. Show that (6.18) is a valid Benders cut when the dual of the subproblem is unbounded.

6.15. Show that the classical Benders algorithm converges if D_x is finite. Assume the dual of the subproblem is never infeasible.

6.3 General Inequality Constraints

The Lagrangean dual for general inequality constraints provides the basis for domain reduction similar to that developed for LP problems in Section 6.2.3.

Suppose that the inequality-constrained optimization problem

$$\min \; \{f(x) \mid \mathbf{g}(x) \geq 0, \; x \in S\} \tag{6.19}$$

is a relaxation of some problem of interest. Also suppose that a feasible solution with value U is known, so that U is an upper bound on the optimal value of the original problem. Let u^* be an optimal solution of the Lagrangean dual $\max\{\theta(u) \mid u \geq 0\}$ of (6.19), where

$$\theta(u) = \min_{x \in S} \{\theta(u, x)\}, \quad \text{and} \quad \theta(u, x) = f(x) - u\mathbf{g}(x) \tag{6.20}$$

Let $v^* = \theta(u^*)$ be the optimal value of the dual. This means v^* is a lower bound on the value of any solution that is feasible in (6.19), due to weak duality (Lemma 4.1). If x^* solves the minimization problem in (6.20), then $\theta(u^*) = \theta(u^*, x^*)$. Suppose further that constraint i is tight, so that $g_i(x^*) = 0$. If the constraint is changed to $g_i(x^*) \geq \Delta_i$, the function $\theta(u)$ for this altered problem is

$$\theta'(u) = \min_{x \in S} \{\theta'(u, x)\}, \quad \text{where} \quad \theta'(u, x) = f(x) - u\mathbf{g}(x) + u_i \Delta_i$$

Because $\theta'(u^*, x)$ differs from $\theta(u^*, x)$ only by a constant, any x that minimizes $\theta(u^*, x)$ also minimizes $\theta'(u^*, x)$. So

$$\theta'(u^*) = \theta'(u^*, x^*) = \theta(u^*) + u_i^* \Delta_i = v^* + u_i^* \Delta_i$$

is a lower bound on the optimal value of the altered problem, due to weak duality.

The optimal value $v^* + u_i^* \Delta_i$ of the altered relaxation is a lower bound on the value of any solution that is feasible in the altered relaxation. If $v^* + u_i^* \Delta_i > U$, then the optimal value of (6.19) rises above the value of a known feasible solution. So a solution x that is feasible in (6.19) and satisfies $g_i(x) \geq \Delta_i$ can be optimal only if $v^* + u_i^* \Delta_i \leq U$, or equivalently

$$\Delta_i \leq \frac{U - v^*}{u_i^*}$$

An optimal solution x must therefore satisfy

$$g_i(x) \leq \frac{U - v^*}{u_i^*} \tag{6.21}$$

This inequality can now be propagated.

In the example (4.19), suppose that the best known feasible solution is $x = (1, 3)$, which means the upper bound is $U = 13$. The optimal dual solution is $u^* = [u_1^* \ u_2^*] = [\frac{15}{19} \ \frac{14}{19}]$ with $v^* = 7\frac{2}{19}$. The optimal solution $x^* = (0, 0)$ of (6.20) makes both constraints tight. The inequality (6.21) corresponding to the first constraint is

$$30x_1 + 5x_2 \leq 82 \tag{6.22}$$

Bounds propagation using the original constraints yields the domains $x_1, x_2 \in \{1, 2, 3\}$. Propagation of (6.22) reduces the domain of x_1 to $\{1, 2\}$.

6.4 Propositional Logic

The logic of propositions provides a convenient way to express logical relations among variables. Both inference methods and relaxations are well developed for propositional logic. The inference methods can achieve k-completeness and k-consistency, not only for logical propositions, but for general constraint sets with two-valued variables.

Constraints expressed in propositional logic appear as logical formulas in which variables take Boolean values (1 and 0) corresponding to true and false. Formulas are constructed by prefixing $\neg$ (*not*), or by joining subformulas with $\wedge$ (*and*) and $\vee$ (*inclusive or*). Formulas can be defined recursively as consisting of (a) the empty formula, which is false by definition; (b) a single variable x_j, which is an unanalyzed or *atomic* proposition; or (c) expressions of the form $\neg F$, $F \wedge G$, and $F \vee G$, where F and G are formulas. The *material conditional* $F \rightarrow G$ is frequently introduced by defining it to mean $\neg F \vee G$. The *equivalence* $F \equiv G$ means $(F \rightarrow G) \wedge (G \rightarrow F)$.

The truth functions corresponding to the formulas are recursively defined in the obvious way. $\neg F$ is true if and only if F is false, $F \wedge G$ is true if and only if F and G are true, and $F \vee G$ is true if and only if at least one of the formulas F, G is true.

6.4.1 Logical Clauses

It is useful for purposes of both inference and relaxation to convert propositional formulas to *conjunctive normal form* (CNF) or *clausal form*.

A *clause* is a disjunction of zero or more *literals*, each of which is a variable or its negation. Thus, $x_1 \vee \neg x_2 \vee x_3$ is a clause of three literals. A formula in CNF or clausal form is a conjunction of zero or more logical clauses. An attractive property of clauses is that it is easy to tell when one implies another. Clause C implies clause D when C *absorbs* D, meaning that all the literals of C occur in D.

Any propositional formula can be converted to an equivalent formula in clausal form. One way to do so is to bring negations inward by using De Morgan's laws

$$\neg(F \wedge G) \equiv (\neg F \vee \neg G)$$
$$\neg(F \vee G) \equiv (\neg F \wedge \neg G)$$

and to distribute disjunctions using the equivalence

$$(F \vee (G \wedge H)) \equiv ((F \vee G) \wedge (F \vee H))$$

For instance, the formula

$$\neg(x_1 \vee \neg x_2) \vee (x_1 \wedge \neg x_3)$$

can be put in CNF by first applying De Morgan's law

$$(\neg x_1 \wedge x_2) \vee (x_1 \wedge \neg x_3)$$

and then distributing the disjunction :

$$(\neg x_1 \vee x_1) \wedge (\neg x_1 \vee \neg x_3) \wedge (x_2 \vee x_1) \wedge (x_2 \vee \neg x_3)$$

and finally deleting the first clause since it is a *tautology* (necessarily true).

Distribution can lead to an exponential explosion, however. For example, the formula

$$(x_1 \wedge y_1) \vee \cdots \vee (x_n \wedge y_n) \tag{6.23}$$

converts to the conjunction of 2^n clauses of the form $F_1 \vee \cdots \vee F_n$, where each F_j is x_j or y_j. The explosion can be avoided by adding variables as follows. Rather than distribute a disjunction $F \vee G$ (when neither F nor G is a literal), replace it with the conjunction

$$(z_1 \vee z_2) \wedge (\neg z_1 \vee F) \wedge (\neg z_2 \vee G)$$

where z_1, z_2 are new variables, and the clauses $\neg z_1 \vee F$ and $\neg z_2 \vee G$ encode the implications $z_1 \to F$ and $z_2 \to G$, respectively. The conversion requires only linear time and space. Formula (6.23), for example, yields the conjunction

$$(z_1 \vee \cdots \vee z_n) \wedge \bigwedge_{j=1}^{n} (\neg z_j \vee x_j) \wedge (\neg z_j \vee y_j)$$

6.4.2 A Complete Inference Method

A simple inference method, *resolution*, is complete with respect to logical clauses. Given any two clauses for which exactly one variable x_j occurs positively in one clause and negatively in the other, one can infer the *resolvent* of the clauses, which consists of all the literals in the clauses except x_j and $\neg x_j$. For instance, the two clauses

$$\begin{array}{cc} x_1 \vee x_2 \vee x_3 \\ \neg x_1 \vee x_2 \vee \neg x_4 \end{array} \qquad (6.24)$$

imply their resolvent

$$x_2 \vee x_3 \vee \neg x_4 \qquad (6.25)$$

The resolvent is obtained by "resolving on x_1."

Resolution is a valid inference because it reasons by cases. In the example, x_1 is either false or true. If it is false, then the first clause of (6.24) implies $x_2 \vee x_3$. If it is true, the second clause implies $x_2 \vee \neg x_4$. In either case, (6.25) follows by absorption.

Section 6.1.1 defines what it means to apply an inference method to a constraint set. Thus, to apply the resolution method to a clause set S is to add to S all resolvents from S that are not already absorbed by a clause in S, and to repeat the procedure until no further resolvents can be added. S is unsatisfiable if and only if resolution eventually generates the *empty clause*, (the clause with zero literals) by resolving two *unit clauses* x_j and $\neg x_j$.

Theorem 6.2. *Resolution is a complete inference method with respect to logical clauses.*

Proof. Let S' be the result of applying the resolution method to a clause set S containing variables $x_1, \ldots, x_n$. It suffices to show that any clause implied by S is absorbed by a clause in S'. Suppose to the contrary. Let C be the longest clause with variables in $\{x_1, \ldots, x_n\}$ that is implied by S but absorbed by no clause in S', where length is measured by the number of literals in the clause. One can suppose without loss of generality that no variables in C are negated, because any negated variable x_j can be replaced in every clause of S with $\neg x_j$ without changing the problem. Note first that C cannot contain all the variables $x_1, \ldots, x_n$. If it did, setting all variables to false would violate C and must therefore violate some clause C' of S', because S' implies C. This means that C' contains only positive literals and therefore absorbs C, contrary to the definition of C.

Thus, C must lack some variable x_j. But in this case, the clauses $x_j \vee C$ and $\neg x_j \vee C$, which S implies (because it implies C), must be absorbed, respectively, by clauses in S', say D and $\bar{D}$ (because $x_j \vee C$ and $\neg x_j \vee C$ are longer than C). Furthermore, D must contain x_j, because otherwise it would absorb C, contrary to C's definition, and similarly $\bar{D}$ must contain $\neg x_j$. So, the resolvent of D and $\bar{D}$ absorbs C, which means that some clause in S' absorbs C, which is again inconsistent with C's definition. $\square$

If all absorbed clauses are deleted at each step of the resolution algorithm, the clause set S' that remains contains the *prime implications* of the original clause set S. They are the undominated clauses implied by S; that is, the implications of S that are absorbed by no implication of S. An example appears in Section 2.4.

The resolution algorithm has exponential complexity in the worst case and can explode in practice if there are too many atomic propositions. There are cases in which resolution is well worth a substantial time investment, however, as when each node of the branching tree incurs a high computational cost, perhaps due to complex nonlinearities in the problem. The resolution algorithm may fix some of the logical variables and avoid expensive branches that would otherwise be explored.

6.4.3 Unit Resolution and Horn Clauses

When the full resolution algorithm is too slow, one can use weaker
variants of resolution that run more rapidly and may nonetheless fix
some of the logical variables. The simplest variant is *unit resolution*,
introduced in Section 5.2.4. This is ordinary resolution restricted to
the case in which one of the clauses resolved is a unit clause.

The running time of unit resolution on a clause set S is proportional to the number of literals in S, because it is essentially a form of
back substitution. An efficient implementation of unit resolution uses
watched literals, described in Section 5.2.4.

Unit resolution can be used to check the satisfiability of an important
class of clauses—renamable Horn clauses. A clause is *Horn* if it contains
at most one positive literal. A Horn clause with exactly one positive
literal, such as $x_1 \vee \neg x_2 \vee \neg x_3 \vee \neg x_4$, can be viewed as a conditional
statement or *rule* whose consequent is a unit clause: $(x_2 \wedge x_3 \wedge x_4) \to x_1$.
Such propositions are called *definite clauses* and are the norm in practical rule bases. A clause set is *renamable Horn* if all its clauses become
Horn after some set of zero or more variables x_j are replaced by their
negations $\neg x_j$.

Unit resolution is not a complete inference method even for Horn
clauses. For example, it fails to derive $x_2 \vee \neg x_3$ from $\neg x_1 \vee x_2$ and
$x_1 \vee \neg x_3$. However, the following is easily shown.

Theorem 6.3. *A renamable Horn clause set S is unsatisfiable if and
only if the unit resolution algorithm derives the empty clause from S.*

6.4.4 Domain Consistency and k-Completeness

A clause set S is k-complete if its projection onto any set of k variables
is described by some set of clauses in S containing only those variables.
One can therefore achieve k-completeness by computing the projection
of S onto every subset of k variables.

The projection of S onto a given subset $\{x_1, \ldots, x_k\}$ of variables can
be computed by applying the resolution method to S and selecting the
resulting clauses that contain only variables in $\{x_1, \ldots, x_k\}$. In fact, a
slight restriction of the resolution method accomplishes the same task.

Theorem 6.4. *A restricted resolution method that resolves only on variables in $\{x_{k+1}, \ldots, x_n\}$ derives all clauses that describe the projection of a clause set onto variables $x_1, \ldots, x_k$.*

Proof. It is convenient to begin by proving a lemma. Suppose that clause set S implies clause C, and let S_C be the result of removing from S all clauses that contain a variable with sign opposite to its sign in C. Then S_C also implies C. To show this, suppose to the contrary. Then some truth assignment $v = (v_1, \ldots, v_n)$ satisfies every clause in S_C, violates C, and violates some clause $C' \in S$. Since $C' \notin S_C$, C' contains some variable that occurs in C with the opposite sign. But in this case it is impossible that v violate both C and C'.

Now, take any clause C implied by S that contains only variables in $\{x_1, \ldots, x_k\}$. It suffices to show that the restricted form of resolution derives from S a clause that absorbs C. But S_C implies C, due to the lemma. Since full resolution is complete, it derives from S_C a clause that absorbs C. This derivation never resolves on a variable in $\{x_1, \ldots, x_k\}$, because these variables occur in S_C always with the same sign. It therefore derives C from S using only restricted resolution. $\square$

The theorem can be illustrated by projecting the following clause set onto x_1, x_2:

$$
\begin{aligned}
& x_1 \vee \ x_2 \\
& \neg x_1 \qquad\quad \vee \ x_3 \\
& \neg x_1 \vee \neg x_2 \vee \neg x_3 \\
& x_1 \qquad\qquad \vee \ x_3 \vee \ x_4 \\
& x_2 \vee \ x_3 \vee \neg x_4
\end{aligned}
\tag{6.26}
$$

Only one resolvent on x_3 or x_4 that is not already absorbed by a clause can be generated, namely $\neg x_1 \vee \neg x_2$. The projection is therefore described by the clauses $x_1 \vee x_2$ and $\neg x_1 \vee \neg x_2$.

Unfortunately, full resolution is needed to achieve k-completeness. Since the projection must be computed for each subset of k variables, it is necessary to resolve on every variable, which achieves k-completeness for each k. Even simple domain consistency requires full resolution. Strong k-consistency is easier to achieve, however, as will be seen in the next section.

Resolution can achieve k-completeness not only for logical clauses but for any constraint set S that contains only Boolean variables. If the two values in each variable's domain are arbitrarily identified with *true* and *false*, then one can, at least in principle, identify the logical clauses implied by a constraint in $\mathcal{S}$. Let $\mathcal{S}'$ consist of all clauses implied by some constraint in $\mathcal{S}$. Then $\mathcal{S}$ and $\mathcal{S}'$ are equivalent, and one can achieve k-completeness for all k by applying the resolution algorithm to $\mathcal{S}'$

6.4.5 Strong k-Consistency

Let k-*resolution* be the resolution method modified so that only resolvents with fewer than k literals are generated. When full resolution is too slow, but unit resolution is ineffective, k-resolution for small k may be a useful alternative. It can reduce and even eliminate backtracking, due to the following theorem:

Theorem 6.5. *The k-resolution method achieves strong k-consistency.*

Proof. It suffices to show, for any $i \leq k$, that i-resolution applied to a clause set $\mathcal{S}$ achieves i-consistency. Let $\mathcal{S}'$ be the set that results from applying i-resolution to $\mathcal{S}$, which means that no further i-resolvents can be generated for $\mathcal{S}'$. Suppose, contrary to the claim, that $\mathcal{S}'$ is not i-consistent. Then there is an assignment

$$(x_{j_1}, \ldots, x_{j_{i-1}}) = (v_1, \ldots, v_{j_{i-1}}) \tag{6.27}$$

that violates no clause in $\mathcal{S}'$, but for which the extended assignments

$$(x_{j_1}, \ldots, x_{j_i}) = (v_1, \ldots, v_{j_{i-1}}, 0)$$
$$(x_{j_1}, \ldots, x_{j_i}) = (v_1, \ldots, v_{j_{i-1}}, 1)$$

respectively violate two clauses in $\mathcal{S}'$. These clauses must contain only variables in $\{x_{j_1}, \ldots, x_{j_i}\}$ and must therefore have a resolvent R (on x_{j_i}) that contains fewer than i variables. No clause in $\mathcal{S}'$ absorbs R because (6.27) violates no clause in $\mathcal{S}'$. But in this case, i-resolution generates R, which contradicts the assumption that no further resolvents can be generated. $\square$

Strong 3-consistency can be achieved for the clause set (6.26) by generating all resolvents with fewer than three literals. The first round of 3-resolution yields the new resolvents

$$\neg x_1 \vee \neg x_2$$
$$x_2 \vee x_3$$
$$x_3 \vee x_4$$

after which no further 3-resolvents can be added to the clause set. Adding the above to (6.26) therefore achieves strong 3-consistency.

It was observed in the previous section that resolution can achieve k-completeness for any constraint set S with Boolean variables. Resolution can also achieve strong k-consistency for S. Let S' consist of all logical clauses implied by some constraint in S. Apply k-resolution to S', and let S'' be the result. Then, by Theorem 6.5, $S \cup S''$ is strongly k-consistent.

6.4.6 Parallel Resolution

Parallel resolution, a restricted form of resolution, is a key element of partial-order dynamic backtracking and its special cases (Section 5.2.7). This section establishes some key properties of parallel resolution.

Recall that partial-order dynamic backtracking enumerates partial assignments and generates a nogood clause for each. The next partial assignment is obtained by finding a conforming solution of the set $\mathcal{N}$ of nogoods. Parallel resolution is important because if it is applied to $\mathcal{N}$, a solution (if one exists) can be found by a greedy algorithm, without backtracking. This, in turn, is because parallel resolution has a completeness property in the context of partial-order dynamic backtracking. Specifically, it is a complete inference method with respect to conforming clauses.

Partial order dynamic backtracking designates a variable in each clause as the *last* variable, and the remaining variables in the clause as *penultimate* variables. The algorithm is designed so that when the penultimate variables in each clause are viewed as preceding the last variable in the clause, the resulting precedence relations determine a well-defined partial order on the variables (i.e., there are no cycles). It also ensures that any given variable always has the same sign whenever it occurs in a penultimate literal.

Parallel resolution is similar to ordinary resolution except that it resolves only on last variables. A clause D *parallel-absorbs* clause C if D is the empty clause, or $D = C$, or some penultimate literal of C is last in D. Each step of the parallel resolution algorithm identifies a pair of clauses that have a resolvent R on a variable that is last in both clauses, where R is parallel-absorbed by no clause. Then all clauses that are parallel-absorbed by R are removed, and R is added to the clause set. The process continues until no such pair of clauses exists.

A partial assignment *conforms* to a clause set if each variable that occurs penultimately is assigned a value opposite to the sign in which it occurs. Thus, a partial assignment

$$(x_{j_1}, \ldots, x_{j_r}) = (v_{j_1}, \ldots, v_{j_r}) \tag{6.28}$$

conforms to a nogood set $\mathcal{N}$ if for each literal x_j that occurs penultimately in $\mathcal{N}$, x_j is one of the variables x_{j_i} and $v_{j_i} = F$, and for each literal $\neg x_j$ that occurs penultimately in $\mathcal{N}$, x_j is one of the variables x_{j_i} and $v_{j_i} = T$. Conformity is well defined because x_j has the same sign whenever it occurs penultimately (see Exercises).

The negation of a partial assignment (6.28) is the clause C given by

$$L_1 \vee \cdots \vee L_r$$

where $L_i = \neg x_{j_i}$ if $v_{j_i} = T$ and $L_i = x_{j_i}$ if $v_{j_i} = F$. One can say that clause C conforms to $\mathcal{N}$ when it is the negation of a partial assignment (6.28) that conforms to $\mathcal{N}$. Then C conforms to $\mathcal{N}$ when every penultimate literal in $\mathcal{N}$ occurs in C.

A basic property of parallel absorption is the following.

Lemma 6.6 *Suppose that clause C conforms to $\{D\}$. Then if C is absorbed by some clause that D parallel-absorbs, C is absorbed by D.*

Proof. Suppose that C is absorbed by a clause D' that D parallel-absorbs. If $D = D'$ the lemma is trivial, so suppose $D \neq D'$. Then D' contains a penultimate literal that is last in D. Thus, each penultimate literal of D occurs in C, since C conforms to D, and the last literal of D occurs in C, since D' absorbs C. It follows that D absorbs C. $\square$

The nogood set $\mathcal{N}$ is solved by the following greedy algorithm. The variables $x_{j_1}, \ldots, x_{j_r}$ that occur penultimately in $\mathcal{N}$ are first assigned values (6.28) that conform to $\mathcal{N}$. Then each remaining variable x_j is assigned a value in such a way that it, along with the assignments already made, does not violate any clause in $\mathcal{N}$. For this procedure to find a solution without backtracking (assuming a solution exists), it suffices that any partial assignment that conforms to $\mathcal{N}$ and falsifies no clause of $\mathcal{N}$ can be extended to a solution of $\mathcal{N}$.

This is equivalent to a completeness property. A partial assigment falsifies a clause D if and only if its negation is absorbed by D. Thus, the desired property is this: any clause conforming to $\mathcal{N}$ that is implied by $\mathcal{N}$ is also absorbed by some clause in $\mathcal{N}$. In other words, $\mathcal{N}$ is complete with respect to clauses that conform to it. It can be shown that $\mathcal{N}$ has this property after parallel resolution is applied.

Theorem 6.7. *Parallel resolution applied to a clause set S is complete with respect to clauses that conform to S.*

Proof. Let S' be the result of applying parallel resolution to S. The claim is that any clause that conforms to S and is implied by S is absorbed by some clause of S'. Suppose to the contrary, and let C be the longest clause that conforms to S and is implied by S but is absorbed by no clause of S'. Also suppose without loss of generality that all the literals of C are positive. As in the proof of Theorem 6.2, C must omit at least one variable x_j. Then the clauses $C \vee x_j$ and $C \vee \neg x_j$, which S implies, must be absorbed respectively by clauses in S', say D and $\bar{D}$, where D contains x_j and $\bar{D}$ contains $\neg x_j$. Clauses D and $\bar{D}$ therefore have a resolvent R on x_j that absorbs C. Furthermore, R is a parallel resolvent because x_j must be the last variable in D and $\bar{D}$. If x_j were not last in D, conformity would require C to contain x_j, which it does not, and the same is true if x_j were not last in $\bar{D}$. Now because R is a parallel resolvent, S' must contain some clause R' that parallel absorbs R, because S' was obtained by applying parallel resoluton to S. This means that C conforms to R'. Clearly $R' \neq R$, because otherwise R absorbs C, contrary to assumption. Thus, each literal of R' is either penultimate in R' or, if it is last in R', penultimate in R. So, Lemma 6.6 and the fact that R absorbs C imply that R' absorbs C, contrary to assumption. The theorem follows. $\square$

It can be shown that parallel resolution has polynomial complexity when applied to the nogood set S in any step of partial-order dynamic backtracking. In fact, the running time is linear in the summed lengths of the clauses of S.

Exercises

6.16. Write the formula $(a \wedge b) \rightarrow (c \wedge d)$ in CNF, first without adding variables, and then by adding two new variables.

6.17. Use the resolution algorithm to compute the prime implications of the formulas $\delta_{10} \vee \delta_{11}$, $\delta_{20} \vee \delta_{21} \vee \delta_{22}$, and $\delta_{11} \rightarrow \neg \delta_{22}$ in the example of Section 2.4.

6.18. It is impossible to put three pigeons in two pigeon holes with at most one pigeon per hole. Prove this by the resolution algorithm. Let x_{ij} be true when pigeon i is placed in hole j and formulate the problem in clausal form. The resolution proof for pigeon hole problems explodes exponentially with the number of pigeons.

6.19. Write the inequalities in Exercise 6.6 as clauses. Use 3-resolution to achieve strong 3-consistency.

6.20. Prove Theorem 6.3. *Hints:* Prove the theorem first for Horn clauses. Then note that a unit resolution proof before renaming is a unit resolution proof after renaming.

6.21. A 2-satisfiability (2SAT) problem is a CNF problem with at most two literals in each clause. Show that 3-resolution, which has polynomial complexity, is complete for 2SAT. Describe a faster method to check for satisfiability by formulating the problem on a graph. Create a vertex for every variable and its negation and two directed edges for each clause.

6.22. Formulate the problem of checking whether a clause set is renamable Horn as a 2SAT problem. For each of the n variables x_j, let y_j be true when x_j is renamed. The number of clauses is quadratic in n, but one can add variables to make it linear in n. How?

6.23. Suppose that constraint set S contains only Boolean variables, and suppose that every clause that is implied by a constraint in S belongs to clause set S'. Show that if S' contains no other clauses, S and S' are equivalent.

6.24. Two families are meeting for party. One family consists of Jane, Robert, and Suzie, and the other consists of Juan, Maria, and Javier. Since the house is small, three will congregate outdoors and three indoors, with both families represented in each location. Jane is shy and insists on being in the same location as another family member, and similarly for Juan. Suzie wants to be with Juan or Maria, and Juan wants to be with Robert or Suzie. Since the choice for each family member is Boolean, the feasibility problem is naturally modeled in propositional logic. Draw the dependency graph and use k-resolution to achieve the degree of consistency necessary to solve the problem without backtracking.

6.25. Let a *multivalent clause* have the form $\bigvee_j (x_j \in S_j)$ where each x_j has a finite domain D_{x_j} and each $S_j \subset D_{x_j}$. Generalize the resolution algorithm to obtain a complete inference method for multivalent clauses, and prove completeness. Define an analog of unit resolution. *Hints:* a resolvent can be derived from several clauses. Resolving on x_j requires taking an intersection. Multivalent resolution can be used in dynamic backtracking methods.

6.26. Let be $\mathcal{N}$ be a clause set for which conformity is well defined. Show that the following are equivalent: (a) any partial assigment that conforms to $\mathcal{N}$ and falsifies no clause in $\mathcal{N}$ can be extended to a solution of $\mathcal{N}$; (b) any clause conforming to $\mathcal{N}$ that is implied by $\mathcal{N}$ is also absorbed by some clause in $\mathcal{N}$.

6.27. Show that at any stage of partial-order dynamic backtracking, any variable x_j that occurs in the nogood set has the same sign whenever it occurs penultimately.

6.5 0-1 Linear Inequalities

0-1 linear inequalities are useful for modeling because they combine elements of both linear inequalities and propositional logic. They have the numerical character of linear inequalities and can therefore deal with costs, counting, and the like. At the same time, they contain 0-1 variables that can express logical relations and, in fact, include logical clauses as a subset.

A second advantage of 0-1 linear inequalities is that they provide a ready-made continuous relaxation, formed simply by replacing the 0-1 variable domains with continuous unit intervals. This relaxation can often be strengthened by the addition of cutting planes to the constraint set, which are implied 0-1 linear inequalities that *cut off* noninteger solutions of the continuous relaxation.

Inference methods for 0-1 linear inequalities can in fact be studied with two purposes in mind—making the constraint set more nearly complete, and making the continuous relaxation tighter. The former goal leads to a theory of inference for 0-1 linear inequalities, which is developed in this section, and the latter inspires cutting-plane theory, which is taken up in Section 7.2.

6.5.1 Implication between Inequalities

A 0-1 linear inequality has the form $ax \geq a_0$, where $x \in \{0,1\}^n$. An inequality $ax \geq a_0$ implies a second 0-1 inequality $cx \geq c_0$ when any $x \in \{0,1\}^n$ that satisfies the first also satisfies the second. Thus, $ax \geq a_0$ implies $cx \geq c_0$ if and only if the following *0-1 knapsack problem*[1] has an optimal value of at least c_0:

$$\min \ cx$$
$$ax \geq a_0, \quad x \in \{0,1\}^n \tag{6.29}$$

This problem can be solved in polynomial time relative to the number of variables, provided the coefficients a_j are bounded in absolute value; in other words, the problem is pseudopolynomial. The problem of checking implication between 0-1 linear inequalities is therefore pseudopolynomial.

To study implication further, it is convenient to write a 0-1 inequality in the form $ax \geq a_0 + n(a)$, where $n(a)$ is the sum of the negative components of a. In this case, a_0 is the *degree* of the inequality. The degree is what the right-hand side would be if all variables with negative coefficients were replaced by their complements. Thus, $-2x_1 + 3x_2 \geq 1$ is written $-2x_1 + 3x_2 \geq 3 - 2$ and has degree 3.

There are two sufficient conditions for implication that can be quickly checked: absorption and reduction. A 0-1 inequality *absorbs* any inequality obtained by strengthening the left-hand side and/or weakening the right-hand side. Thus, if $a, b \geq 0$, $ax \geq a_0$ absorbs $bx \geq b_0$ when $a \leq b$ and $a_0 \geq b_0$. For arbitrary a and b, $ax \geq a_0 + n(a)$ absorbs $bx \geq b_0 + n(b)$ if $|a| \leq |b|$, $a_0 \geq b_0$, and $a_j b_j \geq 0$ for all j, where $|a| = (|a_1|, \ldots, |a_n|)$. For example, $-2x_1 + 3x_2 \geq 3 - 2$ absorbs $-3x_1 + 3x_2 + x_3 \geq 2 - 3$.

Theorem 6.8. *A 0-1 linear inequality implies any 0-1 linear inequality it absorbs.*

Proof. Suppose that $ax \geq a_0 + n(a)$ absorbs $bx \geq b_0 + n(b)$. They may respectively be written

$$\sum_{j \in J_1} a_j x_j - \sum_{j \in J_0} |a_j| x_j \geq a_0 - \sum_{j \in J_0} |a_j| \tag{6.30}$$

and

$$\sum_{j \in J_1} b_j x_j - \sum_{j \in J_0} |b_j| x_j \geq b_0 - \sum_{j \in J_0} |b_j| \tag{6.31}$$

---

[1] Here it is assumed that a 0-1 knapsack problem can have negative coefficients a_j, c_j. If not, the problem is easily converted to a knapsack problem by replacing each variable x_j for which $a_j, c_j < 0$ with its complement $1 - \bar{x}_j$. If $a_j < 0$ and $c_j \geq 0$, then one can set $x_j = 0$ in any optimal solution, and x_j can be dropped from the problem.

where $a_j, b_j \geq 0$ for $j \in J_1$ and $a_j, b_j \leq 0$ for $j \in J_0$. (Since $a_j b_j \geq 0$, a_j and b_j do not differ in sign.) The following inequality

$$\sum_{j \in J_0} (|a_j| - |b_j|) x_j \geq \sum_{j \in J_0} (|a_j| - |b_j|) \tag{6.32}$$

is valid because $|a_j| - |b_j| \leq 0$. Adding (6.32) to (6.30), one obtains

$$\sum_{j \in J_1} a_j x_j - \sum_{j \in J_0} |b_j| x_j \geq a_0 - \sum_{j \in J_1} |b_j|$$

which implies (6.31) because $a_0 \geq b_0$ and $a_j \leq b_j$ for $j \in J_1$. $\square$

Absorption for 0-1 inequalities is a generalization of absorption for logical clauses. A clause such as $x_1 \vee \neg x_2 \vee \neg x_3$ can be written as the 0-1 inequality $x_1 + (1 - x_2) + (1 - x_3) \geq 1$ or $x_1 - x_2 - x_3 \geq 1 - 2$, where 0 and 1 correspond to false and true. A 0-1 inequality that represents a clause is a *clausal inequality* and always has degree 1. Clearly, one clause absorbs another if and only if the same is true of the corresponding clausal inequalities.

A second sufficient condition for implication between 0-1 inequalities is reduction. An inequality $ax \geq a_0$ *reduces to* any inequality obtained by reducing coefficients on the left-hand side and adjusting the right-hand side accordingly. Assuming $a, b \geq 0$, the inequality $ax \geq a_0$ reduces to $bx \geq a_0 - \sum_j (a_j - b_j)$ if $a \geq b$. Thus, $3x_1 + x_2 + x_3 \geq 3$ reduces to $2x_1 + x_2 \geq 1$. More generally, $ax \geq a_0 + n(a)$ reduces to $bx \geq a_0 + n(b) - \sum_j (|a_j| - |b_j|)$ if $|a| \geq |b|$ and $a_j b_j \geq 0$ for all j. For instance, $-3x_1 + 3x_2 + x_3 \geq 4 - 3$ reduces to $-2x_1 + 2x_2 \geq 1 - 2$.

Theorem 6.9. *A 0-1 linear inequality implies any 0-1 linear inequality to which it reduces.*

Proof. Suppose $ax \geq a_0 + n(a)$ reduces to:

$$bx \geq a_0 + n(b) - \sum_j (|a_j| - |b_j|)$$

These two inequalities may respectively be written

$$\sum_{j \in J_1} a_j x_j - \sum_{j \in J_0} |a_j| x_j \geq a_0 - \sum_{j \in J_0} |a_j| \tag{6.33}$$

and

$$\sum_{j \in J_1} b_j x_j - \sum_{j \in J_0} |b_j| x_j \geq$$

$$a_0 - \sum_{j \in J_0} |b_j| - \left(\sum_{j \in J_1} a_j + \sum_{j \in J_0} |a_j| \right) + \left(\sum_{j \in J_1} b_j + \sum_{j \in J_0} |b_j| \right) \tag{6.34}$$

where $a_j, b_j \geq 0$ for $j \in J_1$ and $a_j, b_j \leq 0$ for $j \in J_0$. Note that two terms in (6.34) cancel. The following inequality is valid

$$\sum_{j \in J_1} (b_j - a_j) x_j + \sum_{j \in J_0} (|a_j| - |b_j|) x_j \geq - \sum_{j \in J_1} a_j + \sum_{j \in J_1} b_j, \tag{6.35}$$

because $b_j - a_j \leq 0$ for $j \in J_1$ and $|a_j| - |b_j| \geq 0$ for $j \in J_0$. Adding (6.35) to (6.33) yields (6.34). $\square$

6.5.2 Implication of Logical Clauses

It is easy to check whether a 0-1 linear inequality implies a clause or clausal inequality. The idea can be seen in an example. The inequality

$$x_1 - 2x_2 + 3x_3 - 4x_4 \geq 2 \tag{6.36}$$

implies the clause

$$x_1 \lor \neg x_2 \lor x_5$$

because the falsehood of the clause implies that (6.36) is violated. The clause is false only when $(x_1, x_2, x_3) = (0, 1, 0)$, and if the first two values are substituted into (6.36), the left-hand side is maximized when $(x_3, x_4) = (1, 0)$. This means the maximum value of the left-hand side is 1, and (6.36) is necessarily violated. In general,

Theorem 6.10. *The 0-1 inequality $ax \geq a_0$ implies the logical clause*

$$\bigvee_{j \in J_1} x_j \lor \bigvee_{j \in J_0} \neg x_j$$

if and only if

$$\sum_{\substack{j \notin J_1 \\ a_j > 0}} a_j + \sum_{\substack{j \in J_0 \\ a_j < 0}} a_j < a_0$$

Let succeed = **Dominate**$(\emptyset, 0)$.

Function **Dominate**(J, k).
    If $\sum_{j \notin J} a_j < a_0$ then generate the implied clause $\bigvee_{j \in J} x_j$ and
    return *true*.
    Else if $k = n$ then return *false*.
    Else
        Let *succeed* = *true*.
        For $j = k + 1, \ldots, n$ while *succeed*:
            Let *succeed* = **Dominate**$(J \cup \{j\}, j)$.
        Return *succeed*.

Fig. 6.2 Recursive function procedure for generating all nonredundant clauses implied by a 0-1 inequality $ax \geq a_0$ with $a_1 \geq \cdots a_n \geq 0$. It returns *true* if at least one implied clause exists.

A simple recursive function (Fig. 6.2) generates all nonredundant logical clauses implied by a 0-1 inequality $ax \geq a_0$. It assumes that $a_1 \geq \cdots \geq a_n \geq 0$. When the function is called with arguments (J, k), it determines recursively whether clause $C = \bigvee_{j \in J} x_j$ is implied by $ax \geq a_0$ or can be extended to a clause that is implied (where x_k is the last variable in C). If so, it returns the value *true*, and otherwise *false*. The first step is to check whether C itself is implied by checking whether $\sum_{j \notin J} a_j \leq a_0$. If so, the procedure generates C and returns *true*. If not, it tries adding the next variable x_{k+1} to C, then x_{k+2}, and so on, until the function returns false to indicate failure to find an implied extension of C. If k is already n, and C is not implied, the function immediately returns *false* to indicate failure.

For example, the 0-1 inequality $9x_1 - 6x_2 + 5x_3 + 3x_4 + x_5 \geq 4$ can be written with all positive terms by replacing x_2 with $1 - \bar{x}_2$, where $\bar{x}_2$ represents the $\neg x_2$:

$$9x_1 + 6\bar{x}_2 + 5x_3 + 3x_4 + x_5 \geq 10$$

Algorithm 6.2 now derives the clauses

$$x_1 \vee \neg x_2$$
$$x_1 \vee x_3 \vee x_4$$
$$\neg x_2 \vee x_3 \vee x_4 \vee x_5$$

6.5.3 Implication of Cardinality Clauses

A *cardinality clause* is a generalization of a logical clause. It states that at least a certain number of literals must be true. Thus, a 0-1 inequality $ax \geq a_0 + n(a)$ is a cardinality clause when each $a_j \in \{0, 1, -1\}$ and a_0 is a nonnegative integer. For instance, $x_1 + x_2 - x_3 \geq 2 - 1$ is a cardinality clause and states that at least 2 of the literals $x_1, x_2, \neg x_3$ must be true.

Cardinality clauses deserve attention because they preserve the counting ability of 0-1 inequalities, and yet inference is much easier for them. In particular, it is easy to determine whether a 0-1 linear inequality implies a cardinality clause, and whether one cardinality clause implies another.

Suppose, without loss of generality, that $a \geq 0$ in a given 0-1 inequality $ax \geq a_0 + n(a)$ (if some $a_j < 0$, replace x_j with its complement). This inequality, which can be written $ax \geq a_0$, implies a cardinality clause $bx \geq b_0 + n(b)$ if and only if it implies the clause without its negative terms; that is, if and only if it implies

$$\sum_{j=1}^{k} x_j \geq b_0 \tag{6.37}$$

where $b_1, \ldots, b_k$ are the positive coefficients in b. But $ax \geq a_0$ implies (6.37) if and only if the $\min\{b_0, k\}$ largest coefficients in $\{a_1, \ldots, a_k\}$ are required to accumulate a sum of at least a_0 after $x_{k+1}, \ldots, x_n$ are set equal to one.

Theorem 6.11. *Let $b_j > 0$ for $j \leq k$ and $b_j \leq 0$ for $j > k$ in cardinality clause $bx \geq b_0 + n(b)$. Then if $a \geq 0$ and $a_1 \geq \cdots \geq a_k$, the 0-1 inequality $ax \geq a_0$ implies $bx \geq b_0 + n(b)$ if and only if*

$$\sum_{j < \min\{b_0, k\}} a_j + \sum_{j > k} a_j < a_0$$

For example, the 0-1 inequality

$$4x_1 + 3x_2 + 2x_3 + x_4 + 5x_6 \geq 13$$

implies the cardinality clause

$$x_1 + x_2 + x_3 + x_4 + x_5 - x_7 \geq 3 - 1$$

because $a_1 + a_2 + a_6 + a_7 = 12$ is less than 13 (note that $a_5 = a_7 = 0$). Implication between cardinality clauses is also easy to check.

Theorem 6.12. *Consider the cardinality clauses $ax \geq a_0 + n(a)$ and $bx \geq b_0 + n(b)$. Let $\Delta = \sum_j |a_j| \max\{a_j b_j, 0\}$, which means Δ is the number of terms in the first clause that do not appear in the second clause. The following are equivalent:*

(i) $ax \geq a_0 + n(a)$ *implies* $bx \geq b_0 + n(b)$.
(ii) $\Delta \leq a_0 - b_0$.
(iii) $ax \geq a_0 + n(a)$ *reduces to a cardinality clause that absorbs* $bx \geq b_0 + n(b)$.

Before proving the theorem, consider the following example:

$$
\begin{aligned}
x_1 - x_2 + x_3 - x_4 + x_5 \quad &\geq 4 - 2 \\
x_1 - x_2 \quad &\geq 1 - 1 \\
x_1 - x_2 - x_3 + x_4 \quad + x_6 &\geq 1 - 2
\end{aligned}
$$

The first cardinality clause implies the third. The second is a clause that is a reduction of the first and that absorbs the third, as predicted by the theorem. Also $\Delta = 3 \leq 4 - 1$, as required by the theorem.

Proof of Theorem 6.12. Part (iii) of the theorem implies (i) by virtue of Theorems 6.8 and 6.9.

To show that (i) implies (ii), assume that $a \geq 0$. The proof is easily generalized to arbitrary a by complementing variables. The first cardinality clause may therefore be written $ax \geq a_0$. It suffices to show that for any set J of a_0 indices j with $a_j = 1$, setting $x_j = 1$ for $j \in J$ is enough to satisfy $bx \geq b_0 + n(b)$. But this assignment does satisfy $bx \geq b_0 + n(b)$ if $b_j \neq 1$ for at most $a_0 - b_0$ of the indices in J. This means that $\Delta \leq a_0 - b_0$.

To show that (ii) implies (iii), construct a third cardinality clause

$$cx \geq a_0 - \Delta + n(c) \tag{6.38}$$

by setting

$$c_j = \begin{cases} a_j & \text{if } a_j = b_j \\ 0 & \text{otherwise} \end{cases}$$

This is a cardinality clause because $\Delta \leq a_0 - b_0$ and $b_0 \geq 1$ imply $a_0 - \Delta \geq 1$. In addition, Δ is the number of terms removed from the left-hand side of $ax \geq a_0 + n(a)$ to obtain (6.38), which means that (6.38) is a reduction of $ax \geq a_0 + n(a)$. Finally, (6.38) absorbs $bx \geq b_0 + n(b)$ by construction of c, and the fact that $b_0 \leq a_0 - \Delta$. $\square$

6.5.4 0-1 Resolution

The resolution method for logical clauses can be generalized to a complete inference method for 0-1 linear inequalities, which might be called *0-1 resolution*.

0-1 resolution is analogous to the well-known Chvátal–Gomory cutting-plane procedure, to be presented in Section 7.2.1, in that both procedures generate all implied 0-1 inequalities. 0-1 resolution relies on the ability to recognize when one inequality implies another, but it achieves completeness by using only two special cases of the cutting planes generated by the Chvátal–Gomory procedure.

Neither 0-1 resolution nor the Chvátal–Gomory procedure is a practical method for solving general 0-1 problems. However, they help provide a theoretical foundation for the theory of 0-1 inference in one case and cutting-plane theory in the other.

The 0-1 resolution procedure consists of two repeated operations. One is essentially clausal resolution. The other exploits a diagonal pattern in coefficients and might be called *diagonal summation*. It is assumed that the coefficients and right-hand sides are integers. If they are rational numbers, they can be converted to integers by multiplying each inequality by an appropriate number.

A set S of 0-1 linear inequalities has a *resolvent R* when R is the resolvent of two clausal inequalities C_1 and C_2 such that C_1 is implied by some inequality in S, and similarly for C_2. For example, the inequalities

$$\begin{aligned} 3x_1 - 2x_2 \qquad\quad + 3x_4 \qquad\;\; &\geq 3 - 2 \\ -x_1 - 3x_2 + 3x_3 \qquad\quad + 4x_5 &\geq 5 - 4 \end{aligned} \qquad (6.39)$$

respectively imply the clausal inequalities

$$\begin{aligned} x_1 - x_2 \qquad + x_4 &\geq 1 - 1 \\ -x_1 - x_2 + x_3 \qquad &\geq 1 - 2 \end{aligned}$$

which have the resolvent $-x_2 + x_3 + x_4 \geq 1 - 1$. So, an inequality set containing the inequalities (6.39) has the resolvent $-x_2 + x_3 + x_4 \geq 1 - 1$.

An inequality $cx \geq \delta + n(c)$ is a *diagonal sum* of the system

$$c^i x \geq \delta - 1 + n(c^i), \quad i \in J \qquad (6.40)$$

where $J = \{j \mid c_j \neq 0\}$ and

$$c_j^i = \begin{cases} c_j - 1 \text{ if } c_j > 0 \\ c_j + 1 \text{ if } c_j < 0 \\ 0 \qquad \text{ if } c_j = 0 \end{cases}$$

Furthermore, $cx \geq \delta + n(c)$ is a diagonal sum of a 0-1 inequality set $\mathcal{S}$ if it is a diagonal sum of a system (6.40) in which each inequality is implied by some inequality in $\mathcal{S}$.

For example, inequality (d) below is a diagonal sum of the first three inequalities.

$$\begin{array}{ll} 2x_1 - 2x_2 + 4x_3 \geq 2 - 2 & (a) \\ 3x_1 - x_2 + 4x_3 \geq 2 - 1 & (b) \\ 3x_1 - 2x_2 + 3x_3 \geq 2 - 2 & (c) \\ 3x_1 - 2x_2 + 4x_3 \geq 3 - 2 & (d) \end{array} \qquad (6.41)$$

Note that the diagonal coefficients of (a)–(c) are reduced by one in absolute value, and that diagonal summation raises the degree by one. Inequality (d) is also a diagonal sum of the system

$$\begin{array}{l} 4x_3 + 2x_4 \geq 4 \\ 3x_1 - 3x_2 + 3x_3 + x_4 \geq 4 - 3 \end{array} \qquad (6.42)$$

because (a) and (b) of (6.41) are implied by the first inequality of (6.42), and (c) is implied by the second.

Each iteration of the 0-1 resolution method consists of the following. If some resolvent R of $\mathcal{S}$ is implied by no inequality in $\mathcal{S}$, R is added to $\mathcal{S}$. Then, if some diagonal sum D of $\mathcal{S}$ is implied by no inequality in $\mathcal{S}$, D is added to $\mathcal{S}$. The iterations continue until no such resolvent and no such diagonal sum can be generated.

It is first shown below that 0-1 resolution is complete with respect to clausal inequalities. This will lead to a proof that it is complete with respect to all 0-1 linear inequalities.

Theorem 6.13. *The 0-1 resolution method is complete with respect to clausal inequalities.*

Proof. Given a 0-1 system $Ax \geq b$, let $\mathcal{S}$ be the set of clausal inequalities implied by some inequality in $Ax \geq b$. Then $\mathcal{S}$ is equivalent to $Ax \geq b$. 0-1 resolution generates all clausal inequalities that are generated by applying classical resolution to $\mathcal{S}$, or inequalities that imply them. Thus, by Theorem 6.2, 0-1 resolution is complete with respect to clausal inequalities. $\square$

Theorem 6.14. *The* 0-1 *resolution method is complete with respect to* 0-1 *linear inequalities.*

Proof. Let S' be the result of applying 0-1 resolution to a set S of 0-1 inequalities. It suffices to show that any 0-1 inequality $cx \geq \delta + n(c)$ implied by S is implied by an inequality in S'. The proof is by induction on the degree δ.

First, suppose that $\delta = 1$. Note that any inequality $cx \geq 1 + n(c)$ is equivalent to the clausal inequality $c'x \geq 1 + n(c')$, where $c'_j = 1$ if $c_j > 0$, -1 if $c_j < 0$, and 0 if $c_j = 0$. Now Theorem 6.13 implies that $c'x \geq 1 + n(c')$, and therefore $cx \geq 1 + n(c)$, is implied by an inequality in S'.

Assuming now that the theorem is true for all inequalities of degree $\delta - 1$, it can be shown that it is true for all inequalities of degree δ. Suppose otherwise, and let $cx \geq \delta + n(c)$ be an inequality that is implied by S but by no inequality in S'. But $cx \geq \delta + n(c)$ is a diagonal sum of the system (6.40). It suffices to show that each inequality in (6.40) is implied by an inequality in S', since in this case S' contains an inequality that implies $cx \geq \delta + n(c)$, contrary to hypothesis. To show this, note first that $c^i x \geq \delta - 1 + n(c^i)$ is a reduction of $cx \geq \delta + n(c)$ and is therefore implied by S. Because $c^i x \geq \delta - 1 + n(c^i)$ has degree $\delta - 1$, the induction hypothesis ensures that it is implied by some inequality in S', as claimed. $\square$

A slightly stronger statement can be made. The above theorem states that 0-1 resolution generates an inequality I that implies any given implication $cx \geq \delta + n(c)$ of S. An extra step of diagonal summation generates $cx \geq \delta + n(c)$ itself. Since each inequality in (6.40) is implied by $cx \geq \delta + n(c)$, and therefore by I, $cx \geq \delta + n(c)$ is another diagonal sum that can be obtained.

6.5.5 Domain Consistency and k-Completeness

Section 6.4.4 pointed out that k-completeness can be achieved for any constraint set with Boolean variables by applying resolution to the logical clauses implied by the constraints. This idea can be applied to 0-1 linear inequalities.

Given a 0-1 system $Ax \geq b$, the algorithm of Fig. 6.2 can be used to generate all clausal inequalities implied by each inequality in $Ax \geq b$.

If the implied clauses are collected in a set $\mathcal{S}$, $Ax \geq b$ is equivalent to $\mathcal{S}$. One can now achieve k-completeness for $\mathcal{S}$ by computing the projection of $\mathcal{S}$ onto every subset of k variables. Using Theorem 6.4, this is accomplished by applying to $\mathcal{S}$ a restricted resolution algorithm that resolves only on $x_{k+1}, \ldots, x_n$.

If one wishes to achieve full k-completeness rather than simply project $Ax \geq b$ onto a given set of k variables, it is necessary to resolve on all variables. This, of course, achieves k-completeness for all k. In particular, it achieves domain consistency ($k = 1$).

Consider, for example, the 0-1 system

$$2x_1 - x_2 + 3x_3 - 2x_4 \geq 4 - 3 \quad (a)$$
$$-x_1 + 2x_2 - x_3 + 3x_4 \geq 3 - 4 \quad (b) \qquad (6.43)$$
$$3x_1 + 2x_2 - 2x_3 + x_4 \geq 4 - 2 \quad (c)$$

The feasible solutions are $(1,0,1,1)$, $(1,1,0,0)$, and $(1,1,1,1)$.

Each of the three inequalities (a), (b), and (c) implies the clauses listed below:

$$\begin{aligned}
(a): \quad x_1 - x_2 \qquad\quad - x_4 &\geq 1 - 2 \\
x_1 \qquad + x_3 \qquad &\geq 1 \\
x_3 - x_4 &\geq 1 - 1 \\
(b): -x_1 \qquad - x_3 + x_4 &\geq 1 - 2 \\
x_2 \qquad\quad + x_4 &\geq 1 \\
(c): \quad x_1 + x_2 \qquad\qquad &\geq 1 \\
x_1 \qquad - x_3 \qquad &\geq 1 - 1 \\
x_2 - x_3 + x_4 &\geq 1 - 1
\end{aligned} \qquad (6.44)$$

To project (6.43) onto x_1, x_2, it is enough to resolve only on x_3, x_4, which yields the following after deleting clauses that are absorbed by others:

$$\begin{aligned}
x_1 \qquad\qquad\qquad &\geq 1 \\
x_2 + x_3 \qquad &\geq 1 \\
x_2 \qquad + x_4 &\geq 1 \\
x_3 - x_4 &\geq 1 - 1 \\
-x_3 + x_4 &\geq 1 - 1
\end{aligned} \qquad (6.45)$$

The only clause that contains variables in $\{x_1, x_2\}$ is the unit clause x_1, which describes the projection $\{(1,0), (1,1)\}$ onto x_1, x_2. To achieve domain consistency or k-completeness for $k > 1$, it is necessary to resolve on all the variables. As it happens, this obtains no further resolvents from (6.45). Domain consistency is therefore achieved by fixing $x_1 = 1$, and k-completeness is achieved for all k by adding (6.45) to the constraint set (6.43). One can solve (6.43) without backtracking by solving (6.45) with 1-step lookahead.

6.5.6 Strong k-Consistency

Strong k-consistency for a 0-1 system $Ax \geq b$ can be achieved in a manner analogous to k-completeness. First, use the algorithm of Fig. 6.2 to create the set S of clausal inequalities that are implied by inequalities in S. Then apply the k-resolution algorithm to S by generating resolvents with fewer than k literals. Using Theorem 6.5, $Ax \geq b$ becomes strongly k-consistent after augmenting it with the resulting clauses in S. Note that it is necessary to include in S only clausal inequalities with k or fewer terms, which can accelerate the process considerably.

For example, the 0-1 inequality set (6.43) can be made 2-consistent by first generating implied clauses with two or fewer terms:

$$\begin{aligned}
x_1 \quad\quad + x_3 \quad\quad &\geq 1 \\
x_3 - x_4 &\geq 1 - 1 \\
x_2 \quad\quad + x_4 &\geq 1 \\
x_1 + x_2 \quad\quad\quad &\geq 1 \\
x_1 \quad\quad - x_3 \quad\quad &\geq 1 - 1
\end{aligned}$$

Application of 2-resolution yields the following after dropping redundant clauses:

$$\begin{aligned}
x_1 \quad\quad\quad\quad\quad &\geq 1 \\
x_2 + x_3 \quad\quad &\geq 1 \\
x_2 \quad\quad + x_4 &\geq 1 \\
x_3 - x_4 &\geq 1 - 1
\end{aligned} \tag{6.46}$$

The original constraint set (6.43) therefore becomes strongly 2-consistent after augmenting it with the clauses in (6.46).

Obtaining strong k-consistency can be practical if k is small or if the individual inequalities in $Ax \geq b$ contain only a few variables and therefore do not imply a large number of nonredundant clauses.

Exercises

6.28. Show that one clause absorbs another if and only the same is true of the corresponding clausal inequalities.

6.29. Show by counterexamples that neither absorption nor reduction is a necessary condition for implication between 0-1 linear inequalities.

6.30. Let a *roof point* for a 0-1 inequality $ax \geq a_0$ (with $a \geq 0$) be a minimal satisfier of $ax \geq a_0$; that is, a point $\bar{x}$ that satisfies $ax \geq a_0$ but fails to do so when any one $\bar{x}_j$ is flipped from 1 to 0. For any roof point $\bar{x}$, $\{j \mid \bar{x}_j = 1\}$ is a *roof set*. A *satisfaction set* for $bx \geq b_0$ is an index set J such that $bx \geq b_0$ whenever $x_j = 1$ for all $j \in J$. Show that the 0-1 inequality $ax \geq a_0$ (with $a \geq 0$) implies another 0-1 inequality $bx \geq b_0$ if and only if all roof sets of the former are satisfaction sets of the latter. Use this fact to prove Theorems 6.8 and 6.9.

6.31. Use the algorithm of Fig. 6.2 to derive the four nonredundant clauses implied by $10x_1 + 8x_2 + 4x_4 + 3x_5 + 2x_6 \geq 12$.

6.32. Show that the algorithm of Fig. 6.2 generates all nonredundant clauses implied by a 0-1 inequality $ax \geq a_0$ with $a_1 \geq \cdots \geq a_n \geq 0$.

6.33. What is the smallest a_0 for which $5x_1 + 4x_2 + 3x_3 + 3x_4 + 3x_5 \geq a_0$ implies the cardinality clause $x_1 + x_2 + x_3 + x_4 - x_5 \geq 3 - 1$?

6.34. Prove Theorem 6.11 by considering two cases: $b_0 \leq k$ and $b_0 > k$.

6.35. Show that the cardinality clause $x_1 + x_2 - x_3 - x_4 + x_5 \geq 4 - 2$ implies $-x_1 + x_2 - x_3 + x_4 + x_5 \geq 2 - 2$ by exhibiting a cardinality clause to which the former reduces and that absorbs the latter. Without looking at the proof of Theorem 6.12, indicate how such an intermediate clause can be identified in general.

6.36. Show that the resolvent of two clausal inequalities C_1, C_2 is the result of taking a nonnegative linear combination of C_1, C_2 and bounds $0 \leq x_j \leq 1$ and rounding up any fractions that result. Thus, clausal resolution is a special case of a Chvátal–Gomory cut. *Hint:* Assign multiplier $\frac{1}{2}$ to C_1 and C_2.

6.37. Show that the diagonal sum $cx \geq \delta + n(c)$ of a system (6.40) is the result of taking a nonnegative linear combination of the inequalities of (6.40) and rounding up any fractions that result. Thus, diagonal summation is a special case of a Chvátal–Gomory cut. *Hint:* Assign each inequality $c^i \geq \delta - 1 + n(c^i)$ the multiplier $|c_i|/(\sum_j |c_j| - 1)$.

6.38. The proof of Theorem 6.14 states that "$cx \geq \delta + n(c)$ is a diagonal sum of the system (6.40)." For what set J in (6.40) is this true?

6.39. Verify by resolution that (6.45) is the set of prime implications of (6.44). Solve (6.45) by 1-step lookahead.

6.40. Show that 0-1 resolution terminates after finitely many iterations.

6.6 Integer Linear Inequalities

Inference methods for integer programming problems provide the basis for both logic-based Benders methods and constraint-directed search. The subadditive and branching duals can generate Benders cuts for integer programming. The LP dual of the continuous relaxation can deliver enumerative nogoods for constraint-directed branching. The latter technique is now being incorporated in state-of-the-art mixed-integer programming solvers.

6.6.1 Benders Cuts from the Subadditive Dual

Because the subadditive and branching duals are inference duals, both can provide the basis for constraint-directed search. In particular, they provide a Benders method when applied to problems of the form

$$\min \ f(x) + cy$$
$$g(x) + Ay \geq b \tag{6.47}$$
$$x \in D_x, \ y \geq 0 \text{ and integral}$$

The problem becomes an integer programming subproblem when the variables x are fixed, and solution of its subadditive or branching dual can yield Benders cuts.

An important special case arises when the entire problem (6.47) is an integer programming problem but simplifies when x is fixed, perhaps by separating into a number of smaller problems. This occurs in stochastic integer programming, for instance, in which there are many scenarios or possible outcomes and a set of constraints associated with each. When certain variables are fixed, the problem separates into smaller problems corresponding to the scenarios. Benders decomposition is an attractive alternative for such problems.

Benders cuts based on the subadditive dual are straightforward. If $\bar{x}$ is the solution of the previous master problem, the subproblem is

$$\min \; f(\bar{x}) + cy$$
$$Ay \geq b - g(\bar{x}) \tag{6.48}$$
$$y \geq 0 \text{ and integral}$$

The subadditive dual of (6.48) is

$$\max_{h \in H} \{ h(b - g(\bar{x})) \mid h(A) \leq c \}$$

where H is the set of subadditive, homogeneous, nondecreasing functions. If a Chvátal function h solves the dual, then

$$f(\bar{x}) + h(b - g(\bar{x}))$$

is the optimal value of the subproblem. Since h remains a feasible dual solution for any x, by weak duality $f(x) + h(b - g(x))$ is a lower bound on the optimal value of (6.47) for any x. This yields the Benders cut

$$z \geq f(x) + h(b - g(x))$$

Such cuts have the practical shortcoming, however, that the dual is hard to solve and the Chvátal function that solves it is typically very complicated. It could be difficult to solve a master problem that contains them.

6.6.2 Benders Cuts from the Branching Dual

A more promising Benders approach is based on the branching dual. In this case, the inference dual of the subproblem (6.48) is solved by the branch-and-bound tree that solves (6.48) itself. The issue for generating a Benders cut is how one can bound the optimal value of (6.47) when $\bar{x}$ is replaced by some other value of x. Fortunately, this change only affects the right-hand sides, which simplifies the analysis.

Suppose, as in Section 4.7.3, that (6.48) is a 0-1 programming problem, and let z^* be its optimal value. The branching tree that solves (6.48) continues to prove a lower bound on the optimal value, provided the infeasible nodes remain infeasible. Using the results of Section 4.7.3, this occurs when the variable x satisfies

$$u^i(b - g(x)) > B_i \tag{6.49}$$

where

$$B_i = \sum_{j \in J_i} u^i A_j t_{ij} + \sum_{j \notin J_i} \max\{0, u^i A_j\}$$

at every infeasible leaf node i. Here, u^i is the vector of dual multipliers for $Ax \geq b$ in the solution of the relaxation at node i, J_i is the set of indices j for which x_j is fixed by branching down to node i, and t_{ij} is the value to which x_j is fixed.

As long as the infeasible nodes remain infeasible, the optimal value is bounded below by the minimum of the relaxation values at the remaining leaf nodes. Let v_i^* be the current optimal value of the relaxation at node i. When x is changed to some value other than $\bar{x}$, weak duality implies that the quantity

$$v_i^* + u^i \Delta b = v_i^* - u^i (g(x) - g(\bar{x}))$$

remains a valid lower bound on the optimal value at node i. This provides a Benders cut

$$z \geq \begin{cases} \min_{i \in L_2} \{v_i^* - u^i (g(x) - g(\bar{x}))\}, & \text{if } u^i(b - g(x)) > B_i, \text{ all } i \in L_1 \\ -\infty, & \text{otherwise} \end{cases}$$

$$(6.50)$$

where L_1 is the set of infeasible leaf nodes and L_2 the set of the remaining leaf nodes.

If the original problem (6.47) is a linear 0-1 programming problem, one can linearize the cut (6.50). Suppose that (6.47) minimizes $dx + cy$ subject to $Dx + Ay \geq b$, where each $x_j, y_j \in \{0, 1\}$. The cut (6.50) can be written

$$z \geq z_{\min} - M \left(|L_1| - \sum_{i \in L_1} \delta_i \right)$$

$$z_{\min} \geq v_i^* - u^i D(x - \bar{x}) - M(1 - \delta_i), \quad i \in L_2 \qquad (6.51)$$

$$\sum_{i \in L_2} \delta_i = 1$$

$$u^i(b - Dx) \geq B_i + \epsilon - M(1 - \delta_i), \quad i \in L_1$$

where M is a large number and each δ_i is a 0-1 variable. Variable δ_i for $i \in L_1$ takes the value 1 when infeasible leaf node i remains infeasible. Variable $z_{\min}$ represents the minimum of $v_i^* - u^i D(x - \bar{x})$ over all $i \in L_2$.

Suppose, for example, a Benders method is applied to the problem

$$\min 3x_1 + 4x_2 + 5y_1 + 6y_2 + 7y_3$$
$$x_1 + x_2 + 4y_1 + 3y_2 - y_3 \geq 2$$
$$x_1 - x_2 - y_1 + y_2 + 4y_3 \geq 3 \qquad (6.52)$$
$$x_j, y_j \in \{0, 1\}$$

in which x_1, x_2 are the master problem variables. Here

$$g(x) = Dx = \begin{bmatrix} 1 & 1 \\ 1 & -1 \end{bmatrix} \begin{bmatrix} x_1 \\ x_2 \end{bmatrix}$$

The initial master problem has no constraints, and one can arbitrarily choose $\bar{x} = (0,0)$ as the solution. The resulting subproblem (6.48) is the problem (4.47) studied in Section 4.7, with variables y_j rather than x_j. Recall that nodes 2 and 5 of the branching tree are infeasible, while nodes 3 and 6 are feasible (Fig. 4.5). One can check that $(B_2, B_5) = (0,1)$. The Benders cut (6.50) is

$$z \geq \begin{cases} \min \begin{cases} 12 - \frac{7}{4}(x_1 - x_2), \\ 13 - \frac{5}{4}(x_1 + x_2) \end{cases}, & \text{if } \begin{cases} 11 - 5x_1 - 3x_2 > 0 \\ 3 - x_1 + x_2 > 1 \end{cases} \\ -\infty, & \text{otherwise} \end{cases}$$

In this case, the conditions $11 - 5x_1 - 3x_2 > 0$ and $3 - x_1 + x_2 > 1$ are satisfied by all $x_1, x_2 \in \{0, 1\}$, which means that the infeasible nodes (nodes 2 and 5) remain infeasible for all x. The variables δ_2, δ_5 can therefore be dropped from the linearization (6.51), which simplifies to

$$z \geq z_{\min}$$
$$z_{\min} \geq 12 - \frac{7}{4}(x_1 - x_2) - M(1 - \delta_3)$$
$$z_{\min} \geq 13 - \frac{5}{4}(x_1 + x_2) - M(1 - \delta_6)$$
$$\delta_3 + \delta_6 = 1$$

Solution of the master problem yields $z = 10\frac{1}{4}$ and $(\bar{x}_1, \bar{x}_2) = (0, 1)$, which defines the next subproblem.

In practice, a master problem with linearized cuts (6.51) is likely to be too hard to solve, unless it happens to simplify. Each cut introduces a distinct 0-1 variable δ_i for each leaf node of the subproblem branching tree. However, simpler cuts can be derived from the information in (6.50). One simple cut states that the optimal value is bounded by v^* (the optimal value of the subproblem) if $x_j = \bar{x}_j$ for all variables x_j that are fixed at some leaf node. So if J is the union of J_i over all leaf nodes i, the cut is

$$z \geq v^* - M \left(\sum_{\substack{j \in J \\ \bar{x}_j = 0}} x_j + \sum_{\substack{j \in J \\ \bar{x}_j = 1}} (1 - x_j) \right) \tag{6.53}$$

One can also derive cuts heuristically. For example, one might flip the $\bar{x}_j$'s one at a time and determine that the infeasible nodes remain infeasible after certain flips, perhaps after variables x_j for $j \in \bar{J} \subset J$ are flipped. That is, if one defines $\bar{x}^j$ by $\bar{x}_j^j = 1 - \bar{x}_j$ and $\bar{x}_k^j = \bar{x}_j$ for $k \neq j$, one might find that $u^i(b - D\bar{x}^j) > B_i$ for all $i \in L_1$ when $j \in \bar{J}$. Flipping variable x_j yields a lower bound $v^*(j)$ equal to the minimum of the bounds obtained at all the feasible and fathomed nodes:

$$v^*(j) = \min_{i \in L_2} \left\{ v_i^* - u^i D(\bar{x}^j - \bar{x}) \right\}$$

$$= \min \left\{ \min_{\substack{i \in L_2 \\ \bar{x}_j = 0}} \left\{ v_i^* - u^i D_j \right\}, \min_{\substack{i \in L_2 \\ \bar{x}_j = 1}} \left\{ v_i^* + u^i D_j \right\} \right\}$$

where D_j is column j of D. This gives rise to a cut that introduces only one 0-1 variable δ:

$$z \geq \min_{j \in \bar{J}} \left\{ v^*(j) \right\} - M \left(\delta + \sum_{\substack{j \in J \setminus \bar{J} \\ \bar{x}_j = 0}} x_j + \sum_{\substack{j \in J \setminus \bar{J} \\ \bar{x}_j = 1}} (1 - x_j) \right) \tag{6.54}$$

$$\delta \geq \sum_{\substack{j \in \bar{J} \\ \bar{x}_j = 0}} x_j + \sum_{\substack{j \in \bar{J} \\ \bar{x}_j = 1}} (1 - x_j) - 1, \quad \delta \geq 0$$

In the example, the simple nogood cut (6.53) is

$$z \geq 12 - M(x_1 + x_2)$$

It is easily checked that the infeasible nodes remain infeasible when x_1 is flipped from 0 to 1, as well as when x_2 is flipped from 0 to 1. So

$$v^*(1) = \min \left\{ 12 - \tfrac{7}{4} \cdot 1, \ 13 - \tfrac{5}{4} \cdot 1 \right\} = 10\tfrac{1}{4}$$
$$v^*(2) = \min \left\{ 12 + \tfrac{7}{4} \cdot 1, \ 13 - \tfrac{5}{4} \cdot 1 \right\} = 11\tfrac{3}{4}$$

The resulting cut (6.54) is

$$z \geq 10\tfrac{1}{4} - M\delta, \quad \delta \geq x_1 + x_2 - 1, \quad \delta \geq 0$$

One might also observe that the infeasible nodes remain infeasible when certain variables, perhaps the variables x_j for $j \in \hat{J} \subset J$, are flipped individually or in combination, and no other variables are flipped. Let $X(\hat{J})$ be the set of all values of x that result from these flips. The lower bound is

$$v^*(\hat{J}) = \min_{x \in X(\hat{J})} \left\{ \min_{i \in L_2} \{ v_i^* - u^i D(x - \bar{x}) \} \right\}$$

and the linearized cut is

$$z \geq v^*(\hat{J}) - M \left(\sum_{\substack{j \in J \setminus \hat{J} \\ \bar{x}_j = 0}} x_j + \sum_{\substack{j \notin J \setminus \hat{J} \\ \bar{x}_j = 1}} (1 - x_j) \right) \tag{6.55}$$

In the example, one can set $\hat{J} = \{1, 2\}$ and note that $v^*(\hat{J}) = 10\tfrac{1}{4}$. The resulting cut (6.55) is simply $z \geq 10\tfrac{1}{4}$.

6.6.3 Constraint-Directed Branching

Constraint-directed branching derives enumerative nogoods from a dual solution of the relaxation at leaf nodes (Section 5.2.3). This idea is readily applied to integer linear programming, because the relaxation is an LP whose dual solution is easily obtained. The same idea can be extended to mixed-integer/linear programming and specialized to 0-1 programming.

Given an integer programming problem $\min\{cx \mid Ax \geq b, \ x \geq 0, \ x \in \mathbb{Z}^n\}$, the continuous relaxation problem solved at leaf node i of a search tree has the form.

$$
\begin{aligned}
\min \ & cx \\
& Ax \geq b \quad (u^i) \\
& x \geq p \quad (\sigma^i) \\
& -x \geq -q \quad (\tau^i)
\end{aligned}
\tag{6.56}
$$

Here, p and q are lower and upper bounds imposed by branching or appearing in the original problem, where $p \geq 0$. If $p_j = q_j$, then variable x_j has been fixed and is effectively eliminated from the subproblem.

Dual variables u^i, σ^i, τ^i are associated with the constraints in (6.56) as shown. If the (6.56) is infeasible, then the solution of the dual is a ray (u^i, σ^i, τ^i) that proves infeasibility because $u^i A + \sigma^i - \tau^i \leq 0$ and

$$u^i b + \sigma^i p - \tau^i q > 0 \qquad (6.57)$$

Given a set of branching bounds p, q, the same dual solution (u^i, σ^i, τ^i) proves infeasibility of the resulting LP relaxation if the bounds satisfy (6.57). So a given set of bounds p, q are consistent with feasibility only if

$$u^i b + \sigma^i p - \tau^i q \leq 0 \qquad (6.58)$$

This fact can guide future branching. Suppose p, q are the branching bounds at a subsequent node of the search tree, and one wishes to branch on a variable x_j with fractional value $\hat{x}_j$ in the relaxation at this node. A left branch defined by the bound $x_j \leq \lfloor \hat{x}_j \rfloor$ is feasible only if setting $q_j = \lfloor \hat{x}_j \rfloor$ satisfies (6.58), which is to say

$$\lfloor \hat{x}_j \rfloor \geq Q_{ij} = \sigma^i p + u^i b - \sum_{k \neq j} \tau_k^i q_k$$

Thus, the left branch can be eliminated if $\lfloor \hat{x}_j \rfloor < Q_{ij}$. A right branch defined by $\lceil \hat{x}_j \rceil \leq x_j$ is feasible only if

$$\lceil \hat{x}_j \rceil \leq P_{ij} = \tau^i q - u^i b - \sum_{k \neq j} \sigma_k^i p_k$$

and the branch can be eliminated if $\lceil \hat{x}_j \rceil > P_{ij}$. These tests can be applied at the current node for each infeasible leaf node i that has been encountered so far.

One can also generate enumerative nogood constraints in terms of x. If x satisfies the inequality $ub + \sigma^i x - \tau^i x > 0$, then the bounds p, q satisfy (6.58) when one sets $p = q = x$. Thus, an LP relaxation with bounds p, q is infeasible. But because x lies within these bounds, x can be feasible only if it satisfies the nogood constraint

$$u^i b + (\sigma^i - \tau^i) x \leq 0 \qquad (6.59)$$

Note that σ_j and τ_j will not both be positive in a basic solution, because the corresponding columns of the dual are linearly dependent.

The nogood constraint (6.59) is particularly useful in 0-1 programming, because it implies logical clauses as described in Section 6.5.2.

These clauses can then be used as nogood constraints to direct the search, much as conflict clauses are used in propositional satisfiability algorithms (Section 5.2.4).

Consider, for example, the 0-1 programming problem

$$\begin{aligned}
\min\ & 10x_1 + 2x_2 + 9x_3 + 4x_4 + 5x_5 \\
& 5x_1 - x_2 + 5x_3 + x_4 - 2x_5 \geq 4 \\
& 6x_1 - x_2 + 4x_3 - x_4 + 2x_5 \geq 5 \\
& x_j \in \{0, 1\},\ i = 1, \ldots, 5
\end{aligned} \tag{6.60}$$

If the variables (x_1, x_2, x_3) are fixed to $(0, 1, 0)$ at a leaf node i, the LP relaxation of (6.60) is infeasible. An extreme ray solution of the dual is $u^i = (1, 1)$, $\sigma^i = (0, 2, 0)$, $\tau^i = (11, 0, 9)$. The nogood constraint (6.59) becomes $11x_1 - 2x_2 + 9x_3 \geq 9$. Using the algorithm of Section 6.5.2, the constraint is found to imply the clauses $x_1 \vee \neg x_2$ and $x_1 \vee x_3$. All subsequent branching must satisfy these clauses to obtain a feasible solution.

A similar principle can be applied to general integer programming by introducing 0-1 variables that indicate which branching bounds are enforced. Suppose that the bounds on x in the original problem are $0 \leq x \leq d$. Then (6.58) can be written

$$u^i b - \tau^i d + \sum_j \sigma^i_j p_j + \sum_j \tau^i_j (d_j - q_j) \leq 0 \tag{6.61}$$

Let the 0-1 variable $\delta_{x_j \geq p_j}$ be 1 when $x_j \geq p_j$ is enforced, and similarly for $\delta_{x_j \leq q_j}$. Then for any given set of bounds p, q, the constraint (6.61) implies

$$u^i b - \tau^i d + \sum_j \sigma^i_j p_j \delta_{x_j \geq p_j} + \sum_j \tau^i_j (d_j - q_j) \delta_{x_j \leq q_j} \leq 0 \tag{6.62}$$

This inequality implies logical clauses that can be used to prune the search tree.

Suppose, for example, that the original problem contains bounds $0 \leq x_j \leq 3$ for $j = 1, \ldots, 4$, and (6.58) at a given infeasible leaf node is

$$5p_1 - 2q_2 + 3p_3 \leq 5$$

The bounds p, q can be set to any integer in the interval $[0, 3]$, but a natural choice is to set them to their values at the current leaf node, perhaps $p = (1, 0, 2, 1)$ and $q = (2, 1, 3, 3)$. Then (6.62) becomes

$$5\delta_{x_j \geq 1} + 4\delta_{x_2 \leq 1} + 6\delta_{x_3 \geq 2} \leq 11$$

which implies a single clause,

$$\neg\delta_{x_j \geq 1} \vee \neg\delta_{x_2 \leq 1} \vee \neg\delta_{x_3 \geq 2}$$

The relaxation at a subsequent node can be feasible only if $x_j \geq 1$, $x_2 \leq 1$, or $x_3 \geq 2$ is not enforced by branching bounds. That is, the branching bounds must satisfy $p_1 = 0$, $q_2 \geq 2$, or $p_3 \leq 1$.

Enumerative nogoods can be learned at fathomed nodes as well as infeasible nodes. At a fathomed node i, the optimal value of the relaxation (6.56) equals or exceeds the current upper bound U obtained from the incumbent solution. Thus, $u^i b + \sigma^i p - \tau^i q \geq U$. This means that a set of bounds p, q can result in an optimal solution only if they satisfy

$$u^i b + \sigma^i p - \tau^i q < U \tag{6.63}$$

This implies that a left branch $x_j \leq \lfloor \hat{x}_j \rfloor$ can lead to an optimal solution only if $\lfloor \hat{x}_j \rfloor > Q_{ij} - U$, and a right branch $\lceil \hat{x}_j \rceil \leq x_j$ only if $\lceil \hat{x}_j \rceil < P_{ij} + U$, where P_{ij}, Q_{ij} are defined as above.

Returning to the example (6.60), suppose that variables (x_1, x_4, x_5) are fixed to $(0, 1, 1)$ at a leaf node, and the incumbent solution has value $U = 10$. The relaxation (6.56) is feasible, with optimal value 18 and dual solution $u = (1.8, 0)$, $v = (0, 0, 0, 2.2, 8.6)$, $w = (0, 0, 0, 0, 0)$. The nogood constraint (6.63) becomes $2.2x_4 + 8.6x_5 < 2.8$. This implies the single clause $\neg x_5$, which means that subsequent branching must set $x_5 = 0$ to obtain an optimal solution.

Exercises

6.41. Suppose that the problem

$$\min x_1 + x_2 + 3y_1 + 4y_2$$
$$2x_1 + x_2 - y_1 + 3y_2 \geq 0$$
$$x_1 + 2x_2 + 2y_1 + y_2 \geq 5$$
$$x_1, x_2, y_1, y_2 \geq 0 \text{ and integral}$$

is to be solved by Benders decomposition with x_1, x_2 in the master problem. Suppose that the initial solution of the master problem is $(x_1, x_2) = (0, 0)$, so that the initial subproblem is (4.38) but with variables y_1, y_2. Write a Benders cut based on the subadditive dual.

6.42. Complete the solution of (6.52) by Benders decomposition based on the branching dual.

6.7 The Element Constraint

The element constraint implements variable indexing, which is a central modeling feature of constraint programming and integrated problem solving. A variable index is an index or subscript whose value depends on one or more variables. The existence of effective filters and good relaxations for the element constraint contributes substantially to the efficiency of integrated algorithms.

The simplest form of the element constraint is

$$\text{element}(y, z \mid (a_1, \ldots, a_m)) \tag{6.64}$$

where y is a discrete variable and each a_i is a constant. The constraint says that z must take the yth value in the list $a_1, \ldots, a_m$. Expressions of the form a_y in an integrated model can be implemented by replacing each occurrence of a_y with z and adding (6.64) to the constraint set. One can then apply filtering algorithms and generate relaxations for (6.64).

Another form of the element constraint is

$$\text{element}(y, (x_1, \ldots, x_m), z) \tag{6.65}$$

where each x_i is a variable. It sets z equal to the yth variable in the list $x_1, \ldots, x_m$. This constraint implements expressions of the form x_y, which are common in channeling constraints that connect two formulations of the same problem. This was illustrated in Section 2.5.

Element constraints can also be multidimensional. For example,

$$\text{element}((y_1, y_2), z \mid A) \tag{6.66}$$

selects the element in row y_1 and column y_2 of matrix A and assigns that value to z. For purposes of domain filtering, a constraint of this sort can be treated as a single-dimensional constraint. Thus, if A is $m \times n$, (6.66) can be converted to $\text{element}(y, z \mid a)$ and filtered in that form, where $y = m(y_1 - 1) + y_2$ and $a = (A_1, \ldots, A_n)$. Here A_i is a tuple of the elements in row i of A.

A constraint like (6.66) implements doubly-indexed expressions, as in the traveling salesman problem. If A_{ij} is the cost of traveling from point i to point j, the salesman's objective is to minimize $\sum_{ij} A_{y_i y_{i+1}}$ subject to the constraint that $y_1, \ldots, y_n$ take different values (and where y_{n+1} is identified with y_1).

It is sometimes advantageous to analyze a specially structured element constraint. For example, variably indexed coefficients in expressions of the form $a_y x$ are quite common and are used, for example, in Sections 2.7 and 2.8. They are implemented by the *indexed linear element* constraint

$$\text{element}(y, x, z \mid (a_1, \ldots, a_m)) \qquad (6.67)$$

where x is now a single variable. The constraint sets z equal to the yth term in the list $a_1 x, \ldots, a_m x$. It is possible to exploit the structure of (6.67) when designing filters and relaxations. This is more efficient than implementing x_y with (6.65) and then adding the constraints $x_i = a_i x$ for $i = 1, \ldots, m$. There is also a vector-valued version

$$\text{element}(y, x, z \mid (A_1, \ldots, A_m))$$

in which z and each A_k are tuples of the same length.

6.7.1 Domain Consistency

Filtering algorithms for element constraints can achieve full domain consistency with a modest amount of computation.

It is a trivial matter to filter domains for an element constraint (6.64) that contains a list of constants. Let D_z, D_y be the current domains of z and y, respectively, and let D'_z, D'_y be the new, reduced domains. It is assumed that D_y and D_z are finite. Then D_z can be restricted to a_i's whose indices are in D_y:

$$D'_z = D_z \cap \{a_i \mid i \in D_y\}. \qquad (6.68)$$

D_y can now be restricted to indices of the a_i's that remain in D'_z:

$$D'_y = \{i \in D_y \mid a_i \in D'_z\} \qquad (6.69)$$

This achieves domain consistency.

For example, consider the element constraint

$$\text{element}(y, z \mid (20, 30, 60, 60))$$

where initially the domains are

$$D_z = \{20, 40, 60, 80, 90\}, \quad D_y = \{1, 2, 4\}$$

Rules (6.68)–(6.69) yield the filtered domains

$$D'_z = \{20, 40, 60, 80, 90\} \cap \{20, 30, 60\} = \{20, 60\}$$
$$D'_y = \{1, 2, 4\} \cap \{1, 3, 4\} = \{1, 4\}$$

Achieving domain consistency for an element constraint with variables (6.65) presents a more interesting problem. D_y is assumed finite, but since $x_1, \ldots, x_m$ and z may be continuous variables, their domains may be infinite.

First, D_z must be a subset of the combined domains of the variables x_i whose indices are in D_y:

$$D'_z = D_z \cap \bigcup_{i \in D_y} D_{x_i} \tag{6.70}$$

D_y can now be restricted to indices i for which D'_z intersects D_{x_i}:

$$D'_y = \{i \in D_y \mid D'_z \cap D_{x_i} \neq \emptyset\} \tag{6.71}$$

Finally, D_{x_i} can be restricted if i is the only index in D'_y:

$$D'_{x_i} = \begin{cases} D'_z & \text{if } D'_y = \{i\} \\ D_{x_i} & \text{otherwise.} \end{cases} \tag{6.72}$$

This achieves domain consistency.

Consider, for example, the element constraint

$$\text{element}(y, (x_1, x_2, x_3), z)$$

where initially the domains are

$$D_z = [50, 80], \quad D_y = \{1, 3\}$$
$$D_{x_1} = [0, 20], \quad D_{x_2} = [10, 40], \quad D_{x_3} = [30, 60] \cup [70, 90]$$

Rules (6.70)–(6.72) imply that the reduced domains are

$$D'_z = [50, 80] \cap ([0, 20] \cup [30, 60] \cup [70, 90]) = [50, 60] \cup [70, 80]$$
$$D'_y = \{3\}, \quad D'_{x_1} = D_{x_1}, \quad D'_{x_2} = D_{x_2}, \quad D'_{x_3} = D'_z$$

Thus, y is fixed to 3, which means that $x_3 = z$. The common domain of x_3 and z is the intersection of their original domains.

The indexed linear element constraint (6.67) is processed as follows. D_z can be reduced to its intersection with the set of values that can be obtained by multiplying an a_i by a possible value of x:

$$D'_z = D_z \cap \bigcup_{i \in D_y} \{a_i v \mid v \in D_x\} \tag{6.73}$$

D'_y is the set of indices i for which D'_z contains $a_i v$ for some possible value v of x:

$$D'_y = \{i \in D_y \mid D'_z \cap \{a_i v \mid v \in D_x\} \neq \emptyset\} \tag{6.74}$$

Finally, D'_x can be reduced to values whose multiples by some a_i belong to D'_z:

$$D'_x = \bigcup_{i \in D'_y} \{v \in D_x \mid a_i v \in D'_z\} \tag{6.75}$$

The filtering procedure can be extended to the vector-valued element constraint.

Suppose, for example, that the variables in constraint

$$\text{element}(y, x, z \mid (1, 2, 3))$$

have initial domains

$$D_z = [36, 48], \quad D_y = \{1, 2, 3\}, \quad D_x = [10, 30]$$

Applying (6.73)–(6.75):

$$D'_z = [36, 48] \cap ([10, 30] \cup [20, 60] \cup [30, 90]) = [36, 48]$$
$$D'_y = \{2, 3\}$$
$$D'_x = \emptyset \cup [\tfrac{36}{2}, \tfrac{48}{2}] \cup [\tfrac{36}{3}, \tfrac{48}{3}] = [12, 16] \cup [18, 24]$$

6.7.2 Bounds Consistency

Bounds consistency for an element constraint (6.64) with constants is trivial to achieve. Let $I_z = [L_z, U_z]$ be the current interval domain of z. The updated domain is $[L'_z, U'_z]$, where

$$L'_z = \max\left\{L_z, \min_{i \in D_y}\{a_i\}\right\}, \quad U'_z = \min\left\{U_z, \max_{i \in D_y}\{a_i\}\right\}$$

D_y can also be filtered:

$$D'_y = \{i \in D_y \mid a_i \in [L'_z, U'_z]\}$$

To tighten bounds for an element constraint (6.65) with variables, let $[L_{x_i}, U_{x_i}]$ be the current interval domain of x_i. The domains of z, y can be updated by setting

$$L_z' = \max\left\{L_z, \min_{i \in D_y}\{L_{x_i}\}\right\}, \quad U_z' = \min\left\{U_z, \max_{i \in D_y}\{U_{x_i}\}\right\}$$

$$D_y' = \{i \in D_y \mid [L_{x_i}, U_{x_i}] \cap [L_z', U_z'] \neq \emptyset\}$$

If D_y' is a singleton $\{i\}$, the bounds on x_i can be updated:

$$L_{x_i}' = \max\left\{L_{x_i}, L_z'\right\}, \quad U_{x_i}' = \min\left\{U_{x_i}, U_z'\right\},$$

This procedure achieves bounds consistency if the variables (other than y) have interval domains, but otherwise need not do so. For example, if

$$D_z = [20, 30], \quad D_y = \{1\}, \quad D_{x_1} = [0, 10] \cup [40, 50]$$

then applying the procedure to element$(y, (x_1), z)$ reduces the bounds for x_1 to $[20, 30]$ and has no effect on the bounds for z. However, the constraint is infeasible, and bounds consistency is achieved only by reducing the domain of x_1 and z to an empty interval.

Bounds can be updated for element$(y, x, z \mid (a_1, \ldots, a_m))$ by setting

$$L_z' = \max\left\{L_z, \min_{i \in D_y}\left\{\min\{a_i U_x, a_i L_x\}\right\}\right\}$$

$$U_z' = \min\left\{U_z, \max_{i \in D_y}\left\{\min\{a_i L_x, a_i U_x\}\right\}\right\}$$

and

$$L_x' = \max\left\{L_x, \min_{i \in D_y}\left\{\min\left\{\frac{U_z'}{a_i}, \frac{L_z'}{a_i}\right\}\right\}\right\}$$

$$U_x' = \min\left\{U_x, \max_{i \in D_y}\left\{\min\left\{\frac{L_z'}{a_i}, \frac{U_z'}{a_i}\right\}\right\}\right\}$$

An index i can be deleted from D_y if the reduced interval for z does not overlap the reduced interval for $a_i x$:

$$D_y' = \left\{i \in D_y \mid [\min\{a_i L_x', a_i U_x'\}, \max\{a_i L_x', a_i U_x'\}] \cap [L_z', U_z'] \neq \emptyset\right\}$$

This procedure achieves bounds consistency if the variables other than y have interval domains. For example, if

$$D_z = [30, 40], \quad D_y = \{1, 2\}, \quad D_x = [-10, 10]$$

then the updated bounds for element$(y, x, z \mid (-3, 5))$ are $D_z' = [30, 40]$ and $D_x' = [-5, 5]$. Also, D_y is reduced to $\{2\}$.

Exercises

6.43. Show that formulas (6.68)–(6.69) achieve domain consistency for element$(y, z \,|\, a)$.

6.44. Show that formulas (6.70)–(6.72) achieve domain consistency for element(y, x, z).

6.45. Suppose one is given initial domains $y \in \{1, 2, 3\}$, $x \in \{-1\} \cup [1, 3]$, and $z \in [-2, -1] \cup [1, 2]$. Reduce domains to achieve domain consistency for element$(y, x, z \,|\, (-1, 0, 2))$. Also, reduce domains to achieve bounds consistency for this constraint.

6.46. Extend the filtering procedure (6.70)–(6.72) to the vector-valued indexed linear element constraint.

6.47. Interpret the following expression using the appropriate element constraints, where the variables are y, y_i, x_i, and $w_{i1}, \ldots, w_{im}$ for $i = 1, \ldots, n$:

$$\sum_{i=1}^{n} a_{iy_i} x_i = \sum_{i=1}^{n} b_i w_{iy}$$

6.8 The All-Different Constraint

The all-different (alldiff) constraint is one of the most frequently used global constraints in CP models. It arises in assignment and sequencing problems for which a CP approach is particularly well suited. There are efficient filtering algorithms for achieving domain consistency and bounds consistency. In fact, the matching algorithm commonly used to filter alldiff has inspired a number of similar filtering techniques for other global constraints. There are also continuous relaxations for alldiff, which will be discussed in Section 7.9.

The all-different constraint

$$\text{alldiff}(x_1, \ldots, x_n)$$

requires that the variables $x_1, \ldots, x_n$ all take distinct values, where the domain D_{x_i} of each x_i is finite.

A simple assignment problem involving alldiff might require that each of five workers be assigned one job, and each of six available jobs be assigned to at most one worker. Each worker has the necessary skills to do only certain jobs. The problem is to find a feasible solution for

$$\text{alldiff}(x_1, \ldots, x_5)$$

$$
\begin{aligned}
D_{x_1} &= \{1\} & D_{x_4} &= \{1, 5\} \\
D_{x_2} &= \{2, 3, 5\} & D_{x_5} &= \{1, 3, 4, 5, 6\} \\
D_{x_3} &= \{1, 2, 3, 5\}
\end{aligned}
\tag{6.76}
$$

Thus, worker 1 can do only job 1; worker 2 can do job 2, 3, or 5; and so forth. One can see right away that job 1 must be assigned to worker 1 and can therefore be removed from the domains of x_3, x_4, and x_5. In fact, the domains can be reduced considerably more than this.

6.8.1 Bipartite Matching Formulation

The alldiff constraint can be viewed as a matching problem on a bipartite graph (Section 3.2.5). This allows one to bring graph algorithms to bear to solve the constraint and achieve various kinds of consistency.

Given the constraint $\text{alldiff}(x_1, \ldots, x_n)$, one can construct a bipartite graph G with vertices on one side that correspond to variables $x_1, \ldots, x_n$, and vertices on the other side that correspond to elements in the variable domains. The graph contains edge (x_j, i) whenever $i \in D_{x_j}$. The alldiff constraint is feasible if and only if some matching covers the vertices $x_1, \ldots, x_n$.

The example (6.76) corresponds to the bipartite graph in Fig. 6.3. A matching that covers $x_1, \ldots, x_5$ is shown by heavy lines. It corresponds to a solution of the alldiff constraint, namely $(x_1, \ldots, x_5) = (1, 2, 3, 5, 4)$.

An easy way to check whether an alldiff constraint is feasible is to find a maximum cardinality matching on the associated graph G. The alldiff is feasible if and only if a maximum cardinality matching covers $x_1, \ldots, x_n$. As noted in Section 3.2.5, a matching on G has maximium cardinality if and only if there is no alternating path in G. A simple algorithm, stated in that section, finds a maximum cardinality matching by increasing the cardinality by means of alternating paths until no further alternating paths exist.

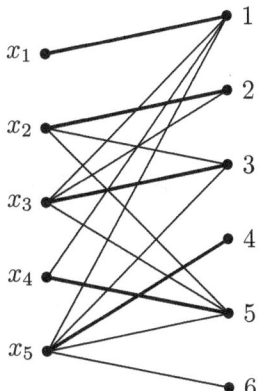

Fig. 6.3 Bipartite graph corresponding to an alldiff constraint. The heavy lines indicate a maximum cardinality matching.

6.8.2 Domain Consistency

A domain-filtering method for alldiff can be derived from properties of a maximum cardinality matching. It is based on the fact that x_j can take the value i if and only the edge (x_j, i) is part of some maximum cardinality matching that covers $x_1, \ldots, x_n$. This can, in turn, be checked by examining the maximum flow model for bipartite matching (Section 3.2.5).

Consider the maximum cardinality matching shown in Fig. 6.3. The maximum flow model adds arcs (s, x_j) from a source s and arcs (i, t) to a sink t, along with a return arc (t, s). The matching shown in the figure corresponds to a maximum flow of 5. Suppose that the capacity bounds $[p_{ts}, q_{ts}]$ for the return arc are fixed to $[5, 5]$, so that the total flow is fixed to 5. Then arc $(x_2, 3)$, for example, belongs to a maximum cardinality matching if and only if the maximum flow on this arc is 1.

Thus, arc $(x_2, 3)$ belongs to no maximum cardinality matching if the current flow of zero is a maximum flow. This can be determined by checking whether there is an augmenting path from 3 to x_2 in the residual graph of Fig. 6.4 (Corollary 3.8). There is such an augmenting path, shown in bold arcs. One therefore cannot remove 3 from the domain of x_2. Similarly, the augmenting path from 6 to x_5 shows that 6 cannot be removed from the domain of 5. On the other hand, there is no augmenting path from 1 to x_3, which means that 1 can be removed from the domain of x_3.

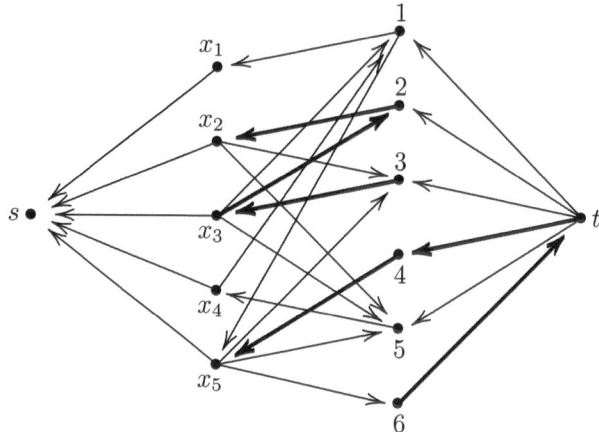

Fig. 6.4 Residual graph for the flow pattern corresponding to the maximum cardinality matching in Fig. 6.3. The heavy arcs represent augmenting paths from 3 to x_2 and from 6 to x_5.

The existence of an augmenting path can be checked by examining only the bipartite graph of Fig. 6.3. Note that the path from 3 to x_2 makes edge $(x_2, 3)$ part of an *alternating cycle* in the bipartite graph. This is a cycle in which every other edge belongs to the matching. Also the existence of the path from 6 to x_5 implies that edge $(x_5, 6)$ belongs to an *even alternating path*. This is a path of even length, one end of which is a vertex that is not covered by the matching. This pattern holds in general.

Corollary 6.15 *Consider any maximum cardinality matching of a graph G and an edge (i, j) that does not belong to the matching. Then (i, j) belongs to some maximum cardinality matching if any only if it is part of an alternating cycle or even alternating path.*

Proof. Let f be a maximum flow in the network model, and fix the capacity of the return arc to be $[f_{ts}, f_{ts}]$. Suppose first that (i, j) is part of an alternating cycle or an even alternating path in G. In the former case, removing (i, j) from the cycle leaves an augmenting path from j to i in the residual graph $R(f)$. In the latter case, suppose that the even alternating path runs from i' to j'. Then if i' is connected to s, there is a path from $j \to j' \to s \to i' \to i$ in $R(f)$. If i' is connected to t, there is a path $j \to j' \to t \to i' \to i$ in $R(f)$. Thus, the maximum flow on (i, j) is one.

For the converse, suppose that (i, j) belongs to a maximum cardinality matching, which means that the maximum flow on (i, j) is one. Then there is a path P from j to i in $R(f)$. If P contains neither s nor t, then adding (i, j) to P creates an alternating cycle in G. Alternatively, assume that P contains s or t. Suppose first that when following P from j, the first arc to a source or sink node is (j', t) for some j'. Then the last arc in P from a source or a sink node to i must be (t, i') for some i'. Because node j' is uncovered, the path from i' to j' is an even alternating path that contains (i, j). Similarly, if (j', s) is the first arc to a source or sink node from j, then by symmetric reasoning (i, j) is again part of an even alternating path. $\square$

Variable domains for alldiff can be now be filtered as follows. Find a maximum cardinality matching on the associated bipartite graph G. If the matching fails to cover the vertices $x_1, \ldots, x_n$ on one side of G, then the alldiff is infeasible. Otherwise, for each uncovered vertex on the other side, mark all edges that are part of an alternating path that starts at that vertex. Also, mark every edge that belongs to some alternating cycle, for the same reason. Now delete all unmarked edges that are not part of the matching. The remaining edges correspond to elements of the filtered domains. This achieves domain consistency.

In the graph of Fig. 6.3, only the vertex labeled 6 is uncovered. The only alternating path starting at that vertex contains also vertices x_5 and 4. Edges $(x_5, 6)$ and $(x_6, 4)$ are therefore marked. The only alternating cycle contains vertices $x_2, 2, x_3$, and 3. Its edges are also marked. One can now delete all edges that are not in the matching except the marked edges $(x_5, 6)$, $(x_2, 3)$ and $(x_3, 2)$. This yields the graph in Fig. 6.5(a). The domains are therefore reduced to

$$
\begin{aligned}
D_{x_1} &= \{1\} & D_{x_4} &= \{5\} \\
D_{x_2} &= \{2, 3\} & D_{x_5} &= \{4, 6\} \\
D_{x_3} &= \{2, 3\}
\end{aligned}
$$

The marked edges can be fund algorithmically by modifying the given bipartite graph G to obtain a directed graph G' on the same vertices. Directed edge (x_j, i) belongs to G' if the same (undirected) edge is part of the initial maximum cardinality matching on G. Also (i, x_j) belongs to G' if it belongs to G but is not part of the matching. Identify the strongly connected components of G'. These are the maximal node-induced subgraphs in which each pair of vertices i, j is connected by a directed path from i to j and a directed path from j

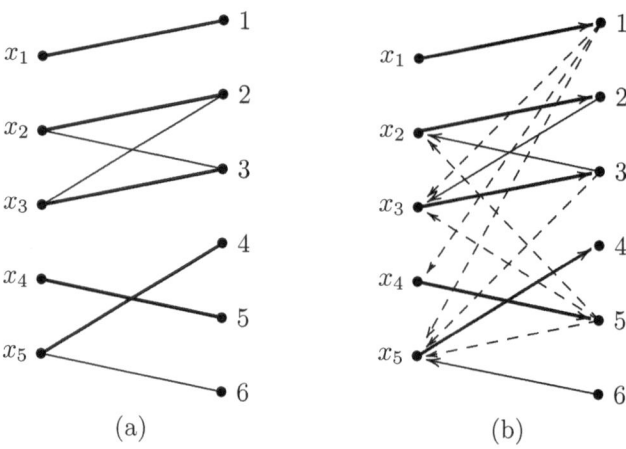

Fig. 6.5 (a) Results of domain filtering for an alldiff constraint. (b) Directed bipartite graph corresponding to Fig. 6.3. Dashed edges can be eliminated.

to i. The strongly connected components can be found in time proportional to $m + n$, where m is the number of edges and n the number of vertices. Mark all edges of G' that are part of the matching or lie on directed paths starting at vertices that are uncovered by the matching. Mark all edges that connect vertices in strongly connected components of G'. The unmarked edges can be deleted.

Figure 6.5(b) shows G' for the graph G of Fig. 6.3. Vertices x_2, x_3, 2, and 3 belong to a strongly connected component, and the connecting edges are marked. Also edges on the path from 6 to x_5 to 4 are marked.

6.8.3 Bounds Consistency

Bounds consistency makes sense for the alldiff constraint when the domain elements have a natural ordering, as for example when they are integers. The matching model achieves bounds consistency more rapidly than domain consistency, because the bipartite graph that represents bounds has a convexity property that allows for faster identification of a maximum cardinality matching.

Let $L_j = \min D_{x_j}$ and $U_j = \max D_{x_j}$ be the endpoints of x_j's domain. Alldiff($x_1, \ldots, x_n$) is bounds consistent if for any variable x_k, $x_k = L_k$ in some solution of the alldiff for which each x_j lies in the interval

$$I_{x_j} = \{L_j, \ldots, U_j\}$$

and $x_k = U_k$ in some solution for which each $x_j \in I_{x_j}$.

The bipartite graph is constructed as before, except that the interval I_j is treated as the domain of x_j. Consider the example on the left below, where the intervals I_{x_j} are shown on the right.

$$
\begin{aligned}
&\text{element}(x_1, \ldots, x_5) \\
&D_{x_1} = \{1, 2, \quad 4 \quad \} && I_{x_1} = \{1, 2, 3, 4 \quad \} \\
&D_{x_2} = \{ \quad 2, 3, \quad 6\} && I_{x_2} = \{ \quad 2, 3, 4, 5, 6\} \\
&D_{x_3} = \{ \quad 3, \quad 5 \quad \} && I_{x_3} = \{ \quad 3, 4, 5 \quad \} \\
&D_{x_4} = \{ \quad 3, 4 \quad \} && I_{x_4} = \{ \quad 3, 4 \quad \} \\
&D_{x_5} = \{ \quad 4, 5 \quad \} && I_{x_5} = \{ \quad 4, 5 \quad \}
\end{aligned}
\tag{6.77}
$$

The corresponding graph appears in Fig. 6.6.

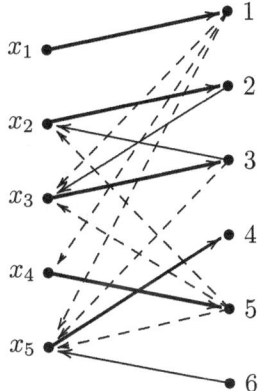

Fig. 6.6 Bipartite graph used to achieve bounds consistency for an all-different constraint. The heavy lines indicate a matching that covers all x_j and therefore satisfies alldiff.

In general, the bipartite graph representing interval domains is *convex*, meaning that if x_j is linked to domain elements i and k for $i < k$, then it is linked to all the elements $i, \ldots, k$. The maximum cardinality matching problem can be solved on a convex graph in linear time relative to the number of variables. This is accomplished as follows. For each vertex $i = 1, \ldots, m$ on the right, let an (x_j, i) be an edge in the matching, where j is the index that minimizes U_j subject to the condition that (x_j, i) is an edge of the graph and x_j is not already covered. If there is no such edge (x_j, i), then i is left uncovered. The matching for problem (6.77) is shown in Fig. 6.6.

The matching obtained as just described covers all the variables x_j if any matching does. This is implied by the following result of graph theory. Let a *perfect* matching be one that covers all vertices.

Theorem 6.16. *If G is a convex bipartite graph, the above algorithm finds a perfect matching for G, if one exists.*

The theorem can be specialized to the present situation as follows.

Corollary 6.17 *If G is a convex graph with vertices $x_1, \ldots, x_n$ on the left, and $m \geq n$ vertices on the right, the above algorithm finds a matching for G that covers $x_1, \ldots, x_n$, if one exists.*

Proof. Construct graph G' from G by adding vertices $x_{n+1}, \ldots, x_m$ to the left and edges from each new vertex to all vertices on the right. If G has a matching that covers $x_1, \ldots, x_n$, then G' has a perfect matching. By Theorem 6.16, the algorithm finds such a matching when applied to G'. Modify the algorithm so that (i) when two or more edges (x_j, i) in G' minimize U_j, the edge with the smaller index j is added to the matching, and (ii) all edges in the matching that are incident to $x_{n+1}, \ldots, x_m$ are dropped. The modified algorithm is the original algorithm applied to G, and it produces a matching that covers $x_1, \ldots, x_n$. $\square$

Once a matching that covers $x_1, \ldots, x_n$ is found, one can mark edges as described in the previous section. For each unmarked edge (x_j, i) that is not part of the matching, remove i from D_{x_j} if $i \in D_{x_j}$. Let D'_{x_j} be the updated domain for each j. Now, update the bounds by setting $L_j = \min D'_{x_j}$ and $U_j = \max D'_{x_j}$ for each j. This achieves bounds consistency.

In the example, edges from x_1 to 3 and 4 are removed, as are edges from x_2 to 3, 4, and 5. This reduces D_{x_1} from $\{1, 2, 4\}$ to $D'_{x_1} = \{1, 2\}$ and reduces D_{x_2} from $\{2, 3, 6\}$ to $\{2, 6\}$, while leaving the other domains unchanged. So, the bounds for x_1 are updated from $(L_1, U_1) = (1, 4)$ to $(1, 2)$, and the remaining bounds are unaffected.

Again, the marked edges can be found by forming the directed graph described in the previous section and identifying edges that (i) lie on directed paths from uncovered vertices, or (ii) connect vertices in strongly connected components. If properly implemented, the running time is linear in n except for the time required to sort the interval endpoints in increasing order.

Exercises

6.48. Use the bipartite matching algorithm to achieve domain consistency for alldiff$(x_1, \ldots, x_5)$, where $x_1 \in \{1, 4\}$, $x_2 \in \{1, 3\}$, $x_3 \in \{3, 6\}$, $x_4 \in \{2, 3, 5\}$, $x_5 \in \{1, 2, 3, 4, 5, 6, 7\}$. Do it once by identifying even alternating cycles and even alternating paths on the graph. Then do it again by identifying strongly connected components on the associated directed graph.

6.49. Use the bipartite matching algorithm for convex graphs to achieve bounds consistency for the alldiff constraint in the previous exercise.

6.50. The alldiffExceptZero(X) constraint requires the variables $x_1, \ldots, x_n$ with nonzero values to take different values. Indicate how to obtain domain consistency for this constraint with a bipartite matching model.

6.9 The Cardinality and Nvalues Constraints

The *cardinality* constraint is an extension of the alldiff constraint that counts how many variables take each of a given set of values. It is also known as the *distribute*, *gcc*, or *generalized cardinality* constraint. The *nvalues* constraint is another extension of alldiff that counts how many different values are assumed by a set of variables. Both are versatile constraints, and both can be efficiently filtered using a network flow model.

6.9.1 The Cardinality Constraint

The cardinality constraint is illustrated in Section 2.5. It can be written

$$\text{cardinality}(X \mid v, \ell, u) \tag{6.78}$$

where $X = \{x_1, \ldots, x_n|$ is a set of variables, $v = (v_1, \ldots, v_m)$ a tuple of values, $\ell = (\ell_1, \ldots, \ell_m)$ a tuple of lower bounds, and $u = (u_1, \ldots, u_m)$ a tuple of upper bounds. The constraint is satisfied when, for each $i \in \{1, \ldots, m\}$, at least ℓ_i and at most u_i variables x_j assume the value v_i.

Consider for example the constraint

$$\text{cardinality}(\{x_1, x_2, x_3, x_4\} \mid (a, b, c), (1, 1, 0), (2, 3, 2)) \qquad (6.79)$$

with domains $D_{x_1} = D_{x_3} = \{a\}$, $D_{x_2} = \{a, b, c\}$, $D_{x_4} = \{b, c\}$. The constraint requires that at least one, and at most two, of the variables $x_1, \ldots, x_4$ take the value a, and analogously for values b and c. Obviously a must be assigned to x_1 and x_3, which means that a cannot be used again and therefore can be removed from the domain of x_2. It will be seen shortly that no other values can be removed from domains.

The alldiff constraint is a special case in which the domain of each x_j is a subset of $\{v_1, \ldots, v_m\}$, $\ell = (0, \ldots, 0)$, and $u = (1, \ldots, 1)$.

6.9.2 Network Flow Model for Cardinality

The network flow model for the cardinality constraint (6.78) associates each variable x_j and each value v with a node of a directed graph G. There is a directed edge from v to x_j in G when v belongs to the domain of x_j. Directed edges also run from a source vertex s to each value v, from each variable x_j to a sink vertex t, and from t to s. The graph corresponding to the example (6.79) appears in Fig. 6.7.

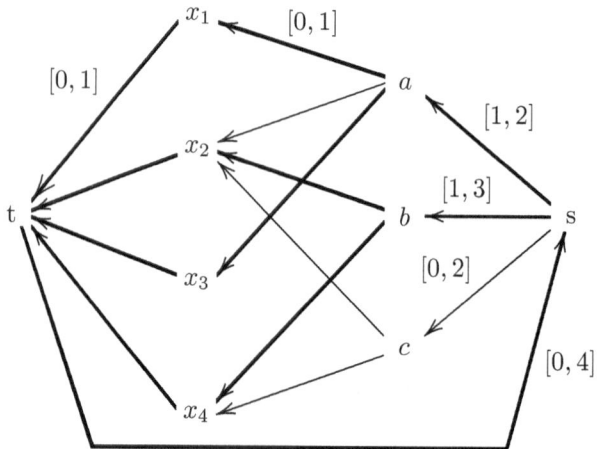

Fig. 6.7 Flow model of a cardinality constraint. One feasible flow pattern is indicated by heavy arrows. The positive flow volumes are all 1 except on edges (s, a) and (s, b), where the flow volume is 2.

Each edge of G has capacity bounds $[\ell, u]$, where ℓ is a lower bound on how much flow the edge can carry, and u an upper bound. A *feasible flow* on G assigns a flow volume to each edge in such a way that the capacity bounds are observed and flow is conserved at each vertex; that is, the total flow into a vertex equals the total flow out. The capacity bounds in the present case are $[\ell_i, u_i]$ for each edge (s, v_i), $[0, 1]$ for each edge (v_i, x_j) and (x_j, t), and $[0, n]$ for (t, s). If there are values in the domains that are not among $\{v_1, \ldots, v_m\}$, they can be represented by a single additional value v in the graph, with capacity $[0, n]$ on the arc (s, v).

A flow of 1 from v_i to x_j is interpreted as assigning value v_i to x_j. Clearly any feasible flow assigns at most one value to each variable and ensures that each value v_i is assigned to at least ℓ_i, and at most u_i, variables. The cardinality constraint (6.78) is feasible if and only if the maximum feasible flow from t to s is n. One maximum flow on the graph corresponding to the cardinality constraint (6.79) appears in Fig. 6.7. It assigns the values $(x_1, \ldots, x_4) = (a, b, a, b)$, which satisfy the cardinality constraint.

A maximum flow from s to t can be found by identifying augmenting paths, as discussed in Section 3.2.4. It is nontrivial to find an initial feasible flow, however, because there are positive lower bounds on arc flows. One approach is to begin with all the bounds on edges (s, v_i) set to $[0, u_i]$, so that an initial flow of zero on all edges is feasible. For $i = 1, \ldots, m$ do the following. If the current flow on (s, v_i) is less than ℓ_i, maximize the flow on (s, v_i) by treating this arc as the return arc, restore the bounds to $[\ell_i, u_i]$, and use this flow as a starting feasible solution for the next max flow problem. If at any point maximizing flow on (s, v_i) yields a flow less than ℓ_i, there is no feasible flow. If the capacity bounds are integral, there is a maximum flow in which the flow on every edge is integral.

6.9.3 Domain Consistency

Domains can be filtered for the cardinality constraint much as for the alldiff constraint. Suppose a maximum flow f of volume n from t to s is found. If f places a flow of one on (v, x_j), then x_j can take value v. If the flow on (v, x_j) is zero, then v can be removed from the domain of x_j if and only if zero is the maximum flow on (v, x_j) when the flow on (t, s) is fixed to zero. Zero is the maximum flow if and only if there is no augmenting path from x_j to v (Corollary 3.8) in the residual graph $R(f)$. But this is the case if and only if v and x_j belong to different strongly connected components of $R(f)$. This demonstrates the following theorem:

Theorem 6.18. *Let f be a feasible flow in the graph G associated with a cardinality constraint (6.78). An element v can be eliminated from the domain of x_j if and only if v and x_j belong to different strongly connected components of the residual graph $R(f)$ of G.*

Thus, one can achieve domain consistency for a cardinality constraint (6.78) by the following procedure:

(a) Apply a maximum flow algorithm to find a flow f on G that maximizes the flow volume on (t, s). This can be done in time proportional to n and the number of edges of G. If the maximum flow on (t, s) is less than n, then (6.78) is infeasible.

(b) Identify the strongly connected components of the residual graph $R(f)$ of G'. This can be done in time that is proportional to $m + n$.

(c) For each value v in the domain of x_i, eliminate v from the domain if f places a flow of zero on edge (v, x_j) of $R(f)$ and the two vertices v, x_j belong to different strongly connected components of $R(f)$.

The residual graph for the flow in Fig. 6.7 appears in Fig. 6.8. The vertices t and x_1 respectively belong to two strongly connected components, and the remaining vertices belong to a third component. Thus, of the three edges (a, x_2), (c, x_2), and (c, x_3) with zero flow volume in f, only (a, x_2) necessarily carries zero flow, because only this edge has endpoints in different strongly connected components. Thus, the domain of x_2 can be reduced from $\{a, b, c\}$ to $\{b, c\}$, and domain consistency is achieved.

6.9.4 The Nvalues Constraint

The *nvalues* constraint is written nvalues$(x \mid \ell, u)$, where x represents a set $\{x_1, \ldots, x_n\}$ of variables, ℓ a lower bound, and ℓ an upper bound. It

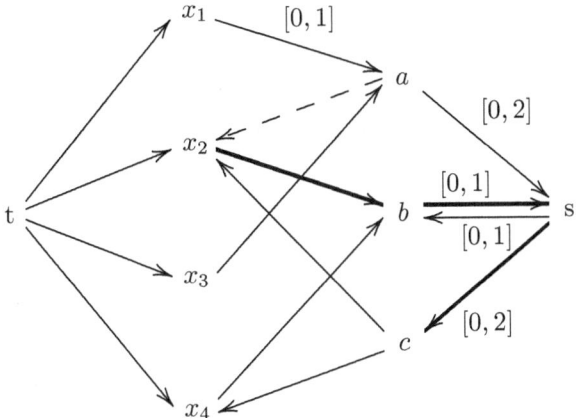

Fig. 6.8 Residual graph for the flow indicated in Fig. 6.7, assuming the flow on (t, s) is fixed to 4. The heavy arrows show an augmenting path from x_2 to c, which indicates that the flow on (c, x_2) can be increased to 1. The dashed edge represents a flow that must be zero, which means a can be eliminated from the domain of x_2. No other values can be eliminated from domains.

is illustrated in Section 2.5. The network flow model presented above is easily modified to provide filtering for the nvalues constraint. This is left as an exercise.

Filtering algorithms have also been developed for versions of nvalues in which ℓ or u is a variable and must be filtered along with x. A max flow model can be adapted to achieve domain consistency for nvalues$(x, \ell \mid u)$, in which ℓ is a variable. However, achieving domain consistency for nvalues$(x, u \mid \ell)$, in which u is a variable, is NP-hard. Incomplete polynomial-time filters have been proposed.

Exercises

6.51. Use the max flow model to filter the constraint

$$\text{cardinality}((x_1, \ldots, x_5) \mid (a, b, c), (1, 1, 1), (2, 2, 2))$$

where $x_1 \in \{a\}$, $x_2 \in \{b\}$, $x_3 \in \{a, c\}$, $x_4 \in \{b\}$, $x_5 \in \{b, c\}$.

6.52. Formulate a flow model that can be used to achieve domain consistency for the constraint nvalues$(x \mid \ell, u)$. *Hint:* Some variables may not receive flow in a feasible solution.

6.10 The Among and Sequence Constraints

The *among* and *sequence* constraints are basic tools for the formulation of sequencing problems. The among constraint places bounds on the number of variables that can take specified values. It might be used to restrict the number of afternoon or night shifts an employee works during a given period, or the number of cars on an assembly line that will receive air conditioning or sun roofs. The sequence constraint applies multiple among constraints to a rolling time horizon.

6.10.1 The Among Constraint

The *among* constraint can be written

$$\text{among}(X \mid V, \ell, u)$$

where $X = \{x_1, \ldots, x_q\}$ is a set of variables, V a set of values, and ℓ, u are nonnegative integers. The constraint requires that at least ℓ and at most u of the variables take a value in the set V. That is, $\ell \leq |\{j \mid x_j \in V\}| \leq u$.

It is straightforward to achieve domain consistency for the constraint. Let q_{in} be the number of variables whose domain is a subset of V, and q_{out} the number of variables whose domain is disjoint from V. The remaining domains are *overlapping* (i.e., they overlap V). Then at least q_{in} variables must take values in V, and at most $q - q_{\text{out}}$ variables can take values not in V. So the constraint is unsatisfiable if $u < q_{\text{in}}$ or $\ell > q - q_{\text{out}}$. If $u = q_{\text{in}}$ or $\ell = q - q_{\text{out}}$, some domain filtering is possible. In the former case, remove elements in V from overlapping domains, and in the latter case, remove elements not in V from overlapping domains. This achieves domain consistency.

Consider, for example, the constraint $\text{among}(X, \{a, b\}, \ell, u)$, where $X = \{x_1, \ldots, x_5\}$ and the domains are

$$D_{x_1} = \{a\}, \ D_{x_2} = \{a, b\}, \ D_{x_3} = \{a, c\}, \ D_{x_4} = \{b, c, d\}, \ D_{x_5} = \{c, d\}$$

Here $q = 5$, $q_{\text{in}} = 2$, and $q_{\text{out}} = 1$. At least two variables (namely, x_1 and x_2) must take values in V, and at most four (all but x_5) can take values in V. So the constraint is infeasible if $u < 2$ or $\ell > 4$. If $u = 2$, remove a from D_{x_3} and b from D_{x_4}. If $\ell = 4$, remove c from D_{x_3} and d from D_{x_4}.

6.10.2 The Sequence Constraint

It is common in scheduling applications to impose among constraints on overlapping subsequences of variables. This gave rise to the *sequence* constraint, which takes the form

$$\text{sequence}\,(x \mid q, V, \ell, u) \tag{6.80}$$

where $x = (x_1, \ldots, x_n)$ and $q \leq n$. It imposes an among constraint on each subsequence of q consecutive variables and is therefore equivalent to the constraints

$$\text{among}(X_j \mid V, \ell, u), \quad j = 1, \ldots, n - q + 1$$

where each $X_j = \{x_j, \ldots, x_{j+q-1}\}$.

The advantage of using the sequence constraint, aside from a more succinct model, is effective domain filtering. Although domain consistency is easily achieved for the individual among constraints, this and propagation do not achieve domain consistency for their conjunction. However, filtering algorithms designed specifically for the sequence constraint achieve domain consistency in polynomial time.

A natural application of the sequence constraint is to assembly line sequencing. Suppose, for example, that several car models are to be manufactured on an assembly line. The models are distinguished by which options are installed, such as air conditioning or a sun roof. Let V_k be the set of models that require option k. Due to the time and equipment required to install option k, it can be installed in at most u_k in every subsequence of q_k consecutive automobiles. There is also an output requirement: of the n cars assembled on a given day, at least ℓ_i must be model i, for each i. The constraints can be written

$$\text{sequence}\,(x \mid V_k, 0, u_k), \quad \text{all } k$$
$$\text{cardinality}\,(X \mid v, \ell, u)$$

where $x = (x_1, \ldots, x_n)$ and $X = \{x_1, \ldots, x_n\}$. Here v is a tuple of model types v_i, ℓ a tuple of production requirements ℓ_i, and $u = (\infty, \ldots, \infty)$.

Figure 6.9 illustrates a small instance of the car sequencing problem with $n = 7$ and models a, b, c, and d. Models b and d require air conditioning, while models c and d require a sun roof. At most, three of every five consecutive cars can receive air conditioning, and at most one of every three a sun roof. The constraints are

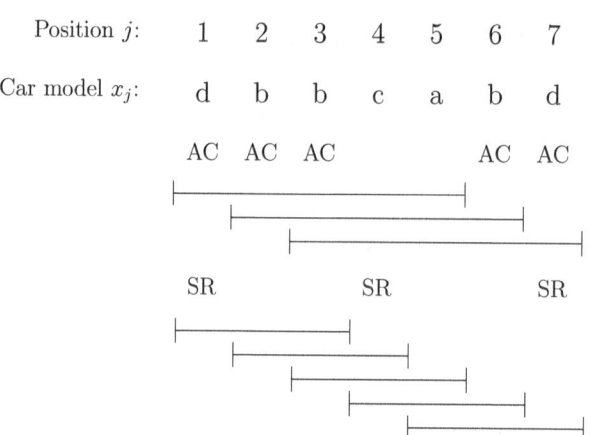

Fig. 6.9 Solution of a small instance of the car sequencing problem, where AC indicates that air conditioning is installed, and SR indicates a sun roof. The brackets correspond to among constraints implied by the two sequence constraints.

$$\text{sequence}\,((x_1,\ldots,x_7),\{b,d\},5,0,3)$$
$$\text{sequence}\,((x_1,\ldots,x_7),\{c,d\},3,0,1)$$
$$\text{cardinality}\,(\{x_1,\ldots,x_7\},(a,b,c,d),(1,3,1,2),(\infty,\infty,\infty,\infty))$$

A feasible solution appears in Fig. 6.9.

6.10.3 Filtering Based on Cumulative Sums

There are several filtering algorithms for the sequence constraint, two of which are presented here. Both achieve domain consistency, run in polynomial time, and are competitive approaches for practical use.

One filter is based on cumulative sums. It is convenient to write the sequence constraint (6.80) in terms of binary variables y_j that take the value 1 when $x_j \in V$:

$$\text{sequence}\,(y \mid q, \{1\}, \ell, u) \tag{6.81}$$

where $y = (y_1, \ldots, y_n)$. The constraint says that each subsequence of length q must contain at least ℓ and at most u ones. The initial domain of y_j contains 1 if and only if D_{x_j} contains a value in V, and it contains 0 if and only if D_{x_j} contains a value not in V.

Now if the domains of the y_j's are filtered to achieve domain consistency, the domains of the x_j's can be adjusted in the obvious way to achieve domain consistency for (6.80). If the filtered domain of y_j does not contain 1, remove from D_{x_j} all elements in V, and if it does not contain 0, remove all elements not in V.

The y_j domains can be filtered by a shaving algorithm. Fix each y_j to each of its domain values v, one at a time, and check whether the constraint (6.81) is feasible. If not, remove v from D_{y_j}. The feasibility of (6.81) is checked by applying an algorithm that can be illustrated as follows.

First define the cumulative sum

$$S_j = \sum_{i=1}^{j} y_i$$

where $S_0 = 0$. Thus, the sequence constraint requires precisely that $\ell \le S_j - S_{j-q} \le u$ for $j = q, \ldots, n$. Consider the constraint

$$\text{sequence}\,((y_1, \ldots, y_6) \mid 4, \{1\}, 2, 2) \tag{6.82}$$

with domains

$$D_{y_1} = \{0, 1\}, \ D_{y_2} = \{1\}, \ D_{y_3} = \{0, 1\}$$
$$D_{y_4} = \{0, 1\}, \ D_{y_5} = \{1\}, \ D_{y_6} = \{0, 1\}$$

To check whether 0 can be removed from D_{x_4}, for example, fix $x_4 = 0$ and check whether (6.82) has a feasible solution.

A solution of (6.82) can be found, if one exists, as shown in Fig. 6.10. First, set each y_j to the smallest value in its domain. This results in the values of S_j shown in Fig. 6.10(a). If this solution satisfies (6.82), the algorithm terminates. However, it violates (6.82) because $S_4 - S_0 = 1 < 2$. To fix this, lift the right end S_4 to 2 by setting $y_4 = 1$, as shown by the arrow. This requires adjusting the values to the right of S_4, because y_5 cannot be 0. It also requires adjusting one value to the left because y_4 cannot be 1.

This results in the solution shown in Fig. 6.10(b). It again violates (6.82), because $S_5 - S_1 = 3 > 2$. To fix this, lift the left end S_1 to 1 by setting $y_1 = 1$. This requires adjusting one value to the right, resulting in Fig. 6.10(c). Now S_6 is lifted because $S_6 - S_2 = 1 < 2$, and the algorithm terminates with the feasible solution of Fig. 6.10(d).

To check whether 1 can be removed from D_{x_4}, set $x_4 = 1$. The initial solution appears in Fig. 6.11(a). Because $S_5 - S_1 = 3 > 2$, S_1 is lifted to 1, resulting in the adjusted solution of Fig. 6.11(b). Now S_0 must be

298

6 Inference

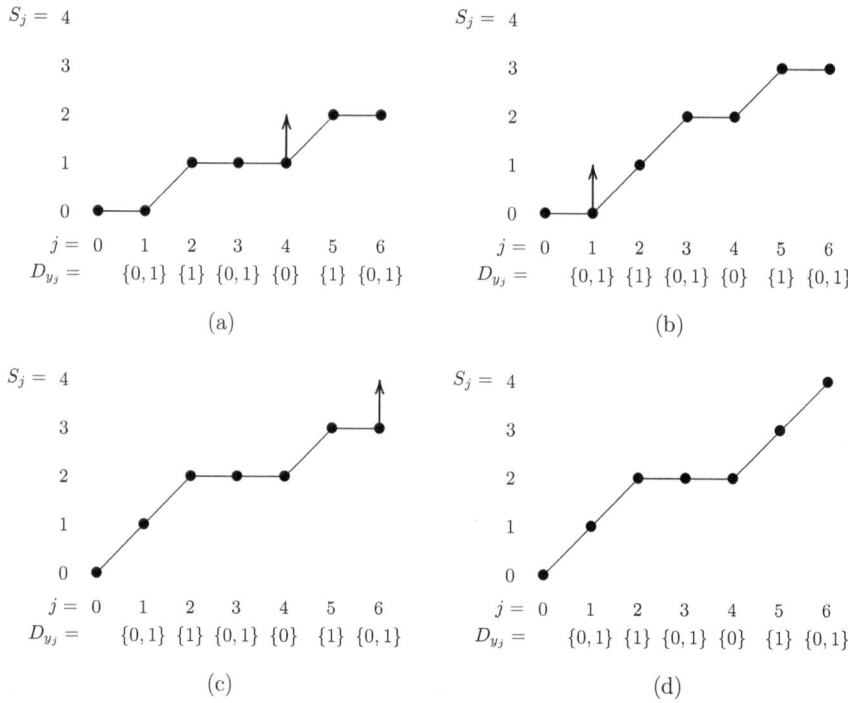

Fig. 6.10 Computing a minimum feasible solution of a sequence constraint.

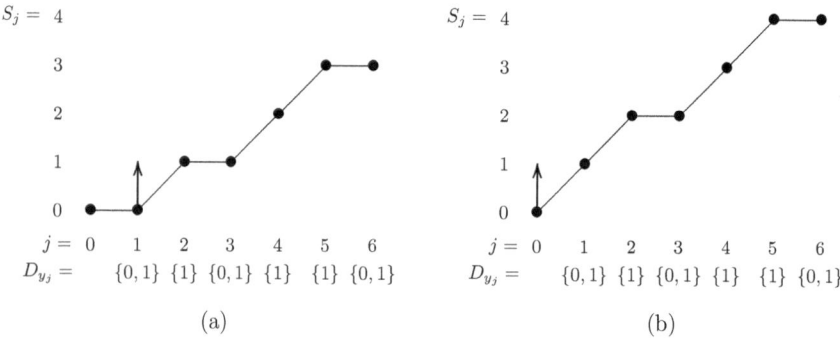

Fig. 6.11 Showing that a sequence constraint has no feasible solution.

lifted, which is impossible. There is no feasible solution, and 1 can be removed from D_{y4}. The precise feasibility checking algorithm appears in Fig. 6.12.

It can be shown (see Exercises) that if S, S' are two feasible solutions of a sequence constraint, then so is $S^* = \min\{S, S\}$, where the minimum is taken componentwise ($S_j^* = \min\{S_j, S_j'\}$ for all j). This

Let $S_0 = 0$ and $S_j = S_{j-1} + \min D_{y_j}$ for $j = 1, \ldots, n$.
While an among constraint $\ell \le S_j - S_{j-q} \le u$ is violated for some j:
    If $\ell > S_j - S_{j-q}$ then
        Perform $\mathbf{lift}(j, \ell - S_j + S_{j-q})$.
    Else
        Perform $\mathbf{lift}(j - q, S_j - S_{j-q} - u)$.
    Terminate with success.

Procedure $\mathbf{Lift}(j, \Delta)$.
    Let $S_j = S_j + \Delta$. If $S_j > j$ then terminate with failure.
    Repair S on the left:
    While $j > 0$ and $(S_j - S_{j-1} > 1$ or $(S_j - S_{j-1} = 1$ and $1 \notin D_{y_{j-1}}))$:
        If $1 \notin D_{y_{j-1}}$ then let $S_{j-1} = S_j$, else let $S_{j-1} = S_j - 1$.
        If $S_{j-1} > j - 1$ then terminate with failure.
        Let $j = j - 1$.
    Repair S on the right:
    While $j < n$ and $(S_j - S_{j+1} > 0$ or $(S_j - S_{j+1} = 0$ and $0 \notin D_{y_j}))$:
        If $0 \notin D_{y_j}$ then let $S_{j+1} = S_j + 1$, else $S_{j+1} = S_j$.
        Let $j = j + 1$.

Fig. 6.12 Algorithm for checking the feasibility of a sequence constraint. If the constraint is feasible, the algorithm finds the minimum feasible solution S, where $y_j = S_j - S_{j-1}$ for $j = 1, \ldots, n$.

implies that there is a unique minimum feasible solution of any feasible sequence constraint. The following can now be shown.

Theorem 6.19. *The algorithm of Fig. 6.12 finds the minimum feasible solution S of a sequence constraint or proves that none exists.*

Proof. The algorithm is finite, because each iteration increases some S_j, and the algorithm terminates if any $S_j > j$. If the constraint is infeasible, then some among constraint $\ell \le S_j - S_{j-q} \le u$ is always violated. The algorithm cannnot terminate with success and therefore terminates with failure.

It remains to show that when the algorithm terminates with success, the solution S obtained is the minimum feasible solution S^*. For this it suffices to show that at any point in the algorithm, $S \le S^*$, so that the algorithm can terminate with success only when $S = S^*$. The proof is by induction on the number of among constraint violations processed in the main loop of the algorithm. The initial S trivially satisfies $S \le S^*$.

For the inductive step, suppose that the current solution S satisfies $S \leq S^*$, and a violated among constraint $\ell \leq S_j - S_{j-q} \leq u$ is processed next. It suffices to show that the resulting solution S' satisfies $S' \leq S^*$. Assume first that $S_j - S_{j-q} < \ell$, so that S'_j is set to $S_{j-q} + \ell$. But $S_{j-q} \leq S^*_{j-q}$ by hypothesis, which implies $S_{j-q} + \ell \leq S^*_{j-q} + \ell \leq S^*_j$ because S^* is feasible, and therefore $S'_j \leq S^*_j$. Now when a value S_i is repaired on the left of S'_j, S_i is increased only when the resulting S'_i is the smallest possible value consistent with S'_j and the domains of $y_{i+1}, \ldots, y_j$. Thus, $S'_j \leq S^*_j$ implies $S'_i \leq S^*_i$. When a value S_i is repaired on the right, S_i is increased only when S'_i is the smallest value consistent with S'_j and the domains of $y_{j+1}, \ldots, y_i$. Thus, $S'_j \leq S^*_j$ implies $S'_i \leq S^*_i$. The reasoning is similar when $S_j - S_{j-q} > u$. $\square$

The complexity of the algorithm is $\mathcal{O}(n^2)$. Because each lifting operation increases an S_j by at least one and each $S_j \leq j$, there are at most $\mathcal{O}(n^2)$ lifting operations. Each operation may adjust other S_i's, but each adjustment replaces one or more lifting operations in this accounting, resulting in an overall complexity of $\mathcal{O}(n^2)$. Domain consistency can therefore be achieved in $\mathcal{O}(n^3)$ time.

The above algorithm is easily modified to achieve domain consistency for the *generalized sequence constraint*, which is an arbitrary combination of among constraints that apply to subsequences of consecutive variables. The generalized constraint is written

$$\text{genSequence}\,(x \mid \mathcal{X}, V, \ell, u)$$

Here $\mathcal{X} = (X_1, \ldots, X_m)$, where each X_i is a subset of consecutive variables occurring in x. Also $\ell = (\ell_1, \ldots, \ell_m)$ and $u = (u_1, \ldots, u_m)$. The generalized sequence constraint is equivalent to

$$\text{among}\,(X_i \mid V, \ell_i, u_i)\,, \ i = 1, \ldots, m$$

The proof of correctness for the algorithm is almost identical to the above proof, and the complexity of achieving domain consistency is again $\mathcal{O}(n^3)$.

6.10.4 Flow-Based Filtering

When the sequence constraint is given an integer programming model, the resulting constraint matrix has special structure that permits the model to be solved by linear programming. In fact, the LP problem is equivalent to a network flow problem, which means that network flow techniques can achieve domain consistency as they do for alldiff and cardinality constraints.

The constraint matrix has the well-known *consecutive ones property* and is therefore *totally unimodular* (Section 7.3.5). This means that that any basic feasible solution of the LP relaxation is integral, provided the right-hand sides are integral. This alone provides a polynomial-time method for achieving domain consistency, because an LP problem can be solved in polynomial time. Furthermore, problems with the consecutive ones property have a network flow formulation, which is particularly convenient for achieving domain consistency.

The idea can be seen in an example. Consider the constraint

$$\text{sequence}\,((y_1, \ldots, y_7) \,|\, 3, \{1\}, \ell, u)$$

An integer programming formulation of the constraint is

$$\ell \leq y_{j-2} + y_{j-1} + y_j \leq u, \quad j = 3, \ldots, 7$$

where each $y_j \in D_{y_j} \subset \{0,1\}$. Adding slack variables z_j and surplus variables w_j, this can be written

$$
\begin{bmatrix}
1 & 1 & 1 & & & & & & & & -1 \\
1 & 1 & 1 & & & & & & & & & 1 \\
 & 1 & 1 & 1 & & & & & & & & & -1 \\
 & 1 & 1 & 1 & & & & & & & & & & 1 \\
 & & 1 & 1 & 1 & & & & & & & & & & -1 \\
 & & 1 & 1 & 1 & & & & & & & & & & & 1 \\
 & & & 1 & 1 & 1 & & & & & & & & & & & -1 \\
 & & & 1 & 1 & 1 & & & & & & & & & & & & 1 \\
 & & & & 1 & 1 & 1 & & & & & & & & & & & & -1 \\
 & & & & 1 & 1 & 1 & & & & & & & & & & & & & 1
\end{bmatrix}
\begin{bmatrix}
y_1 \\
\vdots \\
\\
y_7 \\
w_3 \\
z_3 \\
\vdots \\
\\
w_7 \\
z_7
\end{bmatrix}
=
\begin{bmatrix}
\ell \\
u \\
\ell \\
u \\
\ell \\
u \\
\ell \\
u \\
\ell \\
u
\end{bmatrix}
\quad (6.83)
$$

where zeros are omitted for readability. A 0-1 matrix has the consecutive ones property when the ones in each row occur consecutively. The transpose of the above matrix has this property after the negative columns are multiplied by -1. The matrix is therefore totally unimodular.

Recall from Section 3.2 that the coefficient matrix for a network flow problem contains a 1 and -1 in each column. It is evident that the matrix in (6.83) can be converted to this form by subtracting each row from the next. If a row of zeros is added to the bottom of the matrix before carrying out the row operations, the following equivalent problem results:

$$
\begin{bmatrix}
1 & 1 & 1 & & & & & -1 & & & & \\
 & & & 1 & 1 & & & -1 & -1 & & & \\
-1 & & 1 & & & & & & & & & \\
 & & & & 1 & 1 & & & -1 & -1 & & \\
-1 & & & 1 & & & & & & & & \\
 & & & & & 1 & 1 & & & -1 & -1 & \\
 & -1 & & & 1 & & & & & & & \\
 & & & & & & 1 & 1 & & & -1 & -1 \\
 & & -1 & & & 1 & & & & & & \\
 & & & & & & & 1 & 1 & & & -1 \\
 & & & -1 & 1 & & & & & & & \\
 & & & & -1 & -1 & -1 & & & & & -1
\end{bmatrix}
\begin{bmatrix}
y_1 \\ \vdots \\ y_7 \\ w_3 \\ z_3 \\ \vdots \\ w_7 \\ z_7
\end{bmatrix}
=
\begin{bmatrix}
\ell \\ u-\ell \\ \ell-u \\ u-\ell \\ \ell-u \\ u-\ell \\ \ell-u \\ u-\ell \\ \ell-u \\ u-\ell \\ -u
\end{bmatrix}
\begin{matrix}
(b_3) \\ (a_3) \\ (b_4) \\ (a_4) \\ (b_5) \\ (a_5) \\ (b_6) \\ (a_6) \\ (b_7) \\ (a_7) \\ (b_8)
\end{matrix}
$$

The matrix now describes a network flow problem in which each row corresponds to a node and each column to a directed arc. Figure 6.13 shows the network, in which the node labels correspond to the row labels on the right above. A source node s and sink node t are added, as is a return arc (t, s). The supply at each node a_j is enforced by requiring the flow on arc (s, a_j) to be equal to this supply, and similarly for the demand at each node b_j. Arcs connecting b_3 to b_4, b_5, and b_6 represent y_1, y_2, and y_3; arc (b_4, b_7) represents y_4; and arcs connecting b_5, b_6, b_7 to b_8 represent y_5, y_6, and y_7. The flow on these arcs must

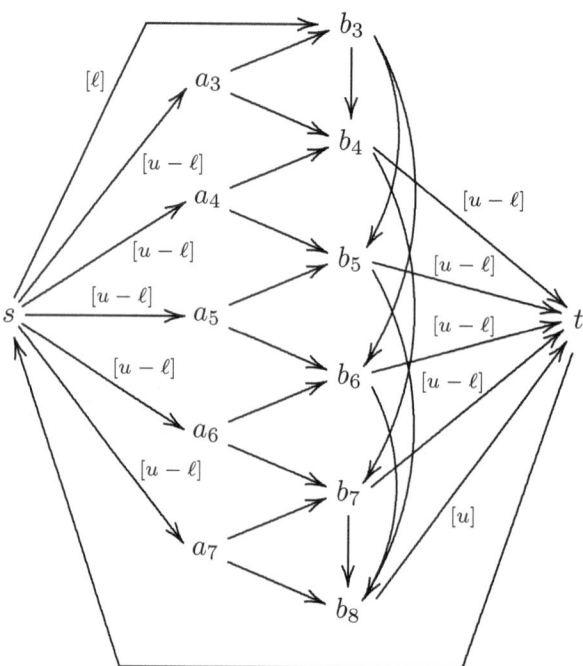

Fig. 6.13 Network flow model of a sequence constraint. Flow is fixed to ℓ, u, or $u - \ell$ on arcs that are so labeled.

lie within the interval $[0, 1]$. It is fixed to 0 if $D_{y_j} = \{0\}$ and to 1 if $D_{y_j} = \{1\}$.

In general, the constraint (6.81) is represented by a network as follows. The nodes are s, t, $a_q, \ldots, a_n$, and $b_q, \ldots, b_{n+1}$. There is an arc (s, b_q) with flow fixed to ℓ, arcs (s, a_j) with flow fixed to $u - \ell$ for $j = q, \ldots, n$, arcs (b_j, t) with flow fixed to $u - \ell$ for $j = q + 1, \ldots, n + 1$, and an arc (t, s). There are also arcs (b_q, b_j) corresponding to y_{j-q} for $j = q + 1, \ldots, 2q$, arcs (b_j, b_{j+q}) corresponding to y_j for $j = q + 1, \ldots, n - q$, and arcs (b_j, b_{n+1}) corresponding to y_j for $j = n - q + 1, \ldots, n$. The flow on these arcs is bounded as noted above. Now 1 can be removed from the domain of y_j if the maximum flow on the corresponding arc is zero, and 0 can be removed if the minimum flow is 1.

The generalized sequence constraint does not necessarily have a network flow model. However, it may be possible to permute the rows so that the transpose of the constraint matrix has the consecutive ones property. This can be checked in $\mathcal{O}(m + n + r)$ time, where r is the number of nonzeros in the matrix. If there is such a permutation, a network flow model can be derived by row operations as described above. If not, there may nonetheless be an equivalent network matrix, which can be checked in $\mathcal{O}(mr)$ time. In any case, linear programming can always achieve domain consistency for genSequence, because the constraint matrix is totally unimodular even when there is no network model. The columns corresponding to $y_1, \ldots, y_n$ form a submatrix that has the consecutive ones property and is therefore totally unimodular, and the remaining columns are unit columns or their negations. The addition of such columns preserves total unimodularity (Section 7.3.5).

Exercises

6.53. Find the minimal feasible solution of the constraint

$$\text{sequence}\,((x_1, \ldots, x_6) \mid 3, \{1\}, 1, 1)$$

using the method of cumulative sums, given domains $D_{x_1} = D_{x_3} = D_{x_6} = \{0, 1\}$ and $D_{x_2} = D_{x_4} = D_{x_5} = \{0\}$. Now use the same method to filter the domains.

6.54. Write the integer programmimg model and draw the network flow model of the constraint in Exercise 6.53.

6.55. Given a 0-1 solution $(y_1, \ldots, y_n)$ of a sequence constraint, let $S = (S_1, \ldots, S_n)$ represent the solution as cumulative sumes. That is, each S_j is the sum $\sum_{i=1}^{j} y_i$. Show that if S, S' are feasible solutions of a sequence constraint, then $S^* = \min\{S, S'\}$ is a feasible solution of the constraint, where $S_j^* = \min\{S_j, S_j'\}$ for all j.

6.11 The Stretch Constraint

The stretch constraint was originally designed for scheduling workers in shifts. It is illustrated in Section 2.5. Analysis of this constraint provides an opportunity to show how dynamic programming can contribute to filtering.

The stretch constraint is written

$$\text{stretch}(x \,|\, v, \ell, u, P)$$

where x is a tuple $(x_1, \ldots, x_n)$ of variables. In typical applications, x_i represents the shift that a given employee will work on day i. Also, v is an m-tuple of possible values of the variables, ℓ an m-tuple of lower bounds, and u an m-tuple of upper bounds. A *stretch* is a maximal sequence of consecutive variables that take the same value. Thus, $x_j, \ldots, x_k$ is a stretch if for some value v, $x_j, \ldots, x_k = v$, $x_{j-1} \neq v$ (or $j = 1$), and $x_{k+1} \neq v$ (or $k = n$).

The stretch constraint requires that for each $j \in \{1, \ldots, m\}$, any stretch of value v_j in x have length at least ℓ_j and at most u_j. In addition P is a set of *patterns*, which are pairs of values $(v_j, v_{j'})$. The constraint requires that when a stretch of value v_j immediately precedes a stretch of value $v_{j'}$, the pair $(v_j, v_{j'})$ must be in P. Thus, the constraint puts bounds on how many consecutive days the employee can work each shift, and which shifts can immediately follow another. For instance, one of the shifts may represent a day off, and it is common to require that the employee never work two different shifts without at least one intervening day off.

Consider, for example, a shift scheduling problem in which there are three shifts (a, b, and c) and the domain of each x_j is listed beneath each variable in Table 6.2. A stretch constraint for these variables might be

$$\text{stretch}(x \,|\, (a, b, c), (2, 2, 2), (3, 3, 3), P) \tag{6.84}$$

where P contains the pairs (a, b), (b, a), (b, c), and (c, b). Thus, a stretch can be at least two and at most three days long, and the worker cannot

Table 6.2 Variable domains in a shift scheduling problem.

x_1	x_2	x_3	x_4	x_5	x_6	x_7
a	a	a		a	a	a
	b	b	b		b	b
c	c		c	c		

change directly between shifts a and c. The constraint allows only two feasible schedules: $aabbaaa$ and $ccbbaaa$.

There is a cyclic version of the constraint, *stretchCycle*, which recognizes stretches that continue from x_n to x_1. It can be used when one wants to have the same schedule every week and allows stretches to extend through the weekend into the next week. The filter described below can be modified for the cyclic constraint.

6.11.1 Dynamic Programming Model

A dynamic programming model can be constructed by viewing a stretch as a control that takes an employee from one state to another. An employee who is just finishing stretch of shift v_j on day i is in state (i, v_j). A stretch of value v_k that starts on day $i+1$ takes the employee to state $(i + \delta, v_k)$, where δ is the length of the stretch, and v_k is a shift that can immediately follow v_j. The stretch must of course be feasible in the sense that $v_k \in D_{x_h}$ for $h = i + 1, \ldots, i + \delta$.

A state transition graph appears in Fig. 6.14, where the edges are labeled by the corresponding controls. Only forward reachable states are shown. Because the time horizon is seven days, a terminal state must be a *finishing* state of the form $(7, v_j)$. The state $(6, a)$ is therefore not feasible as a terminal state and can be deleted.

The state transitions are Markovian because the feasible transitions from a given state (i, v_j) depend only on the state itself, and not how the employee reached the state. The possible length of a stretch of v_j depends only on v_j and i. The shifts to which the employee is allowed to transition depend only on v_j, because the pattern set P contains only pairs. If triples of consecutive stretches could be constrained, the Markovian property would disappear.

Backward reachable states can be identified by traversing the graph backwards from the one feasible terminal state. This allows state $(3, a)$

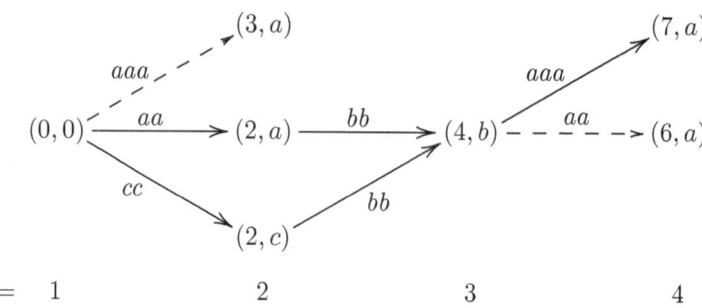

Fig. 6.14 State transition graph for a stretch constraint. Only forward reachable states are shown. Dashed edges lead into states that are not backward reachable.

to be deleted as well. The two paths in the remaining graph correspond to the two feasible solutions. If the control variable is y_k, the solutions are $(y_1, y_2, y_3) = (aa, bb, aaa)$ and (cc, bb, aaa). In the original variables, this corresponds to solutions $(x_1, \ldots, x_7) = (a, a, b, b, a, a, a)$ and (c, c, b, b, a, a, a).

The number K of stages must be large enough to reach all finishing states that are forward reachable. If a finishing state is reached at stage $k < K$, control variables $y_{k+1}, \ldots, y_K$ take null values. In the small instance solved here, all reachable finishing states are reached in stage 4.

The original variables x_i can be used as control variables, but this results in less efficient filtering for the stretch constraint. However, the original varables serve as control variables when the problem is formulated with the more general regular constraint. This formulation is presented in Section 6.12.

6.11.2 Domain Consistency

One way to filter domains is to determine which states are forward and backward reachable in the state transition graph described above. Variable x_i takes value v_j in some feasible solution if $x_i = j$ is part of a stretch whose endpoint is forward and backward reachable. In the example of Table 6.2, $x_3 = b$ is part of a feasible solution because the stretch bb can occur on days 3 and 4, and $(4, b)$ is forward and backward reachable.

It is more efficient, however, to work with *starting* states as well as the *ending states* described above. A starting state (i, v_j) indicates that a stretch of value v_j starts on day i, while an ending state (i, v_j) indicates that a stretch of value v_j ends on day i. A starting state (i, v_j) can transition to a starting state $(i + \delta, v_k)$ if there is a feasible v_j stretch of length δ starting on day i, and $(v_k, v_k) \in P$.

Now x_i takes value v_j in some feasible solution if it is part of a stretch whose starting state is forward reachable and whose ending state is backward reachable. That is, there is a feasible v_j stretch from i' to i'' such that $i' \leq i \leq i''$, (i', v_j) is a forward reachable starting state, and (i'', v_j) is a backward reachable ending state.

Forward reachability is efficiently computed by defining a function $f(i, v_j)$ that denotes the number of days i' on or before i on which starting state (i', v_j) is forward reachable. The function is useful because a state (i, v_j) is forward reachable at some point during a given interval $k, \ldots, k'$ if and only if $f(k', v_j) - f(k - 1, v_j) > 0$.

The forward reachability function can be computed recursively. This is because (i, v_k) is forward reachable when a state (i', v_j) is forward reachable for a day i' prior to i, such that (a) $(v_j, v_k) \in P$, (b) $\ell_k \leq i - i' \leq u_k$, and (c) $v_j \in D_{x_h}$ for $h = i', \ldots, i - 1$. Condition (c) can be checked by precomputing a run length function r_{ij} that measures the longest stretch ending on day i that shift v_j can run. The recursion is

$$f(i+1, v_k) = \begin{cases} f(i, v_k) + 1 & \text{if } f(i_{\max}, v_j) - f(i_{\min}, v_j) > 0 \text{ for some } v_j \\ & \quad \text{for which } i_{\max} \geq i_{\min} \text{ and } (v_j, v_k) \in P \\ f(i, v_k) & \text{otherwise} \end{cases}$$

for $i = 1, \ldots, n$, where

$$i_{\max} = i - \ell_j + 1, \qquad i_{\min} = i - \min\{u_j, r_{ij}\}$$

The boundary condition are set to get the recursion started correctly:

$$f(0, v_j) = 0, \quad f(1, v_j) = r_{1j}, \quad \text{all } j$$

The run length can itself be computed by a simple recursion,

$$r_{ij} = \begin{cases} r_{i-1,j} + 1 & \text{if } v_j \in D_{x_i} \\ 0 & \text{otherwise,} \end{cases} \quad \text{all } j \text{ and } i = 1, \ldots, n$$

where $r_{0j} = 0$ for all j. The values of $f(i, v_i)$ for the example of Table 6.2 are displayed in Table 6.3.

Table 6.3 Forward and backward reachability functions for filtering a stretch constraint.

	v_j	$i = 0$	1	2	3	4	5	6	7
$f(i, v_j)$:	a	0	1	1	1	1	2	2	2
	b	0	0	0	1	2	2	2	3
	c	0	1	1	1	1	2	2	2

	v_j	$i = 1$	2	3	4	5	6	7	8
$b(i, v_j)$:	a	4	3	2	2	2	1	1	0
	b	3	3	2	2	1	1	1	0
	c	3	2	1	1	1	0	0	0

A similar backward reachability function $b(i, v_j)$ denotes the number of days on or after day i on which ending state (i, v_j) is backward reachable. It is recursively computed

$$b(i-1, v_k) = \begin{cases} b(i, v_k) + 1 & \text{if } b(i_{\min}, v_j) - b(i_{\max}, v_j) > 0 \text{ for some } v_j \\ & \text{for which } i_{\min} \le i_{\max} \text{ and } (v_k, v_j) \in P \\ b(i, v_k) & \text{otherwise} \end{cases}$$

for $i = n, \ldots, 1$, where

$$i_{\min} = i + \ell_j - 1, \qquad i_{\max} = i + \min\{u_j, \bar{r}_{ij}\}$$

with boundary conditions

$$b(n, v_j) = \bar{r}_{nj}, \quad b(n+1, v_j) = 0, \quad \text{all } j$$

The run length is measured in a backwards direction,

$$\bar{r}_{ij} = \begin{cases} \bar{r}_{i+1,j} + 1 & \text{if } v_j \in D_{x_i} \\ 0 & \text{otherwise,} \end{cases} \quad \text{all } j \text{ and } i = n, \ldots 1$$

where $\bar{r}_{n+1,j} = 0$ for all j. The function values for the example appear in Table 6.3.

Although the reachability functions can be computed via the above recursions, it is more efficient to compute them using the algorithms of Fig. 6.15 and Fig. 6.16.

Possible values of each x_i can now be determined by examining the forward and backward reachability functions. Consider each possible

For $j = 1, \ldots, m$ let $f(0, v_j) = 0$ and $f(1, v_j) = r_{1j}$.
For $i = 1, \ldots, n$:
    For $j = 1, \ldots, m$ let $f(i + 1.v_j) = f(i, v_j)$.
    For $j = 1, \ldots, m$:
        Let $i_{\max} = i - \ell_j$ and $i_{\min} = i - \min\{u_j, r_{ij}\}$.
        If $i_{\max} \geq i_{\min}$ and $f(i_{\max} + 1, v_j) - f(i_{\min}, v_j) > 0$ then
            For $k = 1, \ldots, m$:
                If $(v_j, v_k) \in P$ then let $f(i + 1, v_k) = f(i, v_k) + 1$.

Fig. 6.15 Algorithm for computing the forward reachability function $f(i, v_j)$.

For $j = 1, \ldots, m$ let $b(n + 1, v_j) = 0$ and $b(n, v_j) = \bar{r}_{nj}$.
For $i = n, \ldots, 1$:
    For $j = 1, \ldots, m$ let $b(i - 1.v_j) = b(i, v_j)$.
    For $j = 1, \ldots, m$:
        Let $i_{\min} = i + \ell_j$ and $i_{\max} = i + \min\{u_j, \bar{r}_{ij}\}$.
        If $i_{\max} \geq i_{\min}$ and $b(i_{\min} - 1, v_j) - b(i_{\max}, v_j) > 0$ then
            For $k = 1, \ldots, m$:
                If $(v_k, v_j) \in P$ then let $b(i - 1, v_k) = b(i, v_k) + 1$.

Fig. 6.16 Algorithm for computing the backward reachability function $b(i, v_j)$.

feasible stretch of each shift that is within the length bounds. For each day i in a stretch of shift v_j, mark v_j as a possible value of x_i if it is forward reachable on the first day of the stretch and backward reachable on the last day of the stretch. All assignments to x_i that remain unmarked after all possible stretches are considered can be deleted from the domain of x_i. This process achieves domain consistency.

More precisely, for each v_j consider every possible stretch $x_k, \ldots, x_{k'}$ of shift v_j for which $\ell_j \leq k' - k + 1 \leq u_j$ and $v_j \in D_{x_i}$ for $i = k, \ldots, k'$. If state (k, v_j) is forward reachable and (k', v_j) is backward reachable, mark v_j as a possible value of x_i for $i = k, \ldots, k'$. State (k, v_j) is forward reachable if $f(k, v_j) - f(k - 1, v_j) > 0$, and (k', v_j) is backward reachable if $b(k', v_j) - b(k' + 1, v_j) > 0$. Each v_j that remains unmarked as a possible value of x_i can be removed from D_{x_i}.

For example, a stretch bb on days 3 and 4 is compatible with the domains shown in Table 6.2. Shift b is forward reachable on day 3

Table 6.4 Reduced variable domains in a shift scheduling problem.

x_1	x_2	x_3	x_4	x_5	x_6	x_7
a	a			a	a	a
		b	b			
c	c					

because $f(3, b) - f(2, b) > 0$ and backward reachable on day 4 because $b(4, b) - b(5, b) > 0$. So, value b is marked to remain in the domains of x_3 and x_4. The reduced domains after completion of the filtering algorithm appear in Table 6.4.

The filtering process can be completed in $\mathcal{O}(nm^2)$ time by maintaining, for each shift, a queue of days that are candidates for the end of a stretch. While examining each day i in reverse order, add i to the back of the queue if i is backward reachable. If i is forward reachable, remove days from the front of the queue that cannot be part of a stretch starting at i, either because the stretch would have the wrong length, or the run length at the front of the queue is too short to reach back to i. Then, add the stretch from i to the front of the queue to a list of possible stretches. The algorithm appears in Fig. 6.17.

Let Q be a queue and $\text{front}(Q)$ the element at the front of Q.
Let L be a list of possible stretches.
For $j = 1, \ldots, m$:
    Let $Q, L = \emptyset$.
    For $i = n, \ldots, 1$:
        If $b(i, v_j) - b(i + 1, v_j) > 0$ then add i to the back of Q.
        If $f(i, v_j) - f(i - 1, v_j) > 0$ then
            Repeat while $Q \neq \emptyset$ and $\min\{u_j, r_{\text{front}(Q)j}\} < \text{front}(Q) - i + 1$:
            Remove $\text{front}(Q)$ from Q.
            If $Q \neq \emptyset$ and $\ell_j \leq \text{front}(Q) - i + 1$ then add $[i, \text{front}(Q)]$ to L.
    For each $[k, k'] \in L$:
        Mark v_j as a feasible value of x_i for $i = k, \ldots, k'$.
For $i = 1, \ldots, n$ and $j = 1, \ldots, m$:
    If v_j is not marked as a feasible value of x_i then remove v_j from D_{x_i}.

Fig. 6.17 Domain consistency algorithm for the stretch constraint.

Exercises

6.56. Suppose that the domains of $x_1, \ldots, x_7$ are as given beneath each variable below:

$$x_1 \ x_2 \ x_3 \ x_4 \ x_5 \ x_6 \ x_7$$

a	a	a		a	a	a
b		b	b	b		b
c	c		c	c	c	c

Draw the state transition graph for

$$\text{stretch}(x \mid (a, b, c), (2, 2, 2), (7, 7, 7), P)$$

where $P = \{(a, b), (b, c)\}$, and use it to filter the domains.

6.57. Use the algorithm of Fig. 6.17 to filter the domains in Table 6.2 on the basis of the recursive function values in Table 6.3.

6.58. Compute the forward and backward reachability functions for the constraint in Exercise 6.56 and use them to filter domains with the algorithm of Fig. 6.17.

6.59. Indicate how dynamic programming can filter a stretchCycle constraint. Draw the appropriate state transition graph for the domains in Table 6.2 and use it to show that the constraint is infeasible. *Hints:* Introduce a third state variable that remembers the first shift taken. Use the state transition graph to solve $\text{stretch}((x_i, \ldots, x_{i+7}) \mid v, \ell, u, P)$ for $i = 1, 2, 3$, where the domains of x_8, x_9 are those of x_1, x_2, respectively.

6.12 The Regular Constraint

Sequencing and scheduling problems frequently impose constraints on possible sequences of operations. For example, workers may require a day off after working a certain number of days, or a machine may not be able to manufacture products A, B, and C consecutively. The *regular constraint* is designed for such situations.

The regular constraint imposes any restriction that can be encoded as a deterministic finite automaton. Putting a constraint in this form in effect creates a dynamic programming model in which the transition function does not depend on the stage. The name *regular* derives from the fact that the set of feasible solutions of the constraint comprises a regular language, as defined in computer science.

Because the regular constraint defines a dynamic programming model, domains can be filtered as for any finite-domain dynamic programmimg problem. Simply eliminate states that are neither forward nor backward reachable and then observe which controls x_i are applied in each stage i of the remaining state transition graph. These are the possible values of variable x_i.

In fact, one can define a more general *dynamic programming constraint* that is filtered in exactly the same way. This constraint allows models in which the transition functions $t_i(s_i, x_i)$ depend on the stage i. A dynamic programming constraint conveniently formulates some common sequencing problems that are more difficult to capture in a regular constraint.

The stretch constraint, described in the previous section, encodes certain types of seqencing restrictions—limits on the size of a stretch, and pairwise precedence constraints. It allows more efficient filtering than the regular constraint, but at a cost of less generality.

The regular constraint constraint can also be filtered by decomposing it into a series of simpler constraints that define the dynamic programming recursion. Domain consistency can be achieved by filtering and propagating the simpler constraints in two passes. This suggests a generalization to a regular constraint based on nondeterministic rather than deterministic finite automata, because its dynamic programming model can be decomposed and filtered in the same fashion.

6.12.1 Determistic Finite Automata

A *deterministic finite automaton* $A = (S, D, t, \alpha, F)$ consists of a finite set S of states, a set D of controls, and a partial transition function t that specifies the result $t(s, x)$ of applying control x in state s. Note that the transition function is not indexed by the stage i as in general dynamic programming (Section 3.4). The function is partial in the sense that it may be defined only for certain pairs (s, x). That is, only certain controls x can be applied in state s. Also $t(s, x)$ and $t(s, x')$ are distinct (when defined) for $x \neq x'$. There is an initial state α and a set F of *accepting* (or *final*) states. The automaton *accepts* a sequence of values (*string*) $x_1, \ldots, x_n$ if and only if

$$s_1 = \alpha, \quad s_{n+1} \in F, \quad \text{and} \quad s_{i+1} = t(s_i, x_i) \text{ for } i = 1, \ldots, n \qquad (6.85)$$

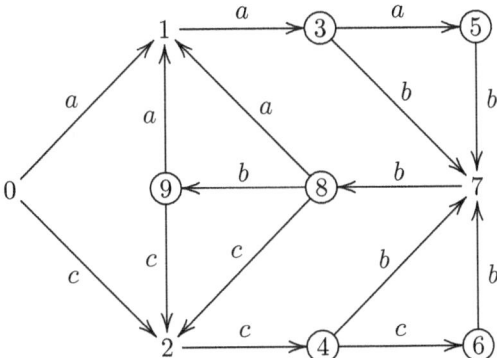

Fig. 6.18 Deterministic finite automaton for a shift scheduling problem, with states shown as vertices and transitions as edges. State 0 is the initial state, and accepting states are circled.

If a *language* is defined to be a set of strings, a *regular* language a set of strings accepted by some finite state automaton.

The regular constraint is written

$$\text{regular}\,(x \mid A)$$

where $x = (x_1, \ldots, x_n)$ and A is a deterministic finite automaton. It requires that A accept $x_1, \ldots, x_n$. Every regular language is specified by some regular constraint.

As with dynamic programming, the key to formulating a problem as a finite automaton is defining the states. This can be illustrated with the shift scheduling problem of Section 6.11. It was formulated as a stretch constraint in that section, but it can also be expressed as a regular constraint.

The states for this problem appear as vertices in Fig. 6.18. The initial state 0 represents the beginning of the week. Shifts a and c are available on day 1, which create transitions to states 1 and 2, respectively. In state 1, the only option is another day in shift a, because the minimum stretch is 2. In state 3, the stretch can be terminated by moving to shift b or extending the stretch one more day, and so forth. The state numbers have no significance other than as arbitrary labels for the states. The circled states can terminate stretches and can therefore serve as accepting states.

It can be checked that strings accepted by the automaton of Fig. 6.18 are precisely those with the desired properties: any stretch occurring in the string has length 2 or 3, and shifts a and c never occur consecutively. The automaton can accept arbitrarily long strings, but only strings $x_1, \ldots, x_7$ of length 7 are of interest here. If A represents the automaton, the shift scheduling problem is therefore captured by the constraint

$$\text{regular}\,((x_1, \ldots, x_7)\,|\,A)$$

and the variable domains (Table 6.2).

It may be more convenient to specify a regular constraint by characterizing the acceptable strings syntactically, rather than by describing the finite automaton. There is a generally accepted notation for specifying the strings belonging to a regular language. It builds acceptable strings as concatenations of structured substrings. For example, ab^* indicates a concatenation of a with zero or more b's, resulting in the set of strings $\{a, ab, abb, \ldots\}$. (The * is known as the *Kleene star*.) A vertical bar indicates alternatives, so that $(a|b)(c|d)$, for example, indicates the set $\{ac, ad, bc, bd\}$. The symbol ϵ denotes the empty string, so that $a(\epsilon|c)$ denotes the set $\{a, ac\}$. The automaton of Fig. 6.18 specifies the regular language described by

$$((aaa^*bbb^*)^*|(ccc^*bbb^*)^*)^*(\epsilon|aaa^*|ccc^*)$$

Here the lower limit of 2 on the stretch length is imposed by expressions of the form aaa^*. The upper limit of 7 is ignored, because it derives from the length of a week rather than shift rules.

6.12.2 Domain Filtering

Domains can be filtered by building a state transition graph, deleting unreachable states, and observing the possible controls in each stage. A state transition graph for the automaton of Fig. 6.18 appears in Fig. 6.19. Only the first seven controls $x_1, \ldots, x_7$ are relevant. Note that the variable domains D_{x_i} are now taken into account. This graph may be contrasted with the state transition graph (Fig. 6.14) used to filter the stretch constraint, where a different dynamic programming model is used. In that model, the controls are stretches rather than individual shifts.

Because state 7 in the last stage of Fig. 6.19 is not an accepting state of the automaton, it is deleted. The backward reachable states can now be identified by traversing the graph backward from the one

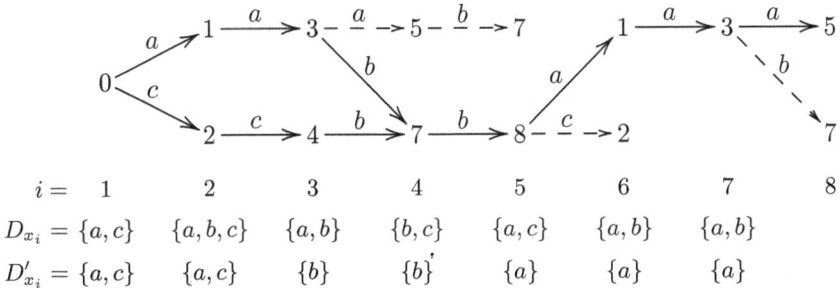

$i =$	1	2	3	4	5	6	7	8
$D_{x_i} =$	$\{a,c\}$	$\{a,b,c\}$	$\{a,b\}$	$\{b,c\}$	$\{a,c\}$	$\{a,b\}$	$\{a,b\}$	
$D'_{x_i} =$	$\{a,c\}$	$\{a,c\}$	$\{b\}$	$\{b\}$	$\{a\}$	$\{a\}$	$\{a\}$	

Fig. 6.19 State transition graph for a shift scheduling problem. Dashed edges lead into states that are not backward reachable. Reduced domains D'_{x_i} are shown.

terminal state that remains. Unreachable states are deleted, and the feasible solutions are precisely the remaining paths from state 0 to a terminal state. In this case, there are two paths. The possible values of x_i in a feasible solution are the values that appear on arcs leaving stage i. The reduced domains D'_{x_i} are indicated in Fig. 6.19. This procedure achieves domain consistency.

In general, a constraint regular$(x \mid A)$ with $A = (S, D, t, \alpha, F)$ is given a dynamic programming model (3.25) by setting $t_i(s_i, x_i) = t(s_i, x_i)$ when $t(s_i, x_i)$ is defined. The set $X_i(s_i)$ of available controls in state s_i is D_{x_i} intersected with the set of values x_i for which $t(s_i, x_i)$ is defined. Also $S_1 = \{\alpha\}$ and $S_{n+1} = F$. Backward reachable states are identified in the state transition graph by traversing it backward from terminal states in F. Unreachable states are then deleted, and each reduced domain D'_{x_i} is the set of values v such that $(s_i, t_i(s_i, v))$ is an edge of the graph that remains. This achieves domain consistency.

6.12.3 Filtering by Decomposition

The variable domains also can be filtered by filtering and propagating the individual state transition constraints (6.85). The propagation scheme is quite simple, consisting of a forward and backward pass, as in dynamic programming. The complexity is the same as computing forward and backward reachability.

Table 6.5 illustrates the process for the above example. The first state transition constraint $s_2 = t(s_1, x_1)$ has feasible solutions $(0, a, 1)$

Table 6.5 Filtering and propagation for a regular constraint that is decomposed into state transition constraints.

Constraint $s_{i+1} = t(s_i, x_i)$	Forward pass D_{s_i}	D_{x_i}	(s_i, x_i, s_{i+1})	Backward pass D_{s_i}	D_{x_i}	(s_i, x_i, s_{i+1})
$s_2 = t(s_1, x_1)$	$\{0\}$	$\{a, c\}$	$(0, a, 1)$ $(0, c, 2)$	$\{0\}$	$\{a, c\}$	$(0, a, 1)$ $(0, c, 2)$
$s_3 = t(s_2, x_2)$	$\{1, 2\}$	$\{a, \not b, c\}$	$(1, a, 3)$ $(2, c, 4)$	$\{1, 2\}$	$\{a, c\}$	$\overline{(1, a, 3)}$ $(2, c, 4)$
$s_4 = t(s_3, x_3)$	$\{3, 4\}$	$\{a, b\}$	$(3, a, 5)$ $(3, b, 7)$ $(4, b, 7)$	$\{3, 4\}$	$\{a, b\}$	$(3, a, 5)$ $(3, b, 7)$ $(4, b, 7)$
$s_5 = t(s_4, x_4)$	$\{5, 7\}$	$\{b, \not c\}$	$(5, b, 7)$ $(7, b, 8)$	$\{\not 5, 7\}$	$\{b\}$	$\overline{(5, b, 7)}$ $(7, b, 8)$
$s_6 = t(s_5, x_5)$	$\{\not 7 8\}$	$\{a, c\}$	$(8, a, 1)$ $(8, c, 2)$	$\{8\}$	$\{a, \not c\}$	$(8, a, 1)$ $\overline{(8, c, 2)}$
$s_7 = t(s_6, x_6)$	$\{1, \not 2\}$	$\{a, \not b\}$	$(1, a, 3)$	$\{1\}$	$\{a\}$	$(1, a, 3)$
$s_8 = t(s_7, x_7)$	$\{3, \not 5\}$	$\{a, \not b\}$	$(3, a, 5)$ $\overline{(3, b, 7)}$			

and $(0, c, 2)$ as shown in column 4 of the table. Nothing can be filtered from domains D_{s_1} and D_{x_1}, and the domain $D_{s_2} = \{1, 2\}$ is passed to the next constraint. The second constraint $s_3 = t(s_2, x_2)$ has the two feasible solutions shown, and b is filtered from D_{x_2} as indicated by the strikeout. The process continues to the end, where solution $(3, b, 7)$ is deleted because state 7 is not an accepting state. The process then continues in reverse. On the backward pass, constraint $s_7 = t(s_6, x_6)$ has only one feasible solution $(1, a, 3)$ because D_{s_7} has been reduced to $\{3\}$. Solution $(8, c, 2)$ is deleted for constraint $s_6 = t(s_5, x_5)$ because D_{s_6} has been reduced to $\{1\}$, and so forth.

The property that permits this scheme is Berge acyclicity of the constraint hypergraph. The *constraint hypergraph* for a constraint satisfaction problem consists of a vertex for each variable and a hyperedge for each constraint, where the hyperedge corresponding to a constraint is the set of variables in the constraint. The constraint hypergraph for the shift scheduling example appears in Fig. 6.20.

A *Berge cycle* in a given hypergraph is a sequence

$$E_1, x_1, E_2, x_2, \ldots, E_m, x_m, E_1$$

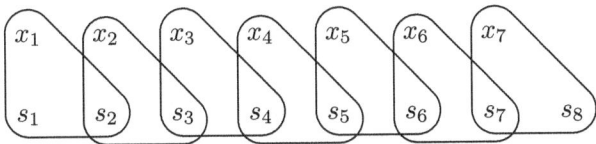

Fig. 6.20 Constraint hypergraph for a dynamic programming formulation of the shift scheduling problem.

where $x_1, \ldots, x_m$ are distinct vertices, $E_1, \ldots, E_m$ are distinct hyperedges, $m \geq 2$, and $x_i \in E_i, E_{i+1}$ for $i = 1, \ldots, m$. A hypergraph is *Berge acyclic* if it contains no Berge cycles. In particular, no pair of hyperedges in a Berge acyclic hypergraph can have more than one vertex in common. The hypergraph of Fig. 6.20 is clearly Berge acyclic.

A two-pass propagation scheme achieves domain consistency for a constraint satisfaction problem that has a Berge acyclic constraint hypergraph. The propagation order is determined by a topological sort on the hypergraph. First, remove a constraint that has a common variable with only one other constraint. Call the deleted constraint C_1 and the common variable x_1. Repeat this step until no constraints remain, obtaining $C_1, \ldots, C_m$ and $x_1, \ldots, x_{m-1}$. Then filter domains for C_1, pass the reduced domain of x_1 to C_2, filter C_2, and so forth to C_m, and similarly in a reverse direction for $C_{m-1}, \ldots, C_1$. It is assumed that domain consistency is obtained for the individual constraint C_i in each step of the procedure.

A simple inductive argument shows that this procedure obtains global domain consistency. The procedure is clearly valid for two constraints C_1 and C_2, because they have only one variable in common. Now suppose it is valid for any set of $m - 1$ constraints that have a Berge acyclic constraint hypergraph. An arbitrary constraint hypergraph for m constraints is obtained by attaching a constraint C_1 to some Berge acyclic hypergraph for $m-1$ constraints $C_2, \ldots, C_m$, where C_1 has a common variable x_1 with only one other constraint C_2. The constraint set $\{C_2, \ldots, C_m\}$ can be viewed as a single constraint that has only on variable x_1 in common with C_1, and the problem reduces to the two-constraint case.

A forward and backward pass achieve domain consistency in a dynamic programming model because the constraint hypergraph is not only Berge acyclic, but also has a natural topological ordering based on the ordering of the stages.

6.12.4 Nondeterministic Finite Automata

The decomposition scheme for filtering of the last section suggests a generalization of the regular constraint that can be filtered in the same fashion. It is based on a *nondeterministic* rather than on a deterministic finite automaton. This extension allows the representation of some constraints to be reduced by an exponential factor.

A nondeterministic finite automaton $N = (S, D, \tau, \alpha, F)$ consists of the same elements as a deterministic automaton, except that the transition function $\tau(s, x)$ maps a control x in state s to a *set* of states. The automaton accepts a string $x_1, \ldots, x_n$ if and only if there is a sequence of states $s_1, \ldots, s_{n+1} \in S$ such that

$$s_1 = \alpha, \quad s_{n+1} \in F, \quad \text{and } s_{i+1} \in \tau(s_i, x_i) \text{ for } i = 1, \ldots, n \qquad (6.86)$$

A generalized regular constraint $\text{regular}(x \mid N)$ requires that the nondeterministic automaton N accept $x_1, \ldots, x_n$. Domain consistency can be achieved by filtering and propagating the individual constraints (6.86) in a forward and then a backward pass.

For example, suppose that a machine can take any of three actions: $r = $ run, $c = $ clean, and $f = $ final shutdown. It can clean in any period, but it is required to clean k periods before final shutdown. A deterministic finite automaton requires at least 2^k states to model the problem, because it is unknown how long the machine will run before shutdown. However, a nondetermistic model has linear size. Let 0 be the initial state, $0'$ a state in which the machine must be cleaned, and $k + 1$ the state after shutdown (the only accepting state). The nondeterministic transition function is $\tau(0, r), \tau(0, c) \in \{0, 0'\}$; $\tau(0', c) \in \{1\}$; and $\tau(i, r), \tau(i, c) \in \{i + 1\}$ for $i = 1, \ldots, k$ (Fig. 6.21). The automaton does not determine how long the machine will run and/or clean before it reaches the final k transitions. Yet it accepts only schedules that clean the machine k periods before shutdown.

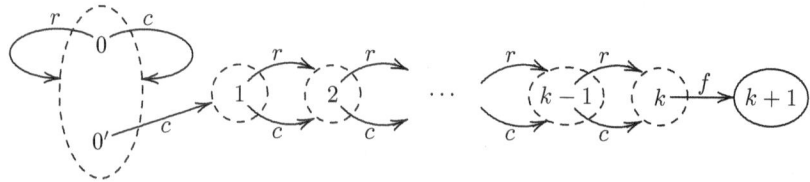

Fig. 6.21 Nondeterministic automaton for a machine scheduling problem. State 0 is the starting state, and $k + 1$ is the accepting state.

6.12.5 Cyclic Regular Constraint

As with the stretch constraint, it is conveninent to define a regular
constraint for cyclic schedules. It can be written

$$\text{regularCycle}\,(x \mid A)$$

where $x = (x_1, \ldots, x_n)$. The constraint requires that automaton A
accept the string $x_1, \ldots, x_n, x_1$.

Filtering for regularCycle can be accomplished by converting the
cyclic constraint to a standard regular constraint $\text{regular}(x \mid A')$, which
is then filtered in the normal fashion. The automaton A' is obtained by
making a copy of the portion of A that follows each initial move, and
ensuring that each copy terminates in the correct state. The automaton
A of Fig. 6.18, for example, becomes the automaton A' of Fig. 6.22.
Strings with length $n + 1$ ending in a appropriate state are desired,
rather than length n as before.

An alternate approach is to carry along the initial control as part of
the state in the dynamic programming model. If $A = (S, D, t, \alpha, F)$,
the state variable is a tuple (s_i, y_i), where y_i remembers the initial
control, and the transition function is

$$t_1(\alpha, x_1) = (t(\alpha, x_1), x_1)$$
$$t_i\,((s_i, y_i), x_i) = (t(s_i, x_i), y_i)\,, \quad i = 1, \ldots, n-1$$
$$t_{n+1}\,((s_{n+1}, y_{n+1}), x_{n+1}) = (t(s_{n+1}, x_{n+1}), x_{n+1}) \text{ for } x_{n+1} = y_{n+1} \text{ only}$$

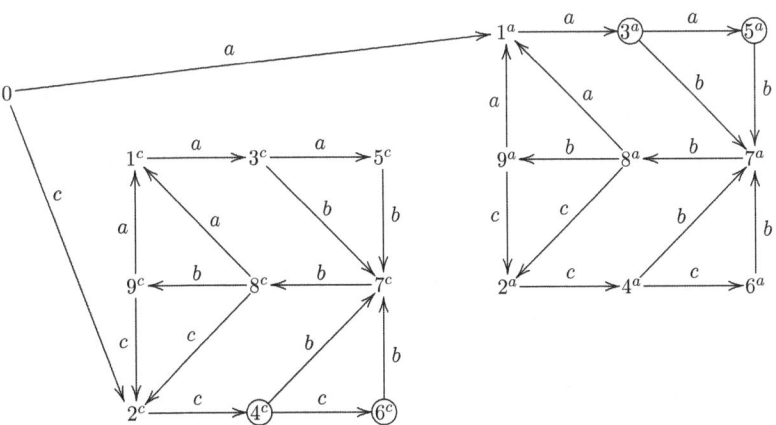

Fig. 6.22 Deterministic automaton for a cyclic shift scheduling problem.
Accepting states are circled.

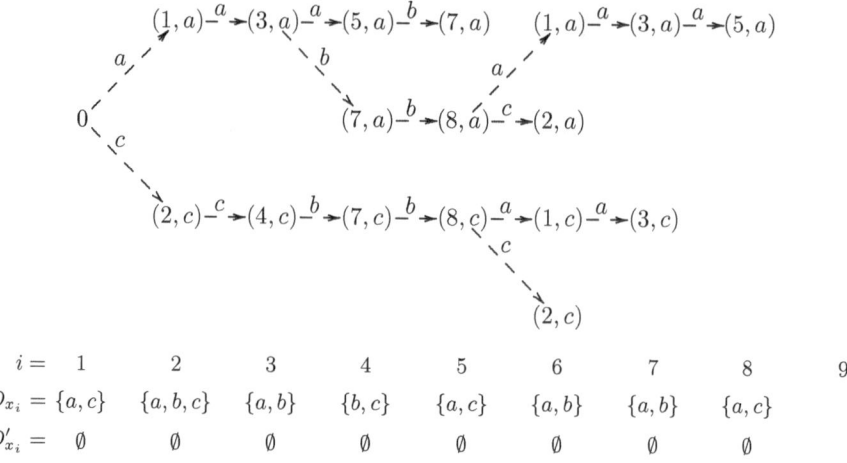

$i =$	1	2	3	4	5	6	7	8	9
$D_{x_i} =$	$\{a,c\}$	$\{a,b,c\}$	$\{a,b\}$	$\{b,c\}$	$\{a,c\}$	$\{a,b\}$	$\{a,b\}$	$\{a,c\}$	
$D'_{x_i} =$	$\emptyset$	$\emptyset$	$\emptyset$	$\emptyset$	$\emptyset$	$\emptyset$	$\emptyset$	$\emptyset$	

Fig. 6.23 State transition graph for a cyclic shift scheduling problem. All edges are dashed because no states are backward reachable (there are no terminal states). All domains reduce to the emptyy set.

The available controls are as before, except in stage $n+1$, where y_{n+1} is the only available control for state (s_{n+1}, y_{n+1}). The resulting state transition graph for the shift scheduling example is shown in Fig. 6.23. Note that no terminal states are reached in stage 9. As a result, no states are backward reachable, and the problem is infeasible.

6.12.6 A Dynamic Programming Constraint

If a regular constraint is filtered by determining forward and backward reachable states, the same filtering mechanism can be applied to a constraint that is expressed directly as a dynamic programming model (3.25). It might be written

$$\text{dynamicProgramming}\,(x, s \mid t, S_1, S_{m+1}, X, c, C)$$

where $x = (x_1, \ldots, x_n)$ are the control variables, $s = (s_1, \ldots, s_m)$ the state variables, $t = (t_1, \ldots, t_m)$ the transition functions, S_1 the set of feasible starting states, and S_{m+1} the set of feasible terminal states. Also $X = (X_1, \ldots, X_m)$, where $X_i(s_i)$ is the set of available controls in state s_i. Optional arguments $c = (c_1, \ldots, c_m)$ and C indicate a tuple of cost functions $c_i(s_i, x_i)$ and a maximum cost C. The constraint requires that x, s satisfy the constraints of the dynamic programming model (3.25), and if costs are present, it requires that the optimal value of (3.25) be less than or equal to C.

The dynamic programming constraint is more general than the regular constraint because it allows the transition function $t_i(s_i, x_i)$ to be indexed by the stage i. For example, it can express the knapsack example of Section 3.4, while the regular constraint cannot. The constraint can also be filtered by decomposition, simply by propagating the state transition constraints $x_{i+1} = t_i(s_i, x_i)$. It can therefore be generalized to nondeterministic transitions. Note, however, that this is not necessary for the machine scheduling example of the previous section. Because the transitions are indexed, one need only specify $D_{x_{m-k}} = \{c\}$ and $D_{x_i} = \{r, c\}$ for all $i \neq m - k, m + 1$

Exercises

6.60. Four drugs are available for a five-day medication regimen. The regimen must begin wigth drug a ior drug b on Monday. Drug c can immediately follow only a, and d can immediately follow only b. Each drug can follow only a drug that precedes it in alphabetical order. No drug can be taken more than two days in a row. Formulate the problem using a regular constraint, draw the deterministic finite automaton, and draw the state transition diagram than can be used for filtering. What are the feasible solutions? What are the reduced domains of each variable? *Hint:* There are three accepting states.

6.61. Describe syntactically the strings belonging to the regular language specified by the automaton of the previous exercise.

6.62. Solve the problem in Exercise 6.60 by decomposition.

6.63. Formulate the problem in Exercise 6.60 using the dynamic programming constraint, with transitions a, aa, b, bb, c, and cc. Note that there is no need to use transition c or d before stage 3.

6.13 The Circuit Constraint

The circuit constraint is similar to the alldiff constraint in that it requires a set of variables to indicate a permutation, but the variables encode the permutation in a different way.

In the constraint alldiff$(x_1, \ldots, x_n)$, each variable x_i indicates the ith item in a permutation of $1, \ldots, n$. In the circuit constraint

$$\text{circuit}(x_1, \ldots, x_n) \tag{6.87}$$

each variable x_i denotes which item follows i. Thus, (6.87) requires that $y_1, \ldots, y_n$ be a permutation of $1, \ldots, n$, where each $y_{i+1} = x_{y_i}$ (and y_{n+1} is identified with y_1).

Cyclic permutations of $y_1, \ldots, y_n$ represent the same solution of (6.87). For example, if each $x_i \in \{1, 2, 3\}$, then $\text{circuit}(x_1, x_2, x_3)$ has only two solutions, namely $(x_1, x_2, x_3) = (2, 3, 1)$ and $(3, 1, 2)$. The permutations $(y_1, y_2, y_3) = (1, 2, 3)$, $(2, 3, 1)$, and $(3, 1, 2)$ correspond to the first solution, while $(3, 2, 1)$, $(2, 1, 3)$, and $(1, 3, 2)$ coresond to to the second. In general, (6.87) has $(n - 1)!$ solutions for domains $x_i \in \{1, \ldots, n\}$, while $\text{alldiff}(x_1, \ldots, x_n)$ has $n!$ solutions.

The circuit constraint can be viewed as describing a *Hamiltonian cycle* on a directed graph. The elements $1, \ldots, n$ may be viewed as vertices of a directed graph G that contains an edge (i, j) whenever $j \in D_{x_i}$. An edge (i, j) is selected when $x_i = j$, and (6.87) requires that the selected edges form a Hamiltonian cycle, which is a path through vertices $y_1, \ldots, y_n, y_1$ for which $y_1, \ldots, y_n$ are all distinct. An edge is Hamiltonian when it is part of some Hamiltonian cycle. An element j can be deleted from D_{x_i} if and only if (i, j) is a non-Hamiltonian edge, which means that domain consistency can be achieved by identifying all non-Hamiltonian edges.

Achieving domain consistency is more difficult for the circuit constraint than for alldiff and is in fact an NP-hard problem. On the other hand, there may be more potential for domain reduction because the circuit constraint is stronger than alldiff when each variable domain is $\{1, \ldots, n\}$. Any $x = (x_1, \ldots, x_n)$ that satisfies (6.87) also satisfies $\text{alldiff}(x_1, \ldots, x_n)$, whereas the reverse is not true. There are also strong relaxations for circuit, some based on cutting planes that have been developed for the traveling salesman problem (Section 7.11).

6.13.1 Modeling with Circuit

The circuit constraint is naturally suited to modeling situations in which costs or constraints depend on which item immediately follows another in a permutation. This occurs, for instance, in the traveling salesman problem, in which a salesman wishes to visit each of n cities once and return home while minimizing the distance traveled. If c_{ij} is the distance from city i to city j, the problem is

$$\min \sum_i c_{ix_i}$$

$$\text{circuit}(x_1, \ldots, x_n)$$ (6.88)

where x_i is the city visited immediately after city i. If certain cities cannot be visited immediately after city i, this can be reflected in the domain of x_i.

On the other hand, if one wishes to assign workers to jobs, and the cost of assigning worker i to job j is c_{ij}, then the alldiff constraint provides the natural formulation:

$$\min \sum_i c_{iy_i}$$

$$\text{alldiff}(y_1, \ldots, y_n)$$

Here, y_i is the job assigned to worker i, and the objective is to minimize total cost. This is the classical assignment problem. If there is a restriction on which jobs may be assigned to worker i, this can be reflected in the initial domain of y_i.

The traveling salesman problem can be modeled with alldiff as well as circuit, but the alldiff formulation provides no natural way to constrain which cities may follow a given city in the salesman's tour. The alldiff model is

$$\min \sum_i c_{y_i y_{i+1}}$$

$$\text{alldiff}(y_1, \ldots, y_n)$$

where y_i is the ith city visited and city $n + 1$ is identified with city 1. If the salesman is allowed to visit only cities in D_i after visiting i, then one must add the channeling constraints

$$y_{i+1} = x_{y_i}, \quad i = 1, \ldots, n - 1$$
$$y_1 = x_{y_n}$$ (6.89)

Now the city immediately following city i can be constrained by specifying the domain of x_i.

The assignment problem can be modeled with circuit as well as alldiff, but only by introducing variables y_i that indicate the job assigned to worker i:

$$\min \sum_i c_{iy_i}$$

$$\text{circuit}(x_1, \ldots, x_n)$$

$$\text{constraints (6.89)}$$

The domain of y_i can be restricted to indicate which cities may be assigned to salesman i.

The circuit constraint is well suited for scheduling problems that involve sequence-dependent setup times or costs. Suppose, for example, that when job i immediately precedes job j, the time required to process job i and then set up job j is c_{ij}. The domain of each x_i contains the jobs that can immediately follow job i. If the jobs are to be sequenced so as to minimize makespan (total time required to complete the jobs), the problem can be written as (6.88). This formulation is valid if job n is interpreted as a dummy job that immediately follows the last job processed. Then $c_{in} = c_{ni} = 0$ for $i = 1, \ldots, n - 1$, and x_1 is the first job to run.

6.13.2 Elementary Filtering Methods

Checking a circuit constraint for feasibility is equivalent to checking whether a directed graph has a Hamiltonian cycle, which is an NP-hard problem. Achieving domain consistency for circuit is therefore NP-hard. There are useful incomplete filtering methods, however, that run in polynomial time.

Two elementary filtering methods for circuit are based on alldiff filtering and vertex-degree filtering. The alldiff filtering methods of Section 6.8 can be applied because the variables in $\text{circuit}(x_1, \ldots, x_n)$ must take different values.

Vertex-degree filtering is based on the fact that the in-degree and out-degree of every vertex in a Hamiltonian cycle is one. Let the inverse domain of j be $D_j^{-1} = \{i \mid j \in D_{x_i}\}$, which is the index set of all variables whose domain contains j. The vertex-degree filtering algorithm cycles through two steps until no further domain reduction is possible:

1. If $D_{x_i} = \{j\}$ for some i, remove j from D_{x_k} for all $k \neq i$.
2. If $D_j^{-1} = \{i\}$ for some j, reduce D_{x_i} to $\{j\}$.

Alldiff filtering is slower than vertex-degree filtering but strictly dominates it with respect to values removed. Domination is easy to show, and strict domination is demonstrated by the example of Fig. 6.24. Because $D_{x_1} = D_{x_2} = \{3, 4\}$, alldiff filtering removes 3 and 4 from the other two domains. Vertex-degree filtering has no effect, however, because the in-degree and out-degree of every vertex is 2.

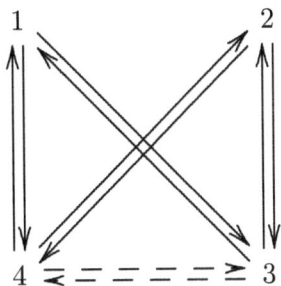

Fig. 6.24 Circuit problem in which alldiff filtering removes the dashed edges, but vertex-degree filtering has no effect.

6.13.3 Filtering Based on Separators

A more thorough filtering method than alldiff or vertex-degree filtering can be obtained by identifying one or more vertex separators of the associated graph G. A vertex separator is a set of vertices that, when removed, separate G into two or more connected components. By defining a certain kind of labeled graph on a vertex separator, one can state a necessary condition for an edge of that graph to be Hamiltonian in G. This allows one to filter domains by analyzing a graph that may be much smaller than G.

Let $G = (V, E)$ be a directed graph with vertex set V and edge set E. A subset S of V is a *(vertex) separator* of G if $V \setminus S$ induces a subgraph with two or more connected components.

The *separator graph* for a separator S of G is a graph $G_S = (S, E_S)$ in which E_S contains *labeled* and *unlabeled* edges. Edge (i, j) is an unlabeled edge of G_S if $(i, j) \in G$. Let C be any connected component of the subgraph of G induced by $V \setminus S$. Edge (i, j) is an edge of G_S with *label C* if (i, c_1) and (c_2, j) are edges of G for some pair of vertices c_1, c_2 of C (possibly $c_1 = c_2$).

Consider, for example, the graph G of Fig. 6.25. Vertex set $S = \{1, 2, 3\}$ separates G into three connected components that may be labeled A, B and C, each of which contains only one vertex. The separator graph G_S contains the three edges that connect its vertices in G plus four labeled edges (shown in dashes). For example, there is an edge $(1, 2)$ labeled A, which can be denoted $(1, 2)^A$, because there is an edge $(1, 4)$ from 1 to a vertex in component A, an there is an edge $(4, 2)$ from a vertex in A to 2.

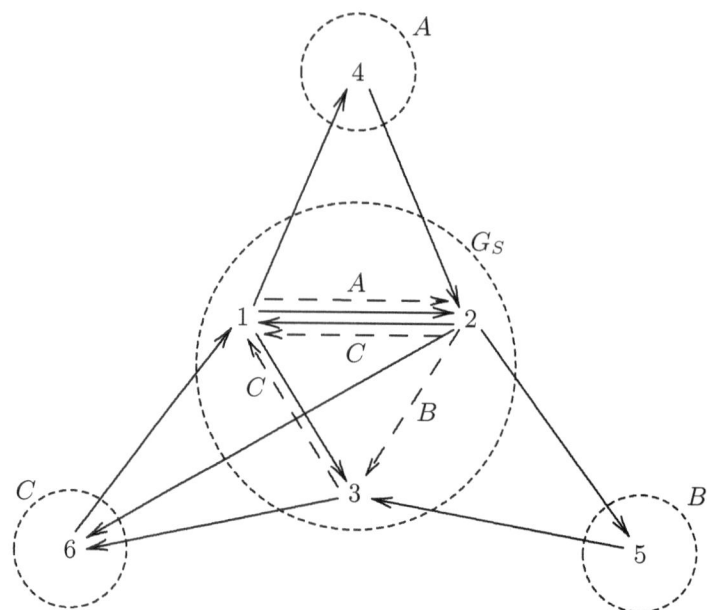

Fig. 6.25 Graph G on vertices $\{1,\ldots,6\}$ contains all the solid edges, and the separator graph G_S on $S = \{1,2,3\}$ contains the solid and dashed edges within the larger circle. The small circles surround connected components of the separated graph.

A Hamiltonian cycle of the separator graph G_S is *permissible* if it contains at least one edge bearing each label. Thus, the edges $(1,2)^A$, $(2,3)^B$, and $(3,1)^C$ form a permissible Hamiltonian cycle in Fig. 6.25. The Hamiltonian cycle $(1,2)$, $(2,3)^B$, and $(3,1)^C$ is nor permissible because the label A is missing. In general, a permissible cycle can contain unlabeled edges, but every label must be present on some edge.

Theorem 6.20. *If S is a separator of directed graph G, then G contains a Hamiltonian cycle only if G_S contains a permissible Hamiltonian cycle. Furthermore, an edge of G connecting vertices in S is Hamiltonian only if it is part of a permissible Hamiltonian cycle of G_S.*

Proof. The first task is to show that if H is an arbitrary Hamiltonian cycle of $G = (V, E)$, then one can construct a permissible Hamiltonian cycle H_S for G_S. Consider the sequence of vertices in H and remove those that are not in S. Let $v_1, \ldots, v_m, v_1$ be the remaining sequence of vertices. H_S can be constructed on these vertices as follows. For any pair v_i, v_{i+1} (where v_{m+1} is identified with v_1), if they are adjacent in H then (v_i, v_{i+1}) is an unlabeled edge of G_S and connects v_i and v_{i+1} in H_S. If v_i, v_{i+1} are not adjacent in H, then all vertices in H between v_i and v_{i+1} lie in the same connected component C of the subgraph of G induced by $V \setminus S$. This means (v_i, v_{i+1}) is an edge of G_S with label C, and $(v_i, v_{i+1})^C$ connects v_i and v_{i+1} in H_S. Since H passes through all connected components, every label must occur on some edge of H_S, and H_S is permissible.

The second task is to show that if (i, j) with $i, j \in S$ is an edge of a Hamiltonian cycle H of G, then (i, j) is an edge of a permissible Hamiltonian cycle of G_S. But in this case (i, j) is an unlabeled edge of G_S and, by the above construction, (i, j) is part of H_S. $\square$

As noted earlier, the separator graph G_S in Fig. 6.25 contains a permissible Hamiltonian cycle with edges $(1, 2)^A$, $(2, 3)^B$, and $(3, 1)^C$. Because this is the only permissible Hamiltonian cycle, none of the unlabeled edges $(1, 2)$, $(2, 1)$, and $(1, 3)$ are part of a permissible Hamiltonian cycle. Thus, by Theorem 6.20, these edges are non-Hamiltonian in the original graph G. The domain D_{x_1} can be reduced from $\{2, 3, 4\}$ to $\{4\}$, and D_{x_2} from $\{1, 5, 6\}$ to $\{5, 6\}$.

The application of Theorem 6.20 requires that one find one or more separators of the original graph G. One way to find them is to use a simple breadth-first-search heuristic. Let vertices i, j be *neighbors* if (i, j) or (j, i) is an edge of G. Arrange the vertices of G in levels as follows. Arbitrarily select a vertex i of G as a *seed* and let level 0 contain i alone. Let level 1 contain all neighbors of i in G. Let level k (for $k \geq 2$) contain all vertices j of G such that (a) j is a neighbor of some vertex on level $k - 1$, and (b) j does not occur in levels 0 through $k - 1$. If $m \geq 2$, the vertices on any given level k ($0 < k < m$) form a separator of G. Thus, the heuristic yields $m - 1$ separators.

The heuristic can be run several times as desired, each time beginning with a different vertex on level 0. In the example, using vertex 6 as a seed produces the separator of Fig. 6.25 on vertices 1, 2, and 3. Using vertex 4 as a seed, however, yields a separator graph on vertices 1 and 2 only, and so forth.

6.13.4 Network Flow Model

The task that remains is to identify edges of G_S that are part of no permissible Hamiltonian cycle in G_S, or *nonpermissible* edges for short. Rather than attempt to find all nonpermissible edges, which could be computationally expensive even for relatively small separator graphs, one can identify edges that satisfy a weaker condition. One approach is to construct a network flow model that enforces a relaxation of the condition in Theorem 6.20, as well as a vertex-degree constraint. The flow model is similar to that presented for the cardinality constraint in Section 6.9.1.

First, construct a network $N(G, S)$ as follows. There is a source node s, a node C for every label, a node U, a node for every ordered pair (i, j) for which at least one edge connects i to j in G_S, a node for every vertex of G_S, and a sink node t. The arcs of $N(G, S)$ are as follows:

- an arc (s, C) with capacity range $[1, \infty)$ for each label C
- an arc (s, U) with capacity $[0, \infty)$
- an arc $(C, (i, j))$ with capacity $[0, 1]$ for each edge $(i, j)^C$ of G_S
- an arc $(U, (i, j))$ with capacity $[0, 1]$ for each unlabeled edge (i, j) of G_S
- an arc $((i, j), i)$ with capacity $[0, 1]$ for each node (i, j) of $N(G, S)$
- an arc (i, t) with capacity $[0, 1]$ for each vertex i if G_s
- a return arc (t, s) with capacity $[m, m]$.

The network $N(G, S)$ for the separator graph G_S of Fig. 6.25 appears in Fig. 6.26.

Every permissible Hamiltonian cycle on G_S describes a flow pattern on $N(G, S)$, in which a flow of 1 from C to (i, j) indicates that $(i, j)^C$ is part of the cycle, and a flow of 1 from U to (i, j) means that the unlabeled edge (i, j) is part of the cycle. Since the cycle must contain all m vertices of G_S, the flow on the return arc must be m. Since each label must occur on the cycle, the flow on each arc (s, C) must be at least 1. Since the out-degree of each vertex i of the cycle is 1, the flow on each (i, t) can be at most 1. In addition:

Theorem 6.21. *An edge (i, j) of G is non-Hamiltonian if there is a separator S of G for which the maximum flow on edge $(U, (i, j))$ of $N(G, S)$ is zero.*

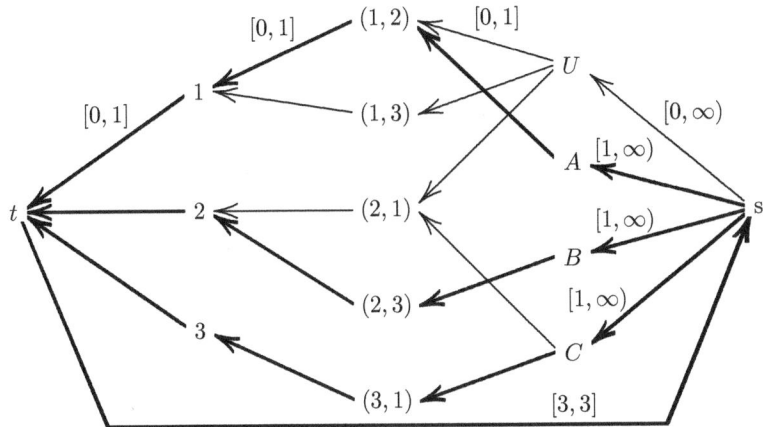

Fig. 6.26 Flow model for simultaneous cardinality and out-degree filtering of non-Hamiltonian edges. Heavy lines show the only feasible flow. Since the maximum flow on edges $(U, (1, 2))$, $(U, (1, 3))$, and $(U, (2, 1))$ is zero, edges $(1, 2)$, $(1, 3)$, and $(2, 1)$ in the original graph are non-Hamiltonian.

As in Section 3.2.4, this can be checked by computing a feasible flow f on $N(G, S)$. The maximum flow on $(U, (i, j))$ is zero if and only if it is zero in f and there is no augmenting path from (i, j) to U (Corollary 3.8).

For example, the zero flow on edges $(U, (1, 2))$, $(U, (1, 3))$ and $(U, (2, 1))$ of Fig. 6.26 is maximum in each case. So, the three edges $(1, 2)$, $(1, 3)$, and $(2, 1)$ of G_S are non-Hamiltonian.

A similar test can be devised to combine cardinality filtering with in-degree filtering. It is unclear how to combine cardinality filtering with both out-degree and in-degree filtering in the same network model.

Exercises

6.64. Assume there is an oracle that can quickly tell whether a graph is Hamiltonian. Describe how to use this oracle to check quickly whether a particular edge in a graph is Hamiltonian.

6.65. Apply Theorem 6.20 to the graph in Fig. 6.25 when the separator is $S = \{1, 2\}$ and when the separator is $S = \{1, 2, 3, 4\}$. Does the theorem identify all non-Hamiltonian edges in the subgraph induced by S? For each separator graph use the flow model to detect nonpermissible edges. Does it identify all nonpermissible edges?

6.66. Consider the graph with directed edges $(1, 2), (1, 4), (1, 5), (2, 3), (2, 7),$ $(3, 4), (3, 7), (3, 8), (4, 1), (4, 5), (5, 6), (6, 2), (6, 5), (7, 3), (8, 1), (8, 4)$ and the separator $S = \{1, 2, 3, 4\}$. Use Theorem 6.20 to filter as many edges as possible. Does this filter remove any edges that are not removed by vertex-degree filtering? Does it remove any that are not removed by alldiff filtering? Does it remove all non-Hamiltonian edges connecting vertices of S?

6.67. Use the flow model to detect nonpermissible edges in the separator graph constructed in Exercise 6.66. Does it identify all nonpermissible edges?

6.68. Show by counterexample that filtering based on Theorem 6.20 is incomplete, even when all separators are used.

6.14 Disjunctive Scheduling

Disjunctive scheduling is the problem of scheduling jobs that must run one at a time, subject to a release time and deadline for each job. The processing time of each job is fixed. In *preemptive scheduling*, one job can be interrupted to start processing another job, while this is not allowed in *nonpreemptive scheduling*. The focus here is on nonpreemptive scheduling.

The basic disjunctive scheduling constraint is

$$\mathrm{noOverlap}(s \,|\, p)$$

where $s = (s_1, \ldots, s_n)$ is a tuple of variables s_j indicating the start time of job j. The parameter $p = (p_1, \ldots, p_n)$ is a tuple of processing times p_j for each job. The constraint requires that for any pair of jobs, one must finish before the other starts. That is, it enforces the disjunctive condition

$$(s_i + p_i \leq s_j) \vee (s_j + p_j \leq s_i)$$

for all i, j with $i \neq j$. The constraint programming literature often refers to the noOverlap constraint as a *unary resource* constraint because each job requires one unit of resource while it is running, and only one unit of resource is available at any one time.

Each job j is associated with an earliest start time E_j and a latest completion time L_j. Initially, these are the release time and deadline of the job, respectively, but they may be updated in the course of the solution algorithm. Thus, the current domain of s_j is the interval $[E_j, L_j - p_j]$, and the release time and deadline of each job is indicated by the initial domain of s_j. Constraint programming systems treat the domain of s_j as a sequence of consecutive integers, but none of the techniques described here presuppose that the domain elements be integral. If all the problem data are integral, however, E_j should always be rounded up and L_j rounded down.

The filtering task for the noOverlap constraint is to reduce the domains of the s_j's as much as possible. Achieving full bounds consistency is an NP-hard problem, since checking for the existence of a feasible schedule is NP-hard. Yet filtering algorithms that stop short of full bounds consistency can be very valuable in practice.

The most popular filtering methods are based on the edge-finding principle and the not-first/not-last principle. The former finds jobs that must precede or follow others, and the latter finds jobs that cannot be first or last in a given subset of jobs. Either can allow one to shrink some of the time windows $[E_j, L_j]$.

6.14.1 Edge Finding

Edge finding is the best-known filtering method for disjunctive scheduling, and it plays an important role in practical solvers. It identifies subsets of jobs that must all precede, or all follow, a particular job. The name *edge finding* derives from the fact that the procedure finds new edges for the precedence graph, which is a graph in which directed edges indicate which jobs must precede other jobs.

It is helpful to introduce some notation that is specialized to scheduling analysis. For a subset J of jobs, let $E_J = \min_{j \in J}\{E_j\}$, $L_J = \max_{j \in J}\{L_j\}$, and $p_J = \sum_{j \in J} p_j$. Also, $i \gg J$ means that job i starts after every job in J has finished, and $i \ll J$ means that job i finishes before any job in J starts.

The edge-finding principle can be expressed in two symmetrical rules:

$$\text{If } L_J - E_{J \cup \{i\}} < p_i + p_J, \text{ then } i \gg J \quad (a)$$
$$\text{If } L_{J \cup \{i\}} - E_J < p_i + p_J, \text{ then } i \ll J \quad (b)$$

$$(6.90)$$

Rule (a) is based on the fact that if job i does not follow the jobs in J, then some job in J must run last. This means that all the jobs in

$J \cup \{i\}$ must be performed between their earliest start time $E_{J \cup \{i\}}$ and the latest finish time L_J for jobs in J alone. If the total time $p_i + p_J$ required for this will not fit in this interval, then job i must follow all the jobs in J. Rule (b) is based on the same reasoning in reverse direction.

If it is found that job i must follow the jobs in J, then job i cannot start until all the jobs in J finish. A lower bound on this start time is the maximum of $E_{J'} + p_{J'}$ over all subsets J' of J. Similarly, if job i must precede the jobs in J, then i must finish before any of the jobs in J start. An upper bound on this finish time is the minimum of $L_{J'} - p_{J'}$ over all $J' \subset J$. In summary,

$$\text{If } i \gg J, \text{ then update } E_i \text{ to max} \left\{ E_i, \max_{J' \subset J} \{E_{J'} + p_{J'}\} \right\}.$$

$$\text{If } i \ll J, \text{ then update } L_i \text{ to min} \left\{ L_i, \min_{J' \subset J} \{L_{J'} - p_{J'}\} \right\}.$$

(6.91)

As an example consider the four-job scheduling problem described in Table 6.6 and illustrated in Fig. 6.27. Since

$$L_{\{1,2\}} - E_{\{1,2,4\}} = 6 - 0 < 3 + (3 + 1) = p_4 + (p_1 + p_2)$$

rule (6.90a) implies that job 4 must follow jobs 1 and 2, or $4 \gg \{1, 2\}$. So, E_4 is updated to

$$\max \left\{ E_4, \max\{E_{\{1,2\}} + p_{\{1,2\}}, E_1 + p_1, E_2 + p_2\} \right\}$$
$$= \max \{0, \max\{1 + 4, 2 + 1, 1 + 3\}\} = 5$$

Note that, although $2 \ll \{3\}$, edge finding does not deduce this fact.

The practicality of edge finding rests on the fact that one need not examine all subsets J of jobs to find all the bound updates that can be established by the edge-finding rules. In fact, the following polynomial-time algorithm suffices. It runs in time proportion to n^2, where n is the

Table 6.6 Data for a four-job disjunctive scheduling problem.

j	p_j	E_j	L_j
1	1	2	5
2	3	1	6
3	1	3	8
4	3	0	9

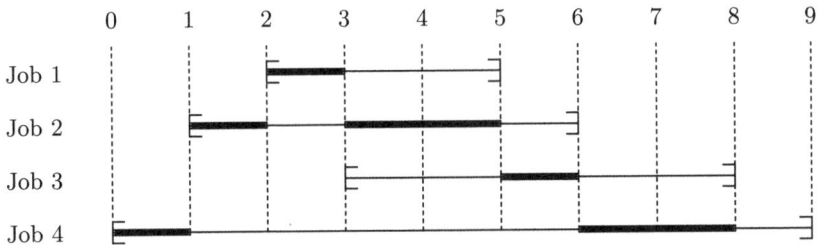

Fig. 6.27 Time windows (horizontal lines) for the four-job disjunctive scheduling problem of Table 6.6. The heavy lines show the Jackson preemptive schedule.

number of jobs. The fastest known algorithm runs in time proportional to $n \log n$, but it requires much more complex data structures.

The first step of the algorithm is to compute the *Jackson preemptive schedule* (JPS) for the given instance. Let us say that a job j is *available* at a given time t if $t \in [E_j, L_j]$ and job j has not completed. Then, as one moves forward in time, the job in process at each time t should be the job j that has the smallest L_j among the jobs available (if any) at t. A JPS is illustrated in Fig. 6.27.

Now, for each job i, do the following. Let J_i be the set of jobs that are not finished at time E_i in the JPS. Let $\bar{p}_j$ be the processing time left for job j at time E_i in the JPS. Finally, let J_{ik} be the jobs in J_i, other than i, that have deadlines at or before job k's deadline:

$$J_{ik} = \{j \in J_i \setminus \{i\} \mid L_j \le L_k\}$$

Examine the jobs $k \in J_i$ ($k \ne i$) in decreasing order of deadline L_k, and select the first job for which

$$L_k - E_i < p_i + \bar{p}_{J_{ik}} \qquad (6.92)$$

Then conclude that $i \gg J_{ik}$ and update E_i to JPS(i, k), which is the latest completion time in the JPS of the jobs in J_{ik}.

This algorithm updates the earliest start times E_i. The same algorithm is run with the direction of time reversed to update the latest completion times L_i. Table 6.7 details the execution of the algorithm for the example of Fig. 6.27. It identifies the one precedence that is established by edge finding, $4 \gg \{1, 2\}$, and updates E_1 from 0 to 5, which is the latest finish time of jobs 1 and 2 in the JPS shown in Fig. 6.27.

Table 6.7 Execution of an algorithm for edge finding.

i	J_i	$\bar{p}$	k	J_{ik}	$L_k - E_i$	$p_i + \bar{p}_{J_{ik}}$
1	$\{1,2,3,4\}$	$(1,2,1,2)$	4	$\{2,3,4\}$	$9-2$	$1+5$
			3	$\{2,3\}$	$8-2$	$1+3$
			2	$\{2\}$	$6-2$	$1+3$
2	$\{1,2,3,4\}$	$(1,3,1,2)$	4	$\{1,3,4\}$	$9-1$	$3+4$
			3	$\{1,3\}$	$8-1$	$3+2$
			1	$\{3\}$	$5-1$	$3+1$
3	$\{2,3,4\}$	$(0,2,1,2)$	4	$\{2,4\}$	$9-3$	$1+4$
			2	$\{2\}$	$6-3$	$1+2$
4	$\{1,2,3,4\}$	$(1,3,1,3)$	3	$\{1,2,3\}$	$8-0$	$3+5$
			2	$\{1,2\}$	$6-0$	$3+4$
	Conclude that $4 \gg \{1,2\}$ and update E_4 from 0 to 5					

Theorem 6.22. *The above edge-finding algorithm is valid and identifies all updated domains that can be deduced from the edge-finding rules (6.90)–(6.91).*

Proof. First show that the algorithm is valid. It suffices to show that any precedence $i \gg J_{ik}$ discovered by the algorithm is valid, because once it is given that $i \gg J_{ik}$, updating E_i to JPS(i, k) is clearly valid (the argument is similar for updating L_i). Suppose, then, that the algorithm derives that $i \gg J_{ik}$, which means that J_{ik} satisfies (6.92). To verify that $i \gg J_{ik}$, it suffices to show that J_{ik} satisfies the edge-finding rule (6.90a):

$$L_{J_{ik}} - E_{J_{ik} \cup \{i\}} < p_i + p_{J_{ik}} \tag{6.93}$$

This can be deduced as follows from (6.92). By definition of J_{ik}, $L_k = L_{J_{ik}}$. Thus, it suffices to show $E_i - E_{J_{ik} \cup \{i\}} = p_{J_{ik}} - \bar{p}_{J_{ik}}$. But this follows from the fact that none of the jobs in J_{ik} are finished at time E_i in the JPS.

Now show that for any valid update E_i' that can be obtained from the edge-finding rules (6.90)–(6.91), the algorithm obtains an update $E_i'' \geq E_i'$ (the argument is similar for any valid update L_i'). It is given that

$$L_J - E_{J \cup \{i\}} < p_i + p_J \tag{6.94}$$

for some J and

$$E_i' = E_{J'} + p_{J'} \tag{6.95}$$

for some $J' \subset J$. Let k be a job in J with the largest L_k. It will first be shown that

$$L_k - E_i < p_i + \bar{p}_{J_{ik}} \tag{6.96}$$

Clearly, $L_k = L_J$. So, if $\Delta = E_i - E_{J \cup \{i\}}$, then by (6.94) it suffices to show

$$p_J - \bar{p}_{J_{ik}} \le \Delta \tag{6.97}$$

Let p^* be the total JPS processing time of jobs in J between $E_{J_{ik} \cup \{i\}}$ and E_i. Also let $\bar{p}_{J_{ik} \setminus J}$ be the total JPS processing time that remains for jobs in $J_{ik} \setminus J$ at time E_i. Then,

$$\bar{p}_{J_{ik}} = p_J + \bar{p}_{J_{ik} \setminus J} - p^* \tag{6.98}$$

But, since the jobs running between $E_{J_{ik} \cup \{i\}}$ and E_i run one at a time, $p^* \le \Delta$. Thus, (6.98) implies $\bar{p}_{J_{ik}} \ge p_J - \Delta$, which implies (6.97).

Since (6.96) holds, the algorithm will discover the precedence $i \gg J_{ik'}$ for some k' for which $L_k \le L_{k'}$ (perhaps $k = k'$), and therefore for which $J_{ik} \subset J_{ik'}$. The algorithm therefore obtains the update $E_i'' = \text{JPS}(i, k')$. Thus

$$E_i'' \ge E_i + \bar{p}_{J_{ik'}} \ge E_i + \bar{p}_{J'} \ge E_{J'} + \hat{p} + \bar{p}_{J'} = E_{J'} + p_{J'} = E_i' \tag{6.99}$$

The first inequality is due to the fact that the time interval between E_i and $\text{JPS}(i, k')$ is at least the total processing time $\bar{p}_{J_{ik'}}$ that remains for the jobs in $J_{ik'}$. The second inequality is due to the fact that all the jobs in J' that are unfinished in the JPS at time E_i belong to $J_{ik} \subset J_{ik'}$. The third inequality holds if $\hat{p}$ is defined to be the total JPS processing time between $E_{J'}$ and E_i of jobs in J', since these jobs must run one at a time. The first equation is due to the definition of $\bar{p}_{J'}$. Thus, $E_i'' \ge E_i'$, which completes the proof. $\square$

6.14.2 Not-First/Not-Last Rules

Edge finding identifies jobs i that must occur first or last in a set $J \cup \{i\}$ of jobs. A complementary type of rule identifies jobs i that cannot occur first or cannot occur last in $J \cup \{i\}$:

$$\begin{aligned} &\text{If } L_J - E_i < p_i + p_J, \text{ then } \neg(i \ll J) \quad (a) \\ &\text{If } L_i - E_J < p_i + p_J, \text{ then } \neg(i \gg J) \quad (b) \end{aligned} \tag{6.100}$$

Rule (a) is based on the fact that if job i occurs first in $J \cup \{i\}$, then there must be enough time between E_i and L_J to run all the jobs. If there is not enough time, then i cannot be first. Similarly, rule (b) is based on the fact that if job i occurs last, then there must be enough time between E_J and L_i to run all of the jobs. If there is not, then i cannot be last.

If a not-first or not-last position is established, then the updating of bounds is simpler than in the case of edge finding:

$$\text{If } \neg(i \ll J), \text{ then update } E_i \text{ to } \max\left\{E_i, \min_{j \in J}\{E_j + p_j\}\right\} \quad (a)$$

$$\text{If } \neg(i \gg J), \text{ then update } L_i \text{ to } \min\left\{L_i, \max_{j \in J}\{L_j - p_j\}\right\} \quad (b)$$

$$(6.101)$$

It is simpler because the inner min or max is over all jobs in J rather than all subsets of J.

Returning to the example of Fig. 6.27, rule (6.100a) deduces the fact that $\neg(4 \ll \{1, 2\})$ because

$$L_{\{1,2\}} - E_4 = 6 - 0 < 3 + (1 + 3) = p_4 + (p_1 + p_2)$$

This allows E_4 to be updated from 0 to $\min\{E_1 + p_1, E_2 + p_2\} = \min\{3, 4\} = 3$. Actually, this result is dominated by the fact that $4 \gg \{1, 2\}$, discovered by edge finding, which updates E_4 to 5. Rule (6.100a) implies that $\neg(3 \ll \{2\})$, however, and rule (6.100b) implies that $\neg(2 \gg \{3\})$. That is, $2 \gg \{3\}$, a fact not deduced by edge finding. The conclusion that $\neg(3 \ll \{2\})$ allows E_3 to be updated from 3 to 4, although $\neg(2 \gg \{3\})$ has no effect on L_2.

As in the case of edge finding, the not-first/not-last rules can be applied without examining all subsets J of jobs. The following algorithm runs in time proportional to n^2, where n is the number of jobs. One part of the algorithm identifies all updates that result from the not-first rule (6.100a)–(6.101a); the procedure for the not-last rule is similar. A running time proportional to $n \log n$ can be obtained by introducing a binary tree data structure.

It is assumed that the jobs are indexed in nondecreasing order of deadlines, so that $j \le k$ implies $L_j \le L_k$. Let $\bar{J}_{jk}$ be the set of jobs with deadlines no later than L_k whose earliest finish time is no earlier than job j's earliest finish time. That is,

$$\bar{J}_{jk} = \{\ell \le k \mid E_j + p_j \le E_\ell + p_\ell\}$$

Also let LST_{jk} be the following upper bound on the latest time at which the jobs in $\bar{J}_{jk}$ can start,

$$\mathrm{LST}_{jk} = \min_{\ell \le k} \left\{ L_\ell - p_{\bar{J}_{j\ell}} \right\}$$

where, by default, $p_\emptyset = -\infty$. For each j, the quantity LST_{jk} can be computed recursively for $k = 1, \ldots, n$.

The not-first part of the algorithm goes as follows. Let E_i' be the updated release time for each job i, which is initialized to E_i. For each job j do the following. For $i = 1, \ldots, n$ $(i \ne j)$ let $E_i' = \max\{E_i', E_j + p_j\}$ if one of the two conditions below is satisfied:

(i) $E_i + p_i < E_j + p_j$ and $E_i + p_i > \mathrm{LST}_{jn}$
(ii) $E_i + p_i \ge E_j + p_j$ and either $E_i + p_i > \mathrm{LST}_{j,i-1}$ or $E_i > \mathrm{LST}_{jn}$

Table 6.8 shows how the not-first algorithm is applied to the example of Fig. 6.27. Supporting data appear in Table 6.9. Note in Table 6.8 that entries appear in the column labeled $E_i + p_i < E_j + p_j$ when this condition is satisfied, and they otherwise appear in the columns headed $E_i + p_i \ge E_j + p_j$. The algorithm updates E_3 from 3 to 4, and E_4 from 0 to 3.

It may be assumed that $E_\ell + p_\ell \le L_\ell$ for all ℓ, because otherwise it is trivial to check that there is no feasible schedule.

Table 6.8 Execution of a not-first algorithm.

j	E_j+p_j	LST_{j4}	i	$E_i+p_i < E_j+p_j$ $E_i+p_i > \mathrm{LST}_{j4}$	$E_i+p_i \ge E_j+p_j$ $\mathrm{LST}_{j,i-1}$	$E_i+p_i >$ $\mathrm{LST}_{j,i-1}$	$E_i >$ LST_{j4}	Update
1	3	1	2		4	no	no	
			3		2	yes	yes	$E_3' = 3$
			4		2	yes	no	$E_4' = 3$
2	4	3	1	no				
			3		3	yes	no	$E_3' = 4$
			4	no				
3	4	7	1	no				
			2		∞	no	no	
			4	no				
4	3	6	1				no	
			2	no				
			3	no				

Table 6.9 Data for execution of the not-first algorithm shown in Table 6.8.

i	E_i	$E_i + p_i$	$\bar{J}_{jk}$	$k=1$	2	3	4
1	2	3	$j=1$	$\{1\}$	$\{1,2\}$	$\{1,2,3\}$	$\{1,2,3,4\}$
2	1	4	2	$\emptyset$	$\{2\}$	$\{2,3\}$	$\{2,3\}$
3	3	4	3	$\emptyset$	$\emptyset$	$\{3\}$	$\{3\}$
4	0	3	4	$\emptyset$	$\emptyset$	$\emptyset$	$\{4\}$

$L_k - p_{\bar{J}_{jk}}$	$k=1$	2	3	4
$j=1$	4	2	3	1
2	∞	3	4	5
3	∞	∞	7	8
4	∞	∞	∞	6

LST_{jk}	$k=1$	2	3	4
$j=1$	4	2	2	1
2	∞	3	3	3
3	∞	∞	7	7
4	∞	∞	∞	6

Theorem 6.23. *The above not-first/not-last algorithm is valid and identifies all updated domains that can be deduced from the not-first/not-last rules (6.100)–(6.101).*

The not-first part of the theorem follows from Lemmas 6.25 and 6.26 below, which in turn rely on Lemma 6.24. The argument is similar for the not-last part of the theorem.

Lemma 6.24 *If the not-first rule of (6.100a)–(6.101a) updates E_i to $E_j + p_j$ for some set J, then for some $k \geq j$ the rule yields this same update when $J = \bar{J}_{jk} \setminus \{i\}$.*

Proof. It is given that

$$L_J - E_i < p_i + p_J \tag{6.102}$$

and that

$$E'_i = E_j + p_j = \min_{\ell \in J} \{E_\ell + p_\ell\} \tag{6.103}$$

Let k be the largest index in J. It suffices to show

$$L_{\bar{J}_{jk}\setminus\{i\}} - E_i < p_i + p_{\bar{J}_{jk}\setminus\{i\}} \tag{6.104}$$

and

$$E_j + p_j = \min_{\ell \in \bar{J}_{jk}\setminus\{i\}} \{E_\ell + p_\ell\} \tag{6.105}$$

Because the jobs are indexed in nondecreasing order of deadline, $L_{\bar{J}_{jk}} = L_J$ by definition of k. Also, $J \subset \bar{J}_{jk} \setminus \{i\}$ due to (6.103) and the fact that $i \notin J$. This means $p_J \leq p_{\bar{J}_{jk}\setminus\{i\}}$, and so (6.104) follows from (6.102). Finally, (6.105) is true by definition of $\bar{J}_{jk}$ and the fact that $i \neq j$. $\square$

Lemma 6.25 *If* $E_i + p_i < E_j + p_j$, *the not-first rule updates* E_i *to* $E_j + p_j$ *if and only if* $E_i + p_i > LST_{jn}$.

Proof. Suppose first that the not-first rule updates E_i to $E_j + p_j$. Then, by Lemma 6.24, there is a $k \geq j$ for which (6.104) holds, which implies

$$L_k - E_i < p_i + p_{\bar{J}_{jk}\setminus\{i\}} \tag{6.106}$$

since $L_k = L_{\bar{J}_{jk}\setminus\{i\}}$. First, note that $E_i + p_i < E_j + p_j$ implies $i \notin \bar{J}_{jk}$. Now,

$$E_i + p_i > L_k - p_{\bar{J}_{jk}\setminus\{i\}} = L_k - p_{\bar{J}_{jk}} \geq LST_{jn}$$

where the first inequality is due to (6.106), the equation is due to the fact that $i \notin \bar{J}_{jk}$, and the last inequality is due to the definition of LST_{jn}.

Now suppose that $E_i + p_i > LST_{jn}$. So, $LST_{jn} < \infty$, which means $LST_{jn} = L_k - p_{\bar{J}_{jk}}$ for some $k \geq j$ for which $\bar{J}_{jk}$ is nonempty. Now,

$$E_i + p_i > LST_{jn} = L_k - p_{\bar{J}_{jk}} = L_k - p_{\bar{J}_{jk}\setminus\{i\}} \tag{6.107}$$

where the second equation is due to the fact that $i \notin \bar{J}_{jk}$. But (6.107) implies (6.106), and so the not-first rule updates E_i to $E_j + p_j$. $\square$

Lemma 6.26 *If* $E_i + p_i \geq E_j + p_j$, *the not-first rule updates* E_i *to* $E_j + p_j$ *if and only if either* $E_i + p_i > LST_{j,i-1}$ *or* $E_i > LST_{jn}$.

Proof. First suppose that the not-first rule updates E_i to $E_j + p_j$. Then, by Lemma 6.24 there is a $k \geq j$ for which (6.106) holds. There are two cases, corresponding to $k < i$ and $i < k$. If $k < i$, then $i \notin \bar{J}_{jk}$ and from (6.106),

$$E_i + p_i > L_k - p_{\bar{J}_{jk}\setminus\{i\}} \geq LST_{j,i-1}$$

If $i < k$, then $i \in \bar{J}_{kj}$ and $p_{\bar{J}_{jk}} = p_i + p_{\bar{J}_{jk}\setminus\{i\}}$. So it follows from (6.106) that $L_k - E_i < p_{\bar{J}_{jk}}$, which implies

$$E_i > L_k - p_{\bar{J}_{jk}} \geq LST_{jn}$$

For the converse, first suppose that $E_i + p_i > \mathrm{LST}_{j,i-1}$. Then, since $\mathrm{LST}_{j,i-1} < \infty$, there is a $k \le i$ for which $\mathrm{LST}_{j,i-1} = L_k - p_{\bar{J}_{jk}}$ and $\bar{J}_{jk}$ is nonempty. So,

$$E_i + p_i > \mathrm{LST}_{j,i-1} = L_k - p_{\bar{J}_{jk}} = L_k - p_{\bar{J}_{jk} \setminus \{i\}}$$

where the second equation is due to $i \notin \bar{J}_{jk}$. But this implies (6.106), which means that the not-first rule updates E_i to $E_j + p_j$. Now suppose that $E_i + p_i \le \mathrm{LST}_{j,i-1}$ and $E_i > \mathrm{LST}_{jn}$. Then there is a $k \ge j$ for which $\mathrm{LST}_{jn} = L_k - p_{\bar{J}_{jk}}$. Also $k \ge i$, since otherwise $\mathrm{LST}_{jn} = \mathrm{LST}_{j,i-1} < E_i$, which contradicts the assumption that $E_i + p_i \le \mathrm{LST}_{k,i-1}$. But $k \ge i$ implies $i \in \bar{J}_{jk}$. However, $\bar{J}_{jk} \ne \{i\}$, because otherwise

$$E_i > \mathrm{LST}_{jn} = L_k - p_{\bar{J}_{jk}} = L_k - p_i \ge L_i - p_i$$

which contradicts the assumption that $E_\ell + p_\ell \le L_\ell$ for all ℓ. So $\bar{J}_{jk} \setminus \{i\}$ is nonempty and satisfies (6.106), which means that the not-first rule updates E_i to $E_j + p_j$. $\square$

6.14.3 Benders Cuts

Disjunctive scheduling constraints commonly arise in the context of planning and scheduling problems. For instance, it may be necessary to assign jobs to facilities as well as schedule the jobs on the facilities to which they are assigned. One such problem is discussed in Section 2.8.

Problems of this kind are often suitable for logic-based Benders decomposition (Section 5.2.2). A master problem assigns jobs to facilities, and separable subproblems schedule the jobs assigned to each facility. Generic Benders cuts can be developed for these problems, based on the nature of the objective function. The cuts can normally be strengthened when information from the scheduling algorithm is available, or when the subproblem is re-solved a few times with different job assignments. Section 2.8, for example, illustrates how the cuts can exploit information obtained from edge-finding and branching procedures.

Minimizing Cost

The simplest type of objective minimizes the fixed cost of assigning jobs to facilities. Thus, if f_{ij} is the cost of assigning job j to facility i, and variable x_j is the facility assigned to job j, the basic planning and scheduling problem may be written

$$\text{linear: } \min \sum_j f_{x_j j}$$

$$\text{subproblem: } \begin{cases} \{x_1, \ldots, x_n\} \\ \text{linear: } s_j + p_{x_j j} \leq d_j, \text{ all } j \\ \text{noOverlap: } ((s_j \mid x_j = i) \mid (p_{ij} \mid x_j = i)), \text{ all } i \end{cases}$$

$$\text{domains: } s_j \in [r_j, \infty), \ x_j \in \{1, \ldots, m\}, \text{ all } j$$

(6.108)

Here, r_j and d_j are the release time and deadline for job j, and variable s_j is the job's start time. As in Section 2.8, the notation $(s_j \mid x_j = i)$ refers to the tuple of start times s_j such that $x_j = i$, and similarly for $(p_{ij} \mid x_j = i)$. The noOverlap constraints require that the jobs on each facility be scheduled sequentially. The subproblem constraint allows the argument list of noOverlap to depend on variables x_j, because they are treated as constants inside the scope of the subproblem constraint.

The master problem can be written

$$\min \sum_j f_{x_j j}$$

optional relaxation of subproblem   (6.109)

Benders cuts

$$x_j \in \{1, \ldots, n\}, \text{ all } j$$

Section 7.13.3 discusses how the subproblem can be relaxed. If the solution of (6.109) in iteration k assigns each job j to machine x_j^k, the subproblem separates into the following scheduling problem for each facility i:

$$\min \sum_{\substack{j \\ x_j^k = i}} f_{ij}$$

$$s_j + p_{ij} \leq d_j, \text{ all } j \text{ with } x_j^k = i$$

(6.110)

$$\text{noOverlap}\left((s_j \mid x_j^k = i) \mid (p_{ij} \mid x_j^k = i)\right)$$

$$s_j \in [r_j, \infty), \text{ all } j \text{ with } x_j^k = i$$

Note that the objective function is a constant. If the scheduling problem on some facility i is infeasible, a Benders cut can be generated to rule out assigning those jobs to machine i again:

$$\bigvee_{j \in J_{ik}} x_j \neq i \tag{6.111}$$

Here, J_{ik} is the set of jobs assigned to facility i in iteration k, so that $J_{ik} = \{j \mid x_j^k = i\}$. The cut (6.111) is added to the master problem for facility i on which the scheduling problem is infeasible.

The cut (6.111) can be strengthened by identifying a proper subset of the jobs assigned to facility i that suffice to create infeasibility. The set J_{ik} can then be this proper subset. A smaller J_{ik} can be identified through an analysis of the scheduling algorithm (i.e., and analysis of the solution of the inference dual), as suggested in Section 2.8, or by a sampling procedure that re-solves the scheduling problem several times with different sets of jobs assigned to facility i. One simple procedure goes as follows. Initially, let $J_{ik} = \{j_1, \ldots, j_p\}$ contain all the jobs assigned to facility i. For $\ell = 1, \ldots, p$, do the following: try to schedule the tasks in $J_{ik} \setminus \{j_\ell\}$ on facility i, and if there is no feasible schedule, remove j_ℓ from J_{ik}. It can be well worth the effort of re-solving the subproblem a few times in order to obtain a stronger cut.

In practice, the master problem is often formulated as a 0-1 programming problem in which the decision variables x_{ij} are 1 when job i is assigned to machine j, and 0 otherwise. In this case, the master problem (6.109) becomes

$$\min \sum_{ij} f_{ij} x_{ij}$$

optional relaxation of subproblem

Benders cuts

$$x_{ij} \in \{0, 1\}, \text{ all } j$$

The Benders cut (6.111) must now be formulated as a 0-1 knapsack inequality:

$$\sum_{j \in J_{ik}} (1 - x_{ij}) \geq 1 \tag{6.112}$$

Minimizing Makespan

The Benders cuts are less straightforward when the objective is to minimize makespan. In this case, the problem is

linear: min M

$$
\text{subproblem:}
\begin{cases}
\{x_1, \ldots, x_n\} \\
\text{linear:}
\begin{cases}
M \geq s_j + p_{x_j j}, & \text{all } j \\
s_j + p_{x_j j} \leq d_j, & \text{all } j
\end{cases} \\
\text{noOverlap: } ((s_j \mid x_j = i) \mid (p_{ij} \mid x_j = i)), \quad \text{all } i
\end{cases}
$$

domains: $s_j \in [r_j, \infty)$, $x_j \in \{1, \ldots, m\}$, all j

$$(6.113)$$

The master problem is

$$
\begin{aligned}
& \min v \\
& \text{optional relaxation of subproblem} \\
& \text{Benders cuts} \\
& x_j \in \{1, \ldots, n\}, \text{ all } j
\end{aligned}
$$

$$(6.114)$$

Given a solution x^k of the master problem in iteration k, the subproblem separates into the following min makespan problem for each machine i:

$$
\begin{aligned}
& \min M_i \\
& M_i \geq s_j + p_{ij}, \ s_j + p_{ij} \leq d_j, \text{ all } j \text{ with } x_j^k = i \\
& \text{noOverlap: } \left((s_j \mid x_j^k = i) \mid (p_{ij} \mid x_j^k = i) \right) \\
& s_j \in [r_j, \infty), \text{ all } j \text{ with } x_j^k = i
\end{aligned}
$$

If M_{ik}^* is the minimum makespan on machine i, the overall minimum makespan is $\max_i \{M_{ik}^*\}$.

The most obvious Benders cut is the one presented in Section 2.8. Again, let J_{ik} be the set of jobs assigned to facility i in iteration k. Then a simple Benders cut requires the makespan to be at least M_{ik}^* whenever the jobs in J_{ik} are assigned to facility i:

$$(\{j \mid x_j = i\} \subset J_{ik}) \to (v \geq M_{ik}^*) \qquad (6.115)$$

The cut can be linearized as follows when the master problem is a 0-1 programming problem:

$$v \geq M_{ik}^* \left(\sum_{j \in J_{ik}} x_{ij} - |J_{ik}| + 1 \right) \qquad (6.116)$$

The cut (6.115) or (6.116) is added to the master problem for each facility i. As before, one may be able to identify a smaller set J_{ik} for which these cuts remain valid, either by an analysis of the scheduling algorithm or a sampling procedure.

A difficulty with the cuts (6.115) and (6.116) is that they become useless when even one of the jobs in J_{ik} is not assigned to facility i. The cuts can be improved by considering the effect on minimum makespan when jobs are removed. The simplest case is that in which all the release times r_j are the same. The improved cuts are based on a simple fact.

Lemma 6.27 *Consider a minimum makespan problem in which jobs $1, \ldots, n$ with identical release times are scheduled in facility i. If M^* is the minimum makespan, then the minimum makespan for the same problem with jobs $1, \ldots, s$ removed is*

$$M^* - \sum_{j=1}^{s} p_{ij}$$

A valid Benders cut is therefore

$$v \geq M_{ik}^* - \sum_{\substack{j \in J_{ik} \\ x_j \neq i}} p_{ij} \qquad (6.117)$$

The cut is easily linearized:

$$v \geq M_{ik}^* - \sum_{j \in J_{ik}} (1 - x_{ij}) p_{ij} \qquad (6.118)$$

Minimizing Tardiness

Tardiness can be measured by the number of late jobs or the total time by which the completion times exceed the corresponding deadlines. A model that minimizes the number of late jobs is linear: $\min \sum_{j} \delta_j$

subproblem: $\begin{cases} \{x_1, \ldots, x_n\} \\ \text{conditional: } (T_j > 0) \to (\delta_j = 1), \text{ all } j \\ \text{linear: } T_j \geq s_j + p_{x_j j} - d_j, \text{ all } j \\ \text{noOverlap: } ((s_j \mid x_j = i) \mid (p_{ij} \mid x_j = i)), \text{ all } i \end{cases}$

domains: $s_j \in [r_j, \infty), \; x_j \in \{1, \ldots, m\}, \text{ all } j$

$$(6.119)$$

Again, the master problem is (6.114). Given a solution x^k of the master problem, the subproblem on each facility i is

$$\min_{\substack{j \\ x_j^k = i}} \sum \delta_j$$

$(T_j > 0) \to (\delta_j = 1), \; T_j \geq s_j + p_{ij} - d_j, \text{ all } j \text{ with } x_j^k = i \qquad (6.120)$

$\text{noOverlap: } \left((s_j \mid x_j^k = i) \mid (p_{ij} \mid x_j^k = i)\right), \text{ all } i$

$s_j \in [r_j, \infty), \; \delta_j \in \{0, 1\}, \text{ all } j \text{ with } x_j^k = i$

If L_{ik}^* is the minimum number of late jobs on facility i, then $\sum_i L_{ik}^*$ is the minimum overall.

A trivial Benders cut forces the number of late jobs to be at least L_{ik}^* on each facility i to which all jobs in J_{ik} are assigned. Thus, there is a single Benders cut per iteration that consists of several constraints:

$$v \geq \sum_{i} L_i$$

$$(6.121)$$

$L_i \geq 0, \; (\{j \mid x_j = i\} \subset J_{ik}) \to (L_i \geq L_{ik}^*), \text{ all } i$

The linearization is:

$$v \geq \sum_{i} L_i$$

$$(6.122)$$

$L_i \geq 0, \; L_i \geq L_{ik}^* - L_{ik}^* \sum_{j \in J_{ik}} (1 - x_{ij}), \text{ all } i$

If tardiness is measured by the total amount by which finish times exceed deadlines, the objective function in (6.119) becomes $\sum_j T_j$ and the conditional constraint is dropped. The subproblem (6.120) is similarly modified. If T_{ik}^* is the minimum tardiness on facility i, then $\sum_i T_{ik}^*$ is the overall minimum.

A trivial Benders cut analogous to (6.115) or (6.116) can again be written, but a sampling procedure can strengthen such a cut significantly. One procedure goes as follows. Let $T_i(J)$ be the minimum tardiness on facility i that results when the jobs in J are assigned to the facility, so that $T_i(J_{ik}) = T_{ik}^*$. Let Z_{ik} be the set of jobs in J_{ik} that can be removed, one at a time, without reducing the minimum tardiness:

$$Z_{ik} = \{j \in J_{ik} \mid T_i(J_{ik} \setminus \{j\}) = T_{ik}^*\}$$

Now let T_{ik}^0 be the minimum tardiness that results from removing all the jobs in Z_{ik} at once, so that $T_{ik}^0 = T_i(J_{ik} \setminus Z_{ik})$. This leads to the Benders cut consisting of

$$v \geq \sum_i T_i$$

$$T_i \geq 0, \text{ all } i$$

(6.123)

and the conditional constraints

$$(\{j \in J_{ik} \mid x_j \neq i\} = \emptyset) \rightarrow (v \geq T_{ik}^*), \text{ all } i$$

$$(\{j \in J_{ik} \mid x_j \neq i\} \subset Z_{ik}) \rightarrow (v \geq T_{ik}^0), \text{ all } i$$

(6.124)

The second conditional constraint can be omitted for facility i when $T_{ik}^0 = T_{ik}^*$. The constraints (6.124) are linearized as follows:

$$v \geq T_{ik}^* - T_{ik}^* \sum_{j \in J_{ik}} (1 - x_{ij}), \text{ all } i$$

$$v \geq T_{ik}^0 - T_{ik}^0 \sum_{j \in J_{ik} \setminus Z_{ik}} (1 - x_{ij}), \text{ all } i$$

(6.125)

Again, the second constraint can be dropped when $T_{ik}^* = T_{ik}^0$. Generation of these cuts requires n additional calls to the scheduler in each Benders iteration, where n is the number of jobs.

When all the release times are equal, a second type of Benders cut can be derived from the following lemma.

Lemma 6.28 *Consider a minimum tardiness problem P in which tasks $1, \ldots, n$ have release time 0, due dates $d_1, \ldots, d_n$, and are to be scheduled on a single facility i. Let T^* be the minimum tardiness for this problem, and $\hat{T}$ the minimum tardiness for the problem $\hat{P}$ that is identical to P except that tasks $1, \ldots, s$ are removed. Then,*

$$T^* - \hat{T} \leq \sum_{j=1}^{s} \left(\sum_{\ell=1}^{n} p_{i\ell} - d_j \right)^+$$

(6.126)

Proof. Consider any optimal solution $\hat{S}$ of $\hat{P}$. One may assume that the makespan of $\hat{S}$ is at most $M = \sum_{\ell=s+1}^{n} p_{i\ell}$, because if it is greater than M, at least one task can be moved to an earlier time without increasing the total tardiness of $\hat{S}$. To obtain a feasible solution S for P, schedule tasks $1, \ldots, s$ sequentially after M. That is, for $j = 1, \ldots, s$ let task j start at time $M + \sum_{\ell=1}^{j-1} p_{i\ell}$. The tardiness of task j in S is at most

$$\left(M + \sum_{\ell=1}^{s} p_{i\ell} - d_j \right)^{+} = \left(\sum_{\ell=1}^{n} p_{i\ell} - d_j \right)^{+}$$

The total tardiness of S is therefore at most

$$\hat{T} + \sum_{k=1}^{s} \left(\sum_{\ell=1}^{n} p_{i\ell} - d_j \right)^{+}$$

from which (6.126) follows. $\square$

This leads to the Benders cut

$$v \geq \sum_i T_i'$$

$$T_i' \geq 0, \quad T_i' \geq T_{ik}^* - \sum_{\substack{j \in J_{ik} \\ x_j \neq i}} \left(\sum_{\ell \in J_{ik}} p_{i\ell} - d_j \right)^{+}, \quad \text{all } i \qquad (6.127)$$

which has the linearization

$$v \geq \sum_i T_i'$$

$$T_i' \geq 0, \quad T_i' \geq T_{ik}^* - \sum_{j \in J_{ik}} \left(\sum_{\ell \in J_{ik}} p_{i\ell} - d_j \right)^{+} (1 - x_{ij}), \quad \text{all } i \qquad (6.128)$$

Exercises

6.69. In the example of Table 6.6, verify that edge finding does not deduce the valid precedence $2 \gg \{3\}$.

6.70. Consider the 3-machine disjunctive scheduling problem in which

$$(E_1, E_2, E_3) = (2, 1, 0), \quad (L_1, L_2, L_3) = (6, 5, 8), \quad p = (3, 2, 2)$$

Use the edge-finding conditions (6.90) to check for valid precedences, and update the bounds accordingly. Does edge finding identify all valid precedences?

6.71. Suppose that in the problem of Table 2.5, jobs 2, 3, and 4 are assigned to machine A. Use the edge-finding rules to find jobs that must precede, or follow, subsets of jobs, and update the bounds accordingly. Note that when bounds have been updated, it may be possible to find additional edges. For example, initially one cannot deduce that job 3 must follow 2 (so that E_3 is not updated), even though one can deduce that job 2 must precede $\{3, 4\}$ (which updates L_1). However, after L_1 is updated, one can deduce that 3 follows 2 and update E_3. In this case, edge finding identifies all possible bound updates, but this is not true in general.

6.72. What is the minimum makespan on machine A in the Exercise 6.71? What jobs play a role in deriving the minimum? Trace the algorithm that computes J_j^L and J_j^E to verify this.

6.73. Write a Benders cut that corresponds to the minimum makespan solution of Exercise 6.72.

6.74. Exhibit a disjunctive scheduling problem in which edge finding fails to discover all precedence relations.

6.75. Apply the polynomial-time edge-finding algorithm to the problem of Exercise 6.70.

6.76. Write the polynomial-time edge-finding algorithm for updating E_i's in pseudocode.

6.77. State a polynomial-time edge-finding algorithm for updating L_i's and apply it to the problem of Table 6.6.

6.78. The proof of Theorem 6.22 omits the updating of L_is. State the argument for this case.

6.79. Apply the not-first and not-last rules to the example of Exercise 6.70 and update the bounds accordingly.

6.80. Apply the polynomial-time not-first algorithm to the example of Exercise 6.70.

6.81. State a polynomial-time not-last algorithm and apply it to the problem of Table 6.6.

6.82. Suppose that a minimum makespan planning and scheduling problem is to be solved by logic-based Benders decomposition. In the first iteration, jobs 1, 2 and 3 are assigned to machine A, on which $(p_{A1}, p_{A2}, p_{A3}) = (2, 3, 2)$. Use release times $(E_1, E_2, E_3) = (0, 2, 2)$ and deadlines $(L_1, L_2, L_3) = (3, 5, 7)$. Write the appropriate Benders cut (6.116).

6.83. Suppose in the problem Exercise 6.82 that all the release times are zero. Write the Benders cut (6.118).

6.84. Prove Lemma 6.27.

6.85. Give a counterexample to show that (6.118) may not be a valid cut if the release times are different.

6.86. Suppose that a Benders method is applied to a minimum total tardiness planning and scheduling problem. In the first iteration, jobs 1, 2, 3 and 4 are assigned to machine A, on which $(p_{A1}, p_{A2}, p_{A3}, p_{A4}) = (1, 1, 2, 2)$. The release times are $(E_1, \ldots, E_4) = (0, 0, 2, 0)$ and the due dates are $(L_1, \ldots, L_4) = (2, 2, 3, 5)$. Write the resulting Benders cut (6.125).

6.87. Show by example that $T_i(J_{ik} \setminus Z_{ik}) < T_i(J_{ik})$ is possible in a disjunctive scheduling problem. *Hint:* See Exercise 6.86.

6.88. Show that $T_i(J_{ik} \setminus Z_{ik}) = T_i(J_{ik})$ in a disjunctive scheduling problem when all the release dates are the same.

6.15 Cumulative Scheduling

Cumulative scheduling differs from disjunctive scheduling in that several jobs may run simultaneously. Each job consumes a certain amount of resource, however, and the rate of resource consumption at any one time must not exceed a limit. Cumulative scheduling may therefore be seen as a form of resource-constrained scheduling.

There may be one resource, or multiple resources with a different limit for each one. A resource limit may be constant or variable over time, and schedules may be preemptive or nonpreemptive. Disjunctive scheduling is a special case of cumulative scheduling in which there is one resource, each job consumes one unit of it, and the limit is always one.

Cumulative scheduling is one of the more successful application areas for constraint programming, due in part to sophisticated filtering algorithms that have been developed for the associated constraints. The most important filtering methods include edge finding, extended edge finding, not-first/not-last rules, and energetic reasoning.

Attention is restricted here to the most widely used cumulative scheduling constraint, which schedules nonpreemptively subject to a single resource limit that is constant over time. The constraint is

$$\text{cumulative}(s \mid p, c, C) \tag{6.129}$$

where $s = (s_1, \ldots, s_n)$ is a tuple of variables s_j representing the start time of job j. The remaining arguments are parameters: $p = (p_1, \ldots, p_n)$ is a tuple of processing times for each job, $c = (c_1, \ldots, c_n)$ is a tuple of resource consumption rates, and C is the limit on total resource consumption at any one time. The constraint requires the following:

$$\sum_{\substack{j \\ s_j \le t \le s_j + p_j}} c_j \le C, \quad \text{for all times } t$$

That is, the total rate of resource consumption of the jobs underway at any time t is at most C.

As with disjunctive scheduling, each job j is associated with an earliest start time E_j and a latest completion time L_j. Initially, these are the release time and deadline of the job, respectively, but they may be updated in the course of the solution algorithm. Thus, the current domain of s_j is the interval $[E_j, L_j - p_j]$, and the release time and deadline of each job is indicated by the initial domain of s_j. If all the problem data are integral, E_j should always be rounded up and L_j rounded down.

A small instance of a cumulative scheduling problem is presented in Table 6.10, and a feasible schedule is illustrated in Fig. 6.28.

6.15.1 Edge Finding

Edge finding for disjunctive scheduling can be generalized to cumulative scheduling. The key concept that makes generalization possible is the *energy* e_j of a job j, which is the product $p_j c_j$ of the processing time and resource consumption rate. In Fig. 6.28, the energy of each

Table 6.10 A small instance of a cumulative scheduling problem. The resource limit is $C = 4$.

j	p_j	c_j	E_j	L_j
1	5	1	0	5
2	3	3	0	5
3	4	2	1	7

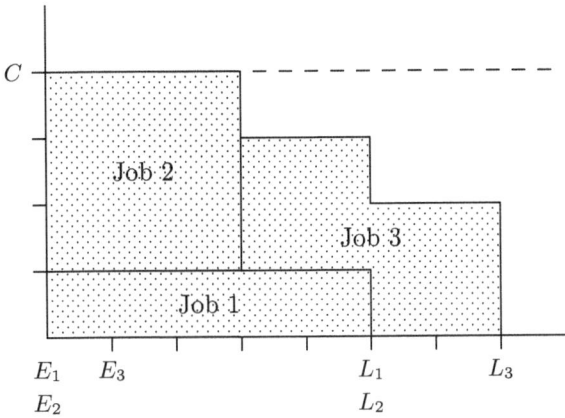

Fig. 6.28 A feasible solution of a small cumulative scheduling problem. The horizontal axis is time and the vertical axis is resource consumption.

job is the area it consumes in the chart. While disjunctive edge finding checks whether the total processing time of a set of jobs fits within a given time interval, cumulative edge finding checks whether the total energy demand exceeds the supply, which is the product of the time interval and the resource limit C.

Unlike disjunctive edge finding, cumulative edge finding does not deduce that one job must start after certain others finish. Rather, it deduces that it must *finish* after the others finish. Nonetheless, cumulative edge-finding algorithms are quite useful for reducing domains.

The notation is similar to that of the previous section. For a set J of jobs, let $p_J = \sum_{j \in J} p_j$, $e_J = \sum_{i \in J} e_j$, $E_J = \min_{j \in j}\{E_j\}$, and $L_J = \max_{j \in J}\{L_j\}$. The notation $i > J$ indicates that job i must finish after all the jobs in J finish, and $i < J$ indicates that job i must start before any job in J starts.

The edge-finding rules are based on the principle that if a set J of jobs requires total energy e_J, then the time interval $[t_1, t_2]$ in which they are scheduled must have a length of at least e_J/C. Thus, one must have $e_J \leq C \cdot (t_2 - t_1)$. In Fig. 6.28, the jobs have a total energy of 22 and therefore require a time interval of at least $\lceil 22/4 \rceil = \lceil 5.5 \rceil = 6$ (rounding up because the data are integral). In fact, they require a time interval of 7 because the jobs cannot be packed perfectly into a rectangular area.

The edge-finding rules may be stated

$$\text{If } e_i + e_J > C \cdot (L_J - E_{J \cup \{i\}}), \text{ then } i > J. \ (a)$$
$$\text{If } e_i + e_J > C \cdot (L_{J \cup \{i\}} - E_J), \text{ then } i < J. \ (b) \qquad (6.130)$$

Rule (a) is based on the fact that if job i does not finish after all the jobs in J finish, then the time interval from $E_{J \cup \{i\}}$ to L_J must cover the energy demand $e_i + e_J$ of all the jobs. If there is not enough energy, then job i must finish after the other jobs finish. The reasoning for rule (b) is analogous. In Fig. 6.28, job 3 must finish after the other jobs finish (i.e., $3 > \{1, 2\}$) because

$$e_3 + e_{\{1,2\}} = 22 > 4 \cdot (5 - 0) = C \cdot (L_{\{1,2\}} - E_{\{1,2,3\}})$$

When it is established that $i > J$, it may be possible to update the release time E_i of job i, as illustrated in Fig. 6.29. If the total energy e_J of the jobs in J exceeds the energy available between E_J and L_J within a resource limit of $C - c_i$, then at some time in the schedule, the jobs in J must consume more resource than $C - c_i$. Since the excess energy is

$$R(J, c_i) = e_J - (C - c_i)(L_J - E_J)$$

the jobs in J must consume more resource than $C - c_i$ for a period of at least $R(J, c_i)/c_i$. None of this excess resource is consumed after job i finishes, because $i > J$. It must be consumed before job i starts, which means that job i can start no earlier than $E_J + R(J, c_i)/c_i$.

On the other hand, suppose that job i must start before the jobs in J start (i.e., $i < J$). The excess resource $R(J, c_i)$ cannot be consumed

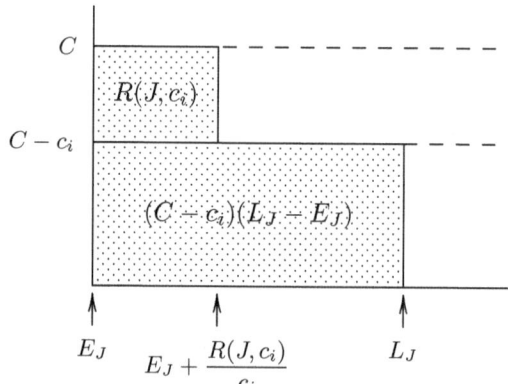

Fig. 6.29 Updating of E_i to $E_i' = E_J + R(J, c_i)/c_i$. The area of each rectangle is indicated. The entire shaded area has area e_J.

before job i starts. It must be consumed after job i finishes, which means that job i can finish no sooner than $E_J - R(J, c_i)/c_i$. Thus, the reasoning is similar when $i < J$. This leads to the update rules:

If $i > J$ and $R(J, c_i) > 0$, update E_i to $\max\left\{E_i, E_J + \dfrac{R(J, c_i)}{c_i}\right\}$.

If $i < J$ and $R(J, c_i) > 0$, update L_i to $\min\left\{L_i, L_J - \dfrac{R(J, c_i)}{c_i}\right\}$.

The domains can in general be reduced further by doing a similar update for all subsets J' of J, because $i > J$ implies $i > J'$ and $i < J$ implies $i < J'$. Thus if $i > J$, E_i can be updated to $\max\{E_i, E_i'(J)\}$, where

$$E_i'(J) = \max_{\substack{J' \subset J \\ R(J', c_i) > 0}} \left\{E_{J'} + \frac{R(J', c_i)}{c_i}\right\} \tag{6.131}$$

and if $i < J$, L_i can be updated to $\min\{L_i, L_i'\}$, where

$$L_i'(J) = \min_{\substack{J' \subset J \\ R(J', c_i) > 0}} \left\{L_{J'} - \frac{R(J', c_i)}{c_i}\right\} \tag{6.132}$$

If this is done for all subsets J of jobs, one can update E_i to $\max\{E_i, E_i'\}$, where

$$E_i' = \min_{\substack{J \\ i \notin J \\ C \cdot (L_J - E_{J \cup \{i\}}) < e_i + e_J}} \left\{E_i'(J)\right\} \tag{6.133}$$

and update L_i to $\min\{L_i, L_i'\}$, where

$$L_i' = \max_{\substack{J \\ i \notin J \\ C \cdot (L_J - E_{J \cup \{i\}}) < e_i + e_J}} \left\{L_i'(J)\right\} \tag{6.134}$$

In the example of Fig. 6.28, the only precedence discovered by the edge-finding rule (6.130a) is $3 > \{1, 2\}$. Since $R(\{1\}, c_3) = -5$, $R(\{2\}, c_3) = -1$, and $R(\{1, 2\}, c_3) = 4$, one has

$$E_3'(\{1, 2\}) = E_{\{1,2\}} + \frac{R(\{1, 2\}, c_3)}{c_3} = 2$$

Thus, $E_3 = 1$ can be updated to $E_3' = 2$.

Figure 6.30 presents an algorithm that computes updates (6.133) in time proportional to n^2. The algorithm assumes that the jobs are indexed in nondecreasing order of release times, so that $E_i \leq E_j$ when $i < j$. For each value of k in the outer loop, the algorithm identifies jobs i for which $i > J_k$, where $J_k = \{j \mid L_j \leq L_k\}$ is the subset of jobs with deadlines no later than job k.

The task of the first i-loop is to compute updates E_i'' based on subsets of J_k in which the jobs have release dates no earlier than E_i. Define $J_{ki} = \{j \in J_k \mid j \geq i\}$, recalling that $j \geq i$ implies $E_j \geq E_i$. Then it can be shown that E_i'', as computed by the algorithm, is equal to

$$\max_{\substack{J' \subset J_{ki} \\ R(J', c_i) > 0}} \left\{ E_{J'} + \frac{R(J', c_i)}{c_i} \right\} \tag{6.135}$$

Let $E_j' = E_j$ for all j.
For $k = 1, \ldots, n$:
    Let $e_J = 0$, $e_J^* = \infty$, $\Delta e = \infty$, and $E^* = L_k$.
    For $i = n, n-1, \ldots, 1$:
        If $L_i \leq L_k$ then
            Let $e_J = e_J + e_i$.
            If $e_J/C > L_k - E_i$ then exit (no feasible schedule).
            If $e_J + CE_i > e_J^* + CE^*$ then let $e_J^* = e_J$ and $E^* = E_i$.
        Else
            Let $R = e_J^* - (C - c_i)(L_k - E^*)$.
            If $R > 0$ then let $E_i'' = E^* + R/c_i$.
            Else let $E_i'' = -\infty$.
    For $i = 1, \ldots, n$:
        If $L_i \leq L_k$ then
            Let $\Delta e = \min\{\Delta e, C \cdot (L_k - E_i) - e_J\}$.
            Let $e_J = e_J - e_i$.
        Else
            If $C \cdot (L_k - E_i) < e_J + e_i$ then let $E_i' = \max\{E_i', E_i''\}$.
            If $\Delta e < e_i$ then
                Let $R = e_J^* - (C - c_i)(L_k - E^*)$.
                If $R > 0$ then let $E_i' = \max\{E_i', E^* + R/c_i\}$.

Fig. 6.30 Edge-finding algorithm for computing updated release times E_i' in time proportional to n^2. The algorithm assumes that jobs are indexed in nondecreasing order of release time E_j.

To see this, let j^* be the index that maximizes $e_{J_{kj}} + CE_j$ over $j \in J_{ki}$. Then, e_j^* as computed in the algorithm is $e_{J_{kj^*}}$, and E^* is E_{j^*}. Also, if $e_{J_{kj^*}} - (C - c_i)(L_k - E_{j^*}) > 0$, then E_i'' as computed is

$$E_i'' = E_{j^*} + \frac{e_{J_{kj^*}} - (C - c_i)(L_k - E_{j^*})}{c_i}$$

Because $j = j^*$ maximizes $e_{J_{kj}} + CE_j$ over $j \in J_{ki}$, however, it also maximizes

$$E_j + \frac{e_{J_{kj}} - (C - c_i)(L_k - E_j)}{c_i} = \frac{e_{J_{kj}} + CE_j}{c_i} + \text{constant}$$

over $j \in J_{ki}$, which implies that E_i'' is equal to (6.135).

The *else* portion of the second i-loop has two functions. The first line updates E_i' to E_i'' when $i > J_{ki}$. The remaining lines identify precedences $i > J$ for subsets J of J_k containing at least one job with release date before E_i. To accomplish this, the *if* part of the second i-loop computes

$$\Delta e = \min_{\substack{j \in J_k \\ j < i}} \left\{ C \cdot (L_k - E_j) - e_{J_{kj}} \right\} \tag{6.136}$$

Thus, Δe is also the minimum of $C \cdot (L_k - E_J) - e_J$ over all subsets J of J_k containing at least one j with $E_j < E_i$. If j' is the minimizing index in (6.136), then $\Delta e \leq e_i$ implies that

$$e_i + e_{J_{kj'}} > C \cdot (L_J - E_{J_{kj'} \cup \{i\}})$$

which implies $i > J_{kj'}$ by the edge-finding rule. Now E_i can be updated to

$$\max \left\{ E_i', E_{j'} + \frac{e_{J_{kj'}} - (C - c_i)(L_k - E_{j'})}{c_i} \right\}$$

But since j^* maximizes

$$E_j + \frac{e_{J_{kj}} - (C - c_i)(L_k - E_j)}{c_i}$$

over $j \in J_{ki} \subset J_{kj'}$, E_i can be updated to

$$\max \left\{ E_i', E_{j^*} + \frac{e_{J_{kj^*}} - (C - c_i)(L_k - E_{j^*})}{c_i} \right\}$$

$$= \max \left\{ E_i', E^* + \frac{e_J^* - (C - c_i)(L_k - E^*)}{c_i} \right\}$$

when $E^* + (e_J^* - (C - c_i)(L_k - E^*))/c_i > 0$, as is done in the algorithm.

Based in part on the above reasoning, one can prove the following result.

Theorem 6.29. *The algorithm of Fig. 6.30 computes all updates of the form (6.133), and a similar algorithm computes all updates of the form (6.134).*

When the algorithm is applied to the problem of Fig. 6.28, the update $E_3' = 2$ is discovered when $k = 1$ and $i = 3$ in the second i-loop. The first i-loop has already computed $e_J^* = 14$ and $E^* = 0$, because the index j that maximizes $e_{J_{kj}} + CE_j$ over $j \in J_{ki} = J_{13} = \{1, 2\}$ is $j^* = 1$. So $e_J^* = e_{J_{kj^*}} = e_{\{1,2\}} = 14$, and $E^* = E_{j^*} = 0$. Also, $\Delta e = 6$, because the index j that minimizes $C \cdot (L_k - e_j) - e_{J_{kj}}$ over $j \in J_k$ and $j < i$ (i.e., over $j = 1, 2$) is $j' = 1$, so that $\Delta e = C \cdot (L_k - E_{j'}) - e_{J_{kj'}} = 6$. Since $\Delta e < e_i$, the algorithm has established that $i > J_{kj'}$ (i.e., $3 > \{1, 2\}$). Now since $R = e_J^* - (C - c_i)(L_k - E^*) = 4 > 0$, it updates $E_3' = 1$ to $E^* + R/c_3 = 2$.

6.15.2 Extended Edge Finding

A weakness of the edge-finding rules is that they may fail to detect that job i must finish after the jobs in J finish when job i has an earlier release time than the other jobs. In such cases, the total time available $L_J - E_{J \cup \{i\}}$ may provide the required energy if it were available to all the jobs, but the period between $E_{J \cup \{i\}}$ and E_J is not available to the jobs in J. If job i cannot finish by E_J, it may have to finish after the other jobs finish. A similar situation can occur when job i has a later deadline than the other jobs. An extended version of edge finding corrects this problem, albeit at the cost of more computation.

Edge finding is extended by adding the rules below:

$$
\begin{aligned}
&\text{If } E_i \leq E_J < E_i + p_i \text{ and} \\
&\quad e_J + c_i(E_i + p_i - E_J) > C \cdot (L_J - E_J), \text{ then } i > J \quad (a) \\
&\text{If } L_i - p_i < L_J \leq L_i \text{ and} \\
&\quad e_J + c_i(L_J - L_i + p_i) > C \cdot (L_J - E_J), \text{ then } i < J \quad (b)
\end{aligned}
\tag{6.137}
$$

The reasoning behind rule (a) is as follows. It is supposed that job i has release time no later than E_J but cannot finish by E_J. If job i

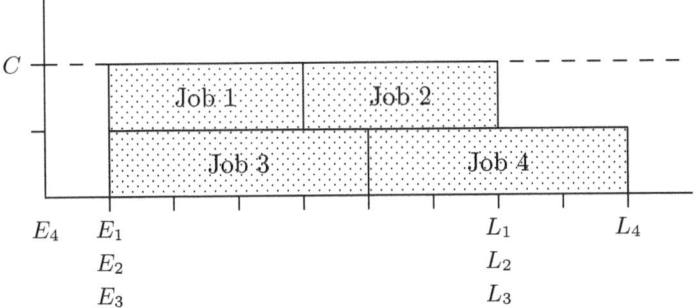

Fig. 6.31 A feasible solution of a small cumulative scheduling problem. Extended edge finding deduces that $4 > \{1, 2, 3\}$, but ordinary edge finding does not.

finishes before some job in J finishes (or at the same time), then it must finish by L_J. Thus, the energy required between E_J and L_J is at least the energy e_J of the jobs in J and the energy $c_i(E_i + p_i - E_J)$ of the portion of job i that must run during this period. If this exceeds the available energy during the period, then job i must finish after all the jobs in J finish. The reasoning is similar for rule (b).

Rule (a) is illustrated in the example of Fig. 6.31. Here, job 4 must finish after the others finish, because $E_4 < E_{\{1,2,3\}} < E_4 + p_4$ and

$$e_{\{1,2,3\}} + c_4(E_4 + p_4 - E_{\{1,2,3\}}) = 10 + 1 \cdot (4 - 1) = 13$$
$$> 12 = 2 \cdot (7 - 1) = C \cdot (L_{\{1,2,3\}} - E_{\{1,2,3\}})$$

Note that ordinary edge finding does not deduce that $4 > \{1, 2, 3\}$, because

$$e_4 + e_{\{1,2,3\}} = 14 \leq 2 \cdot (7 - 0) = C \cdot (L_{\{1,2,3\}} - E_{\{1,2,3,4\}})$$

If $i > J$, then as with ordinary edge finding, E_i can be updated to $\max\{E_i, E'_i(J)\}$, where $E'_i(J)$ is given by (6.131). Similarly, if $i < J$ then L_i can be updated to $\min\{L_i, L'_i(J)\}$. Thus, E_i can be updated to the maximum of $E'_i(J)$ over all J satisfying the extended edge-finding rule (6.137a), and L_i can be updated to the minimum of $L'_i(J)$ over all J satisfying (6.137b). The updates appear to require more computation than the original edge-finding updates. At least one algorithm, not presented here, computes them in time proportional to n^3.

In the example, $R(J, c_4) > 0$ when J is $\{1, 2\}$, $\{2, 3\}$ or $\{1, 2, 3\}$. Thus, $E_4 = 0$ is updated to $E'_4(\{1, 2, 3\}) = \max\{1 + \frac{1}{2}, 1 + \frac{1}{2}, 1 + \frac{4}{2}\} = 3$.

6.15.3 Not-First/Not-Last Rules

As in the case of disjunctive scheduling, one can sometimes deduce in a cumulative scheduling context that a job i cannot be scheduled first, or cannot be scheduled last, in a set $J \cup \{i\}$ of jobs. One must be careful about what *not first* and *not last* mean, however. In disjunctive scheduling, a job i is *not first* when $\neg(i \ll J)$, but in cumulative scheduling *not first* does not mean $\neg(j < J)$. Rather, it means the same thing as in disjunctive scheduling: job i starts after some job in J finishes. Similarly, job i is not last when it finishes before some job in J starts.

Let $F_J = \min_{j \in J}\{E_j + p_j\}$ be the minimum earliest finish time of the jobs in J, and let $S_J = \max_{j \in J}\{L_j - p_j\}$ be the maximum latest start time of the jobs in J. If job i is not first, then it cannot start earlier than F_J, which means that E_i can be updated to $\max\{E_i, F_J\}$. If job i is not last, then it cannot finish later than S_J, and L_i can be updated to $\min\{L_i, S_J\}$. In view of this, the not-first/not-last rules can be stated:

$$\begin{aligned} &\text{If } E_J \leq E_i < F_J \text{ and} \\ &e_J + c_i\left(\min\{E_i + p_i, L_J\} - E_J\right) > C \cdot (L_J - E_J), \qquad (a) \\ &\qquad \text{then update } E_i \text{ to } \max\{E_i, F_J\} \\ &\text{If } S_J < L_i \leq L_J \text{ and} \\ &e_J + c_i\left(L_J - \max\{L_i - p_i, E_J\}\right) > C \cdot (L_J - E_J), \qquad (b) \\ &\qquad \text{then update } L_i \text{ to } \min\{L_i, S_J\} \end{aligned} \qquad (6.138)$$

Rule (a) can be proved as follows. Suppose job i and set J satisfy the conditions of the rule but job i starts at some time t before F_J. Thus, $E_i \leq t < F_J$. Since, by hypothesis, no job in J can finish by t, a resource capacity of c_i must remain unused during the period from E_J to t. This means that the total energy required between E_J and L_J by the jobs in $J \cup \{i\}$, including the energy that cannot be used, is

$$e_J + c_i\left(\min\{t + p_i, L_J\} - E_J\right)$$

But since $t \geq E_i$, this quantity is greater than or equal to

$$e_J + c_i\left(\min\{E_i + p_i, L_J\} - E_J\right)$$

which, by hypothesis, exceeds the available energy $C \cdot (L_J - E_J)$. So job i can start no earlier than F_J. The argument for rule (b) is analogous.

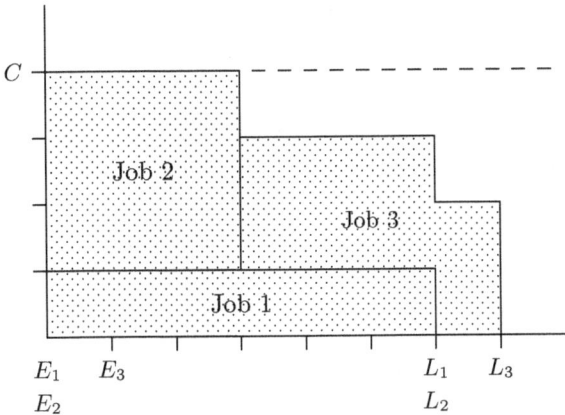

Fig. 6.32 A feasible solution of a small cumulative scheduling problem. It can be deduced that job 3 is not first, and E_3 can therefore be updated to 3. Neither edge finding nor extended edge finding can update E_3.

The example of Fig. 6.28 is modified in Fig. 6.32 to illustrate the not-first principle. Here, $F_{\{1,2\}} = \min\{0+3, 0+6\} = 3$. One can deduce that job 3 is not first because $E_{\{1,2\}} \leq E_3 < F_{\{1,2\}}$ and

$$e_{\{1,2\}} + c_3 \cdot \left(\min\{e_3 + p_3, L_{\{1,2\}}\} - E_{\{1,2\}}\right) =$$
$$15 + 2 \cdot (\min\{5, 6\} - 0) = 25 > 4 \cdot (6 - 0) = C \cdot (L_{\{1,2\}} - E_{\{1,2\}})$$

E_3 is therefore updated to $\max\{E_3, F_{\{1,2\}}\} = \max\{1, 3\} = 3$. Note that although $3 \gg \{1, 2\}$, this cannot be deduced by any of the edge-finding rules discussed earlier. Thus, the not-first rule discovers an update that is missed by edge finding, even when extended.

The not-first/not-last rules (6.138) can be used to update each E_i to the maximum of $\max\{E_i, F_J\}$ over all sets J satisfying the conditions in rule (a), and to update L_i to the minimum of $\min\{L_i, S_J\}$ over all sets J satisfying the conditions in (b). It is possible to compute these updates in time proportional to n^3.

6.15.4 Energetic Reasoning

The concept of energy plays a role in all of the filtering methods presented here for the cumulative scheduling constraint. A sharper analysis of the energy required by a set of jobs in a given time interval $[t_1, t_2]$, however, provides an additional technique for reducing domains.

The analysis is based on *left shifting* and *right shifting* jobs. A job i is left-shifted if it is scheduled as early as possible (i.e., if it starts at E_i) and is right-shifted if it is scheduled as late as possible (i.e., starts at $L_i - p_i$). Suppose that job i runs $p_i^\ell(t_1)$ time units past t_1 when it is left-shifted and starts $p_i^r(t_2)$ time units before time t_2 when right-shifted. Thus

$$p_i^\ell(t_1) = \max\{0, p_i - \max\{0, t_1 - E_i\}\}$$
$$p_i^r(t_2) = \max\{0, p_i - \max\{0, L_i - t_2\}\}$$

The amount of time job i runs during $[t_1, t_2]$ is at least the minimum of $p_i^\ell(t_1)$, $p_i^r(t_2)$, and $t_2 - t_1$. So the minimum energy consumption of job i during $[t_1, t_2]$ is

$$e_i(t_1, t_2) = c_i \min\left\{p_i^\ell(t_1), p_i^r(t_2), t_2 - t_1\right\}$$

The minimum total energy consumption during interval $[t_1, t_2]$ is therefore $e(t_1, t_2) = \sum_i e_i(t_1, t_2)$, where the sum is taken over all jobs.

There is clearly no feasible schedule if $e(t_1, t_2) > C \cdot (t_2 - t_1)$ for some interval $[t_1, t_2]$. Fortunately, it is not necessary to examine every interval to check whether such an interval exists. To see this, let

$$\Delta e(t_1, t_2) = C \cdot (t_2 - t_1) - e(t_1, t_2)$$

be the excess energy capacity during $[t_1, t_2]$. Then there is no feasible schedule if $\Delta e(t_1, t_2) < 0$ for some interval $[t_1, t_2]$. If $F_i = E_i + p_i$ is the earliest finish time of job i, and $S_i = L_i - p_i$ the latest start time, the following can be shown:

Lemma 6.30 *All local minima of the function $\Delta e(t_1, t_2)$ occur at points (t_1, t_2) belonging to the set T^* defined to be the union of the following sets:*

$$\{(t_1, t_2) \mid t_1 \in T_1,\ t_2 \in T_2,\ t_1 < t_2\}$$
$$\{(t_1, t_2) \mid t_1 \in T_1,\ t_2 \in T(t_1),\ t_1 < t_2\}$$
$$\{(t_1, t_2) \mid t_2 \in T_1,\ t_1 \in T(t_2),\ t_1 < t_2\}$$

where

$$T_1 = \{E_i, F_i, S_i \mid i = 1, \ldots, n\}$$
$$T_2 = \{F_i, S_i, L_i \mid i = 1, \ldots, n\}$$
$$T(t) = \{E_i + L_i - t \mid i = 1, \ldots, n\}$$

Due to this fact, one can check whether $\Delta e(t_1, t_2)$ goes negative for some interval by evaluating it only for $(t_1, t_2) \in T^*$.

Energy reasoning can be extended to domain reduction, since even when $\Delta e(t_1, t_2)$ is nonnegative, it may be small enough to restrict when some jobs can run. The set of all jobs other than i requires energy of at least $e(t_1, t_2) - e_i(t_1, t_2)$ in the interval $[t_1, t_2]$. Thus, the energy available to run job i in the interval $[t_1, t_2]$ is at most

$$\Delta e_i(t_1, t_2) = C \cdot (t_2 - t_1) - e(t_1, t_2) + e_i(t_1, t_2)$$

If job i is left-shifted, it requires energy $c_i p_i^\ell(t_1)$ after t_1. If $c_i p_i^\ell(t_1) > \Delta e_i(t_1, t_2)$, then

$$c_i p_i^\ell(t_1) - \Delta e_i(t_1, t_2)$$

is the amount of job i's energy consumption that must be moved outside $[t_1, t_2]$ to the right, beyond t_2. This means that job i must run for a time of at least $p_i^\ell(t_1) - \Delta e_i(t_1, t_2)/c_i$ past t_2, and job i's earliest finish time F_i can be updated to

$$\max\left\{ F_i, \ t_2 + p_i^\ell(t_1) - \frac{\Delta e_i(t_1, t_2)}{c_i} \right\}$$

This implies the following:

Theorem 6.31. *If $c_i p_i^\ell(t_1) > \Delta e_i(t_1, t_2)$, then E_i can be updated to*

$$\max\left\{ E_i, \ t_2 + p_i^\ell(t_1) - p_i - \frac{\Delta e_i(t_1, t_2)}{c_i} \right\}$$

Similarly, if $c_i p_i^r(t_2) > \Delta e_i(t_1, t_2)$, then L_i can be updated to

$$\min\left\{ L_i, \ t_1 - p_i^r(t_2) + p_i - \frac{\Delta e_i(t_1, t_2)}{c_i} \right\}$$

One can apply Theorem 6.31 to all jobs i and all pairs $(t_1, t_2) \in T^*$ in time proportional to n^3. It is apparently an open question whether this necessarily detects all updates that can be obtained from the theorem.

6.15.5 Benders Cuts

Like the disjunctive constraint, the cumulative constraint commonly occurs in planning and scheduling problems are amenable to the logic-based Benders decomposition. The Benders cuts are similar to those developed in Section 6.14.3 for disjunctive scheduling.

Minimizing Cost

Consider first a planning and cumulative scheduling problem that minimizes the fixed cost of assigning jobs to facilities. The model is identical to the model (6.108) stated earlier for planning and disjunctive scheduling, except that the disjunctive constraint is replaced by

$$\text{cumulative}((s_j \mid x_j = i) \mid (p_{ij} \mid x_j = i), (c_{ij} \mid x_j = i), C_i), \text{ all } i$$

where c_{ij} the rate of resource consumption of job j in facility i, and C_i is the maximum resource consumption rate in facility i.

The Benders master problem is identical to the master problem (6.109) for disjunctive scheduling, aside from the subproblem relaxation, which becomes more complicated for minimum cost as well as other objectives when one moves to cumulative scheduling. Relaxations for the cumulative scheduling subproblem are presented in Section 7.13.3. The subproblem is the same as before, except that the disjunctive constraint in (6.110) is replaced with a cumulative constraint. The Benders cuts are again (6.111) and have the linearized version (6.111). As before, the cuts can be strengthened by analyzing the solution of the scheduling subproblem (or more precisely, the solution of its inference dual) and identifying which jobs play a role in the proof of infeasibility.

Minimizing Makespan

If the objective is to minimize makespan, the model is (6.113) with the disjunctive constraint replaced by a cumulative constraint. The simple Benders cut (6.116) is still valid. However, Lemma 6.27 does not hold for cumulative scheduling, and the cut (6.118) is not valid. However, a weaker form of the lemma can be proved and a cut written on that basis.

Lemma 6.32 *Consider a minimum makespan problem P in which jobs $1, \ldots, n$, with release time 0 and deadlines $d_1, \ldots, d_n$, are to be scheduled on a single facility i. Let M^* be the minimum makespan for P, and $\hat{M}$ the minimum makespan for the problem $\hat{P}$ that is identical to P except that jobs $1, \ldots, s$ are removed. Then,*

$$M^* - \hat{M} \leq \Delta + \max_j \{d_j\} - \min_j \{d_j\} \tag{6.139}$$

where $\Delta = \sum_{j=1}^{s} p_{ij}$. In particular, when all the deadlines are the same, $M^* - \hat{M} \leq \Delta$.

Proof. Consider any optimal solution of $\hat{P}$ and extend it to a solution S of P by scheduling jobs $1, \ldots, s$ sequentially after $\hat{M}$. That is, for $k = 1, \ldots, s$ let job k start at time $\hat{M} + \sum_{j=1}^{k-1} p_{ij}$. The makespan of S is $\hat{M} + \Delta$. If $\hat{M} + \Delta \leq \min_j\{d_j\}$, then S is clearly feasible for P, so that $M^* \leq \hat{M} + \Delta$ and the lemma follows. Now, suppose $\hat{M} + \Delta > \min_j\{d_j\}$. This implies

$$\hat{M} + \Delta + \max_j\{d_j\} - \min_j\{d_j\} > \max_j\{d_j\} \qquad (6.140)$$

Since $M^* \leq \max_j\{d_j\}$, (6.140) implies (6.139), and again the lemma follows. $\square$

The bound $M^* - \hat{M} \leq \Delta$ need not hold when the deadlines differ. Consider, for example, an instance with three jobs where $(r_1, r_2, r_3) = (0, 0, 0)$, $(d_1, d_2, d_3) = (2, 1, \infty)$, $(p_{i1}, p_{i2}, p_{i3}) = (1, 1, 2)$, $(c_{i1}, c_{i2}, c_{i3}) = (2, 1, 1)$, and $C = 2$. Then, if $s = 1$, $M^* - \hat{M} = 4 - 2 > \Delta = p_{i1} = 1$.

To write a Benders cut, suppose first that all the deadlines are the same. Then by Lemma 6.27, each job removed from facility i reduces the minimum makespan at most p_{ij}:

$$v \geq M_{ik}^* - \sum_{\substack{j \in J_{ik} \\ x_j \neq i}} p_{ij} \qquad (6.141)$$

The cut is easily linearized:

$$v \geq M_{ik}^* - \sum_{j \in J_{ik}} (1 - x_{ij}) p_{ij} \qquad (6.142)$$

When the deadlines differ, the cut (6.141) becomes

$$v \geq M_{ik}^* - \left(\sum_{\substack{j \in J_{ik} \\ x_j \neq i}} p_{ij} + \max_{j \in J_{ik}}\{d_j\} - \min_{j \in J_{ik}}\{d_j\} \right) \qquad (6.143)$$

and its linearization is

$$v \geq M_{ik}^* - \left(\sum_{j \in J_{ik}} p_{ij}(1 - x_{ij}) + \max_{j \in J_{ik}}\{d_j\} - \min_{j \in J_{ik}}\{d_j\} \right) \qquad (6.144)$$

Minimizing Tardiness

When the objective is to minimize the number of late jobs, the model is (6.119) with the disjunctive constraint replaced by a cumulative constraint. The Benders cuts are again (6.121)–(6.122).

Finally, when minimizing total tardiness, one can again use the Benders cuts (6.123)–(6.125). Lemma 6.28 remains valid, but the resulting Benders cuts (6.127)–(6.128) can be quite weak. The reason is that the proof of Lemma 6.28 assumes that the minimum makespan $\hat{M}$ of problem $\hat{P}$ is at most $M = \sum_{j=s+1}^{n} p_{ij}$. The true makespan is likely to be much less than M because some jobs can be scheduled concurrently.

A generally stronger cut could be obtained by assuming $\hat{M}$ is at most the makespan M^* of problem P's minimum tardiness solution. Unfortunately, this is not a valid assumption, because in exceptional cases $\hat{M} > M^*$ even though $\hat{P}$ has fewer jobs than P. Suppose, for example, that $n = 4$ and $s = 1$, with $r = (0, 0, 0, 0)$, $d = (5, 3, 3, 6)$, $p = (1, 2, 2, 4)$, $c = (2, 1, 1, 1)$, and $C = 2$. An optimal solution of P puts $t = (4, 0, 2, 0)$ with tardiness $T^* = 1$ and makespan $M^* = 5$, but the only optimal solution of the smaller problem $\hat{P}$ puts $(t_2, t_3, t_4) = (0, 0, 2)$ with tardiness $\hat{T} = 0$ and makespan $\hat{M} = 6 > M^*$.

It is therefore probably advisable to use Benders cuts (6.123)–(6.125) rather than (6.127)–(6.128) when the subproblem involves cumulative scheduling.

Exercises

6.89. Consider a cumulative scheduling problem with four jobs in which $p = (3, 3, 4, 3)$, $c = (2, 2, 1, 1)$, and the resource limit is 3. All release times are zero, and the deadlines are $(L_1, \ldots, L_4) = (6, 6, 6, 7)$. Apply the edge-finding rules and update the bounds accordingly.

6.90. State the argument for the second edge-finding rule (6.130b).

6.91. Apply the algorithm of Fig 6.30 to the problem of Exercise 6.89.

6.92. State the argument for the second extended edge-finding rule (6.137b).

6.93. Consider a cumulative scheduling problem with four jobs in which $p = (5, 3, 3, 3)$, $c = (1, 2, 2, 1)$, and the resource limit is 3. The release times are $(E_1, \ldots, E_4) = (0, 2, 2, 2)$, and the deadlines are $(L_1, \ldots, L_4) = (9, 8, 8, 8)$. Apply the extended edge-finding rules and update the bounds accordingly.

Does extended edge finding identify a valid precedence that is not identified by ordinary edge finding?

6.94. State the argument for the not-last rule (6.138b).

6.95. Consider a cumulative scheduling problem with four jobs in which $p = (3,3,4,3)$, $c = (2,2,1,1)$, and the resource limit is 3. The release times are $(E_1, \ldots, E_4) = (1,1,1,0)$, and all the deadlines are 7. Apply the not-first/not-last rules and update the bounds accordingly. Does this identify any precedences not identified by edge finding?

6.96. Show by counterexample that Lemma 6.27 is not valid for cumulative scheduling.

6.97. Suppose that a Benders method is applied to a minimum makespan planning and cumulative scheduling problem. In the first iteration, jobs 1, 2, and 3 are assigned to machine A, on which the resource capacity is 2, $(p_{A1}, p_{A2}, p_{A3}) = (1,1,2)$, and $(c_{A1}, c_{A2}, c_{A3}) = (2,1,1)$. All the release times are zero, and all the deadlines are 4. Write the Benders cut (6.142).

6.98. Change the deadlines in the problem of Exercise 6.97 to $(L_1, L_2, L_3) = (2,1,4)$. Write the Benders cut (6.143). Show that the stronger cut (6.142) is not valid in this case.

6.99. Show by example that $T_i(J_{ik} \setminus Z_{ik}) < T_i(J_{ik})$ is possible for cumulative scheduling, even when all the release dates are the same.

6.16 Bibliographic Notes

Section 6.1. The concept of domain consistency was originally developed for binary (two-variable) constraints under the name of arc consistency [343, 360], because constraints were identified with arcs in a dependency graph. The concept was generalized to multiple-variable constraints in [359] and implicitly in [160]. The concept is still called hyperarc consistency, or generalized arc consistency, in recognition of the network model.

Bounds consistency, also known as *interval consistency*, has arisen in many contexts.

An extension of domain consistency to partial assignments yields k-completeness, which is stronger than the concept of k-consistency that originated in the constraints community [211]. The fundamental theorem relating backtracking with strong k-consistency and width of the dependency graph (Theorem 6.1) is due to [212]. This and other forms of consistency are studied in [265, 482].

When the dependency graph has small *induced width*, its structure can be exploited by nonserial dynamic programming, which has been known in operations research for more than 30 years [82]. Essentially the same idea has surfaced in a number of contexts, including Bayesian networks [332], belief logics [451, 455], pseudoboolean optimization [152], location theory [130], k-trees [20, 21], and bucket elimination [169].

The domain store can be viewed as a relaxation of the feasible set, albeit a weak relaxation that contains the entire Cartesian product of the variable domains. This raises the question as to whether some richer structure can provide a stronger relaxation and propagate more information from one constraint to the next. One recent proposal is a *relaxed multivalued decision diagram* (MDD) [13], which is based on the binary decision diagrams used for circuit verification and configuration problems [10, 12, 113]. The maximum *width* of MDDs can be adjusted to result in relaxations of the desired quality, with width 1 corresponding to the domain store. Relaxed MDDs have seen success in some problem areas [81, 252, 270] but have not yet been implemented in CP or optimization solvers.

Section 6.2. The idea for Fourier–Motzkin elimination appears in Fourier's work from the 1820s [209] (English translation in [322]). Motzkin's formulation appeared in the 1930s [362] (English translation in [363]). The same idea was proposed in the 1850s by Boole [106] as a solution method for what is now called linear programming (see [253] for a historical treatment).

Classical Benders decomposition is due to [73] and is extended to nonlinear subproblems (generalized Benders decomposition) in [224].

Section 6.4. The resolution method for propositional logic and the completeness theorem (Theorem 6.2) are due to [407, 408], where the method is applied to problems in disjunctive normal form, rather than the conjunctive normal form used here. In this context, resolution is often called *consensus*. Resolution is extended to first-order predicate logic in [427] and to clauses with multivalued variables in [295]. The connection between Horn clauses and unit resolution (Theorem 6.3) is observed in [181]. The completeness of parallel resoution (Theorem 6.7) as well as its polynomial complexity in partial-order dynamic backtracking are proved in [279].

Section 6.5. The domination theorems for 0-1 inequalities (Theorems 6.8 and 6.9) are proved in [273, 279]. The algorithm for generating all nonredundant clauses implied by a 0-1 linear inequality (Fig. 6.2) is generalized to nonlinear 0-1 inequalities in [241, 242]. The implication theorems for cardinality clauses (Theorems 6.11 and 6.12) are proved in [273, 279]. Additional domination results are proved in [44], which states an efficient algorithm for deriving all cardinality clauses implied by a 0-1 inequality. This algorithm is used as the basis for a 0-1 programming solver OPBDP. The completeness proofs for 0-1 resolution (Theorems 6.13 and 6.14) are due to [275].

Section 6.6. Logic-based Benders cuts for mixed-integer programming are introduced in [279, 296], based on the branching dual. Other schemes for generating mixed-integer Benders cuts are presented in [131, 140]. The use of nogoods (constraint-directed branching) in branch-and-bound search for mixed-integer programming was proposed in [279], and related schemes are given with computational results in [3, 433].

Section 6.7. The element constraint was introduced by [266]. Filtering for domain and bounds consistency is straightforward. Filtering algorithms are widely implemented, except those for the multidimensional and specially structured versions of the constraint.

Section 6.8. The alldiff constraint first appeared in [331]. The matching-based filtering method presented here is due to [150, 416], and the filtering method for achieving bounds consistency is due to [353]. The convex graph result on which it is based, Theorem 6.16, is proved in [229].

Section 6.9. The network flow model given here for filtering the cardinality constraint is due to [418]. There is a bounds consistency algorithm based on flows [311] that exploits convexity of the graph. Filters for the nvalues constraint are discussed under that entry in Chapter 8.

Section 6.10. The among and sequence constraints were introduced in [61]. Incomplete filtering algorithms for sequence appear in [53] and [425]. The latter presents a filter for sequence and cardinality in the car sequencing problem discussed here. The filter based on cumulative sums, due to [488], was the first to achieve domain consistency in polynomial time. Complete filters based in alternate encodings of the constraint are presented in [110]. The network flow model for sequence is introduced in [344]. The conversion of a consecutive-ones matrix (also known as an *interval matrix*) to a network matrix is a classical technique [494]; see also [8]. A procedure that checks whether a permutation of columns reveals the consecutive ones property appears in [107]. Algorithms for testing whether a matrix (perhaps without the consecutive ones property) has a network equivalent are given in [94, 442].

Section 6.11. The stretch constraint originated in [391]. The filtering algorithm presented here is due to [264]. Domain filtering in a dynamic programming context is discussed in [479].

Section 6.12. The idea of representing a CP constraint with a finite automaton appears as early as 1992 in [495]. Filtering for automata-based constraints was introduced by [56, 120]. The regular constraint was proposed by [390, 392].

Filtering by decomposition is discussed in [55, 56, 405, 406]. The basic propagation result for Berge acyclic constraint hypergraphs appears in a paper on relational databases [51] (a constraint can be viewed as a relation, i.e., a set of tuples belong to the Cartesian product of the variable domains). The result states that achieving domain consistency for each pair of linked constraints achieves global domain consistency. The two-pass property used here, along with others, is discussed in [59, 330]. It is an interesting question in general when filtering a decomposed constraint achieves domain consistency; recent work includes [89, 90, 92, 405].

The extension of automata-based constraints to nondeterministic automata appears in [55, 56, 406]. The cyclic regular constraint is briefly discussed in [406], and its reformulation by dynamic programming proposed here. A regular language constraint can be extended to other types of languages in the Chomsky hierarchy, and this is explored in [447].

Section 6.13. The circuit constraint was first formulated as such by [331], but Hamiltonian cycles have been studied by Hamilton, Kirkman, and others at least since the mid-nineteenth century. Elementary filtering methods have been used by [122, 457]. The filtering algorithm based on separators, presented here, appears in [220]. Filtering can also be based on sufficient conditions for nonhamiltonity of a graph, some of which appear in [134, 135, 137]. Given an edge (i, j) in directed graph G, let G_{ij} be the graph that results from inserting a vertex in edge (i, j). Then the edge (i, j) belongs to a Hamiltonian path if and only if G_{ij} is Hamiltonian. Checking whether nonhamiltonicity conditions are met for each edge may require too much computation for a practical filter, however.

Section 6.14. An early study of disjunctive scheduling is [116], and edge finding is introduced in [117]. The $O(n^2)$ edge-finding algorithm described here is that of [118]. Another $O(n^2)$ algorithm appears in [39, 375, 378]. An algorithm that achieves $O(n \log n)$ complexity with complex data structures is given in [119], and an $O(n^3)$ algorithm that allows incremental updates in [121]. Extensions that take setup times into account are presented in [112, 206]. The propagation algorithm for not-first/not-last rules given here appears in [38], while others appear in [180, 477]. A comprehensive treatment of disjunctive and cumulative scheduling is provided by [39].

The Benders cuts for minimizing cost and makespan are used in [279, 300]. They are strengthened by analyzing the edge-finding process in [284] and by re-solving the subproblem in [291]. The cuts for minimizing the number of late jobs and total tardiness are developed in [285] and strengthened in [291].

Section 6.15. The cumulative scheduling constraint originated in [6]. The $O(n^2)$ edge-finding algorithm presented here, which appears in [39], is an improvement of one in [376, 377]. Another algorithm appears in [121]. The extended edge-finding algorithm given here is based on [39, 375]. Not-first/not-last propagation appears in [376, 377] and energetic reasoning in [186, 187].

The Benders cuts for minimizing cost and makespan appear in [279] and are strengthened in [284, 291]. The cuts for minimizing the number of late jobs and total tardiness are developed in [285] and strengthened in [291].

Chapter 7
Relaxation

The ideal problem relaxation is both easy to solve and in some sense *tight*, meaning that it closely resembles the original problem. The solution of a tight relaxation is more likely to be feasible in the original problem, or if not, to provide a good bound on the optimal value of the original problem.

The most widely used medium for formulating a relaxation is a system of linear inequalities in continuous variables. This is due largely to the fact that the mathematical programming field has focused on inequality-constrained problems and has made continuous linear relaxation one of its primary tools for solving them. A relaxation of this sort is easy to solve because it is a linear programming problem. It can also be tight if the problem structure is carefully analyzed and the inequalities selected wisely.

There are at least two ways to obtain linear inequality-constrained relaxations. One is to reformulate the constraints with a mixed integer linear programming (MILP) model, and then to take a continuous relaxation of the resulting model. An MILP model consists of linear inequalities with additional 0-1 or integer-valued variables that capture the discrete elements of the problem. Experience has taught that integer variables allow a sufficiently clever modeler to formulate a wide range of constraints. A continuous relaxation is obtained simply by dropping the integrality constraint on the variables, and the relaxation is strengthened by the addition of valid inequalities (cutting planes), an intensely studied topic of mathematical programming.

A second strategy is to design a relaxation directly, without formulating an MILP model, but using only the original variables and exploiting the specific structure of the constraints. This approach received relatively little attention before the recent rise of interest in integrated methods, because the community most interested in continuous relaxation—the mathematical programming community— normally formulated the problem as an MILP from the start.

The choice between the strategies is not either/or. One can relax some metaconstraints directly and relax others using an MILP model, and then pool the individual relaxations into a single relaxation for the entire problem. The pooled relaxation is solved to obtain a bound on the optimal value of the original problem. This constraint-oriented approach is very much in the spirit of integrated methods, because it allows one to take advantage of the peculiar characteristics of each metaconstraint.

In some cases, it is advantageous to formulate not one relaxation but many. If the relaxations are sufficiently easy to solve, one can examine the space of relaxations in search of a tight one. The problem of finding the tightest relaxation is the *relaxation dual*, already discussed in Section 4.2. Specific relaxation duals include Lagrangean, surrogate, subadditive, and branching duals, also introduced in Chapter 4.

This chapter makes essential use of the basic concepts of linear optimization, which are presented in Section 3.1. The chapter begins with a brief description of how to relax piecewise linear functions without using a mixed-integer formulation. Piecewise linear functions not only occur frequently in their own right, but can be used to approximate some nonlinear problems that would otherwise be difficult to relax.

The relaxation of MILP models is then taken up. This requires an extended discussion of cutting-plane theory, because cutting planes are often necessary to obtain a tight continuous relaxation. General-purpose cutting planes are developed for 0-1, integer, and mixed-integer constraint sets. None of this is applicable unless an MILP model can be written in the first place, and a section is therefore devoted to MILP modeling.

Much of the cutting plane discussion is concerned with *separating cuts*, which are generated only after the continuous relaxation is solved. These cutting planes are therefore not part of the original linear relaxation that might be generated for a metaconstraint. They are added to the pooled relaxation only after it is solved.

The remainder of the chapter is devoted to identifying relaxations for specific metaconstraints—disjunctions of linear and nonlinear inequality systems, formulas of propositional logic, and several popular global constraints from the constraint programming field (disjunctions are discussed before MILP modeling in order to provide necessary background). The global constraints canvassed include the element, all-different, cardinality, and circuit constraints, along with disjunctive and cumulative scheduling constraints. Some constraints are relaxed directly, some are given MILP formulations, and some are treated in both ways. Chapter 8 cites relaxations given in the literature for several additional metaconstraints.

Convex nonlinear relaxations can be useful for relaxing continuous nonlinear constraints for which a good linear relaxation is difficult to identify. This possibility is only briefly addressed here by presenting convex relaxations for a disjunction of convex nonlinear systems.

7.1 Piecewise Linear Functions

A natural generalization of linear inequality constraints is to inequalities or equations with a sum of piecewise linear functions on the left-hand side. Each function takes one variable as an argument.

Constraints of this sort are a highly versatile modeling device, partly because they can approximate separable nonlinear functions. They also represent an excellent application of integrated methods. Although piecewise linear functions can be given mixed-integer models using *type II special ordered sets* of variables, this requires additional variables. By replacing a piecewise linear function with an appropriate meta-constraint, a convex hull relaxation can be quickly generated without adding variables. Branching is also effective and results in substantial propagation. "Gaps" in the function are accommodated without effort, whereas they require still more auxiliary variables in a mixed-integer model.

A piecewise linear inequality constraint has the form

$$\sum_j g_j(x_j) \geq b \tag{7.1}$$

where each g_j is a piecewise linear function. In general a piecewise linear function $g(x)$ is defined on a finite union of closed intervals and can be written

$$g(x) = \frac{U_k - x}{U_k - L_k} c_k + \frac{x - L_k}{U_k - L_k} d_k \quad \text{for } x \in [L_k, U_k], \quad k = 1, \ldots, m$$

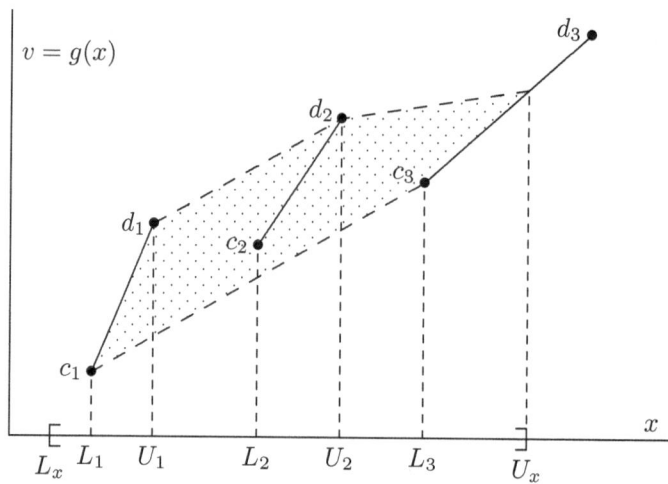

Fig. 7.1 Graph of a piecewise linear function with gaps, and its convex hull (shading). The interval $[L_x, U_x]$ is the current domain of x.

where intervals $[L_k, U_k]$ intersect in at most a point. A sample function is shown in Fig. 7.1.

A separable nonlinear function $\sum_j h_j(x_j)$ can be approximated by a piecewise linear function to a high degree of accuracy by replacing each nonlinear function $h_j(x)$ with a piecewise linear function having sufficiently many linear pieces. The overhead of doing so is small, because a convex hull relaxation can be generated quickly for each function without requiring additional variables.

Figure 7.1 illustrates how a piecewise linear function $g(x)$ with gaps can be approximated with a convex hull relaxation. Note that the domain of x can be used to reduce the interval $[L_3, U_3]$ to $[L_3, U_x]$. Techniques from computational geometry use a divide-and-conquer approach to compute the convex hull in time proportional to $m \log m$.

In general, a constraint (7.1) can be relaxed by replacing it with $\sum_j v_j \geq b$, where each v_j is constrained by a system of linear inequalities in v_j, x_j that describes the convex hull of the graph of $v_j = g_j(x_j)$.

Branching on x can tighten the domains and relaxations considerably. Branching is called for when the solution value $(\bar{v}, \bar{x})$ of (v, x) in the current relaxation of the problem does not satisfy $g(\bar{x}) = \bar{v}$. If $\bar{x}$ is between U_1 and U_2, for example, the domain of x is split into the intervals $[L_x, U_1]$ and $[L_2, U_x]$, resulting in the convex hull relaxations of Fig. 7.2. If $\bar{x}$ lies in the interval $[L_2, U_2]$, one can split the domain

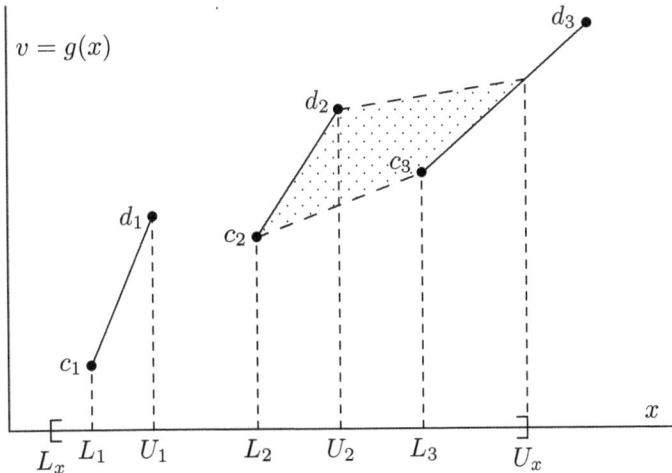

Fig. 7.2 Convex hull relaxations after branching on x with a value between U_1 and L_2.

of x into three intervals, $[L_x, U_1]$, $[L_2, U_2]$, and $[L_3, U_x]$. In this small example, the resulting convex hull relaxations become exact.

Exercises

7.1. Let $h(x) = \sum_{j=1}^{n} h_j(x_j)$ be a separable function, where each x_j has domain D_{x_j}. If

$$S_j = \{(v_j, x_j) \mid v_j = h_j(x_j),\ x_j \in D_{x_j}\}$$

for each j, let $\mathcal{C}_j$ be a constraint set that describes the convex hull of S_j. Show that the feasible set of

$$\mathcal{C} = \left\{ v = \sum_{j=1}^{n} v_j \right\} \cup \bigcup_{j=1}^{n} \mathcal{C}_j$$

projected onto (v, x) is the convex hull of

$$S = \{(v, x) \mid v = h(x),\ x \in D\}$$

where $D = D_{x_1} \times \cdots \times D_{x_n}$. Thus one can model the convex hull of h by modeling the convex hull of each h_j.

7.2. Let each $h_j(x_j)$ in Exercise 7.1 be a fixed-charge function

$$h_j(x_j) = \begin{cases} 0 & \text{if } x_j = 0 \\ f_j + c_j x_j & \text{if } x_j > 0 \end{cases}$$

where each $x_j \in [0, U_j]$. In this case, $h_j(x_j)$ is a lower semicontinuous function. The convex hull of S_j is described by

$$\frac{f_j + c_j U_j}{U_j} x_j \le v_j \le f_j + c_j x_j, \quad x_j \ge 0$$

and is illustrated by the light shaded area in Fig. 7.14(b) of Section 7.6.2. Describe the convex hull of S using only the variables v and $x = (x_1, \ldots, x_n)$.

7.2 0-1 Linear Inequalities

A *0-1 linear inequality* is a linear inequality with 0-1 variables. It has the form $ax \ge a_0$, where $x = (x_1, \ldots, x_n)$ and $x \in \{0, 1\}^n$. General integer linear inequalities, in which the value of each x_j must be a nonnegative integer, are discussed in Section 7.3. It is assumed throughout this section that the coefficients a_j are integral. This entails negligible loss of generality, because an inequality with rational coefficients can be multiplied by a suitable positive scalar to covert the coefficients to integers.

A continuous relaxation of a system 0-1 linear inequalities

$$Ax \ge b, \quad x \in \{0, 1\}^n \tag{7.2}$$

can be obtained by replacing $x_j \in \{0, 1\}$ with $0 \le x_j \le 1$ for each j. A widely used technique for strengthening this continuous relaxation is to generate *cutting planes*. These are inequalities that are implied by (7.2) but are *not* implied by the continuous system $Ax \ge b$. That is, if $cx \ge c_0$ is a cutting plane, then any $x \in \{0, 1\}^n$ that satisfies $Ax \ge b$ also satisfies $cx \ge c_0$, but this is not true of any $x \in [0, 1]^n$. Cutting planes are also known as *valid inequalities* or *valid cuts*.

An example appears in Fig. 7.3. The small circles represent the 0-1 solutions of the system

$$2x_1 + x_2 \ge 1$$
$$x_1 + 2x_2 \ge 1$$

The shaded area is the feasible set of the continoous relaxation. The cutting plane $x_1 + x_2 \ge 1$ cuts off part of this feasible set, leaving the darker shaded region.

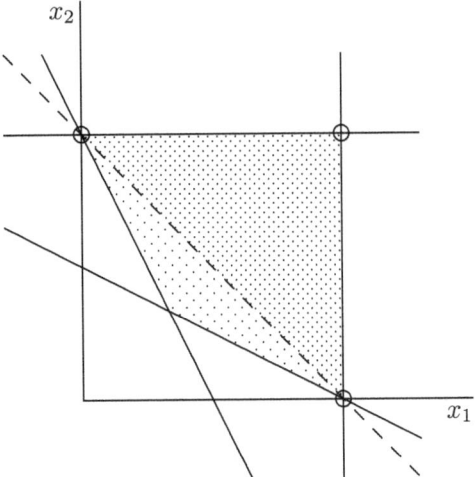

Fig. 7.3 Feasible solutions of a 0-1 programming problem (small circles), feasible set of a continuous relaxation (total shaded area), and feasible set that that remain after adding a valid cut (darker shading).

A 0-1 system (7.2) can therefore be relaxed by combining the continuous linear system $Ax \geq b$ with cutting planes and bounds $0 \leq x_j \leq 1$ for each x_j. Since the cutting planes are not redundant of the *continuous* system $Ax \geq b$, they strengthen the continuous relaxation by *cutting off* points that would otherwise satisfy the relaxation.

Cutting planes are often used to help solve *0-1 linear programming problems*, in which all the constraints are 0-1 linear inequalities. Such problems can be written

$$\min cx$$
$$Ax \geq b, \quad x \in \{0,1\}^n \tag{7.3}$$

where $x = (x_1, \ldots, x_n)$. Cutting planes can also be generated for systems with general integer and continuous variables, which are discussed in Section 7.3.

Cutting planes can be generated by a general-purpose method, or by a method that is specialized to inequality sets with a particular type of structure. Some general-purpose methods are presented in this section, including the Chvátal–Gomory procedure, knapsack cut generation, and lifting. The next section presents Gomory cuts, which are general-purpose cuts for general integer as well as 0-1 inequalities, and mixed-integer rounding cuts for inequalities with both integral and continuous variables. Specialized cutting planes will be discussed in later sections in connection with specific types of constraints.

7.2.1 Chvátal–Gomory Cuts

A relatively simple algorithm, known as the Chvátal–Gomory procedure, can in principle generate any valid cut for a system of linear inequalities with integer-valued variables. It also reveals an interesting connection between logic and cutting plane theory, because the Chvátal–Gomory procedure for 0-1 linear inequalities is related to the resolution method of inference.

The Chvátal–Gomory procedure can be applied to any system of integer linear inequalities, but it is convenient to begin with a 0-1 system (7.2). $Ax \geq b$ is understood to include bounds $0 \leq x \leq e$, where e is a vector of ones. This imposes no restriction but is convenient for generating cuts.

The procedure is based on the idea of an *integer rounding cut*. Given an inequality $ax \geq a_0$ where x is a nonnegative integer, an integer rounding cut is obtained by rounding up any nonintegers among the coefficients and right-hand side. The cut therefore has the form $\lceil a \rceil x \geq \lceil a_0 \rceil$ and is clearly valid for $ax \geq a_0$. The coefficients a on the left-hand side can be rounded up, without invalidating the inequality, because $x \geq 0$. Now that the left-hand side is integral, the right-hand side a_0 can be rounded up as well.

Each step of the Chvátal–Gomory procedure generates an integer rounding cut for a surrogate of $Ax \geq b$ and adds this cut to the system $Ax \geq b$. Recall from Section 4.3 that a surrogate of $Ax \geq b$ is a nonnegative linear combination $uAx \geq ub$. The generated cut therefore has the form $\lceil uA \rceil x \geq \lceil ub \rceil$ where $u \geq 0$. Any inequality generated by a finite number of such steps is a *Chvátal–Gomory cut* for (7.2). A cut that can be generated in k steps, but not $k-1$ steps, is a cut of *rank k*. It is a fundamental result of cutting-plane theory that every valid cut for (7.2) is a Chvátal–Gomory cut.

As an example, consider the 0-1 system:

$$x_1 + 3x_2 \geq 2$$
$$6x_1 - 3x_2 \geq -1$$
$$0 \leq x_j \leq 1, \quad j = 1, 2$$
$$x \in \{0, 1\}^2$$

It has the valid rank 2 cut $2x_1 + 3x_2 \geq 5$. There are many Chvátal–Gomory derivations of this cut, such as the following two-step derivation. The cuts $x_1 \geq 1$ and $x_2 \geq 1$ are generated first:

$$
\begin{array}{ll}
x_1 + 3x_2 \geq 2 \quad (\tfrac{1}{7}) & \qquad x_1 + 3x_2 \geq 2 \quad (\tfrac{1}{3}) \\
6x_1 - 3x_2 \geq -1 \; (\tfrac{1}{7}) & \qquad -x_1 \qquad\quad \geq -1 \; (\tfrac{1}{3}) \\
\hline
x_1 \qquad\quad \geq \tfrac{1}{7} \Rightarrow x_1 \geq 1 & \qquad x_2 \geq \tfrac{1}{3} \Rightarrow x_2 \geq 1
\end{array}
$$

The nonzero multipliers u_i are shown in parentheses, and the surrogate $uAx \geq ub$ appears below the line along with the cut $\lceil uA \rceil x \geq \lceil ub \rceil$. The constraint $-x_1 \geq -1$ reflects the bound $x_1 \leq 1$. A linear combination of these two cuts yields $2x_1 + 3x_2 \geq 5$ in the second step.

A longer derivation of $2x_1 + 3x_2 \geq 5$, outlined in Fig. 7.4, points the way to a proof of the Chvátal–Gomory theorem. The derivation starts with the weaker inequality $2x_1 + 3x_2 \geq 0$, which can be derived from bounds $x_j \geq 0$. From this it obtains, in a manner described in the proof below, the four Chvátal–Gomory cuts at the leaf nodes of the enumeration tree. These four combine in pairs to obtain the cuts at the middle level. For instance, the first two combine as follows:

$$
\begin{array}{ll}
(2x_1 + 3x_2) + x_1 + \qquad\;\; x_2 \geq 1 \; (\tfrac{1}{2}) \\
(2x_1 + 3x_2) + x_1 + (1 - x_2) \geq 1 \; (\tfrac{1}{2}) \\
\hline
(2x_1 + 3x_2) + x_1 \qquad\qquad\;\; \geq \tfrac{1}{2} \Rightarrow (2x_1 + 3x_2) + x_1 \geq 1
\end{array}
$$

This inference is closely parallel to a resolution step (Section 6.4.2). The "logical" parts of the two inequalities are

$$
\begin{array}{l}
x_1 + x_2 \qquad\;\; \geq 1 \\
x_1 + (1 - x_2) \geq 1
\end{array}
$$

These correspond to logical clauses $x_1 \lor x_2$ and $x_1 \lor \neg x_2$, which have the resolvent x_1 (i.e., $x_1 \geq 1$). So the Chvátal–Gomory operations in Fig. 7.4 are parallel to a resolution proof of the empty clause. The conclusion, however, is not the empty clause, but $2x_1 + 3x_2 \geq 1$. This becomes the premise for another round of cuts, which yields $2x_1 + 3x_2 \geq 2$, and so forth, until $2x_1 + 3x_2 \geq 5$ is obtained.

Theorem 7.1. *Every valid cut for a 0-1 linear system is a Chvátal–Gomory cut.*

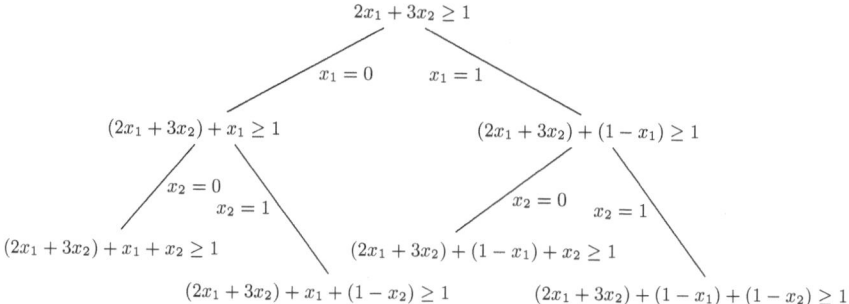

Fig. 7.4 Illustration of the proof of the Chvátal–Gomory theorem for 0-1 inequalities.

Proof. Let $cx \geq c_0$ be a valid cut for (7.2). Let $c_0 - \Delta$ be the smallest possible value of cx in the unit box defined by $0 \leq x \leq e$. Thus, the Chvátal–Gomory cut

$$cx \geq c_0 - \Delta \qquad (7.4)$$

is obtained from a linear combination of bounds (namely, $x_j \geq 0$ with weight c_j when $c_j \geq 0$ and $-x_j \geq -1$ with weight $-c_j$ when $c_j < 0$).

There is a sufficiently large number M such that the following is also a Chvátal–Gomory cut

$$cx + M \sum_{j \in P} x_j + M \sum_{j \in N} (1 - x_j) \geq c_0 \qquad (7.5)$$

for any partition P, N of $\{1, \dots, n\}$. This is shown as follows. First, (7.5) is valid for the polyhedron defined by $Ax \geq b$ and $0 \leq x \leq e$. To see this, it suffices to observe that (7.5) is satisfied at all vertices of this polyhedron. It is satisfied at 0-1 vertices because $cx \geq c_0$ is valid for the 0-1 system. If $\alpha > 0$ is the smallest value of $\sum_{j \in P} x_j + \sum_{j \in N}(1 - x_j)$ over all nonintegral vertices and all partitions, then (7.5) is satisfied at any noninteger vertex when $M = \Delta/\alpha$, due to (7.4). Now, since (7.5) is valid for the polyhedron, by Corollary 4.5 it is dominated by a surrogate of $Ax \geq b$ and $x \leq e$ and is therefore a Chvátal–Gomory cut.

To complete the proof, it suffices to show that if

$$cx \geq c_0 - \delta - 1 \qquad (7.6)$$

is a Chvátal–Gomory cut and $cx \geq c_0 - \delta$ is valid, then $cx \geq c_0 - \delta$ is a Chvátal–Gomory cut.

First, for any partition P, N of $\{1, \ldots, n\}$, the inequality

$$cx + \sum_{j \in P} x_j + \sum_{j \in N} (1 - x_j) \geq c_0 - \delta \qquad (7.7)$$

is a Chvátal–Gomory cut. This is seen by combining (7.5) and (7.6) with weights $1/M$ and $(M-1)/M$, respectively.

Next, suppose the following are Chvátal–Gomory cuts

$$cx + \sum_{j \in P' \cup \{i\}} x_j + \sum_{j \in N'} (1 - x_j) \geq c_0 - \delta$$

$$cx + \sum_{j \in P'} x_j + \sum_{j \in N' \cup \{i\}} (1 - x_j) \geq c_0 - \delta$$

where P', N' and $\{i\}$ partition a subset of $\{1, \ldots, n\}$. Combining these cuts with a weight of $1/2$ each yields the "resolvent"

$$cx + \sum_{j \in P'} x_j + \sum_{j \in N'} (1 - x_j) \geq c_0 - \delta \qquad (7.8)$$

which is therefore a Chvátal–Gomory cut. By applying this operation repeatedly, starting with (7.7), the Chvátal–Gomory cut $cx \geq c_0 - \delta$ is obtained. $\square$

7.2.2 0-1 Knapsack Cuts

It is frequently useful to derive cutting planes for individual 0-1 knapsack inequalities as well as systems of 0-1 inequalities. These cutting planes are known as *knapsack cuts* or *cover inequalities*. Knapsack cuts can often be strengthened by a *lifting* process that adds variables to the cut. Two lifting procedures are described in the following two sections.

Recall that 0-1 knapsack packing inequalities have the form $ax \leq a_0$ and knapsack covering inequalities have the form $ax \geq a_0$, where $a \geq 0$ and each $x_j \in \{0, 1\}$. The nonnegativity restriction on a incurs minimal loss of generality, because one can always obtain $a \geq 0$ by replacing each x_j that has a negative coefficient with $1 - x_j$. Cuts are developed here for packing inequalities $ax \leq a_0$. Covering inequalities can be accommodated by obtaining cuts for $-ax \leq -a_0$.

Define a *cover* for $ax \leq a_0$ to be an index set $J \in \{1, \ldots, n\}$ for which $\sum_{j \in J} a_j > a_0$. A cover is *minimal* if no proper subset is a cover. If J is a cover, the *cover inequality*

$$\sum_{j \in J} x_j \leq |J| - 1 \qquad (7.9)$$

is obviously valid for $ax \leq a_0$. Only minimal covers need be considered, because nonminimal cover inequalities are redundant of minimal ones.

For example, $J = \{1, 2, 3, 4\}$ is a minimal cover for the inequality

$$6x_1 + 5x_2 + 5x_3 + 5x_4 + 8x_5 + 3x_6 \leq 17 \qquad (7.10)$$

and gives rise to the cover inequality

$$x_1 + x_2 + x_3 + x_4 \leq 3 \qquad (7.11)$$

7.2.3 Sequential Lifting

A cover inequality can often be strengthened by adding variables to the left-hand side; that is, by *lifting* the inequality into a higher dimensional space. *Sequential lifting*, in which terms are added one at a time, is presented first. The resulting cut depends on the order in which terms are added. There are also techniques for adding several terms simultaneously, and one of these is described in the next section.

Given a cover inequality (7.9), the first step of sequential lifting is to add a term $\pi_k x_k$ to the left-hand side to obtain

$$\sum_{j \in J} x_j + \pi_k x_k \leq |J| - 1 \qquad (7.12)$$

where π_k is the largest coefficient for which (7.12) is still valid. Thus, one can set

$$\pi_k = |J| - 1 - \max_{\substack{x_j \in \{0,1\} \\ \text{for } j \in J}} \left\{ \sum_{j \in J} x_j \;\middle|\; \sum_{j \in J} a_j x_j \leq a_0 - a_k \right\} \qquad (7.13)$$

For example, (7.11) can be lifted to

$$x_1 + x_2 + x_3 + x_4 + 2x_5 \leq 3 \tag{7.14}$$

because in this case π_5 is

$$3 - \max \{x_1 + x_2 + x_3 + x_4 \mid 6x_1 + 5x_1 + 5x_2 + 5x_4 \leq 17 - 8\} = 2$$

At this point, (7.12) can be lifted further by adding another new term, and so forth. At any stage in this process, the current inequality is

$$\sum_{j \in J} x_j + \sum_{j \in J'} \pi_j x_j \leq |J| - 1$$

The lifting coefficient for the next x_k, where $k \notin L = J \cup J'$, is

$$\pi_k = |J| - 1 - \max_{\substack{x_j \in \{0,1\} \\ \text{for } j \in L}} \left\{ \sum_{j \in J} x_j + \sum_{j \in J'} \pi_j x_j \;\middle|\; \sum_{j \in L} a_j x_j \leq a_0 - a_k \right\} \tag{7.15}$$

For example, if one attempts to lift (7.14) further by adding x_6, the inequality is unchanged because π_6 is

$$3 - \max \left\{ \sum_{j=1}^{4} x_j + 2x_5 \;\middle|\; 6x_1 + 5x_2 + 5x_3 + 5x_4 + 8x_5 \leq 14 \right\} = 0$$

The order of lifting can affect the outcome; if x_6 is added before x_5, the resulting cut is $x_1 + x_2 + x_3 + x_4 + x_5 + x_6 \leq 3$.

The computation of lifting coefficients π_k by (7.13) in effect requires solving a 0-1 programming problem and can be time consuming. Fortunately, the coefficients can be computed recursively. An algorithm for computing the initial lifting coefficient π_k will be given first, and then a different recursion for computing the remaining coefficients.

This approach actually requires one to compute more than π_k; one must compute the function

$$\pi_J(u) = |J| - 1 - g^*(u)$$

for $u \geq 0$, where

$$g^*(u) = \max_{\substack{x_j \in \{0,1\} \\ \text{for } j \in J}} \left\{ \sum_{j \in J} x_j \;\middle|\; \sum_{j \in J} a_j x_j \leq a_0 - u \right\} \tag{7.16}$$

Then, in particular, $\pi_k = \pi_J(a_k)$.

The function $\pi_J(u)$ can be computed directly by solving a dynamic programming problem for each value of u, but it is more efficient to solve a dual problem that exchanges the constraint and objective function in (7.16), and then recover $\pi_J(u)$. The dual problem is to compute the following for $t = 0, \ldots, |J|$:

$$h^*(t) = \min \left\{ \sum_{j \in J} a_j x_j \ \middle| \ \sum_{j \in J} x_j \geq t \right\}$$

Then given any integer $t \in \{0, \ldots, |J| - 1\}$, one can deduce that $g^*(u) = t$ for all u satisfying

$$a_0 - h^*(t+1) < u \leq a_0 - h^*(t)$$

Also, $g^*(u) = |J|$ for $u \geq h^*(|J|)$. The function $h^*(t)$ can be computed using the recursion

$$h_k(t) = \min \{a_k + h_{k-1}(t-1), h_{k-1}(t)\}, \quad t = 1, \ldots, |J| \qquad (7.17)$$

with the boundary conditions $h_k(0) = 0$ for all k and $h_0(t) = \infty$ for $t > 0$. Now $h^*(t) = h_{|J|}(t)$.

In the example, the first three stages of the recursion (7.17) become

$$\begin{aligned}
h_1(1) &= \min\{a_1 + h_0(0), h_0(1)\} = 6 \\
h_2(1) &= \min\{a_2 + h_1(0), h_1(1)\} = 5 \\
h_2(2) &= \min\{a_2 + h_1(1), h_1(2)\} = 11 \\
h_3(1) &= 5, \quad h_3(2) = 10, \quad h_3(3) = 16
\end{aligned}$$

with $h_k(t) = \infty$ for $t > k$. The fourth and last stage yields

$$\begin{aligned}
h^*(0) &= h_4(0) = 0 \\
h^*(1) &= h_4(1) = 5 \\
h^*(2) &= h_4(2) = 10 \\
h^*(3) &= h_4(3) = 15 \\
h^*(4) &= h_4(4) = 21
\end{aligned}$$

From this one can read off the function $g^*(u)$

$$g^*(u) = \begin{cases}
3 & \text{for } 0 \leq u \leq 2 \\
2 & \text{for } 2 < u \leq 7 \\
1 & \text{for } 7 < u \leq 12 \\
0 & \text{for } 12 < u \leq 17 \\
-\infty & \text{for } 17 < u
\end{cases}$$

and $\pi_J(u) = 3 - g^*(u)$. Thus, the lifting coefficient for x_5 is $\pi_5 = \pi_J(a_5) = \pi_J(8) = 2$.

The remaining sequential lifting coefficients are determined as follows. If coefficients π_j have been derived for $j \in J'$, then the next coefficient π_k is $\pi_L(a_k)$, where $L = J \cup J'$ and

$$\pi_L(u) = |J| - 1 - \max_{\substack{x_j \in \{0, 1\} \\ \text{for } j \in L}} \left\{ \sum_{j \in J} x_j + \sum_{j \in J'} \pi_j x_j \;\middle|\; \sum_{j \in L} a_j x_j \leq a_0 - u \right\}$$

The function $\pi_L(u)$ can be computed recursively when $\pi_J(u)$ is known:

$$\pi_{L \cup \{k\}}(u) = \min \{\pi_L(u), \pi_L(u + a_k) - \pi_L(a_k)\} \qquad (7.18)$$

where the first option inside the min corresponds to setting $x_k = 0$, and the second to $x_k = 1$.

In the example, if π_6 is to be computed after π_5, then

$$\pi_6 = \pi_{J \cup \{5\}}(a_6) = \min \{\pi_J(3), \pi_J(6) - \pi_J(3)\} = \min\{1, 1 - 1\} = 0$$

as stated earlier.

The computation of optimal lifting coefficients π_k can be practical when there are not too many variables involved or the problem has special structure. In other cases, one may wish to use a heuristic algorithm to obtain a π_k that may be smaller than the optimal coefficient.

7.2.4 Sequence-Independent Lifting

It is possible to lift all the missing variables in the cover inequality (7.9) simultaneously. This can be done by constructing a function that is analogous to $\pi_J(u)$ in the previous section, but that provides valid lifting coefficients for all the missing variables when they are added as a group. The resulting cut may not be as strong as one obtained by optimal sequential lifting, but it requires much less computation. Sequence-independent lifting has therefore become a standard feature of integer programming solvers.

Recall that function ρ is superadditive if $\rho(u_1 + u_2) \geq \rho(u_1) + \rho(u_2)$ for all real numbers u_1, u_2. The lifting procedure is based on the following fact.

Theorem 7.2. *Suppose that $\rho(u) \leq \pi_J(u)$ for all real numbers u, $\rho(u)$ is superadditive, and J is a cover for the 0-1 knapsack inequality $ax \leq a_0$. Then, the following lifted inequality is valid for $ax \leq a_0$:*

$$\sum_{j \in J} x_j + \sum_{j \notin J} \rho(a_j) x_j \leq |J| - 1 \tag{7.19}$$

Proof. It suffices to show that any 0-1 vector $\bar{x}$ satisfying $ax \leq a_0$ also satisfies (7.19). Let $J' = \{j \notin J \mid \bar{x}_j = 1\}$. Then

$$\sum_{j \in J} \bar{x}_j + \sum_{j \notin J} \rho(a_j) \bar{x}_j = \sum_{j \in J} \bar{x}_j + \sum_{j \in J'} \rho(a_j)$$

$$\leq \sum_{j \in J} \bar{x}_j + \rho\left(\sum_{j \in J'} a_j\right) \leq \sum_{j \in J} \bar{x}_j + \pi_J\left(\sum_{j \in J'} a_j\right)$$

where the first inequality is due to the superadditivity of ρ and the second is due to the fact that π_J bounds ρ. It therefore suffices to show that

$$\sum_{j \in J} \bar{x}_j + \pi_J\left(\sum_{j \in J'} a_j\right) \bar{y} \leq |J| - 1 \tag{7.20}$$

with $\bar{y} = 1$. Let $J = \{1, \ldots, p\}$, and note that $(\bar{x}_1, \ldots, \bar{x}_p, \bar{y})$ satisfies

$$\sum_{j \in J} a_j x_j + \left(\sum_{j \in J'} a_j\right) y \leq a_0 \tag{7.21}$$

Since

$$\sum_{j \in J} x_j + \pi_J\left(\sum_{j \in J'} a_j\right) y \leq |J| - 1 \tag{7.22}$$

is a valid lifted cut for (7.21), $(\bar{x}_1, \ldots, \bar{x}_p, \bar{y})$ satisfies (7.22). Therefore, (7.20) holds. □

A function ρ that satisfies the conditions of Theorem 7.2 can be obtained as follows. Let $\Delta = \sum_{j \in J} a_j - a_0$ and $J = \{1, \ldots, p\}$ with $a_1 \geq \cdots \geq a_p$. Let $A_j = \sum_{k=1}^{j} a_k$ with $A_0 = 0$. Then define

$$\rho(u) = \begin{cases} j & \text{if } A_j \leq u \leq A_{j+1} - \Delta \ \& \ j \in \{0, \ldots, p-1\} \\ j + (u - A_j)/\Delta & \text{if } A_j - \Delta \leq u < A_j \ \& \ j \in \{1, \ldots, p-1\} \\ p + (u - A_p)/\Delta & \text{if } A_p - \Delta \leq u \end{cases}$$

$$\tag{7.23}$$

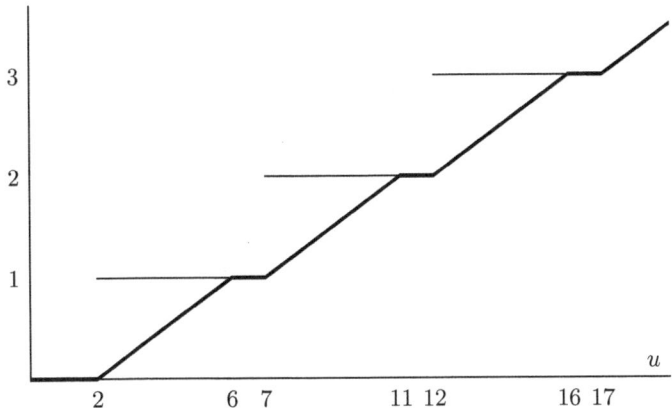

Fig. 7.5 Graph of $\pi_J(u)$ (horizontal line segments) and $\rho(u)$ (heavy line) for a knapsack lifting problem.

It can be shown that $\rho(u)$ is superadditive and bounded above by $\pi_J(u)$. Note that it is easy to compute the lifting coefficients $\rho(a_j)$ using (7.23).

The function $\rho(u)$ for example (7.10) appears in Fig. 7.5. Since $\rho(a_5) = \rho(8) = \frac{5}{4}$ and $\rho(a_6) = \rho(3) = \frac{1}{4}$, the result of lifting the cover inequality (7.11) is

$$x_1 + x_2 + x_3 + x_4 + \tfrac{5}{4}x_5 + \tfrac{1}{4}x_6 \le 3$$

7.2.5 Set-Packing Inequalities

The cover inequalities described in the previous three sections are derived from individual knapsack constraints. In many cases, stronger inequalities can be obtained by considering several knapsack constraints simultaneously. One way to do this is to deduce valid inequalities for the collection of cover inequalities derived from several knapsack constraints. Mixed-integer solvers typically focus on cover inequalities with a right-hand side of one. A collection of these inequalities define a *set-packing* problem, for which several classes of valid inequalities have been studied.

A set-packing problem asks, for a given collection of sets, whether one can choose k sets that are pairwise disjoint. When written in 0-1 form, the problem is to find a feasible solution for

$$Ax \le e, \quad x_j \in \{0, 1\}, \quad j = 1, \ldots, n \tag{7.24}$$

and $\sum_j x_j \ge k$, where A is a 0-1 matrix and e is a tuple of ones. The columns of A correspond to sets and the rows to elements of the sets. Set j contains element i if $A_{ij} = 1$, and set j is selected if $x_j = 1$.

Two families of valid inequalities can be derived right away. They are most easily described by defining the *intersection graph* G that corresponds to (7.24). G has a vertex for each variable x_j and an undirected edge (x_j, x_k) whenever $A_{ij} = A_{ik} = 1$ for some row i. An *odd cycle* of G is a cycle with an odd number of edges.

Theorem 7.3. *Let G be the intersection graph for the set-packing problem (7.24). If C is a subset of vertices of G defining an odd cycle, then the* odd cycle inequality

$$\sum_{j \in C} x_j \le \frac{|C| - 1}{2} \tag{7.25}$$

is valid for (7.24).

It can also be shown that if C defines an odd hole, then (7.25) is facet defining. An odd cycle is an *odd hole* if it has no chords (i.e., no edges other than the edges in the cycle that connect two vertices of the cycle). An inequality is *facet defining* for (7.24) if it describes a facet, or $(n-1)$-dimensional face, of the convex hull of the feasible set. Facet-defining inequalities are desirable as valid inequalities because all valid inequalities are dominated by surrogates of the facet-defining inequalities.

Consider, for example, the small set-packing problem

$$
\begin{aligned}
x_1 + x_2 & & & & \le 1 \\
x_2 + x_3 & & & & \le 1 \\
& x_3 + x_4 & & & \le 1 \\
& & x_4 + x_5 & \le 1 \\
x_1 & & & + x_5 & \le 1
\end{aligned}
$$

The intersection graph is a pentagon on the vertices $x_1, \ldots, x_5$. Since this is an odd hole, the valid inequality $x_1 + x_2 + x_3 + x_4 + x_5 \le 2$ defines a facet of the convex hull of the feasible set.

Clique inequalities are also widely used in practical solvers. A subset C of vertices of a graph G induce a *clique* if every two vertices of C are connected by an edge in G.

Theorem 7.4. *Let G be the intersection graph for the set-packing problem (7.24). If a subset C of vertices of G induce a clique of G, then the clique inequality*

$$\sum_{j \in C} x_j \le 1 \qquad (7.26)$$

is valid for (7.24).

It can also be shown that (7.26) is facet defining if C defines a maximal clique (i.e., no proper superset of C induces a clique).

For example, the intersection graph of the set-packing problem

$$
\begin{aligned}
x_1 + x_2 + x_3 &\le 1 \\
x_1 \qquad\qquad + x_4 &\le 1 \\
x_2 \quad + x_4 &\le 1 \\
x_3 + x_4 &\le 1
\end{aligned}
$$

is itself a clique. The inequality $x_1 + x_2 + x_3 + x_4 \le 1$ is therefore valid and facet defining.

Exercises

7.3. Use the procedure in the proof of Theorem 7.1 to show that the inequality $(2x_1 + 3x_2) + x_1 + x_2 \ge 1$ at the leftmost leaf node in Fig. 7.4 is a Chvátal–Gomory cut. In particular, what are Δ, α, and M in this case?

7.4. Identify a minimal cover for (7.10) other than $\{1, 2, 3, 4\}$, and write the corresponding cover inequality.

7.5. Show that cover inequalities corresponding to nonminimal covers are redundant of those corresponding to minimal covers.

7.6. Suppose that (7.10) is part of a larger problem whose continuous relaxation has the solution $\bar{x} = (0, 0, \frac{4}{5}, \frac{4}{5}, \frac{7}{8}, \frac{2}{3})$. Find a minimal cover of (7.10) for which $\bar{x}$ violates the corresponding cover inequality. Such an inequality is a *separating cut* that cuts off the solution $\bar{x}$.

7.7. Suppose that the 0-1 knapsack packing inequality $ax \le a_0$ is part of a larger problem whose continuous relaxation has the solution $\bar{x}$. Show that a cover inequality $\sum_{j \in J} x_j \le |J| - 1$ cuts off the solution $\bar{x}$ if and only if

$$\sum_{j=1}^{n}(1 - \bar{x}_j)y_j > 1, \quad \sum_{j \in J} a_j y_j \geq a_0 + 1$$

where $y_j = 1$ when $j \in J$ and $y_j = 0$ otherwise. Describe a 0-1 knapsack problem one can solve to find the cover inequality that is most violated by $\bar{x}$. This is the separation problem for $ax \leq a_0$.

7.8. The 0-1 knapsack inequality $5x_1 + 4x_2 + 4x_3 + 4x_4 + 9x_5 + 5x_6 \leq 16$ has the minimal cover inequality $x_1 + x_2 + x_3 + x_4 \leq 3$, among others. Use the lifting formula (7.13) to find a lifting coefficient for x_5. Then compute a lifting coefficient for x_6 using (7.15).

7.9. Compute the lifting coefficient for x_5 in Exercise 7.8 by recursively computing $h^*(t)$. Then, compute the lifting coefficient for x_6 using the recursive formula (7.18).

7.10. Show that the recursive formula (7.18) is correct.

7.11. Plot $\rho(u)$ against u for the cover inequality in Exercise 7.8 using the formula (7.23). What are the sequence-independent lifting coefficients? Note that the resulting cut is, in this case, the same as the one obtained from sequential lifting.

7.12. Show that $\rho(u)$ as defined by (7.23) is superadditive and bounded above by $\pi_J(u)$.

7.13. Derive all valid odd cycle and clique inequalities from the 0-1 system $Ax \leq e$ that are not already in the system, where

$$A = \begin{bmatrix} 1 & 1 & 0 & 0 & 0 \\ 1 & 0 & 1 & 0 & 0 \\ 1 & 1 & 0 & 1 & 0 \\ 0 & 1 & 1 & 0 & 0 \\ 1 & 0 & 1 & 1 & 0 \\ 1 & 0 & 0 & 1 & 1 \end{bmatrix}$$

7.14. Prove Theorem 7.3.

7.15. Prove Theorem 7.4.

7.16. Prove that if C defines a maximal clique in Theorem 7.4, then (7.26) is facet defining. *Hints:* It suffices to exhibit n affinely independent feasible points that satisfy $\sum_{j \in C} x_j \leq 1$ as an equation. Let e^j be the jth unit vector (i.e., a vector of zeros except for a 1 in the jth place) and consider the points e^j for $j \in C$, as well as the points $e^j + e^{i_j}$ for $j \notin C$. Here, $i_j \in C$ is selected so that no row of A contains a 1 in columns j and i_j (show that i_j exists).

7.17. Prove that an odd cycle inequality in Theorem 7.3 that corresponds to an odd hole is facet-defining.

7.3 Integer Linear Inequalities

Perhaps the most popular general-purpose cutting planes are Gomory cuts and mixed-integer rounding cuts. Gomory cuts can be generated equally well for 0-1 and general integer linear inequalities. Mixed integer rounding cuts are designed for mixed-integer/linear systems, in which some of the variables are continuous, and others are required to take 0-1 or general integer values.

The Chvátal–Gomory theorem, proved in the previous section for 0-1 linear inequalities, can also be extended to general integers. In fact, Gomory cuts are actually a special case of rank 1 Chvátal–Gomory cuts, distinguished by the fact that they are *separating cuts*. Suppose that solution $\bar{x}$ of a continuous relaxation of the integer system $Ax \geq b$ is infeasible because it has some nonintegral components. A separating cut $cx \geq c_0$ *cuts off* $\bar{x}$ in the sense that $c\bar{x} < c_0$. The cut is *separating* in that it defines a hyperplane that separates $\bar{x}$ from the convex hull of feasible solutions. There is always a Gomory cut that cuts off $\bar{x}$ when it has some nonintegral components. Typically, one generates several Gomory cuts and adds them to the relaxation to strengthen it.

The motivation for using separating cuts is it provides a principle for selecting relevant cuts. In general, there are a large number of non-redundant valid cuts, and adding them all to the continuous relaxation would be impractical. Separating cuts are relevant in the sense that they exclude the solution of the current relaxation so that when the relaxation is re-solved after adding the cuts, a different solution will be obtained. If desired, one can generate separating cuts for this new solution and add them to the relaxation, and so forth through several rounds.

In some cases, the continuous relaxation of an integer linear system describes an integral polyhedron, meaning that the vertices of the polyhedron have all integral coordinates. In such cases, one can solve the system by finding a vertex solution of its continuous relaxation—for example, by using the simplex method. One sufficient condition for an integral polyhedron is that the coefficient matrix of the problem be totally unimodular. Total unimodularity can, in turn, be characterized by a necessary and sufficient condition that is sometimes useful for showing that certain classes of problems have integral polyhedra.

7.3.1 Chvátal–Gomory Cuts

The Chvátal–Gomory procedure is a complete inference method for general integer inequalities as well as for 0-1 inequalities. It is applied to an inequality system

$$Ax \geq b$$
$$x \text{ integral} \tag{7.27}$$

$Ax \geq b$ is understood to include bounds $0 \leq x_j \leq h_j$, which imposes no practical restriction and avoids the necessity of studying the more difficult unbounded case.

In the proof, the resolution pattern is replaced by a more complex combination of inequalities. Suppose again the goal is to derive an inequality $cx \geq c_0$. As in the 0-1 case, a weaker inequality $cx \geq \delta_0$ (for sufficiently small δ_0) can be derived from variable bounds $0 \leq x \leq h$. It therefore suffices to show that whenever $cx - \delta \geq -1$ is a Chvátal–Gomory cut and $cx - \delta \geq 0$ is valid, then $cx - \delta \geq 0$ is a Chvátal–Gomory cut.

The derivation pattern is illustrated in Fig. 7.6. It can be assumed without loss of generality that each upper bound h_j is the same, namely h_0. Each node on level k of the enumeration tree corresponds to setting variables $x_1, \ldots, x_k$ to certain values $v_1, \ldots, v_k$. The inequality $T(v_1, \ldots, v_k) \geq 0$ is associated with this node, where

$$T(v_1, \ldots, v_k) = \prod_{i=1}^{k}(h_0 + 1 - v_i)(cx - \delta) + \sum_{i=1}^{k} \prod_{j=i+1}^{k} (h_0 + 1 - v_j)x_i$$

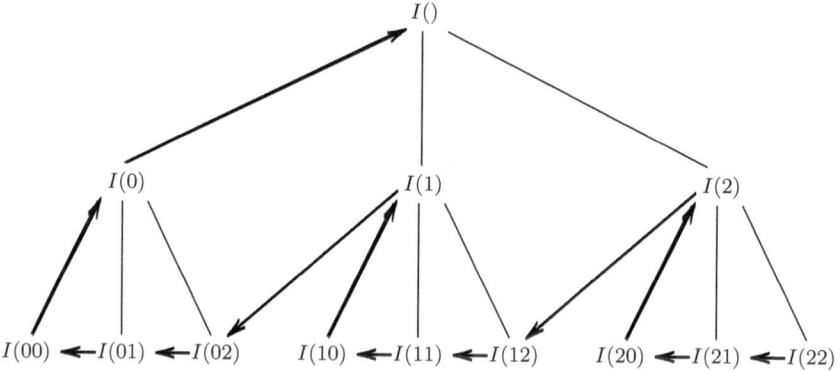

Fig. 7.6 Illustration of the proof of the Chvátal–Gomory theorem for general integer inequalities. $I(v_1v_2)$ stands for the inequality $T(v_1, v_2) \geq 0$.

The inequalities associated with the nodes of Fig. 7.6, beginning at the lower right, are

$$T(2,2) = (cx - \delta) + x_1 + x_2 \geq 0$$
$$T(2,1) = 2(cx - \delta) + 2x_1 + x_2 \geq 0$$
$$T(2,0) = 3(cx - \delta) + 3x_1 + x_2 \geq 0$$
$$T(2) = (cx - \delta) + x_1 \qquad \geq 0$$
$$T(1,2) = 2(cx - \delta) + x_1 + x_2 \geq 0$$
$$T(1,1) = 4(cx - \delta) + 2x_1 + x_2 \geq 0$$
$$T(1,0) = 6(cx - \delta) + 4x_1 + x_2 \geq 0$$
$$T(1) = 2(cx - \delta) + x_1 \qquad \geq 0$$
$$T(0,2) = 3(cx - \delta) + x_1 + x_2 \geq 0$$
$$T(0,1) = 6(cx - \delta) + 2x_1 + x_2 \geq 0$$
$$T(0,0) = 9(cx - \delta) + 3x_1 + x_2 \geq 0$$
$$T(0) = 3(cx - \delta) + x_1 \qquad \geq 0$$
$$T() = (cx - \delta) \qquad\qquad \geq 0$$

Each inequality at a nonleaf node is derived (in part) from its left-most immediate successor, as indicated by arrows in the figure. The remaining arrows show (in part) how the inequalities at leaf nodes are derived.

Theorem 7.5. *Every valid inequality for a bounded integer system is a Chvátal–Gomory cut.*

Proof. It suffices to show that if

$$cx - \delta \geq -1 \tag{7.28}$$

is a Chvátal–Gomory cut, and $cx - \delta \geq 0$ is valid, then $T() \geq 0$ is a Chvátal–Gomory cut. In fact, it will be shown that $T(v_1, \ldots, v_k) \geq 0$ is a Chvátal–Gomory cut for all $v_1, \ldots, v_k$ and all $k \leq n$.

First, it is easy to check that each inequality $T(v_1, \ldots, v_k) \geq 0$ for $k < n$ can be derived from the inequality $T(v_1, \ldots, v_k, 0) \geq 0$ and the bound $-x_{k+1} \geq -1$ by using $1/(h_0 + 1)$ as the multiplier for each.

Now consider the inequalities $T(v_1, \ldots, v_n) \geq 0$. It will be shown inductively that each of these inequalities, beginning with the inequality $T(h_0, \ldots, h_0) \geq 0$, can be derived from inequalities that have already been shown to be Chvátal–Gomory cuts, namely from

$$cx - \delta \geq -1$$

$$x_i \geq 0 \qquad \text{for all } i = 1, \ldots, n \text{ with } v_i = h_0 \qquad (7.29)$$

$$T(v_1, \ldots, v_{i-1}, v_i + 1) \geq 0 \text{ for all } i = 1, \ldots, n \text{ with } v_i < h_0$$

Note first that $T(v_1, \ldots, v_n) \geq -1$ is the sum of the inequalities (7.29). It will be shown that $T(v_1, \ldots, v_n) \geq -1$ cannot be satisfied at equality, which implies that $T(v_1, \ldots, v_n) \geq \epsilon - 1$ is valid for some $\epsilon > 0$. Rounding up the right-hand side yields that $T(v_1, \ldots, v_n) \geq 0$ is a Chvátal–Gomory cut, and the theorem follows.

Suppose then that $T(v_1, \ldots, v_n) \geq -1$ is satisfied at equality. Since each inequality in (7.29) is valid, $T(v_1, \ldots, n_n)$ can be -1 only if $cx - \delta = -1$. But $cx - \delta$ can be -1 and the second inequality in (7.29) satisfied only if $x_1 = v_1$. Given that $x_1 = v_1$, $cx - \delta$ can be -1 and the third inequality in (7.29) satisfied only if $x_2 = v_2$, and so forth. Thus, $(x_1, \ldots, x_n) = (v_1, \ldots, v_n)$ satisfies $cx - \delta = -1$, which means $cx - \delta \geq 0$ is violated by an integer point. This is impossible, because it is given that $cx - \delta \geq 0$ is valid for all integer solutions of $Ax \geq b$. Thus, $T(v_1, \ldots, v_n) \geq -1$ cannot be satisfied at equality. $\square$

7.3.2 Gomory Cuts

There is a systematic and easily implemented method for the generation of separating Chvátal–Gomory cuts. These separating cuts are popularly known as *Gomory cuts* because one of the earliest algorithms for integer programming, invented by Ralph Gomory, is based on them. Gomory originally proposed a pure cutting-plane algorithm that repeatedly re-solves the continuous relaxation after adding inequalities that cut off the fractional solution of the last relaxation. Such an algorithm is rarely efficient, but Gomory cuts can be very effective in a branch-and-cut method and are widely used in commercial MILP solvers today.

To generate Gomory cuts for an integer programming problem

$$\min \bar{c}\bar{x}$$
$$\bar{A}\bar{x} \geq b \qquad (7.30)$$
$$\bar{x} \geq 0 \text{ and integral}$$

where $\bar{x} = (x_1, \ldots, x_n)$, the problem is first converted to an equality-constrained problem by adding surplus variables $x_{n+1}, \ldots, x_{n+m}$:

$$\min \ cx$$
$$Ax = b, \ \ x \geq 0 \text{ and integral} \tag{7.31}$$

where $x = (x_1, \ldots, x_{n+m})$, $A = [\bar{A} \ \ -I]$, and $c = [\bar{c} \ \ 0]$. The surplus variables can be restricted to be integral if it is supposed (with very little loss of generality) that $\bar{A}$ and b have integral components.

Gomory cuts are based on a simple idea from modular arithmetic. Let $\text{frac}(\alpha) = \alpha - \lfloor \alpha \rfloor$ be the fractional part of a real number α. Then clearly, $\text{frac}(\alpha k) \leq \text{frac}(\alpha)k$ for any nonnegative integer k.

Now consider any noninteger basic solution $(x_B, x_N) = (x_B, 0)$ of the continuous relaxation of (7.31). That is, some component x_i of x_B is not an integer. The aim is to find a valid inequality for (7.31) that cuts off this solution.

As indicated in Section 3.1, any solution of $Ax = b$ has the form (x_B, x_N) where $x_B = \hat{b} - \hat{N}x_N$ (recall that $\hat{b} = B^{-1}b$ and $\hat{N} = B^{-1}N$). Thus in particular,

$$x_i = \hat{b}_i - \hat{N}_i x_N \tag{7.32}$$

where $\hat{N}_i$ is row i of $\hat{N}$. Since x_i must be integer in any feasible solution of (7.31), the two terms of (7.32) must have the same fractional part:

$$\text{frac}(\hat{N}_i x_N) = \text{frac}(\hat{b}_i) \tag{7.33}$$

Because x_N consists of nonnegative integers in any feasible solution of (7.31), $\text{frac}(\hat{N}_i x_N) \leq \text{frac}(\hat{N}_i)x_N$. This and (7.33) imply the *Gomory cut*

$$\text{frac}(\hat{N}_i)x_N \geq \text{frac}(\hat{b}_i) \tag{7.34}$$

This obviously cuts off the solution $(x_B, x_N) = (x_B, 0)$, since the right-hand side is strictly positive.

The Gomory cut (7.34) can be written

$$(\hat{N}_i - \lfloor \hat{N}_i \rfloor)x_N \geq \hat{b}_i - \lfloor \hat{b}_i \rfloor$$

Subtracting (7.32) from this yields an alternate form of the cut that contains x_i:

$$x_i + \lfloor \hat{N}_i \rfloor x_N \leq \lfloor \hat{b}_i \rfloor \tag{7.35}$$

As an example, consider the following problem, which was discussed in Section 4.6.1:

$$\min \ 2x_1 + 3x_2$$
$$x_1 + 3x_2 \geq 3 \qquad (a)$$
$$4x_1 + 3x_2 \geq 6 \qquad (b) \tag{7.36}$$
$$x_1, x_2 \geq 0 \text{ and integral}$$

It can be written as an equality-constrained problem by introducing surplus variables x_3, x_4:

$$
\begin{aligned}
\min \ & 2x_1 + 3x_2 \\
& x_1 + 3x_2 - x_3 = 3 \qquad (a) \\
& 4x_1 + 3x_2 - x_4 = 6 \qquad (b) \\
& x_1, \ldots, x_4 \geq 0 \text{ and integral}
\end{aligned}
\qquad (7.37)
$$

The continuous relaxation has the optimal solution

$$
F(x_B, x_N) = (x_1, x_2, x_3, x_4) = (1, \tfrac{2}{3}, 0, 0)
$$

which is illustrated in Fig. 7.7. At this solution,

$$
B = \begin{bmatrix} 1 & 3 \\ 4 & 3 \end{bmatrix}, \qquad
B^{-1}N = \hat{N} = \begin{bmatrix} \tfrac{1}{3} & -\tfrac{1}{3} \\ -\tfrac{4}{9} & \tfrac{1}{9} \end{bmatrix}, \qquad
B^{-1}b = \hat{b} = \begin{bmatrix} 1 \\ \tfrac{2}{3} \end{bmatrix}
$$

Since x_2 is noninteger, a Gomory cut can be formulated to cut off this solution. The cut (7.34) is

$$
\mathrm{frac}(-\tfrac{4}{9})x_3 + \mathrm{frac}(\tfrac{1}{9})x_4 \geq \mathrm{frac}(\tfrac{2}{3}) \qquad \text{or} \qquad \tfrac{5}{9}x_3 + \tfrac{1}{9}x_4 \geq \tfrac{2}{3} \qquad (7.38)
$$

The cut in form (7.35) is

$$
x_2 + \lfloor -\tfrac{4}{9} \rfloor x_3 + \lfloor \tfrac{1}{9} \rfloor x_4 \leq \lfloor \tfrac{2}{3} \rfloor \qquad \text{or} \qquad x_2 - x_3 \leq 0 \qquad (7.39)
$$

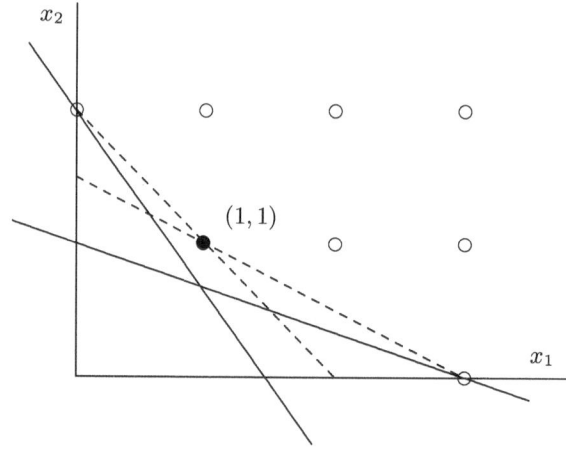

Fig. 7.7 An integer programming problem with two Gomory cuts (dashed lines). The small open circles show some of the feasible solutions, and the solid circle is the optimal solution.

Note that either cut excludes the solution $(1, \frac{2}{3}, 0, 0)$.

The relaxation is now re-solved with the additional constraint (7.39) written as an equality constraint $-x_2 + x_3 - x_5 = 0$, where x_5 is a new surplus variable. The solution remains fractional, with $x_B = (x_1, x_2, x_3) = (\frac{3}{5}, \frac{6}{5}, \frac{6}{5})$. Now, the cut (7.34) is

$$\tfrac{3}{5}x_4 + \tfrac{3}{5}x_5 \geq \tfrac{3}{5} \tag{7.40}$$

and in form (7.35) is

$$x_1 - x_4 \leq 0$$

When this is added to the relaxation in equality form $-x_1 + x_4 - x_6 = 0$, the solution is integral with $(x_1, x_2) = (1, 1)$ and optimal value 5.

Gomory cuts can be written in the original variables $x_1, \ldots, x_n$ by eliminating any surplus variables that occur in cut (7.34). Simply replace any surplus variable x_j in (7.34) with $\bar{A}^{j-n}\bar{x} - b_{j-n}$, where $\bar{A}^{j-n}$ is row $j-n$ of $\bar{A}$. In the example, the two cuts (7.38) and (7.40) can be written in the original variables x_1, x_2 by substituting $x_3 = x_1 + 3x_2 - 3$ and $x_4 = 4x_1 + 3x_2 - 6$. This yields

$$\begin{array}{ll} x_1 + 2x_2 \geq 3 & (c) \\ 3x_1 + 3x_2 \geq 6 & (d) \end{array} \tag{7.41}$$

These cuts are illustrated in Fig. 7.7.

The surplus variable coefficients in (7.38) and (7.40) also indicate how to obtain the cuts (7.41) by taking linear combinations and rounding. Constraints (a) and (b) of the original problem (7.36) are first combined with multipliers $\frac{5}{9}$ and $\frac{1}{9}$ to obtain $x_1 + 2x_2 \geq \frac{21}{9}$ which, after rounding, is (c) above. Thus, cut (c) is a rank 1 Chvátal–Gomory cut for the original system (7.36). Next constraints (a), (b), and (c) are combined with coefficients 0, $\frac{3}{5}$, and $\frac{3}{5}$. Note that surplus variable x_3 has coefficient 0 in (7.40). This yields $3x_1 + 3x_2 \geq \frac{27}{5}$ which, after rounding, is cut (d). In general, one can say the following.

Theorem 7.6. *The Gomory cut (7.34), when expressed in the original variables, is a rank 1 Chvátal–Gomory cut for (7.36).*

Proof. The proof relies on the identity

$$\sum_j \mathrm{frac}(a_j)y_j + \mathrm{frac}\left(-\sum_j a_j y_j\right) = \left\lceil \sum_j \mathrm{frac}(a_j)y_j \right\rceil \tag{7.42}$$

where each y_j is an integer. To prove the theorem, it suffices to exhibit a surrogate of (7.36)'s continuous relaxation that is equivalent to (7.34) after rounding. Let

$$J_1 = \{j \in \{1, \ldots, n\} \mid x_j \text{ is nonbasic}\}$$
$$J_2 = \{j \in \{n+1, \ldots, n+m\} \mid x_j \text{ is nonbasic}\}$$

Assign multiplier $\text{frac}(\hat{N}_{ij})$ to inequality $\bar{A}^{j-n}\bar{x} \geq b_{j-n}$ for each $j \in J_2$ and to bound $x_j \geq 0$ for each $j \in J_1$, and assign a multiplier of zero to all other constraints. The resulting surrogate is

$$\sum_{j \in J_1} \text{frac}(\hat{N}_{ij})x_j + \sum_{j \in J_2} \text{frac}(\hat{N}_{ij})\bar{A}^{j-n}\bar{x} \geq \sum_{j \in J_2} \text{frac}(\hat{N}_{ij})b_{j-n} \quad (7.43)$$

It will be shown that (7.43) with the right-hand side rounded up is equivalent to (7.34). Since $x_j = \bar{A}^{j-n}\bar{x} - b_{j-n}$ for $j \in J_2$, (7.34) expressed in the original variables is

$$\sum_{j \in J_1} \text{frac}(\hat{N}_{ij})x_j + \sum_{j \in J_2} \text{frac}(\hat{N}_{ij})\bar{A}^{j-n}\bar{x} \geq \text{frac}(\hat{b}_i) + \sum_{j \in J_2} \text{frac}(\hat{N}_{ij})b_{j-n}$$
$$(7.44)$$

Since $A = [\bar{A} \;\; -I]$,

$$\hat{b}_i = (B^{-1})^i b = -\hat{N}^i b = -\sum_{j \in J_2} \hat{N}_{ij} b_{j-n}$$

So, (7.44) can be written

$$\sum_{j \in J_1} \text{frac}(\hat{N}_{ij})x_j + \sum_{j \in J_2} \text{frac}(\hat{N}_{ij})\bar{A}^{j-n}\bar{x} \geq$$
$$\text{frac}\left(-\sum_{j \in J_2} \hat{N}_{ij} b_{j-n}\right) + \sum_{j \in J_2} \text{frac}(\hat{N}_{ij})b_{j-n}$$

Due to (7.42) and the fact the b is integral, the right-hand side of this inequality is the result of rounding up the right-hand side of the surrogate (7.43). $\square$

It can be shown that the Gomory cutting-plane algorithm terminates with an optimal solution after a finite number of steps. Since each step generates a Chvátal–Gomory cut (by Theorem 7.6), the Gomory algorithm provides an alternate proof of Theorem 7.5.

7.3.3 Mixed-Integer Rounding Cuts

A *mixed-integer/linear inequality* is a linear inequality in which some variables are continuous and other variables take integer values. Inequalities of this kind are widely used in integer programming, partly because disjunctions of linear constraints and other logical conditions are commonly written in inequality form by introducing 0-1 variables. An integrated solver does not require such formulations in the original model, but mixed-integer/linear inequalities may appear in other contexts. For instance, the solver may generate a mixed-integer model for a constraint to obtain a continuous relaxation by dropping the integrality requirement. It is therefore useful to know how to generate cuts for mixed-integer constraint sets.

The integer rounding cut used in the Chvátal–Gomory procedure can be extended to mixed-integer/linear inequalities, resulting in a *mixed-integer rounding cut*. It is less obvious how to write a rounding cut in the mixed-integer case, but it is derived from a simple observation. Consider the two-variable mixed-integer inequality $y + x \geq b$ where $x \geq 0$, b is nonintegral and y is any integer (possibly negative). One can see from Fig. 7.8 that the rounding cut

$$y + \frac{x}{\mathrm{frac}(b)} \geq \lceil b \rceil \tag{7.45}$$

is a valid inequality for these constraints. More formally,

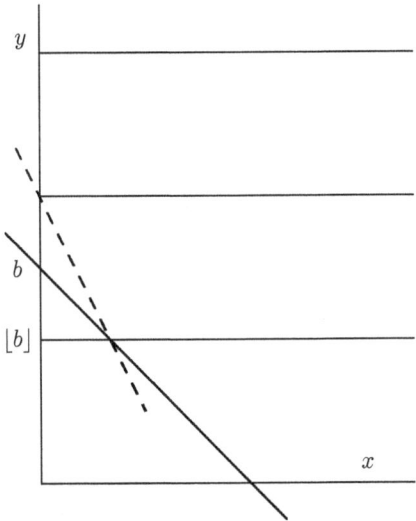

Fig. 7.8 A mixed-integer cut (dashed line) for $y + x \geq b$ (solid line).

Theorem 7.7. *The inequality (7.45) is a valid cut for the constraints* $y + x \geq b$ *and* $x \geq 0$, *where* b *is nonintegral and* y *is integer valued.*

Proof. Because any point (x, y) satisfying the constraints satisfies $y \geq \lceil b \rceil$ or $y \leq \lceil b \rceil - 1$, it suffices to show that (7.45) holds in either case. If $y \geq \lceil b \rceil$, one can take a linear combination of $y \geq \lceil b \rceil$ with multiplier $\mathrm{frac}(b)$ and $x \geq 0$ with multiplier 1 to obtain

$$\mathrm{frac}(b)y + x \geq \mathrm{frac}(b)\lceil b \rceil \tag{7.46}$$

which is equivalent to (7.45). If $y \leq \lceil b \rceil - 1$, one can take a linear combination of $y \leq \lceil b \rceil - 1$ with multiplier $1 - \mathrm{frac}(b)$ and $y + x \geq b$ with multiplier 1 and again obtain (7.46). $\square$

The cut (7.45) is useful because a general mixed-integer/linear inequality can be relaxed to an inequality that has the form $y + x \geq b$. Thus, a valid cut for $y + x \geq b$ leads to a general mixed-integer rounding cut.

A general mixed-integer inequality can be written

$$\sum_{j \in J} a_j y_j + cx \geq b \tag{7.47}$$

where each $y_j \geq 0$ is integer-valued, and each $x_j \geq 0$ is real-valued. Let J_1 be the set of indices j for which $\mathrm{frac}(a_j) \geq \mathrm{frac}(b)$, and let $J_2 = J \setminus J_1$. Then (7.47) can be written

$$\sum_{j \in J_1} a_j y_j + \sum_{j \in J_2} a_j y_j + z_1 - z_2 \geq b \tag{7.48}$$

where the continuous variable $z_1 \geq 0$ is the sum of all terms $c_j x_j$ with $c_j > 0$, and $z_2 \geq 0$ the negated sum of the remaining terms $c_j x_j$. The reason for the partition of J into J_1 and J_2 will become evident shortly.

The inequality (7.48) remains valid if the coefficients a_j for $j \in J_1$ are rounded up. Thus, since $a_j = \lfloor a_j \rfloor + \mathrm{frac}(a_j)$ and $z_2 \geq 0$, (7.48) implies

$$\sum_{j \in J_1} \lceil a_j \rceil y_j + \sum_{j \in J_2} \lfloor a_j \rfloor y_j + \sum_{j \in J_2} \mathrm{frac}(a_j)y_j + z_1 \geq b \tag{7.49}$$

Setting y_0 equal to the sum of the first two terms and z_0 equal to the sum of the last two, (7.49) can be written $y_0 + z_0 \geq b$. Because y_0 is a (possibly negative) integer variable and $z_0 \geq 0$, this inequality has the form required by Theorem 7.7. For nonintegral b, it implies the cut $y_0 + z_0/\text{frac}(b) \geq \lceil b \rceil$. Restoring the expressions for y_0 and z_0 yields a mixed-integer rounding cut for (7.48):

$$\sum_{j \in J_1} \lceil a_j \rceil y_j + \sum_{j \in J_2} \left(\lfloor a_j \rfloor + \frac{\text{frac}(a_j)}{\text{frac}(b)} \right) y_j + \frac{z_1}{\text{frac}(b)} \geq \lceil b \rceil \qquad (7.50)$$

Note that for $j \in J_2$ the coefficient of y_j is less than $\lceil a_j \rceil$ when a_j is noninteger, and therefore strengthens the cut. Partitioning J into J_1 and J_2 avoids using a coefficient of this form when it is greater than $\lceil a_j \rceil$ and would weaken the cut. If $c_j^+ = \max\{c_j, 0\}$, the following has been shown.

Theorem 7.8. *Let* $J_1 = \{ j \in J \mid \text{frac}(a_j) \geq \text{frac}(b) \}$ *and* $J_2 = J \setminus J_1$. *Then the mixed-integer rounding cut*

$$\sum_{j \in J_1} \lceil a_j \rceil y_j + \sum_{j \in J_2} \left(\lfloor a_j \rfloor + \frac{\text{frac}(a_j)}{\text{frac}(b)} \right) y_j + \frac{1}{\text{frac}(b)} \sum_j c_j^+ x_j \geq \lceil b \rceil \quad (7.51)$$

is valid for (7.47) when b is nonintegral, $x_j, y_j \geq 0$, and y_j is integer valued.

For example, the mixed-integer/linear inequality

$$\tfrac{5}{3} y_1 - \tfrac{2}{3} y_2 + \tfrac{1}{4} y_3 + 2x_1 + 3x_2 - 4x_3 \geq \tfrac{3}{2}$$

has $J_1 = \{1\}$ and $J_2 = \{2, 3\}$. The mixed-integer rounding cut (7.51) is

$$2y_1 - \tfrac{1}{3} y_2 + \tfrac{1}{2} y_3 + 4x_1 + 6x_2 \geq 2$$

7.3.4 Separating Mixed-Integer Rounding Cuts

The mixed-integer rounding cuts derived in the previous section lead directly to separating cuts for a mixed-integer system of the form

$$A_1 y + A_2 x = b, \quad x, y \geq 0$$
$$y \in \mathbb{Z}^n \qquad (7.52)$$

If some y_i has a nonintegral value in a basic solution of the continuous relaxation of (7.52), a cut violated by this solution can be derived as follows.

The coefficient matrix $[A_1 \; A_2]$ is first partitioned $[B \; N]$, where B contains the basic columns. Some of the columns of B may correspond to y_j's and others to x_j's, and similarly for N. Let J be the set of indices j for which y_j is nonbasic, and K the set of indices for which x_j is nonbasic. One can solve for y_i in terms of the nonbasic variables:

$$y_i = \hat{b}_i - \sum_{j \in J} \hat{N}_{ij} y_j - \sum_{j \in K} \hat{N}_{ij} x_j \tag{7.53}$$

where $\hat{b} = B^{-1}b$ and $\hat{N} = B^{-1}N$. (7.53) implies the inequality

$$y_i + \sum_{j \in J} \hat{N}_{ij} y_j + \sum_{j \in K} \hat{N}_{ij} x_j \geq \hat{b}_i$$

Applying Theorem 7.7 to this inequality yields the desired cut.

Theorem 7.9. *Let y_i be a noninteger in a basic solution of the continuous relaxation of (7.52), in which N is the matrix of nonbasic columns. Let J be the set of indices j for which y_j is nonbasic, and K the set of indices for which x_j is nonbasic. Then,*

$$y_i + \sum_{j \in J_1} \lceil \hat{N}_{ij} \rceil y_j + \sum_{j \in J_2} \left(\lfloor \hat{N}_{ij} \rfloor + \frac{\mathrm{frac}(\hat{N}_{ij})}{\mathrm{frac}(\hat{b}_i)} \right) y_j + \frac{1}{\mathrm{frac}(\hat{b}_i)} \sum_{j \in K} \hat{N}_{ij}^+ x_j \geq \lceil \hat{b}_i \rceil$$

is a separating mixed-integer rounding cut, where the index sets are $J_1 = \{ j \in J \mid \mathrm{frac}(\hat{N}_{ij}) \geq \mathrm{frac}(\hat{b}_{ij}) \}$ and $J_2 = J \setminus J_1$.

The cut is clearly separating because y_i has the noninteger value $\hat{b}_i$ in the given solution, and all the other variables in the cut are nonbasic and equal to zero.

The cut may be illustrated with the constraint set

$$-6y_1 - 4y_2 + 3x_1 + 4x_2 = 1$$
$$-y_1 - y_2 + x_1 + 2x_2 = 3$$
$$y_1, y_2, x_1, x_2 \geq 0, \quad y_1, y_2 \in \mathcal{Z}$$

In one basic solution, the basic variables are $(y_1, x_1) = (\frac{8}{3}, \frac{17}{3})$. The relevant data appear below.

$$B = \begin{bmatrix} -6 & 3 \\ -1 & 1 \end{bmatrix}, \quad N = \begin{bmatrix} -4 & 4 \\ -1 & 2 \end{bmatrix}, \quad B^{-1} = \begin{bmatrix} -\frac{1}{3} & 1 \\ -\frac{1}{3} & 2 \end{bmatrix}$$

$$\hat{b} = \begin{bmatrix} \frac{8}{3} \\ \frac{17}{3} \end{bmatrix}, \quad \hat{N} = \begin{bmatrix} \frac{1}{3} & \frac{2}{3} \\ -\frac{2}{3} & \frac{8}{3} \end{bmatrix}$$

Since y_1 is nonintegral, Theorem 7.9 provides a separating cut:

$$y_1 + \left(\lfloor \tfrac{1}{3} \rfloor + \tfrac{1/3}{2/3} \right) y_2 + \tfrac{1}{2/3} (\tfrac{2}{3})^+ x_2 \geq \lceil \tfrac{8}{3} \rceil$$

or $y_1 + \tfrac{1}{2} y_2 + x_2 \geq 3$. In this instance, $J_1 = \emptyset$ and $J_2 = \{2\}$.

7.3.5 Integral Polyhedra

A polyhedron is *integral* when every vertex has all integral coordinates. When a set of mixed-integer linear inequalities describes an integral polyhedron, it has the same feasible set as its continuous relaxation. In this happy circumstance, one can solve the constraints by solving their continuous relaxation, which is a much easier problem.

There is no easy rule for recognizing when mixed-integer inequalities define an integral polyhedron, but there is a well-known sufficient condition that is often useful. Namely, $Ax \geq b$, $x \geq 0$ describes an integral polyhedron if b is integral and the matrix A is *totally unimodular,* meaning that every square submatrix of A has a determinant equal to 0, 1, or -1.

Several properties of total unimodularity follow immediately from its definition. Recall that a unit vector e^i consists of all zeros except for a 1 in the ith place.

Theorem 7.10. *Matrix A is totally unimodular if and only if A' is totally unimodular, where A' is the matrix obtained by any of the following operations on A: transposition; swapping two columns; negating any column; or adding a unit column e^i.*

The main property is the following:

Theorem 7.11. *A matrix A with integral components is totally unimodular if and only if $Ax \geq b$, $x \geq 0$ defines an integral polyhedron for any integral b.*

Proof. Suppose A is totally unimodular. Any vertex of the polyhedron $P = \{x \geq 0 \mid Ax \geq b\}$ is a basic feasible solution $(B^{-1}b, 0)$ of $Ax - s = b$, $x, s \geq 0$ for some square submatrix B of $[A \ \ -I]$. But by Theorem 7.10, $[A \ \ -I]$ is totally unimodular, which means B^{-1}, and therefore $B^{-1}b$, are integral.

Conversely, suppose P is integral for each integral b, and let $\bar{B}$ be any square submatrix of A. If $\bar{B}$ is singular then $\det \bar{B} = 0$, and so one may assume $\bar{B}$ is nonsingular. Let B be the following square matrix consisting of a subset of the columns of $[A \ \ -I]$:

$$\begin{bmatrix} \bar{B} & 0 \\ C & -I \end{bmatrix}$$

Note that B is nonsingular. Consider the system $Ax - s = b$, $x, s \geq 0$, where $b = Bz + e^i$ for arbitrary unit vector e^i. Here, z is any integral vector chosen so that $z + (B^{-1})_i \geq 0$, where $(B^{-1})_i$ is column i of B^{-1}. Then $(B^{-1}b, 0) = (z + (B^{-1})_i, 0)$ is a basic feasible solution of $Ax - s = b$, $x, s \geq 0$, which by hypothesis is integral. Since i is arbitrary, B^{-1} and therefore $\bar{B}^{-1}$ are integral. Since $\bar{B}$ is integral (due to the integrality of A), $\det \bar{B}$ and $\det \bar{B}^{-1}$ are integers. Thus, since

$$|\det \bar{B}||\det \bar{B}^{-1}| = |\det(\bar{B}\bar{B}^{-1})| = 1$$

it follows that $|\det \bar{B}| = 1$. $\square$

A necessary and sufficient condition for total unimodularity is the following.

Theorem 7.12. *The $m \times n$ matrix A is totally unimodular if and only if for each $J \subset \{1, \ldots, n\}$, there is a partition $J = J_1 \cup J_2$ such that*

$$\left| \sum_{j \in J_1} A_{ij} - \sum_{j \in J_2} A_{ij} \right| \leq 1 \ \text{ for } i = 1, \ldots, m \tag{7.54}$$

Proof. First, suppose that A is totally unimodular, and let J be any subset of $\{1, \ldots, n\}$. Let $\delta_j = 1$ when $j \in J$ and $\delta_j = 0$ otherwise, and consider the polyhedron defined by

$$P = \left\{ x \mid \left\lfloor \tfrac{1}{2} A\delta \right\rfloor \leq Ax \leq \left\lceil \tfrac{1}{2} A\delta \right\rceil, \ 0 \leq x \leq \delta \right\}$$

Due to Theorems 7.10 and 7.11, P is integral. Also since $\tfrac{1}{2}\delta \in P$, the polyhedron P is nonempty, and one can choose an arbitrary integral point $y \in P$. Since $\delta_j - 2y_j = \pm 1$, one can partition J by letting

$$J_1 = \{ j \in J \mid \delta_j - 2y_j = 1 \}$$
$$J_2 = \{ j \in J \mid \delta_j - 2y_j = -1 \}$$

Now for each i,

$$\sum_{j \in J_1} A_{ij} - \sum_{j \in J_2} A_{ij} = \sum_{j \in J} A_{ij}(\delta_j - 2y_j)$$

$$= \begin{cases} A_i\delta - A_i\delta = 0 & \text{if } A_i\delta \text{ is even} \\ A_i\delta - (A_i\delta \pm 1) = \pm 1 & \text{if } A_i\delta \text{ is odd} \end{cases}$$

For the converse, suppose that for any $J \in \{1, \ldots, n\}$, there is a partition satisfying (7.54). The proof is by induction on the size of J. For $|J| = 1$, (7.54) simply says that the submatrix, and therefore its determinant, is 0 or ± 1. Suppose, then, that the claim is true for any J with $|J| = k - 1 \geq 1$, and let B be an arbitrary $k \times k$ submatrix of A. If one assumes B is nonsingular, it suffices to show $|\det B| = 1$. By Cramer's rule and the induction hypothesis, $B^{-1} = \bar{B} / \det B$ where each $\bar{B}_{ij} \in \{0, \pm 1\}$. Also, $B\bar{B}_1 = |\det B| e^1$, with $\bar{B}_1$ denoting column 1 of $\bar{B}$. Let

$$J_1' = \{i \in J \mid \bar{B}_{i1} = 1\}, \quad J_2' = \{i \in J \mid \bar{B}_{i1} = -1\}$$

with $J = J_1' \cup J_2'$. Note that $J \neq \emptyset$, because otherwise B^{-1} would be singular. Since $B\bar{B}_1 = |\det B| e^1$, for $i = 2, \ldots, n$ one has

$$(B\bar{B}_1)_i = \sum_{j \in J_1'} B_{ij} - \sum_{j \in J_2'} B_{ij} = 0 \tag{7.55}$$

By hypothesis, there is a partition J_1, J_2 of J such that

$$\Delta_i = \left| \sum_{j \in J_1} B_{ij} - \sum_{j \in J_2} B_{ij} \right| \leq 1, \quad i = 1, \ldots, k \tag{7.56}$$

Because partition J_1, J_2 of J can be created by transferring indices from J_1' to J_2', and vice-versa, and because each transfer alters the difference in (7.55) by an even number, (7.55) implies that Δ_i is even for $i = 2, \ldots, k$. Thus, (7.56) implies that $\Delta_i = 0$ for $i = 2, \ldots, k$. Finally, it can be shown as follows that $\Delta_1 = 1$. For if $\Delta_1 = 0$, then due to (7.56) one has $Bz = 0$ when z_j is defined to be 1 for $j \in J_1$, -1 for $j \in J_2$, and 0 otherwise. Since B is nonsingular, this implies $z = 0$, which is impossible since $J \neq \emptyset$. Thus, $\Delta_1 = 1$ and $Bz = \pm e^1$. But since $B\bar{B}_1 = |\det B| e^1$, and since the components of z and $\bar{B}_1$ belong to $\{0, \pm 1\}$, it follows that $\bar{B}_1 = \pm z$ and $|\det B| = 1$. $\square$

As an example, consider the system

$$
\begin{aligned}
x_1 &\geq b_1 \\
x_1 + x_2 &\geq b_2 \\
&\ \vdots \\
x_1 + x_2 + \cdots + x_n &\geq b_n \\
x_j \geq 0, \ \ \text{all } j&
\end{aligned}
\tag{7.57}
$$

The coefficient matrix is totally unimodular, as one can see by placing every other column of J in J_1 and the remaining columns of J in J_2 and applying Theorem 7.11. Thus, any extreme point solution of (7.57) is integral if $b_1, \ldots, b_n$ are integral.

A useful consequence of Theorem 7.11 for $0, \pm 1$ matrices is the following sufficient condition for total unimodularity.

Corollary 7.13 *A matrix A with components in $\{0, \pm 1\}$ is totally unimodular if each column contains no more than two nonzero entries, and any unit column with two nonzeros contains 1 and -1.*

This can be applied to the well-known network flow model. model Let E be the set of arcs (i, j) in a directed network. The net supply s_i of flow is given for each node i, as is the unit cost c_{ij} of flow on arc (i, j). If variable x_{ij} represents the flow from node i to node j, the problem is to find the minimum-cost feasible flow:

$$
\begin{aligned}
\min \ &\sum_{(i,j)\in E} c_{ij} x_{ij} \\
&\sum_{(j,i)\in E} x_{ji} - \sum_{(i,j)\in E} x_{ij} = s_i, \ \ \text{all } i \\
&x_{ij} \geq 0, \ \ \text{all } i, j
\end{aligned}
\tag{7.58}
$$

Since the coefficient matrix satisfies the conditions of Corollary 7.13, it is totally unimodular. This means that the optimal flow is always integral if the net supplies s_i are integral.

Exercises

7.18. In the proof of Theorem 7.5, show that $T(v_1, \ldots, v_k) \geq 0$ for $k < n$ can be derived from $T(v_1, \ldots, v_k, 0) \geq 0$ and the bound $-x_{k+1} \geq -1$.

7.19. In the proof of Theorem 7.5, show that $T(v_1, \ldots, v_n) \geq -1$ is the sum of the inequalities (7.29).

7.20. Consider the integer programming problem

$$
\begin{aligned}
\min \; & 2x_1 + x_2 \\
& 5x_1 + 4x_2 \geq 10 \\
& x_1 + 3x_2 \geq 3 \\
& x_1, x_2 \geq 0 \text{ and integral}
\end{aligned}
\tag{7.59}
$$

If surplus variables x_3, x_4 are inserted, the optimal solution of the continuous relaxation of (7.59) is $x = (0, \frac{5}{2}, 0, \frac{9}{2})$, with

$$
B^{-1} = \begin{bmatrix} \frac{1}{4} & 0 \\ -\frac{3}{4} & 1 \end{bmatrix}
$$

Write two Gomory cuts in terms of the nonbasic variables. Write the same cuts in terms of a basic variable and the nonbasic variables. Finally, write the two cuts in terms of the original variables, x_1, x_2. Show how to obtain these last two cuts as rank 1 Chvátal–Gomory cuts of the constraints in the continuous relaxation of (7.59). Note that in each case, one of the multipliers corresponds to a bound $x_j \geq 0$.

7.21. Prove the identity (7.42).

7.22. Write a mixed-integer rounding cut for $\frac{3}{2}y_1 + \frac{4}{3}y_2 - \frac{5}{3}y_3 + 2x_1 - \frac{5}{2}x_2 \geq \frac{5}{2}$, where y_1, y_2, y_3 are integral. What would the cut be if the indices of integer-valued variables were not partitioned into J_1 and J_2 in the proof of Theorem 7.8?

7.23. Use Theorem 7.7 to verify that (7.50) is a rounding cut for (7.49).

7.24. Prove Theorem 7.9 as a corollary of Theorem 7.8.

7.25. The continuous relaxation of

$$
\begin{aligned}
& y_1 - 4y_2 + 3x_1 + x_2 = 2 \\
& -3y_1 + y_2 + x_1 - 2x_2 = 1 \\
& x_1, x_2 \geq 0, \; y_1, y_2 \geq 0 \text{ and integral}
\end{aligned}
$$

has a basic solution $y = (0, \frac{1}{7})$, $x = (\frac{6}{7}, 0)$, with

$$
B^{-1} = \frac{1}{7} \cdot \begin{bmatrix} -1 & 3 \\ 1 & 4 \end{bmatrix}
$$

Write a separating mixed-integer rounding cut.

7.26. Prove Corollary 7.13.

7.27. A capacitated network flow model has the form (7.58) plus $x_{ij} \leq U_{ij}$ for all $(i, j) \in E$. Show that if each s_i and each U_{ij} is integral, a capacitated network flow problem always has an integral optimal flow.

7.28. The incidence matrix A for an undirected graph G contains a row for every vertex of G and a column for every edge. $A_{ie} = 1$ when vertex i is incident to edge e, and $A_{ie} = 0$ otherwise. Show that A is totally unimodular if and only if G is bipartite.

7.29. An *interval matrix* is a 0-1 matrix that has the consecutive ones property; that is, the ones in every row (if any) occur consecutively. Show that any interval matrix is totally unimodular.

7.4 Disjunctions of Linear Systems

A broad class of constraints can be written as a disjunction of linear inequality systems. In particular, any constraint that can be expressed by mixed-integer linear inequalities can, at least in principle, be written in this form. It is therefore useful to be able to relax disjunctions of linear systems. A variety of relaxation methods are known, including the convex hull relaxation, which is the tightest possible linear relaxation. There are also weaker relaxations that require fewer variables.

A disjunction of linear systems has the form

$$\bigvee_{k \in K} A^k x \geq b^k \tag{7.60}$$

Each system $A^k x \geq b^k$ represents a polyhedron, and the feasible set of (7.60) is a finite union of polyhedra.

7.4.1 Convex Hull Relaxation

The *convex hull* of a set $S \subset \mathbb{R}^n$ is the set $\text{conv}(S)$ of all convex combinations of points in S. A *convex combination* of points $x^1, \ldots, x^m$ is a point of the form $\sum_{k=1}^{m} \alpha_k x^k$, where $\sum_k \alpha_k = 1$ and each $\alpha_k \geq 0$.

If the convex hull of a set is a polyhedron, then the inequalities describing that polyhedron provide the tightest possible linear relaxation of the set. The convex hull of a finite union of polyhedra is clearly a polyhedron. Thus, the tightest possible linear relaxation of a disjunction of linear systems can be obtained by describing the convex hull of its feasible set.

Suppose for the moment that each disjunct of (7.60) is feasible. Then, every point in the convex hull of the feasible set of (7.60) can be written as a convex combination $x = \sum_{k \in K} \alpha_k \bar{x}^k$, where each $\bar{x}^k$ lies in the polyhedron described by $A^k x \geq b^k$. The convex hull is therefore described by

$$x = \sum_{k \in K} \alpha_k \bar{x}^k$$
$$A^k \bar{x}^k \geq b^k, \quad k \in K \tag{7.61}$$
$$\sum_{k \in K} \alpha_k = 1, \quad \alpha_k \geq 0, \ k \in K$$

This is a nonlinear system, but it can be linearized by the change of variable $x^k = \alpha_k \bar{x}^k$:

$$x = \sum_{k \in K} x^k$$
$$A^k x^k \geq \alpha_k b^k, \quad k \in K \tag{7.62}$$
$$\sum_{k \in K} \alpha_k = 1, \quad \alpha_k \geq 0, \ k \in K$$

It will be shown that (7.62) is a *convex hull relaxation* of (7.60), in the sense that the projection of the feasible set of (7.62) onto x is the closure of the convex hull of (7.60).[1] This is true even when not all disjuncts of (7.60) are feasible.

Theorem 7.14. *System (7.62) is a convex hull relaxation of the disjunction (7.60).*

Proof. Let C be the convex hull of (7.60), P the feasible set of (7.62), and P_x the projection of P onto x. The claim is that $cl(C) = P_x$, where $cl(C)$ is the closure of C.

First show that $cl(C) \subset P_x$. Since P_x is closed, it suffices to show that $C \subset P_x$. It may be assumed that at least one disjunct of (7.60)

---

[1] A *closed* set S is a set that contains all of its limit points. Point x is a limit point of S if, for any $\epsilon > 0$, some point in S is no further than ϵ away from x. The closure of S is the result of adding to S all of its limit points. A polyhedron is clearly closed, as is any projection of a polyhedron.

is feasible, because otherwise C is empty and trivially $C \subset P_x$. Let K_1 be the set of indices $k \in K$ for which $A^k x \geq b^k$ is feasible, and let $K_2 = K \setminus K_1$. Then any $x \in C$ can be written as a convex combination $\sum_{k \in K_1} \alpha_k \bar{x}^k$, where $A^k \bar{x}^k \geq b^k$ for each $k \in K_1$. Introduce the change of variable $x^k = \alpha_k \bar{x}^k$ for $k \in K$. Multiplying $A^k \bar{x}^k \geq b^k$ by the nonnegative number α_k, one obtains $A^k x^k \geq b^k \alpha_k$ for $k \in K_1$. Setting $\alpha_k = 0$ and $x^k = 0$ for $k \in K_2$, one has $x = \sum_{k \in K} x^k$ and $A^k x^k \geq b^k \alpha_k$ for all $k \in K$. Thus x, x^k, and α_k for $k \in K$ satisfy (7.62), and $x \in P_x$.

Now show that $P_x \subset cl(C)$. Define $\hat{P}$ to be the set of points x, x^k, and $\alpha_k > 0$ for $k \in K$ satisfying (7.62). Given any such point in $\hat{P}$, let $\bar{x}^k = x^k / \alpha_k$, and note that $x = \sum_{k \in K} \alpha_k \bar{x}^k$ and $A^k \bar{x}^k \geq b^k$ for $k \in K$. Thus $x \in C$, and $\hat{P}_x \subset C$, which implies $cl(\hat{P}_x) \subset cl(C)$. But since $P = cl(\hat{P})$, it follows that $P_x = cl(\hat{P}_x)$ and $P_x \subset cl(C)$. $\square$

Note that the convex hull relaxation requires that each continuous variable x_j be disaggregated into $|K|$ continuous variables x_j^k. Thus, if $x \in \mathbb{R}^n$, the relaxation requires n new variables x_j^k and one new variable α_k for each disjunct. The convex hull relaxation therefore tends to be more useful when there are only a few disjuncts, or when the relaxation simplifies in a fashion that allows variables to be eliminated.

As an example, consider the disjunction

$$\begin{pmatrix} x_1 - 2x_2 \geq -2 \\ 0 \leq x_j \leq 2, \ j = 1,2 \end{pmatrix} \vee \begin{pmatrix} 3x_1 - x_2 \geq 1 \\ 0 \leq x_j \leq 2, \ j = 1,2 \end{pmatrix} \quad (7.63)$$

The feasible set of (7.63) is shown in Fig. 7.9. The convex hull relaxation (7.62) is

$$x_1 = x_1^1 + x_1^2, \quad x_2 = x_2^1 + x_2^2$$
$$x_1^1 - 2x_2^1 \geq -2\alpha_1, \quad 3x_1^2 - x_2^2 \geq \alpha_2$$
$$0 \leq x_j^1 \leq 2\alpha_1, \quad 0 \leq x_j^2 \leq 2\alpha_2, \quad j = 1,2$$
$$\alpha_1 + \alpha_2 = 1, \quad \alpha_1, \alpha_2 \geq 0$$

which can be simplified somewhat by eliminating x_j^2 and α_2. If α_1 is renamed α, and x_j^1 is renamed y_j, this yields

$$y_1 - 2y_2 \geq -2\alpha, \quad 3(x_1 - y_1) - (x_2 - y_2) \geq 1 - \alpha$$
$$0 \leq y_j \leq 2\alpha, \quad x_j \geq y_j, \quad x_j - y_j \leq 2(1 - \alpha), \quad j = 1,2$$
$$0 \leq \alpha \leq 1$$

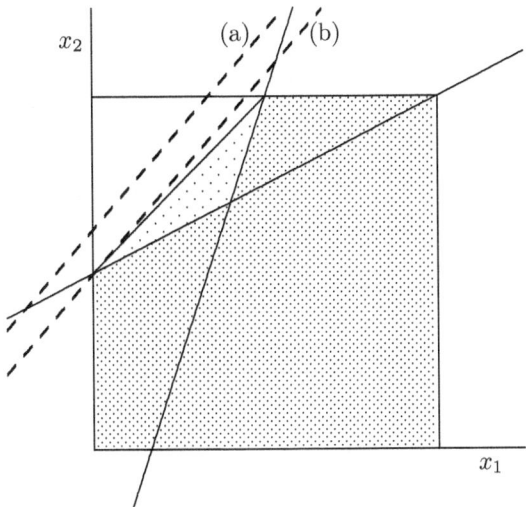

Fig. 7.9 Relaxations of the disjunction (7.63). The darker shaded area is the feasible set, and the entire shaded area is its convex hull. Dashed line (a) represents the inequality $\frac{3}{2}x_1 - \frac{4}{3}x_2 \geq -\frac{5}{3}$, which, with the bounds $0 \leq x_j \leq 2$, describes the projection of the big-M relaxation. Dashed line (b) represents the supporting inequality $\frac{3}{2}x_1 - \frac{4}{3}x_2 \geq -\frac{4}{3}$.

7.4.2 Big-M Relaxation

The big-M relaxation for a disjunction of linear systems (7.60) is generally weaker than the convex hull relaxation, but it requires fewer variables—one additional variable for each disjunct. It is obtained by first formulating a 0-1 model for the disjunction:

$$
\begin{array}{ll}
A^k x \geq b^k - M^k(1 - \alpha_k), & k \in K \quad (a) \\
L \leq x \leq U & (b) \\
\displaystyle\sum_{k \in K} \alpha_k = 1 & (c) \\
\alpha_k \in \{0, 1\}, & k \in K
\end{array}
\qquad (7.64)
$$

The model assumes that each x_j is bounded ($L_j \leq x_j \leq U_j$). Since one α_k is forced to be 1, the constraints (a) force x to lie in at least one of the polyhedra defined by $A^k x \geq b^k$. The components of the vector M^k should be large enough that $a^k x \geq b^k - M^k(1 - \alpha_k)$ does not constrain x when $\alpha_k = 0$. On the other hand, the components of M^k should be as small as possible in order to yield a tighter relaxation. Thus, one can set

$$M_i^k = b_i^k - \min_{L \le x \le U} \{A_i^k x\} = b_i^k - \sum_j \min\{0, A_{ij}^k\} U_j - \sum_j \max\{0, A_{ij}^k\} L_j \tag{7.65}$$

where A_i^k is row i of A^k. The integrality condition in (7.64) is now relaxed to obtain the linear relaxation

$$\begin{aligned}
&A^k x \ge b^k - M^k(1 - \alpha_k), \quad k \in K \\
&L \le x \le U \\
&\sum_{k \in K} \alpha_k = 1, \quad \alpha_k \ge 0, \quad k \in K
\end{aligned} \tag{7.66}$$

For example, the disjunction (7.63) can be viewed as the disjunction

$$(x_1 - 2x_2 \ge -2) \vee (3x_1 - x_2 \ge 1) \tag{7.67}$$

plus bounds $0 \le x_j \le 2$. Since $U_1 = U_2 = 2$, (7.65) sets $M_1^1 = 2$ and $M_1^2 = 3$. So the big-M relaxation (7.64) becomes

$$\begin{aligned}
&x_1 - 2x_2 \ge 2\alpha - 4, \quad 3x_1 - x_2 \ge 1 - 3\alpha \\
&0 \le x_j \le 2, \quad j = 1, 2 \\
&0 \le \alpha \le 1
\end{aligned} \tag{7.68}$$

The projection of this relaxation onto (x_1, x_2) is described by

$$\tfrac{3}{2}x_1 - \tfrac{4}{3}x_2 \ge -\tfrac{5}{3} \tag{7.69}$$

and the bounds $0 \le x_j \le 2$ (Fig. 7.9). It is clearly weaker than the convex hull relaxation.

One way to strengthen the big-M relaxation is to observe that if x does not belong to the polyhedron defined by the kth disjunct, then it must belong to at least one of the other polyhedra. This allows one to set

$$M_i^k = b_i^k - \min_{\ell \ne k} \left\{ \min_{L \le x \le U} \left\{ A_i^k x \mid A^\ell x \ge b^\ell \right\} \right\} \tag{7.70}$$

Using these values in (7.66) generally results in a stronger relaxation. In the example (7.67), one obtains $M_1^1 = 1$ and $M_1^2 = 2$, which yields the relaxation

$$\begin{aligned}
&x_1 - 2x_2 \ge 2\alpha - 3, \quad 3x_1 - x_2 \ge 1 - 2\alpha \\
&0 \le x_j \le 2, \quad j = 1, 2 \\
&0 \le \alpha \le 1
\end{aligned}$$

The projection of this relaxation onto x_1, x_2 is the single inequality

$$\tfrac{3}{2}x_1 - \tfrac{2}{3}x_2 \geq -\tfrac{4}{3} \qquad (7.71)$$

plus the bounds $0 \leq x_j \leq 2$, which, as Fig. 7.9 illustrates, is an improvement over (7.68).

The drawback of using formula (7.70), in general, is that one must solve $|K|$ linear programming problems to compute each M_i^k. It will be shown in the next section, however, that when relaxing a disjunction of single inequalities plus bounds (as is the case here), there is a closed-form expression that yields (7.71).

7.4.3 Disjunctions of Linear Inequalities

The big-M relaxation simplifies considerably when (7.60) is a system of single linear inequalities:

$$\bigvee_{k \in K} a^k x \geq b_k \qquad (7.72)$$

It is again assumed that $L \leq x \leq U$. When projected onto x, the big-M relaxation (7.64) simplifies to a single inequality whose coefficients are trivial to compute.

The big-M relaxation for the disjunction (7.72) is

$$
\begin{aligned}
& a^k x \geq b_k - M_k(1 - \alpha_k), \quad k \in K \ (a) \\
& L \leq x \leq U \qquad\qquad\qquad\qquad\quad (b) \\
& \sum_{k \in K} \alpha_k \geq 1, \ \ \alpha_k \geq 0, \ k \in K \qquad (c)
\end{aligned}
\qquad (7.73)
$$

As before, each M_k is chosen so that it is a lower bound on $a^k x - b_k$, for instance, by using (7.65):

$$M_k = b_k - \min_{L \leq x \leq U}\{a^k x\} = b_k - \sum_j \min\{0, a_j^k\}U_j - \sum_j \max\{0, a_j^k\}L_j$$

$$(7.74)$$

It can be assumed without loss of generality that $M_k > 0$, because otherwise the corresponding inequality is vacuous and can be dropped.

One can now eliminate the variables α_k by taking a linear combination of the inequalities (7.73a), where the kth inequality receives weight $1/M_k$. This yields

$$\left(\sum_{k\in K}\frac{a^k}{M_k}\right)x \geq \sum_{k\in k}\frac{b_k}{M_k} - |K| + 1 \tag{7.75}$$

Theorem 7.15. *If each M_k is given by (7.74), the inequality (7.75) and the bounds $L \leq x \leq U$ describe the projection of the big-M relaxation (7.64) onto x.*

Proof. Let S be the feasible set of (7.64) and $\bar{S}$ the feasible set of (7.75). The claim is that the projection of S onto x is $\bar{S}$. Clearly, the projection of S onto x is a subset of $\bar{S}$, because as just shown, (7.75) is a nonnegative linear combination of (7.64). It therefore suffices to show that for any $\bar{x} \in \bar{S}$ there are $\alpha_k \in [0, 1]$ such that $(\bar{x}, \alpha) \in S$. First "solve" (7.73a) for α_k by setting

$$\bar{\alpha}_k = \frac{1}{M_k}(a^k\bar{x} - b_k + M_k)$$

By construction, $\alpha = (\alpha_1, \ldots, \alpha_{|K|})$ satisfies (7.64a) if $\alpha \leq \bar{\alpha}$. Also,

$$\sum_{k\in K}\bar{\alpha}_k = \left(\sum_{k\in K}\frac{a^k}{M_k}\right)\bar{x} - \sum_{k\in K}\frac{b_k}{M_k} + |K| \geq 1$$

where the inequality follows from (7.75). Further, $\bar{\alpha}_k \geq 0$, because

$$M_k\bar{\alpha}_k = a^k\bar{x} - b_k + M_k = a^k\bar{x} - \sum_j\min\{0, a_j^k\}U_j - \sum_j\max\{0, a_j^k\}L_j \geq 0$$

where the second equality is from (7.74), and the inequality is from the fact that $L \leq x \leq U$. Thus, if one sets

$$\alpha = \frac{\bar{\alpha}}{\displaystyle\sum_{k\in K}\bar{\alpha}_k}$$

then $(\bar{x}, \alpha)$ satisfies (7.64). $\square$

To take an example, the inequality (7.75) for the disjunction (7.67) is precisely (7.69). This inequality (along with bounds $0 \leq x_j \leq 2$) is the projection of the big-M relaxation onto x. The inequality can be strengthened by using the tighter bounds (7.70). As noted earlier, there is a closed-form expression for the relaxation that results from using (7.70) in the special case of a disjunction of single inequalities.

In fact, any valid inequality $cx \geq d$ for the disjunction (7.72) can be strengthened in this way to obtain a *supporting* inequality $cx \geq d^*$ for (7.72). A supporting inequality for a set S is an inequality that is satisfied by every point of S and is satisfied as an equation by at least one point in S. In particular, the inequality (7.75) can be strengthened in this fashion, unless of course it is already a supporting inequality.

The desired right-hand side d^* is the smallest of the minimum values obtained by minimizing cx subject to each of the disjuncts $a^k x \geq b_k$. That is,

$$d^* = \min_{k \in K} d_k^* \tag{7.76}$$

where

$$d_k^* = \min_{L \leq x \leq U} \left\{ cx \mid a^k x \geq b_k \right\}$$

The computation of d_k^* is simplified if $c \geq 0$ and the lower bounds on the variables are zero. To this end, one can introduce the change of variable

$$\hat{x}_j = \begin{cases} x_j - L_j \text{ if } c_j \geq 0 \\ U_j - x_j \text{ otherwise} \end{cases}$$

The strengthened elementary inequality in terms of $\hat{x}$, namely $\hat{c}\hat{x} \geq \hat{d}^*$, can now be computed, where $\hat{c}_j = |c_j|$. The right-hand side of $cx \geq d^*$ can then be recovered from (7.76) by setting

$$d_k^* = \hat{d}_k^* + \sum_{\substack{j \\ c_j > 0}} L_j c_j + \sum_{\substack{j \\ c_j < 0}} U_j c_j \tag{7.77}$$

It remains to compute

$$\hat{d}_k^* = \min_{\hat{x} \geq 0} \left\{ \hat{c}\hat{x} \mid \hat{a}^k \hat{x} \geq \hat{b}_k \right\}, \tag{7.78}$$

where

$$\hat{a}_j^k = \begin{cases} a_j^k \text{ if } c_j \geq 0 \\ -a_j^k \text{ otherwise} \end{cases}$$

and

$$\hat{b}_k = b_k - \sum_{\substack{j \\ c_j > 0}} L_j a_j^k - \sum_{\substack{j \\ c_j < 0}} U_j a_j^k$$

Because $\hat{c} \geq 0$, linear programming duality applied to (7.78) yields

$$\hat{d}_k^* = \min_{\substack{j \\ \hat{a}_j^k > 0}} \left\{ \frac{\hat{c}_j}{\hat{a}_j^k} \right\} \max\{\hat{b}_k, 0\} \tag{7.79}$$

This proves the next theorem.

Theorem 7.16. *A valid inequality* $cx \geq d$ *for the disjunction* (7.72) *is supporting if and only if* $d = d^*$, *where* d^* *is defined by* (7.76), (7.77), *and* (7.79).

To apply this theorem to the disjunction (7.67), note that the inequalities $\hat{a}^k\hat{x} \geq \hat{b}_k$ for $k = 1, 2$ are $\hat{x}_1 + 2\hat{x}_2 \geq 2$ and $3\hat{x}_1 + \hat{x}_2 \geq 3$. Then, $(\hat{d}_1^*, \hat{d}_2^*) = (\frac{4}{3}, \frac{3}{2})$, and $d^* = \min\{d_1^*, d_2^*\} = \min\{-\frac{4}{3}, -\frac{7}{6}\} = -\frac{4}{3}$. This yields the supporting inequality (7.71), which is illustrated in Fig. 7.9.

7.4.4 Disjunctions of Linear Equations

The big-M relaxation for a disjunction of linear equations

$$\bigvee_{k \in K} a^k x = b_k \tag{7.80}$$

has some special structure that will be useful when relaxing the element constraint. Because each $a^k x = b_k$ can be written as a system of two inequalities $a^k x \geq b_k$ and $-a^k x \geq -b_k$, the big-M relaxation of (7.80) is

$$\begin{aligned} a^k x &\geq b_k - M_k'(1 - \alpha_k), \quad k \in K \\ -a^k x &\geq -b_k - M_k''(1 - \alpha_k), \quad k \in K \end{aligned} \tag{7.81}$$

Using (7.70), the big-Ms can be set to

$$\begin{aligned} M_k' &= b_k - \min_{\ell \neq k} \left\{ \min_{L \leq x \leq U} \left\{ a^k x \mid a^\ell x = b_\ell \right\} \right\} \\ M_k'' &= -b_k - \min_{\ell \neq k} \left\{ \min_{L \leq x \leq U} \left\{ -a^k x \mid a^\ell x = b_\ell \right\} \right\} \end{aligned} \tag{7.82}$$

Theorem 7.17. *If* M_k', M_k'' *are as given by* (7.82), *then* (7.81) *and the bounds* $L \leq x \leq U$ *provide a relaxation of* (7.80).

For example, if one relaxes the disjunction

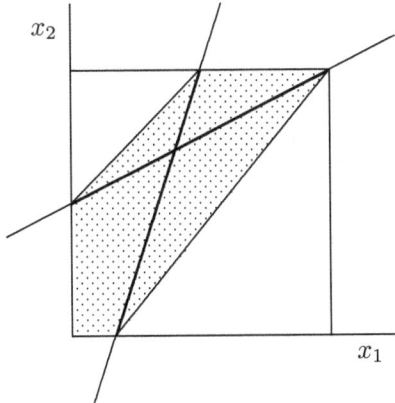

Fig. 7.10 Projected big-M relaxation (shaded area) of the feasible set consisting of the two heavy line segments, which is described by bounds $x_1, x_2 \in [0, 2]$ and the disjunction of the equations $x_1 - 2x_2 = -2$, $3x_1 - x_2 = 1$.

$$(x_1 - 2x_2 = -2) \vee (3x_1 - x_2 = 1)$$

with $x_1, x_2 \in [0, 2]$, then $(M_1', M_2') = (1, 2)$ and $(M_1'', M_2'') = (\frac{7}{3}, 3)$, and (7.81) becomes

$$
\begin{aligned}
x_1 - 2x_2 &\geq -2 - \alpha & 3x_1 - x_2 &\geq 1 - 2(1 - \alpha) \\
-x_1 + 2x_2 &\geq 2 - \tfrac{7}{3}\alpha & -3x_1 + x_2 &\geq -1 - 3(1 - \alpha) \\
0 \leq \alpha &\leq 1
\end{aligned}
$$

plus bounds. Projecting out α, this becomes

$$-x_1 + x_2 \leq 1, \quad 6x_1 - 5x_2 \leq 2$$

plus bounds. As it happens, this is a convex hull relaxation, illustrated in Fig. 7.10.

7.4.5 Separating Disjunctive Cuts

When the convex hull relaxation or big-M relaxation of a disjunction of linear systems has a large number of variables and constraints, it may be advantageous to generate only a separating cut for the disjunction. Unfortunately, the identification of a separating cut requires solving a linear system that is similar in size to the convex hull relaxation. Thus, if there is only one disjunction in the problem, one may as well

put its entire convex hull relaxation into the problem relaxation. But if
there are several disjunctions, their combined convex hull relaxations
can be quite large. It is much faster to generate only a separating cut
for each, because this can be accomplished by solving a separate linear
system for each disjunction, and the resulting relaxation contains only
the separating cuts.

To find a separating cut for a disjunction (7.60) of linear systems,
let $\bar{x}$ be the solution of the current problem relaxation. The goal is
to identify a valid cut $dx \geq \delta$ for (7.60) that $\bar{x}$ violates. This can
be done by recalling from Corollary 4.5 that $dx \geq \delta$ is valid for the
system $A^k x \geq b^k$, $x \geq 0$ if and only if it is dominated by a surrogate
of $A^k x \geq b^k$. If the system is feasible, a dominating surrogate can be
written $uA^i x \geq ub^i$, where $d \geq uA^i$, $\delta \leq ub$, and $u \geq 0$. But $dx \geq \delta$
is valid for the disjunction as a whole if and only if it is valid for each
feasible disjunct.

Theorem 7.18. *The inequality $dx \geq \delta$ is valid for a disjunction (7.60)
of linear systems containing nonnegativity constraints $x \geq 0$ if and
only if for each feasible system $A^k x \geq b^k$ there is a $u^k \geq 0$ such that
$d \geq u^k A^k$ and $\delta \leq u^k b^k$.*

This allows one to write a linear programming problem that finds a
cut $dx \geq \delta$ that $\bar{x}$ violates:

$$
\begin{aligned}
&\max \delta - d\bar{x} &&(a)\\
&\delta \leq u^k b^k, \quad k \in K &&(b)\\
&d \geq u^k A^k, \quad k \in K &&(c)\\
&-e \leq d \leq e &&(d)\\
&u^k \geq 0, \quad k \in K\\
&\delta, d \text{ unrestricted}
\end{aligned}
\qquad (7.83)
$$

The variables in the problem are d, δ, and u^k for $k \in K$. Since a strong
cut is desired, it should be designed so that it is in some sense maximally violated by $\bar{x}$. This is the intent of the objective function (a). If
the maximum value is less than or equal to zero, there is no separating
cut. The constraint (d) is added because, otherwise, the problem would
be unbounded; if (d, δ) is feasible, then any scalar multiple $(\alpha d, \alpha \delta)$ is
feasible. The problem can, in general, be bounded by placing a bound
on some norm of d. One possibility is to bound the L_∞ norm, which
is $\max_j\{|d_j|\}$. This is accomplished by constraint (d), in which e is a

vector of ones. Another possibility is to bound the L_1 norm, which is $\sum_j |d_j|$. This alternative will be taken up shortly.

As an example, consider the disjunction (7.63), and suppose $\bar{x} = (\frac{1}{2}, 2)$ is the solution of the current problem relaxation (Fig. 7.11). To find a separating cut, solve (7.83):

$$
\begin{aligned}
&\max \delta - \tfrac{1}{2}d_1 - 2d_2 \\
&\delta \le -2u_1^1 - 2u_4^1 - 2u_5^1 \\
&\delta \le u_1^2 - 2u_4^2 - 2u_5^2 \\
&u_1^1 + u_2^1 - u_4^1 \le d_1 \\
&-2u_1^1 + u_3^1 - u_5^1 \le d_2 \\
&3u_1^2 + u_2^2 - u_4^2 \le d_1 \\
&-u_1^2 + u_3^2 - u_5^2 \le d_2 \\
&-1 \le d_j \le 1, \quad d = 1, 2 \\
&u_i^k \ge 0, \quad i = 1, \dots, 5, \; k = 1, 2
\end{aligned}
\tag{7.84}
$$

A solution is

$$
d = (1, -1), \; \delta = -1, \; u^1 = (\tfrac{1}{2}, \tfrac{1}{2}, 0, 0, 0), \; u^2 = (\tfrac{1}{3}, 0, 0, 0, \tfrac{2}{3})
$$

This yields the separating cut $d_1 - d_2 \ge -1$, which is the facet-defining cut shown in Fig. 7.11.

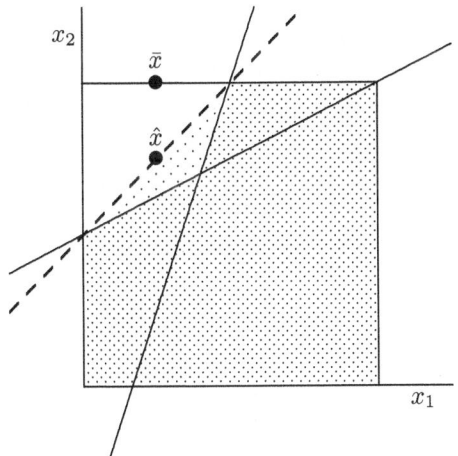

Fig. 7.11 Disjunctive cut (dashed line) that separates $\bar{x} = (\frac{1}{2}, 2)$ from the convex hull (shaded area) of the feasible set (dark shading). The point $\hat{x}$ is a closest point in the convex hull set to $\bar{x}$, as measured by the rectilinear distance.

The dual of (7.83) sheds light on the separation problem. Associating dual variables y_k with (b), x^k with (c), and s, s' with (d), the dual is

$$
\begin{align}
& \min(s + s')e && (a) \\
& \bar{x} - \sum_k x^k = s - s' && (b) \\
& A^k x^k \geq b^k y_k, \quad k \in K && (c) \\
& \sum_k y_k = 1 && (d) \\
& s, s', x^k, y_k \geq 0, \ k \in K
\end{align}
$$

(7.85)

If $s = s'$, the constraint set of (7.85) is the convex hull relaxation of the disjunction (7.60). Thus, (7.85) finds a point $\sum_k x^k$ in the convex hull that has minimum distance to $\bar{x}$, where the distance is measured by $\sum_j (s_j + s'_j)$. This distance measure is the L_1 norm of $\bar{x} - \sum_k x^k$, also known as the rectilinear distance from $\bar{x}$ to $\sum_k x^k$. Thus, the primal problem (7.83) of finding the strongest separating cut is essentially the same as the dual problem (7.85) of finding the closest point $\hat{x}$ in the convex hull to $\bar{x}$. Furthermore,

Theorem 7.19. *The separating cut $dx \geq \delta$ found by the primal problem (7.83) contains a closest point $\hat{x}$ in the convex hull of (7.60) to $\bar{x}$. That is, $d\hat{x} = \delta$.*

Proof. Let δ, d, and $\hat{u}^k$ for $k \in K$ be an optimal solution of the primal problem (7.83), and let $\hat{x}^k, \hat{y}_k$ for $k \in K$ be the corresponding optimal solution of the dual problem (7.85). The closest point $\hat{x}$ is $\sum_k \hat{x}^k$, and the claim is that $d \sum_k \hat{x}^k = \delta$. Apply the complementary slackness principle (Corollary 4.10) to (7.83c) to obtain

$$
\sum_{k \in K} (\hat{u}^k A^k - d)\hat{x}^k = 0
$$

Thus,

$$
d \sum_{k \in K} \hat{x}^k = \sum_{k \in K} \hat{u}^k A^k \hat{x}^k = \sum_{k \in K} \hat{u}^k b^k \hat{y}_k = \sum_{k \in K} \delta \hat{y}_k = \delta
$$

(7.86)

where the second equality is due to complementary slackness applied to (7.85c), the third is due to complementary slackness applied to (7.83b), and the last is due to (7.85d). □

A dual solution for the example problem (7.84) is

$$x^1 = (0, \tfrac{1}{2}), \; x^2 = (\tfrac{1}{2}, 1), \; y = (\tfrac{1}{2}, \tfrac{1}{2}), \; s = (0, 0), \; s' = (0, \tfrac{1}{2})$$

So a closest point in the convex hull to $\bar{x}$ is $\hat{x}^1 + \hat{x}^2 = (\tfrac{1}{2}, \tfrac{3}{2})$. As shown in Fig. 7.11, the separating cut runs through this point.

Another approach to bounding the separating cut problem (7.83) is to bound the L_1 norm rather than the L_∞ norm of d. For example, one can set $\sum_j |d_j| \le 1$. This is accomplished by letting $d = d^+ - d^-$, where $d^+, d^- \ge 0$. Then the bound is $e(d^+ + d^-) \le 1$, where e is a row vector of ones. The problem (7.60) becomes

$$
\begin{aligned}
&\max \delta - (d^+ - d^-)\bar{x} \\
&\delta \le u^k b^k, \quad k \in K \\
&d^+ - d^- \ge u^k A^k, \quad k \in K \\
&e(d^+ + d^-) \le 1 \\
&d^+, d^- \ge 0; \; u^k \ge 0, \; k \in K \\
&\delta \text{ unrestricted}
\end{aligned}
\tag{7.87}
$$

The dual of (7.87) finds a point in the convex hull that is closest to $\bar{x}$, as measured by the L_∞ norm. Again, the separating cut contains this point. Thus, there is a duality of norms. Bounding the L_∞ norm finds a closest point as measured by the L_1 norm, while bounding the L_1 norm finds a closest point as measured by the L_∞ norm.

Exercises

7.30. A common modeling situation requires that an operation either observe certain linear constraints or shut down:

$$[Ax \ge b] \vee [x = 0]$$

Write a convex hull relaxation of this disjunction and simplify it.

7.31. Write a convex hull relaxation of

$$
\begin{bmatrix} x_1 = 0 \\ 0 \le x_2 \le 1 \end{bmatrix} \vee \begin{bmatrix} x_1 \ge 0 \\ x_2 = 0 \end{bmatrix}
$$

Note that the convex hull is not a closed set and that the convex hull relaxation describes its closure.

7.32. Show by example that a convex hull relaxation of

$$[A^1x \geq b^1] \vee [A^2x \geq b^2]$$

combined with $Dx \geq d$ can be weaker than a convex hull relaxation of

$$\begin{bmatrix} A^1x \geq b^1 \\ Dx \geq d \end{bmatrix} \vee \begin{bmatrix} A^2x \geq b^2 \\ Dx \geq d \end{bmatrix}$$

even when $Dx \geq d$ is a simple nonnegativity constraint $x \geq 0$.

7.33. Write a convex hull relaxation for the disjunction

$$[(x_1, x_2) = (0, 1)] \vee [2x_1 - x_2 \geq 2]$$

where each $x_j \in [0, 2]$. Simplify the relaxation so that it contains only the variables x_1, x_2, α where $0 \leq \alpha \leq 1$. Use Fourier–Motzkin elimination to project the relaxation onto x_1, x_2, and verify that it describes the convex hull by drawing a graph of the feasible set. *Hint:* To obtain the convex hull, the bounds on x_j must be included in the disjuncts (see Exercise 7.31). The disjunction becomes

$$\begin{bmatrix} x_1 = 0 \\ x_2 = 1 \end{bmatrix} \vee \begin{bmatrix} 2x_1 - x_2 \geq 2 \\ x_1 \leq 2 \\ x_2 \geq 0 \end{bmatrix}$$

7.34. Write a big-M relaxation of the disjunction in the previous exercise using the big-Ms given by (7.65). Use Fourier–Motzkin elimination to project the relaxation onto x_1, x_2 and draw a graph of the result. Note that the relaxation does not describe the convex hull. *Hint:* Here the bounds on x_j need not be included in the disjuncts. The following disjunction can therefore be relaxed with bounds $x_j \in [0, 2]$:

$$\begin{bmatrix} -x_1 \geq 0 \\ x_2 \geq 1 \\ -x_2 \geq -1 \end{bmatrix} \vee [2x_1 - x_2 \geq 2]$$

7.35. Tighten the relaxation of Exercise 7.34 by using the big-Ms given by (7.70). Project the relaxation onto x_1, x_2 and draw a graph to verify that the relaxation is in fact tighter.

7.36. Use Theorem 7.15 to write a big-M relaxation for

$$[-2x_1 + x_2 \geq 0] \vee [-x_1 + 2x_2 \geq 2]$$

subject to the bounds $x_j \in [0, 2]$. Draw a graph of the relaxation along with the feasible set.

7.37. Use Theorem 7.16 to tighten the relaxation of Exercise 7.36. Draw a graph and note that the tightened inequality supports the feasible set.

7.38. Verify that (7.79) is the optimal value of (7.78) by solving the dual of (7.78).

7.39. Use Theorem 7.17 to write a big-M relaxation of

$$[-x_1 + x_2 = 1] \vee [x_1 - x_2 = 1]$$

subject to the bounds $x_j \in [0, 2]$. Draw a graph of the relaxation along with the feasible set. Note that it is not a convex hull relaxation.

7.40. Solve a linear programming problem to find an optimal separating cut for the point $\bar{x} = (0, 1)$ and the disjunction

$$\begin{bmatrix} x_1 + x_2 \geq 2 \\ x_1, x_2 \geq 0 \end{bmatrix} \vee \begin{bmatrix} x_1 \geq 1 \\ x_2 \geq 0 \end{bmatrix}$$

on the assumption that the L_∞ norm of the vector of coefficients in the cut is bounded. Examine the dual solution to identify the point in the convex hull of the feasible set of the disjunction that is closest to $\bar{x}$, as measured by the L_1 norm.

7.41. Verify the identity (7.86) in the proof of Theorem 7.19.

7.42. Write the dual of (7.87) and indicate how the dual solution is used to identify the point in the convex hull of the feasible set of (7.60) that is closest to $\bar{x}$ as measured by the L_∞ norm.

7.43. Solve the system (7.87) for the problem in Fig. 7.11 and identify the optimal separating cut for the L_1 norm. Note that it is the same as the cut obtained for the L_∞ norm. Examine the dual solution and determine the point in the convex hull of the feasible set that is closest to $\bar{x}$ as measured by the L_∞ norm. *Hint:* The optimal solution of (7.87) has $\delta = -\frac{1}{2}$, $u^1 = (\frac{1}{4}, 0, 0, 0, 0)$, $u^2 = (\frac{1}{6}, 0, 0, 0, \frac{1}{3})$.

7.5 Disjunctions of Nonlinear Systems

When formulating continuous relaxations of nonlinear constraints, the overriding concern is obtaining a convex relaxation. A convex relaxation is one with a convex feasible set (i.e., a feasible set that contains all convex combinations of its points). A convex relaxation is generally much easier to solve, because optimization over a convex feasible set is generally much easier than over more general feasible sets. Most nonlinear programming methods, for instance, are designed to find only a locally optimal solution, which is guaranteed to be optimal only in a small neighborhood surrounding it. Such a solution is globally optimal, however, if one is minimizing a convex function over a convex feasible set (or maximizing a concave function over a convex set).

Fortunately, a disjunction of nonlinear inequality systems can be given a convex relaxation using much the same techniques used earlier to relax disjunctions of linear systems—provided each individual system defines a convex, bounded set. In particular, convex hull relaxations and convex big-M relaxations can be easily formulated for disjunctions of convex nonlinear systems.

7.5.1 Convex Hull Relaxation

The problem is to find a continuous relaxation for the disjunction

$$\bigvee_{k \in K} g^k(x) \leq 0 \tag{7.88}$$

where each $g^k(x)$ is a vector of functions $g_i^k(x)$ with $x \in \mathbb{R}^n$. It is assumed that $x \in [L, U]$, and $g^k(x)$ is bounded when $x \in [L, U]$. It is further assumed that each $g_i^k(x)$ is a convex function on $[L, U]$, meaning that

$$g_i^k((1 - \alpha)x^1 + \alpha x^2) \leq (1 - \alpha)g_i^k(x^1) + \alpha g_i^k(x^2)$$

for all $x^1, x^2 \in [L, U]$ and all $\alpha \in [0, 1]$. This implies that the feasible set of each system $g^k(x) \leq 0$ is convex.

To simplify exposition, it is assumed that every disjunct of (7.88) is feasible. The convex hull of (7.88) consists of all points that can be written as a convex combination of points $\bar{x}^k$ that respectively satisfy the disjuncts of (7.88). Thus,

$$x = \sum_{k \in K} \alpha_k \bar{x}^k$$

$$g^k(\bar{x}^k) \leq 0, \quad \text{all } k \in K$$

$$L \leq \bar{x}^k \leq U, \quad \text{all } k \in K$$

$$\sum_{k \in K} \alpha_k = 1, \quad \alpha_k \geq 0, \quad \text{all } k \in K$$

Using the change of variable $x^k = \alpha_k \bar{x}^k$, the following relaxation is obtained:

$$x = \sum_{k \in K} x^k$$

$$g^k \left(\frac{x^k}{\alpha_k} \right) \leq 0, \quad \text{all } k \in K \tag{7.89}$$

$$\alpha_k L \leq x^k \leq \alpha_k U, \quad \text{all } k \in K$$

$$\sum_{k \in K} \alpha_k = 1, \quad \alpha_k \geq 0, \quad \text{all } k \in K$$

The function $g^k(x^k/\alpha_k)$ is in general nonconvex, but a classical result of convex analysis implies that one can restore convexity by multiplying the second constraint of (7.89) by α_k.

Theorem 7.20. *Consider the set S consisting of all (x, α) with $\alpha \in [0, 1]$ and $x \in [\alpha L, \alpha U]$. If $g(x)$ is convex and bounded for $x \in [L, U]$, then*

$$h(x, \alpha) = \begin{cases} \alpha g(x/\alpha) & \text{if } \alpha > 0 \\ 0 & \text{if } \alpha = 0 \end{cases}$$

is convex and bounded on S.

Proof. To show convexity of $h(x, \alpha)$, arbitrarily choose points (x^1, α_1), $(x^2, \alpha_2) \in S$. Supposing first that $\alpha_1, \alpha_2 > 0$, convexity can be shown by noting that for any $\beta \in [0, 1]$,

$$h\left(\beta x^1 + (1 - \beta)x^2, \beta \alpha_1 + (1 - \beta)\alpha_2\right)$$

$$= (\beta \alpha_1 + (1 - \beta)\alpha_2)\, g\left(\frac{\beta x^1 + (1 - \beta)x^2}{\beta \alpha_1 + (1 - \beta)\alpha_2}\right)$$

$$= (\beta \alpha_1 + (1 - \beta)\alpha_2)\, g\left(\frac{\beta \alpha_1}{\beta \alpha_1 + (1 - \beta)\alpha_2}\frac{x^1}{\alpha_1} + \frac{(1 - \beta)\alpha_1}{\beta \alpha_1 + (1 - \beta)\alpha_2}\frac{x^2}{\alpha_2}\right)$$

$$\leq (\beta \alpha_1 + (1 - \beta)\alpha_2)\left[\frac{\beta \alpha_1}{\beta \alpha_1 + (1 - \beta)\alpha_2}g\left(\frac{x^1}{\alpha_1}\right) + \frac{(1 - \beta)\alpha_1}{\beta \alpha_1 + (1 - \beta)\alpha_2}g\left(\frac{x^2}{\alpha_2}\right)\right]$$

$$= \beta h\left(x^1, \alpha_1\right) + (1 - \beta)h\left(x^2, \alpha_2\right)$$

where the inequality is due to the convexity of $g(x)$. If $\alpha_1 = \alpha_2 = 0$, then

$$h\left(\beta x^1 + (1 - \beta)x^2, \beta \alpha_1 + (1 - \beta)\alpha_2\right)$$
$$= h(0, 0) = \beta h\left(x^1, \alpha_1\right) + (1 - \beta)h\left(x^2, \alpha_2\right)$$

because $\alpha_j L \le x^k \le \alpha_j U$ implies $x^k = 0$. If $\alpha_1 = 0$ and $\alpha_2 > 0$,

$$h\left(\beta x^1 + (1-\beta)x^2, \beta\alpha_1 + (1-\beta)\alpha_2\right)$$
$$= h\left((1-\beta)x^2, (1-\beta)\alpha_2\right) = (1-\beta)g\left(\frac{x^2}{\alpha_2}\right)$$
$$= \beta h(0,0) + (1-\beta)h\left(x^2, \alpha_2\right)$$

Finally, $h(x,\alpha) = \alpha g(x/\alpha)$ is bounded because $\alpha \in [0,1]$, $x/\alpha \in [L,U]$, and $g(x)$ is bounded for $x \in [L,U]$. $\square$

Due to Theorem 7.20, multiplying the second constraint of (7.89) by α_k yields a convex hull relaxation of (7.88):

$$x = \sum_{k \in K} x^k$$

$$\alpha_k g^k\left(\frac{x^k}{\alpha_k}\right) \le 0, \quad \text{all } k \in K \tag{7.90}$$

$$\alpha_k L \le x^k \le \alpha_k U, \quad \text{all } k \in K$$

$$\sum_{k \in K} \alpha_k = 1, \quad \alpha_k \ge 0, \quad \text{all } k \in K$$

The following result can be shown in a manner similar to the proof of Theorem 7.14.

Theorem 7.21. *Suppose each $g^k(x)$ in (7.88) is convex and bounded for $x \in [L,U]$, and each disjunct contains the constraint $L \le x \le U$. Then (7.90) is a convex hull relaxation of (7.88).*

When the boundedness conditions of Theorem 7.21 are violated, the convex hull of a disjunction (7.88) need not be a closed set, even if the disjuncts describe closed sets. For example, the convex hull of

$$\begin{pmatrix} x_1 = 0 \\ x_2 \ge 0 \end{pmatrix} \vee \begin{pmatrix} x_1 \ge 0 \\ x_2 \ge \dfrac{1}{1+x_1} \end{pmatrix}$$

is the nonnegative quadrant except for points $(x_1, 0)$ with $x_1 > 0$. One might say that $x_1, x_2 \ge 0$ is a convex hull relaxation, but only in the sense that it describes the closure of the convex hull.

It is necessary to deal with the fact that α_k can vanish. The simplest approach is to use the constraint

$$(\alpha_k + \epsilon)g^k \left(\frac{x^k}{\alpha_k + \epsilon} \right) \leq 0, \quad \text{all } k \in K \tag{7.91}$$

in place of the second constraint of (7.90), for some small $\epsilon > 0$. The introduction of ϵ preserves convexity. More complex alternatives are available when the small ϵ causes numerical problems (see Section 7.14).

As an example, consider the disjunction

$$\left[x_1^2 + x_2^2 - 1 \leq 0 \right] \vee \left[(x_1 - 2)^2 + x_2^2 - 1 \leq 0 \right] \tag{7.92}$$

with $x_1 \in [-1, 3]$ and $x_2 \in [-1, 1]$. The feasible set for (7.92) is the union of the discs in Fig. 7.12. The convex hull relaxation (7.91) is

$$\begin{bmatrix} x_1 \\ x_2 \end{bmatrix} = \begin{bmatrix} x_{11} \\ x_{21} \end{bmatrix} + \begin{bmatrix} x_{12} \\ x_{22} \end{bmatrix}$$

$$\frac{x_{11}^2 + x_{21}^2}{\alpha + \epsilon} \leq \alpha + \epsilon$$

$$\frac{x_{12}^2 + x_{22}^2}{1 - \alpha + \epsilon} - 4x_{12} + 3(1 - \alpha + \epsilon) \leq 0$$

$$0 \leq \alpha \leq 1$$

(The bounds on x_1, x_2 are redundant and are omitted.) Figure 7.12 shows the projection of the feasible set of this relaxation onto x_1, x_2.

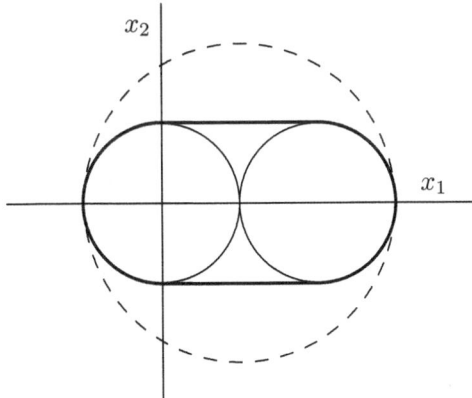

Fig. 7.12 Feasible set of a disjunction of two nonlinear systems (two small discs), the convex hull relaxation (area within heavy boundary), and a big-M relaxation (disc with dashed boundary).

7.5.2 Big-M Relaxation

The big-M relaxation of the disjunction (7.88) introduces a variable
α_k for each $k \in K$, where $\alpha_k = 1$ indicates that the kth disjunct is
enforced. It is assumed that there are bounds $L \leq x \leq U$ on x. The
big-M relaxation is

$$g^k(x) \leq M^k(1 - \alpha_k), \quad \text{all } k \in K$$
$$L \leq x \leq U$$
$$\sum_{k \in K} \alpha_k = 1, \ \alpha_k \geq 0, \quad \text{all } k \in K \tag{7.93}$$

where M^k is a vector of valid upper bounds on the component functions
of $g^k(x)$, given that $L \leq x \leq U$. This relaxation is clearly convex,
assuming that each $g^k(x)$ is convex.

The bounds M^k can be set to

$$M_i^k = \max_{L \leq x \leq U} \left\{ g_i^k(x) \right\} \tag{7.94}$$

but the tightest bound is

$$M_i^k = \max_{\ell \neq k} \left\{ \max_{L \leq x \leq U} \left\{ g_i^k(x) \mid g^\ell(x) \leq 0 \right\} \right\} \tag{7.95}$$

As an example, again consider the disjunction (7.92) with domains
$x_1 \in [-1, 3]$ and $x_2 \in [-1, 1]$. Setting $M^1 = M^2 = 8$ as given by (7.95)
yields the big-M relaxation

$$x_1^2 + x_2^2 - 1 \leq 8(1 - \alpha)$$
$$(x_1 - 2)^2 + x_2^2 - 1 \leq 8\alpha$$
$$0 \leq \alpha \leq 1$$

The large disc in Fig. 7.12 depicts the projection of the relaxation onto
the x-space. The projection is described by $(x_1 - 1)^2 + x_2^2 \leq 4$.

Exercises

7.44. Write a convex hull relaxation for

$$\left[x_1^2 + x_2^2 \leq 1 \right] \vee \left[(x_1, x_2) = (2, 0) \right]$$

and simplify it as much as possible.

7.45. Write a big-M relaxation for the disjunction in the previous exercise, using the big-Ms given by (7.95). Project it onto x_1, x_2, and draw a graph. Is it a convex hull relaxation?

7.46. Show that if $g(x)$ satisfies the conditions of Theorem 7.20, then $\bar{h}(x, \alpha) = (\alpha + \epsilon)g(x/(\alpha + \epsilon))$ is convex for any $\epsilon > 0$. *Hint:* For any convex $f(x)$, it is trivial to show that $\bar{f}(x) = f(x + a)$ is convex.

7.6 Mixed-Integer Modeling

One general method for obtaining a linear relaxation of a constraint set is to reformulate the constraints as a mixed-integer/linear (MILP) model, and then drop the integrality condition on the variables.

Mixed-integer programming provides a highly versatile modeling language if one is sufficiently ingenious. There may be several ways, however, to write an MILP model of the same problem, and some models may be more succinct or have tighter relaxations than others. In fact, the more succinct model is often not the tighter one. Formulating a suitable relaxation is as much an art as a science.

The theory of mixed-integer modeling tells us that it is essentially tantamount to disjunctive and knapsack modeling. A problem can be given an MILP model if and only if its feasible set is a finite union of mixed-integer polyhedra that satisfy a certain technical condition. Each mixed-integer polyhedron is described by a system of knapsack inequalities (or continuous linear inequalities, a special case). This means that the feasible set is described by a disjunction of knapsack systems. The disjunction can then be rewritten as an MILP model using a convex hull or big-M formulation. If there are too many disjunctions, the problem is written as several disjunctions. There may also be a free-standing system of knapsack constraints, which may be viewed as a disjunction consisting of one disjunct. The individual systems of knapsack inequalities are written to capture linear conditions and counting ideas.

This disjunctive approach need not result in the tightest known model, but it often does. It is important, however, to simplify the model as much as possible by eliminating unnecessary variables and constraints.

7.6.1 Mixed-Integer Representability

Representation as an MILP allows the use of auxiliary variables, both continuous and discrete. A subset S of $\mathbb{R}^n \times \mathbb{Z}^p$ is *mixed-integer representable* (MILP representable) if it is the projection onto x of the feasible set of a model of the form

$$Ax + Bu + D\delta \geq b$$
$$x \in \mathbb{R}^n \times \mathbb{Z}^p, \ u \in \mathbb{R}^m, \ y_k \in \{0,1\}, \text{ all } k \qquad (7.96)$$

Some of the auxiliary variables are real-valued (u_j) and some are binary (y_k).[2]

To state the representability theorems, some definitions are necessary. Let a *mixed-integer polyhedron* be the nonempty intersection of any polyhedron in $\mathbb{R}^{n+p}$ with $\mathbb{R}^n \times \mathbb{Z}^p$. Such a polyhedron is illustrated in Fig. 7.13. A vector $r \in \mathbb{R}^{n+p}$ is a *recession direction* of a polyhedron $P \in \mathbb{R}^{n+p}$ if one can go forever in the direction r without leaving P. That is, for any $x \in P$, $x + \alpha r \in P$ for all $\alpha \geq 0$. A rational vector r is a recession direction of a mixed-integer polyhedron Q if it is a recession direction of a polyhedron whose intersection with $\mathbb{R}^n \times \mathbb{Z}^p$ is Q. The *recession cone* of a mixed-integer polyhedron is the set of all its recession directions (Fig. 7.13). The definition is well formed because of the following lemma.

Lemma 7.22 *All polyhedra in $\mathbb{R}^{n+p}$ having the same nonempty intersection with $\mathbb{R}^n \times \mathbb{Z}^p$ have the same recession cone.*

Proof. Let $Q = \{x \in \mathbb{R}^{n+p} \mid Ax \geq b\}$ and $Q' = \{x \in \mathbb{R}^{n+p} \mid A'x \geq b'\}$ be polyhedra, and suppose that $Q \cap (\mathbb{R}^n \times \mathbb{Z}^p) = Q' \cap (\mathbb{R}^n \times \mathbb{Z}^p) = P$, where P is nonempty. It suffices to show that any recession direction d of Q is a recession direction of Q'. Take any $u \in P$. Because $u \in Q$, it follows that $u + \alpha d \in Q$ for any $\alpha \geq 0$. Furthermore, because d is rational, $u + \bar{\alpha} d \in Q \cap (\mathbb{R}^n \times \mathbb{Z}^p)$ for some sufficiently large $\bar{\alpha} > 0$. Now if d is not a recession direction of Q', then because $u \in Q'$, $u + \beta \bar{\alpha} d \notin Q'$ for some sufficiently large integer $\beta \geq 1$. Thus in particular $u + \beta \bar{\alpha} d \notin Q' \cap (\mathbb{R}^n \times \mathbb{Z}^p)$. But because β is integer, $u + \beta \bar{\alpha} d \in Q \cap (\mathbb{R}^n \times \mathbb{Z}^p)$. This violates the assumption that Q, Q' have the same intersection with $\mathbb{R}^n \times \mathbb{Z}^p$. $\square$

---

[2] A broader sense of mixed-integer representability would allow unbounded integer auxiliary variables. For example, the set of even integers x could then be represented by $x = 2y$ for $y \in \mathbb{Z}$.

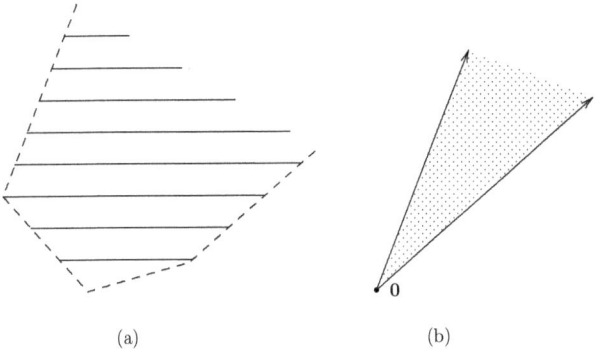

Fig. 7.13 (a) A mixed-integer polyhedron Q (horizontal lines) where $Q = P \cap (\mathbb{R} \times \mathbb{Z})$ and P is bounded by the dashed line. (b) The recession cone of Q.

A necessary and sufficient condition for mixed-integer representability can now be stated.

Theorem 7.23. *A nonempty set $S \subset \mathbb{R}^n \times \mathbb{Z}^p$ is mixed-integer representable if and only if S is the union of finitely many mixed-integer polyhedra in $\mathbb{R}^n \times \mathbb{Z}^p$ having the same recession cone. In particular, S is mixed-integer representable if and only if S is the projection onto x of a mixed-integer formulation of the following form:*

$$
\begin{aligned}
x &= \sum_{k \in K} x^k \\
A^k x^k &\geq b^k y_k, \quad k \in K \\
\sum_{k \in K} y_k &= 1, \quad y_k \in \{0, 1\}, \ k \in K \\
x &\in \mathbb{R}^n \times \mathbb{Z}^p
\end{aligned}
\tag{7.97}
$$

Proof. Suppose first that S is the union of mixed-integer polyhedra P_k, $k \in K$, that have the same recession cone. Each P_k has the form $\{x \mid A^k x^k \geq b^k\} \cap (\mathbb{R}^n \times \mathbb{Z}^p)$. It can be shown as follows that S is represented by (7.97), and is therefore representable, because (7.97) has the form (7.96). Suppose first that $x \in S$. Then x belongs to some P_{k^*}, which means that x is feasible in (7.97) when $y_{k^*} = 1$, $y_k = 0$ for $k \neq k^*$, $x^{k^*} = x$, and $x^k = 0$ for $k \neq k^*$. The constraint $A^k x^k \geq b^k y_k$ is satisfied by definition when $k = k^*$, and it is satisfied for other k's because $x^k = y_k = 0$.

Now suppose that x, y and x^k satisfy (7.97). Let the polyhedron Q_k be $\{x \mid A^k x \geq b^k\}$, so that $P_k = Q_k \cap (\mathbb{R}^n \times \mathbb{Z}^p)$. To show that $x \in S$, note that exactly one y_k, say y_{k^*}, is equal to 1. Then $A^{k^*} x^{k^*} \geq b^{k^*}$ is enforced, which means that $x^{k^*} \in Q_{k^*}$. For other k's, $A^k x^k \geq 0$. Thus, $A^k(\beta x^k) \geq 0$ for all $\beta \geq 0$, which implies that x^k is a recession direction for Q_k. Because by hypothesis all the P_k's have the same recession cone, all Q_k's have the same recession cone. Thus each x^k $(k \neq k^*)$ is a recession direction for Q_{k^*}, which means that $x = x^{k^*} + \sum_{k \neq k^*} x^k$ belongs to Q_{k^*} and therefore to $\bigcup_{k \in K} Q_k$. But because $x \in \mathbb{R}^n \times \mathbb{Z}^p$,

$$x \in \left(\bigcup_{k \in K} Q_k \right) \cap (\mathbb{R}^n \times \mathbb{Z}^p) = \bigcup_{k \in K} (Q_k \cap (\mathbb{R}^n \times \mathbb{Z}^p)) = \bigcup_{k \in K} P_k$$

To prove the converse of the theorem, suppose that S is represented by (7.96). To show that S is a finite union of mixed-integer polyhedra, let $P(\bar{y})$ be the set of all x that are feasible in (7.96) when $y = \bar{y} \in \{0,1\}^{|K|}$. Because S is nonempty, $P(\bar{y})$ is nonempty for at least one $\bar{y}$. Thus we let Y be the set of all $\bar{y}$ for which $P(\bar{y})$ is nonempty. So $P(\bar{y})$ is a mixed-integer polyhedron for all $\bar{y} \in Y$, and $S = \bigcup_{\bar{y} \in Y} P(\bar{y})$. To show that the $P(\bar{y})$'s have the same recession cone, note that

$$P(\bar{y}) = \left\{ x \in \mathbb{R}^n \times \mathbb{Z}^p \;\middle|\; \begin{bmatrix} A & B & D \\ 0 & 0 & 1 \\ 0 & 0 & -1 \end{bmatrix} \begin{bmatrix} x \\ u \\ y \end{bmatrix} \geq \begin{bmatrix} b \\ \bar{y} \\ -\bar{y} \end{bmatrix} \text{ for some } u, y \right\}$$

But x' is a recession direction of $P(\bar{y})$ if and only if (x', u', y') is a recession direction of

$$\left\{ \begin{bmatrix} x \\ u \\ y \end{bmatrix} \in \mathbb{R}^n \times \mathbb{Z}^p \times \mathbb{R}^{m+|K|} \;\middle|\; \begin{bmatrix} A & B & D \\ 0 & 0 & 1 \\ 0 & 0 & -1 \end{bmatrix} \begin{bmatrix} x \\ u \\ y \end{bmatrix} \geq \begin{bmatrix} b \\ \bar{y} \\ -\bar{y} \end{bmatrix} \right\}$$

for some u', y'. The latter is true if and only if

$$\begin{bmatrix} A & B & D \\ 0 & 0 & 1 \\ 0 & 0 & -1 \end{bmatrix} \begin{bmatrix} x' \\ u' \\ y' \end{bmatrix} \geq \begin{bmatrix} 0 \\ 0 \\ 0 \end{bmatrix}$$

This means that the recession directions of $P(\bar{y})$ are the same for all $\bar{y} \in Y$, as desired. $\square$

The theorem says in part that any nonempty mixed-integer representable subset of $\mathbb{R}^n \times \mathbb{Z}^p$ is the feasible set of some disjunction

$$\bigvee_{k \in K} \left(\begin{array}{c} A^k x \geq b^k \\ x \in \mathbb{R}^n \times \mathbb{Z}^p \end{array} \right) \tag{7.98}$$

This and the following lemma provide a technique for writing a convex hull formulation by conceiving the feasible set as a union of mixed-integer polyhedra.

Lemma 7.24 *If each disjunct of (7.98) is a convex hull formulation, then (7.97) is a convex hull formulation of (7.98).*

Proof. It is clear that x satisfies (7.98) if any only if x satisfies (7.97) for some $(x^k, y_k \mid k \in K)$. It remains to show that, given any feasible solution $\bar{x}$, $(\bar{x}^k, \bar{y}_k \mid k \in K)$ of the continuous relaxation of (7.97), $\bar{x}$ belongs to the convex hull of the feasible set of (7.98). But $\bar{x}$ is the convex combination

$$\bar{x} = \sum_{k \in K^+} \bar{y}_k \frac{\bar{x}^k}{\bar{y}_k} \tag{7.99}$$

where $K^+ = \{k \in K \mid \bar{y}_k > 0\}$. Furthermore, each point $\bar{x}^k / \bar{y}_k$ satisfies $A^k(\bar{x}^k / \bar{y}_k) \geq b^k$ because $(\bar{x}^k, \bar{y}_k)$ satisfies $A^k \bar{x}^k \geq b^k \bar{y}_k$. Thus $\bar{x}^k / \bar{y}_k$ satisfies the continuous relaxation of the kth disjunct of (7.98) and so, by hypothesis, belongs to the convex hull of the feasible set of that disjunct. This and (7.99) imply that $\bar{x}$ belongs to the convex hull of the feasible set of (7.98). $\square$

It is important to note that the continuous relaxation of (7.97) is a valid convex hull relaxation of (7.98), due to Theorem 7.14, even if the disjuncts do not describe polyhedra with the same recession cone. Thus, a finite union of polyhedra can be given a convex hull relaxation even when there is no MILP formulation of the union.

An MILP representable subset of $\mathbb{R}^n \times \mathbb{Z}^p$ can also be given a big-M formulation:

$$A^k x \geq b^k - M^k(1 - y_k), \quad k \in K$$
$$x \in \mathbb{R}^n \times \mathbb{Z}^p, \quad \sum_{k \in K} y_k = 1, \quad y_k \in \{0, 1\}, \quad k \in K \tag{7.100}$$

where each M^k is a tuple of finite numbers such that

$$M^k \geq b^k - \min_{\ell \neq k} \left\{ \min_x \left\{ A^k x \mid A^\ell x \geq b^\ell, \ x \in \mathbb{R}^n \times \mathbb{Z}^p \right\} \right\} \tag{7.101}$$

Theorem 7.25. *If set $S \subset \mathbb{R}^n \times \mathbb{Z}^p$ is the union of finitely many mixed-integer polyhedra $P_k = Q_k \cap (\mathbb{R}^n \times \mathbb{Z}^p)$ (for $k \in K$) having the same recession cone, where $Q_k = \{x \mid A^k x \geq b^k\}$, then S is represented by the big-M mixed-integer disjunctive formulation (7.100).*

Proof. System (7.100) clearly represents S if every component of M^k as given by (7.101) is finite. We therefore suppose that some component i of some M^k is infinite, which implies that $\min\left\{A_i^k x \mid A^\ell x \geq b^\ell\right\}$ is unbounded for some $\ell \neq k$. Since P_ℓ is nonempty, this means there is a point $\bar{x} \in P_\ell$ and a rational direction d such that $A_i^k(\bar{x} + \alpha d)$ is unbounded in a negative direction as $\alpha \to \infty$, and such that $\bar{x} + \alpha d \in Q_\ell$ for all $\alpha \geq 0$. This means d is a recession direction of P_ℓ and therefore, by hypothesis, a recession direction of P_k. Thus by Lemma 7.22, d is a recession direction of Q_k. Since P_k is nonempty, there is an x' satisfying $A^k x' \geq b^k$, and for any such x' we have $A^k(x' + \alpha d) \geq b^k$ for all $\alpha \geq 0$. Thus $A^k(\bar{x} + \alpha d) \geq b^k + A^k(\bar{x} - x')$ for all $\alpha \geq 0$, which means that $A_i^k(\bar{x} + \alpha d)$ cannot be unbounded in a negative direction as $\alpha \to \infty$. $\square$

Although a big-M relaxation is generally not as tight as a convex hull relaxation, it may be the better alternative when the convex hull relaxation has a large number of variables. If the big-Ms are large numbers, however, the big-M formulation may provide a very weak relaxation.

In some cases, both the convex hull and big-M relaxations are quite weak, and it is best not to use a relaxation at all. In particular, if one must place large upper bounds on some variables to satisfy the recession cone condition for the convex hull formulation, or to ensure that finite big-Ms exist for a big-M relaxation, then the relaxation that results may be too weak to justify the overhead of solving it. One of the advantages of using a modeling framework more general than MILP is that one is not obliged to incur the overhead of writing an MILP model when it does not provide a useful relaxation.

7.6.2 Example: Fixed-Charge Function

Mixed-integer representability is illustrated by the fixed-charge function, which occurs frequently in modeling. Suppose the cost x_2 of manufacturing quantity x_1 of some product is to be minimized. The cost is zero when $x_1 = 0$ and is $f + cx_1$ otherwise, where f is the fixed cost and c the unit variable cost. In general there would be additional constraints in the problem, but the issue here is how to write MILP constraints that represent feasible pairs (x_1, x_2) for the fixed-charge function.

The problem can be viewed as minimizing x_2 subject to $(x_1, x_2) \in S$, where S is the set depicted in Fig. 7.14(a). S is the union of two polyhedra P_1 and P_2, and the problem is to minimize x_2 subject to the disjunction

$$\begin{bmatrix} x_1 = 0 \\ x_2 \geq 0 \end{bmatrix} \vee \begin{bmatrix} x_2 \geq cx_1 + f \\ x_1 \geq 0 \end{bmatrix}$$

where the disjuncts correspond respectively to P_1 and P_2. In this case, there are no integer variables in the disjuncts $(p = 0)$.

Note that P_1 and P_2 do not have the same recession cone. The recession cone of P_1 is P_1 itself, and the recession cone of P_2 is the set of

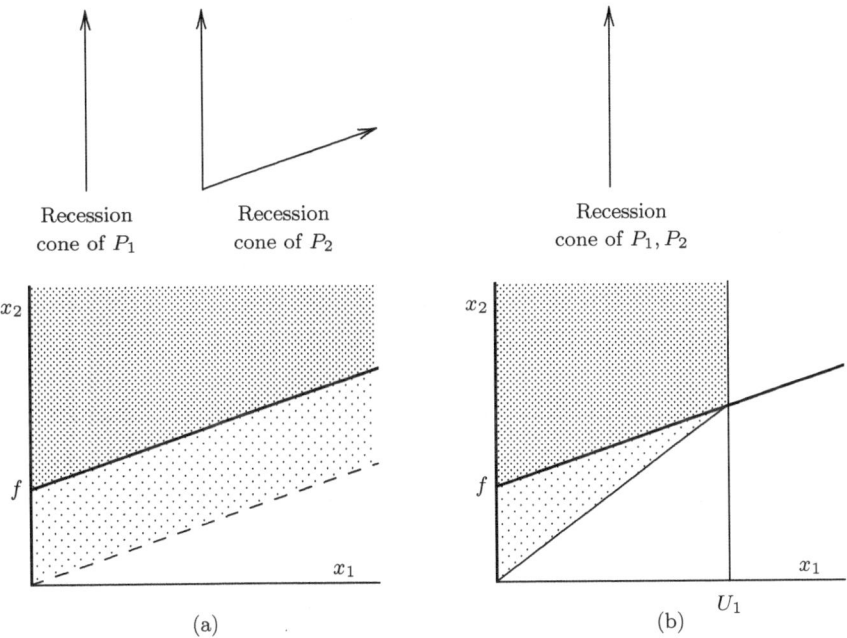

(a)   (b)

Fig. 7.14 (a) Feasible set of a fixed-charge problem, consisting of the union of polyhedra P_1 (heavy vertical line) and P_2 (darker shaded area). The convex hull of the feasible set is the entire shaded area, excluding the dashed line. (b) Feasible set of the same problem with the bound $x_1 \leq L_1$, where P_2' is the darker shaded area. The convex hull of the feasible set is the entire shaded area.

all vectors (x_1, x_2) with $x_2 \geq cx_1 \geq 0$. Thus, by Theorem 7.23, S is not MILP representable, and in particular the convex hull representation (7.97) does not correctly represent S:

$$
\begin{array}{llll}
x_1 = x_1^1 + x_1^2 & x_1^1 \leq 0 & -cx_1^2 + x_2^2 \geq fy_2 & y_1 + y_2 = 1 \\
x_2 = x_2^1 + x_2^2 & x_1^1, x_2^1 \geq 0 & x_1^2 \geq 0 & y_1, y_2 \in \{0, 1\}
\end{array}
\tag{7.102}
$$

This can be seen by simplifying the above model. Only one 0-1 variable appears, which can be renamed y. Also, one can set $x_1^2 = x_1$ (since $x_1^1 = 0$) and $x_2^1 = x_2 - x_2^2$, which yields

$$
x_1 \geq 0, \quad x_2 - x_2^2 \geq 0, \quad x_2^2 - cx_1 \geq fy, \quad y \in \{0, 1\}
$$

Minimizing x_2 subject to this is equivalent to minimizing x_2 subject to

$$
x_1 \geq 0, \quad x_2 - cx_1 \geq fy, \quad y \in \{0, 1\}
$$

The projection onto (x_1, x_2) is the union of the two polyhedra obtained by setting $y = 0$ and $y = 1$. The projection is therefore the set of all points satisfying $x_2 \geq cx_1$, $x_1 \geq 0$, which is clearly different from $P_1 \cup P_2$. The model is therefore incorrect.

Although (7.102) does not correctly model the problem, its continuous relaxation (formed by replacing $y_j \in \{0, 1\}$ with $y_j \in [0, 1]$) is a valid convex hull relaxation. The projection of its feasible set onto x is the closure of the convex hull. As is evident in Fig. 7.14, the convex hull relaxation is quite weak, since x_1 is unbounded.

In practice, one can generally put an upper bound U_1 on x_1 without harm. The problem is now to minimize x_2 subject to

$$
\begin{pmatrix} x_1 = 0 \\ x_2 \geq 0 \end{pmatrix} \vee \begin{pmatrix} x_2 \geq cx_1 + f \\ 0 \leq x_1 \leq U_1 \end{pmatrix}
\tag{7.103}
$$

The recession cone of each of the resulting polyhedra P_1, P_2' (Fig. 7.14b) is the same (namely, P_1), and the feasible set $S' = P_1 \cup P_2'$ is therefore MILP representable. The convex hull formulation (7.97) becomes

$$
\begin{array}{llll}
x_1^1 \leq 0 & -cx_1^2 + x_2^2 \geq fy_2 & x_1 = x_1^1 + x_1^2 & y_1 + y_2 = 1 \\
x_1^1, x_2^2 \geq 0 & 0 \leq x_1^2 \leq L_1 y_2 & x_2 = x_2^1 + x_2^2 & y_1, y_2 \in \{0, 1\}
\end{array}
$$

Again the model simplifies. As before, one can set $y = y_2$, $x_1 = x_1^2$, and eliminate x_2^2, resulting in the model

$$
x_1 \leq L_1 y, \quad x_2 \geq fy + cx_1, \quad x_1 \geq 0, \quad y \in \{0, 1\}
\tag{7.104}
$$

Obviously, y encodes whether the quantity produced is zero or positive, in the former case ($y = 0$) forcing $x_1 = 0$, and in the latter case incurring the fixed charge f. The projection onto (x_1, x_2) is $P_1 \cup P_2$, and the model is therefore correct.

The disjunction (7.103) can also be given a big-M model (7.100). Although x_2 is formally unbounded, M_1, M_2 as given by (7.101) are finite. Specifically, $M_1 = U_1$ and $M_2 = f + U_1$, and the big-M model (7.100) becomes

$$x_1 \le U_1 y, \quad x_2 \ge cx_1 + fy - cU_1(1 - y) \tag{7.105}$$
$$0 \le x_1 \le U_1, \quad y \in \{0, 1\}$$

It is interesting to compare the continuous relaxations of the convex hull model (7.104), shown in Fig. 7.14(b), with the big-M model (7.105), shown in Fig. 7.15. The convex hull model is relaxed by replacing $y \in \{0, 1\}$ with $0 \le y \le 1$. Its projection onto x_1, x_2 is the convex hull of the feasible set, which is described by

$$x_2 \ge cx_1 \tag{7.106}$$

and the bounds on x_1.

Due to Theorem 7.15, the continuous relaxation of the big-M model (7.105), when projected onto (x_1, x_2), is given by (7.75) and the bounds $0 \le x_1 \le U_1, x_2 \ge 0$. The inequality (7.75), in this case, becomes

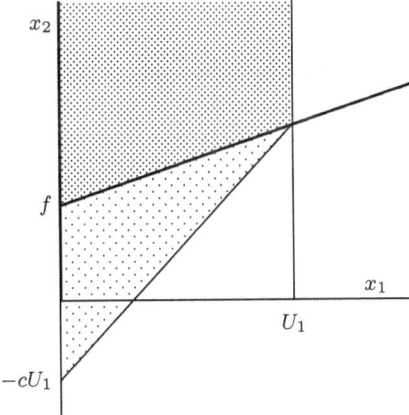

Fig. 7.15 Big-M relaxation of the fixed-charge problem in Fig. 7.14 (entire shaded area).

$$x_2 \geq \left(\frac{f}{U_1} + 2c\right) x_1 - cU_1 \qquad (7.107)$$

This is weaker than the convex hull relaxation (7.106), as is evident in the figures. Since the inequality (7.107) already supports the feasible set, it cannot be further tightened by using Theorem 7.16.

Due to the succinctness of the convex hull relaxation after it is simplified, it is the clear choice for relaxing this particular fixed-charge problem. Other problems, however, may generate so many auxiliary variables in the convex hull relaxation that the big-M relaxation is more practical.

7.6.3 Disjunctive Models

Mixed-integer modeling devices used in practice are generally based on disjunctions, knapsack-like constraints, or combinations of the two. Disjunctive models are considered in this section, and knapsack models in the next.

Disjunctive modeling devices are useful when one must make several discrete choices from two or more alternatives. For example, each choice y_i might be *no* or *yes*, corresponding to 0 or 1 for $i = 1, \ldots, m$. Choosing $y_i = 0$ enforces the first term, and choosing $y_i = 1$ enforces the second term, of the disjunction

$$\left(A^{i0}x \geq b^{i0}\right) \vee \left(A^{i1}x \geq b^{i1}\right) \qquad (7.108)$$

There may also be constraints on which choices $y = (y_1, \ldots, y_m)$ are mutually possible.

The problem can be formulated as a single disjunction by taking the product of the disjunctions (7.108):

$$\bigvee_{y \in Y} \left\{A^{iy_i}x \geq b^{iy_i} \mid i = 1, \ldots, m\right\} \qquad (7.109)$$

where Y is the set of mutually possible choices. This disjunction can, in turn, be given a convex hull or big-M relaxation as described in previous sections. It might be called a *product relaxation* for (7.108). Because (7.109) contains 2^n disjuncts in the worst case, however, the product relaxation may be too large or too complicated for practical use.

Often a more practical approach is to combine individual relaxations of the disjunctions (7.108), along with a relaxation of $y \in Y$, to obtain a relaxation of the entire constraint set. The resulting relaxation, which might be called a *factored relaxation*, is not as tight as a product relaxation in general, but it can be much more succinct.

As a simple illustration of this idea, consider a constraint set consisting of the disjunctions

$$\begin{pmatrix} x_1 = 0 \\ x_2 \in [0,1] \end{pmatrix} \vee \begin{pmatrix} x_2 = 0 \\ x_1 \in [0,1] \end{pmatrix}$$
$$\begin{pmatrix} x_1 = 0 \\ x_2 \in [0,1] \end{pmatrix} \vee \begin{pmatrix} x_2 = 1 \\ x_1 \in [0,1] \end{pmatrix} \tag{7.110}$$

The two disjunctions, respectively, have the convex hull relaxations (a) and (b) below, which are illustrated in Fig. 7.16.

$$x_1 + x_2 \le 1, \quad x_1, x_2 \ge 0 \quad (a)$$
$$x_1 \le x_2, \quad x_1 \ge 0, \quad x_2 \le 1 \ (b) \tag{7.111}$$

The feasible set of the factored relaxation (7.111) is the intersection of the two convex hulls.

The product of the disjunctions in (7.110) is

$$\begin{pmatrix} x_1 = 0 \\ x_2 \in [0,1] \end{pmatrix} \vee \begin{pmatrix} x_1 = 0 \\ x_2 = 1 \\ x_1, x_2 \in [0,1] \end{pmatrix} \vee \begin{pmatrix} x_1 = 0 \\ x_2 = 0 \\ x_1, x_2 \in [0,1] \end{pmatrix} \vee \begin{pmatrix} x_2 = 0 \\ x_2 = 1 \\ x_1 \in [0,1] \end{pmatrix}$$

The feasible set of (7.110), and thus of the product, is the heavy line segment in Fig. 7.16, which is its own convex hull. Thus, the product relaxation is clearly tighter than the factored relaxation (7.111),

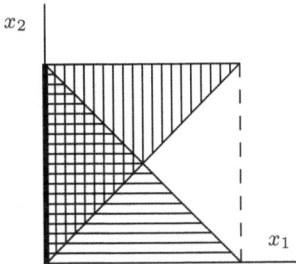

Fig. 7.16 Convex hulls of two disjunctions (vertical and horizontal shading, respectively). Their intersection is the feasible set of the factored relaxation. The heavy vertical line segment is the feasible set of the product relaxation.

although the product relaxation is much more complicated in general—
even if it happens to simplify considerably in this example.

Example: Facility Location

A *capacitated facility location problem* illustrates the above modeling
guidelines. There are m possible locations for factories, and n customers who obtain products from the factories (Fig. 7.17). A factory
installed at location i incurs fixed cost f_i and has capacity C_i. Each
customer j has demand D_j. Goods are shipped from factory i to customer j on trucks, each with capacity K_{ij}, and each incurring a fixed
cost c_{ij}. The problem is to decide which facilities to install, and how
to supply the customers, so as to minimize total cost.

The basic decision, for each location i, is whether to install a factory
at that location. This presents two discrete alternatives that can be
represented as a disjunction. To describe each alternative with knapsack systems, let x_{ij} be the quantity of goods shipped from factory i
to customer j, and let w_{ij} be the number of trucks on which they are
transported. Then if z_i is the total cost incurred at location i, the two
alternatives for location i are represented by the disjunction

$$
\begin{pmatrix}
\sum_j x_{ij} \leq C_i \\
0 \leq x_{ij} \leq K_{ij}w_{ij}, \text{ all } j \\
z_i \geq f_i + \sum_j c_{ij}w_{ij} \\
w_{ij} \in \mathbb{Z}, \text{ all } j
\end{pmatrix}
\vee
\begin{pmatrix}
x_{ij} = 0, \text{ all } j \\
z_i \geq 0
\end{pmatrix}
\qquad (7.112)
$$

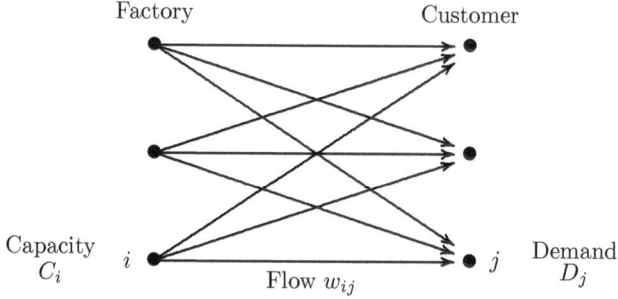

Fig. 7.17 A facility location problem.

The alternative on the left corresponds to installing a factory at location i. The first constraint enforces the factory's capacity limit, and the second does the same for the truck capacities. The third constraint computes the cost incurred at location i. Note that each w_{ij} is integer valued, which means that this disjunct describes a mixed-integer polyhedron. The disjunct on the right corresponds to the case in which no factory is installed at location i.

Customer demand can be satisfied by imposing the constraint $\sum_i x_{ij} \geq D_j$ for each customer j. Each of these constraints can be viewed as a separate disjunction with only one alternative. The objective is to minimize total cost, given by $\sum_i z_i$.

An MILP formulation exists if two disjuncts in (7.112) have the same recession cone. As it happens, they do not. The cone for the first polyhedron is $\{(x_i, w_i, z_i) \mid x_i = 0,\ w_i \geq 0,\ z_i \geq \sum_j c_{ij} w_{ij}\}$ where $x_i = (x_{i1}, \ldots, x_{in})$ and $w_i = (w_{i1}, \ldots, w_{in})$, while the cone for the second is $\{(x_i, w_i, z_i) \mid x_i = 0,\ z_i \geq 0\}$. The cones can be equalized if the innocuous constraints $w_{ij} \geq 0$ and $z_i \geq \sum_j c_{ij} w_{ij}$ are added to the second disjunct. This yields a disjunction that can be given an MILP model:

$$
\begin{pmatrix}
\sum_j x_{ij} \leq C_i \\
0 \leq x_{ij} \leq K_{ij} w_{ij},\ \text{all } j \\
z_i \geq f_i + \sum_j c_{ij} w_{ij} \\
w_{ij} \in \mathbb{Z},\ \text{all } j
\end{pmatrix}
\vee
\begin{pmatrix}
x_{ij} = 0,\ \text{all } j \\
w_{ij} \geq 0,\ \text{all } j \\
z_i \geq \sum_j c_{ij} w_{ij}
\end{pmatrix}
\tag{7.113}
$$

Using (7.97), the convex hull formulation of (7.113) is

$$x_{ij} = x_{ij}^1 + x_{ij}^2, \quad w_{ij} = w_{ij}^1 + w_{ij}^2, \quad z_i = z_i^1 + z_i^2, \quad \text{all } j$$

$$\sum_j x_{ij}^1 \leq C_i y_i \qquad\qquad x_{ij}^2 = 0,\ w_{ij}^2 \geq 0 \ \text{all } j$$

$$\qquad\qquad\qquad\qquad\qquad z_i^2 \geq \sum_j c_{ij} w_{ij}^2$$

$$0 \leq x_{ij}^1 \leq K_{ij} w_{ij}^1,\ \text{all } j$$

$$z_i^1 \geq f_i y_i + \sum_j c_{ij} w_{ij}^1 \qquad y_i \in \{0, 1\},\ w_{ij} \in \mathbb{Z},\ \text{all } j$$

Because the auxiliary 0-1 variables corresponding to the two disjuncts sum to one, they can be written as y_i and $1 - y_i$; the latter does not appear because the right-hand sides in the second disjunct are all zero. The constraints $x_{ij}^2 = 0$ can be dropped (along with the aggregation constraints $x_{ij} = x_{ij}^1 + x_{ij}^2$) if x_{ij}^1 is replaced by x_{ij}, and similarly for the constraints $w_{ij}^2 \geq 0$ and $z_i^2 \geq \sum_j c_{ij} w_{ij}^2$. Because z_i can be replaced by $f_i y_i + \sum_j c_{ij} w_{ij}$ in the objective function $\sum_j z_j$, the complete MILP model becomes

$$\min \sum_i \left(f_i y_i + \sum_j c_{ij} w_{ij} \right)$$

$$\sum_j x_{ij} \leq C_i y_i, \quad \text{all } i$$

$$0 \leq x_{ij} \leq K_{ij} w_{ij}, \quad \text{all } i, j \tag{7.114}$$

$$\sum_i x_{ij} \geq D_j, \quad \text{all } j$$

$$y_i \in \{0, 1\}, \quad w_{ij} \in \mathbb{Z}, \text{ all } i, j$$

Although each disjunction (7.113) is given a convex hull formulation in the MILP model (7.114), the model as a whole is not a convex hull formulation of the problem.

Using (7.100), the big-M model for the disjunction (7.113) is

$$\sum_j x_{ij} \leq C_i + M_{1i}^1(1 - y_i) \qquad\qquad 0 \leq x_{ij} \leq M_{1ij}^2 y_i, \quad \text{all } i, j$$

$$\qquad\qquad\qquad\qquad\qquad\qquad\qquad w_{ij} \geq -M_{2ij}^2 y_i \quad \text{all } i, j$$

$$0 \leq x_{ij} \leq K_{ij} w_{ij} + M_{2ij}^1(1 - y_i), \text{ all } j$$

$$z_i \geq f_i y_i + \sum_j c_{ij} w_{ij} - M_{3i}^1(1 - y_i) \qquad z_i^2 \geq \sum_j c_{ij} w_{ij} - M_{3i}^2 y_i, \quad \text{all } i$$

$$y_i \in \{0, 1\}, \quad w_{ij} \in \mathbb{Z}, \text{ all } j$$

$$\tag{7.115}$$

It can be verified from (7.101) that $M_{1i}^1 = -C_i$, $M_{1ij}^2 = C_i$, and all the other big-Ms are zero in the sharp formulation. The big-M formulation (7.115) therefore reduces to the same model as the convex hull formulation.

Lot Sizing

A *lot-sizing problem with setup costs* illustrates logical variables that interrelate. There is a demand D_t for a product in each period t. No more than C_t units of the product can be manufactured in period t, and any excess over demand is stocked to satisfy future demand. If there is no production in the previous period, then a setup cost of f_t is incurred. The unit production cost is p_t, and the unit holding cost per period is h_t. A starting stock level is given. The objective is to choose production levels in each period so as to minimize total cost over all periods.

Let x_t be the production level chosen for period t and s_t the stock level at the end of the period. In each period t, there are three discrete options to choose from: (1) start producing (with a setup cost), (2) continue producing (with no setup cost), and (3) produce nothing. If v_t is the setup cost incurred in period t, these correspond respectively to the three disjuncts

$$\begin{pmatrix} v_t \geq f_t \\ 0 \leq x_t \leq C_t \end{pmatrix} \vee \begin{pmatrix} v_t \geq 0 \\ 0 \leq x_t \leq C_t \end{pmatrix} \vee \begin{pmatrix} v_t \geq 0 \\ x_t = 0 \end{pmatrix} \qquad (7.116)$$

In addition, there are logical connections between the choices in consecutive periods

$$\begin{array}{l} \text{(2) in period } t \Rightarrow \text{(1) or (2) in period } t-1 \\ \text{(1) in period } t \Rightarrow \text{neither (1) nor (2) in period } t-1 \end{array} \qquad (7.117)$$

The inventory balance constraints are

$$s_{t-1} + x_t = D_t + s_t, \quad s_t \geq 0, \quad t = 1, \ldots, n \qquad (7.118)$$

where s_t is the stock level in period t and s_0 is given. The problem is to minimize

$$\sum_{t=1}^{n} (p_t x_t + h_t s_t + v_t) \qquad (7.119)$$

subject to (7.116)–(7.118).

A convex hull relaxation for (7.116) is

$$\begin{array}{ccc} v_t = \displaystyle\sum_{k=1}^{3} v_t^k, & x_t = \displaystyle\sum_{k=1}^{3} x_t^k, & y_t = \displaystyle\sum_{k=1}^{3} y_{tk} \\[2mm] v_t^1 \geq f_t y_{t1}, & v_t^2 \geq 0, & v_t^3 \geq 0 \\[2mm] 0 \leq x_t^1 \leq C_t y_{t1}, \; 0 \leq x_t^2 \leq C_t y_{t2}, \; x_t^3 = 0 \\[2mm] y_{tk} \geq 0, \quad k = 1, 2, 3 \end{array} \qquad (7.120)$$

To simplify (7.120), define new variables $w_t = y_{t1}$ and $y_t = y_{t2}$ so that $w_t + y_t \leq 1$. Thus, $w_t = 1$ indicates a startup and $y_t = 1$ indicates continued production in period t, while $w_t = y_t = 0$ indicates no production. Since $x_t^3 = 0$, one can set $x_1 = x_t^1 + x_t^2$, which allows the two capacity constraints in (7.120) to be replaced by $0 \leq x_t \leq C_t(w_t + y_t)$. Finally, v_t can replace v_t^1, because v_t is being minimized and v_t^2 and v_t^3 do not appear. The convex hull relaxation (7.120) becomes

$$v_t \geq f_t w_t, \quad w_t \geq 0, \quad y_t \geq 0 \qquad (a)$$
$$0 \leq x_t \leq C_t(w_t + y_t), \quad w_t + y_t \leq 1 \qquad (b)$$

$$(7.121)$$

The logical constraints (7.117) can be formulated

$$y_t \leq w_{t-1} + y_{t-1} \qquad (a)$$
$$w_{t-1} + w_t \leq 1, \quad y_{t-1} + w_t \leq 1 \qquad (b)$$

$$(7.122)$$

The set packing inequalities in (7.122b) and $w_t + y_t \leq 1$ from (7.121b) can be replaced by the clique inequality

$$w_{t-1} + y_{t-1} + y_t \leq 1 \qquad (7.123)$$

The entire problem can now be formulated as minimizing (7.119) subject to (7.119) along with (7.121a), (7.122a), (7.123), and $w_t, y_t \in \{0, 1\}$ for all $t \geq 1$.

7.6.4 Knapsack Models

A large class of mixed-integer models are based on counting ideas, which can be expressed as knapsack inequalities. Recall that a knapsack covering inequality has the form $ax \geq \alpha$, where each $a_j > 0$ and each x_j must take nonnegative integer values, while a knapsack packing inequality has the form $ax \leq \alpha$. If x_j is interpreted as the number of items of type j chosen for some purpose, the left-hand side of a knapsack inequality counts the total number of items selected (perhaps weighting some more than others), and the right-hand side places a bound on the total number or weight. Knapsack inequalities are therefore useful for formulating problems in which some item count must be bounded.

Logic, Covering, and Packing Constraints

Perhaps the simplest problem of this sort is a disjunction of Boolean variables, such as $\bigvee_{j \in J} x_j$, where each $x_j \in \{0, 1\}$. Since this logical clause states that at least one variable is true, it can be written as an inequality $\sum_{j \in J} x_j \geq 1$. This a special case of a knapsack inequality whose left-hand side counts the number of true variables, and whose right-hand side places a lower bound of 1 on the count. Section 7.7 discusses relaxations for this and other types of logical propositions.

A similar pattern occurs in set-covering, set-packing, and set-partitioning problems. A *set-covering problem* begins with a finite collection $\{S_j \mid j \in J\}$ of sets and a finite set $T \subset \bigcup_{j \in J} S_j$. It seeks the smallest subcollection of sets that covers T; that is, the smallest $J' \subset J$ for which $T \subset \bigcup_{j \in J'} S_j$. For example, if one wants to play a certain set T of songs at a party and S_j is the set of songs on compact disk j, then one might wish to buy the smallest number of compact disks that provide the desired songs. Other applications include crew scheduling, facility location, assembly line balancing, and simplification of Boolean functions.

It is easy to formulate a 0-1 model for the set covering problem. Let i index the elements of T, and let $a_{ij} = 1$ when $i \in S_j$, and $a_{ij} = 0$ otherwise. If binary variable $x_j = 1$ when S_j is selected, the problem is to minimize $\sum_j x_j$ (or, if desired, $\sum_j c_j x_j$) subject to the knapsack covering inequality $\sum_j a_{ij} x_j \geq 1$ for each i.

The *set-packing problem* seeks the largest subcollection of sets S_j for which each element in T appears in at most one set. The problem is therefore to maximize $\sum_j c_j x_j$ subject to the knapsack packing inequality $\sum_i a_{ij} x_j \leq 1$ for each i. The *set-partitioning problem* seeks to maximize or minimize $\sum_j c_j x_j$ subject to $\sum_i a_{ij} x_j = 1$, which implies both a covering and a packing inequality.

Other knapsack-like constraints include logical constraints that involve a counting element. A *cardinality clause*, for example, requires that at least k of the Boolean variables in $\{x_j \mid j \in J\}$ be true (Section 6.5.3). It can, of course, be formulated with the knapsack covering inequality $\sum_{j \in J} x_j \geq k$, which along with $0 \leq x_j \leq 1$ provides a convex hull relaxation. A further generalization is the *cardinality conditional*, which states that if k of the variables in $\{x_j \mid i \in I\}$ are true, then at least ℓ of the variables in $\{x_j \mid j \in J\}$ are true. See Chapter 8 for more information on these constraints.

Two obvious opportunities for knapsack formulations are the *knapsack covering* and *knapsack packing* problems discussed in Section 2.3. The *capital budgeting problem* has the same structure as a 0-1 knapsack packing problem. In this problem, there are several possible investments, and each investment j incurs a capital cost of a_j. The return from investment j is c_j, and the objective is to select investments so as to maximize return while staying with the available investment funds α. Thus, the problem is to maximize $\sum_j c_j x_j$ subject to $\sum a_j x_j \leq \alpha$, where $x_j \in \{0, 1\}$ and $x_j = 1$ indicates that investment j is selected.

Example: Package Delivery

A final example illustrates how the approach presented here can result in a formulation that is superior to the standard formulation. A collection of packages are to be delivered by several trucks, and each package j has size a_j. Each available truck i has capacity Q_i and costs c_i to operate. The problem is to decide which trucks to use, and which packages to load on each truck, to deliver all the items at minimum cost.

The problem can be modeled by analyzing it as a combination of knapsack and disjunctive ideas. The decision problem consists of two levels: the choice of which trucks to use, followed by the choice of which packages to load on each truck. The trucks selected must provide sufficient capacity, which leads naturally to a 0-1 knapsack constraint:

$$\sum_{i=1}^{m} Q_i y_i \geq \sum_{j=1}^{n} a_j, \tag{7.124}$$

where each $y_i \in \{0, 1\}$ and $y_i = 1$ when truck i is selected.

The secondary choice of which packages to load on truck i depends on whether that truck is selected. This suggests a disjunction of two alternatives. If the truck i is selected, then a cost c_i is incurred, and the items loaded must fit into the truck (a 0-1 knapsack constraint). If truck i is not selected, then no items can be loaded (another knapsack constraint). The disjunction is

$$\begin{pmatrix} z_i \geq c_i \\ \sum_{j=1}^{n} a_j x_{ij} \leq Q_i \\ 0 \leq x_{ij} \leq 1, \text{ all } j \\ x_{ij} \in \mathbb{Z}, \text{ all } j \end{pmatrix} \vee \begin{pmatrix} z_i \geq 0 \\ x_{ij} = 0, \text{ all } j \end{pmatrix} \tag{7.125}$$

where $z_i \geq 0$ is the fixed cost incurred by truck i, and $x_{ij} = 1$ when package j is loaded into truck i. The feasible set is the union of two mixed-integer polyhedra, which have the same recession cone. If $y_i = 1$ when the first disjunct is enforced, the convex hull formulation of (7.125) is

$$z_i \geq c_i y_i \qquad (a)$$

$$\sum_{j=1}^{n} a_j x_{ij} \leq Q_i y_i \qquad (b) \qquad\qquad (7.126)$$

$$x_{ij} \leq y_i, \text{ all } j \qquad (c)$$

$$y_i, x_{ij} \in \{0,1\}, \text{ all } j$$

Finally, to make sure that each packaged must be shipped, a set of knapsack constraints is imposed:

$$\sum_{i=1}^{m} x_{ij} \geq 1, \text{ all } j \qquad\qquad (7.127)$$

Because (7.124) and (7.127) can be viewed as disjunctions having one disjunct, the problem consists of disjunctions of mixed-integer systems. One can now minimize total cost $\sum_i z_i$ subject to (7.124) and (7.127), as well as (7.126) for all i.

This formulation differs in two ways from a formulation that one might initially write for this problem. One might omit the constraint (7.126c) because the model is correct without it. Yet this constraint makes the relaxation tighter. Also, one might not include (7.124) because, due to (7.127), it is the sum of constraints (7.126b) over all i. Yet the presence of (7.124) allows the solver to deduce lifted knapsack cuts, which create a tighter continous relaxation and result in much faster solution.

Exercises

7.47. Show that the disjunctive representation (7.97) does not correctly model the union of the line segment $\{(0, x_2) \mid 0 \leq x_2 \leq 1\}$ with the ray $\{(x_1, 0) \mid x_1 \geq 0\}$, and explain why. Show that the continuous relaxation of (7.97) nonetheless describes the closure of the convex hull. Truncate the ray to $\{(x_1, 0) \mid 0 \leq x_1 \leq M\}$ in order to make the union MILP representable, and write an MILP model for it.

7.48. Use the disjunctive representation (7.97) to write an MILP model of the union of the cone $\{(x_1, x_2) \mid 0 \leq x_1 \leq x_2\}$ with the cone $\{(x_1, x_2) \mid 1 \leq x_1 \leq x_2 + 1\}$. Note that the model can be simplified so that it contains only the original variables x_1, x_2 and $\alpha \in \{0, 1\}$.

7.49. Use the disjunctive representation (7.97) to write a mixed-integer model of the union of the cone $\{(x_1, x_2) \mid 0 \leq x_1 \leq x_2\}$ with the cone

$\{(x_1, x_2) \mid 1 \le x_1 \le 2x_2 + 1\}$. Show that the model does not correctly represent the union. Why?

7.50. In a scheduling problem, jobs 1 and 2 both have duration 1. Let x_j be the start time of job j. Since the jobs cannot overlap,

$$[x_2 \ge x_1 + 1] \vee [x_1 \ge x_2 + 1]$$

Show that this disjunction has no MILP model. Place an upper bound U on x_1, x_2 and write an MILP representation (7.97) of the resulting disjunction. Draw a graph of the feasible set and its convex hull. Note that because U is typically much larger than 1, the convex hull relaxation is quite weak. The overhead of an MILP formulation may therefore be unjustified.

7.51. Write the tightest possible big-M model of the scheduling disjunction of Exercise 7.50 by using the big-Ms in (7.70). Project its continuous relaxation onto x_1, x_2 and draw a graph. Note that the big-M model is useless for obtaining a relaxation because its continuous relaxation is redundant of the bounds $x_j \in [0, U]$. However, the relaxation is somewhat better if there are time windows. Write the tightest possible big-M relaxation if job j has release time r_j and deadline d_j.

7.52. Let $x \ge 0$ be a continuous variable and y a 0-1 variable. Identify which of the following conditions has an MILP formulation, and explain why. *Hint:* Consider the feasible set in x-y space.

1. $x = 0$ if $y = 0$.
2. $x = 0$ if $y = 0$, and $x \le M$ otherwise.
3. $y = 0$ if $x = 0$.

7.53. There are n manufacturing plants that must produce a total of at least R widgets in a given period. If plant j operates at level A, it produces a_j widgets at cost f_j. If it operates at level B, it produces b_j widgets at cost g_j. The plant can also be shut down, in which case it incurs no cost. Write a factored disjunctive model that minimizes cost subject to the output requirement, and simplify it as much as possible.

7.54. Suppose in Exercise 7.53 that when a plant operates at level A, its output must be in the interval $[a_j, A_j]$, and it incurs cost $f_j + c_j x_j$. When it operates at level B, $x_j \in [b_j, B_j]$, and the cost is $g_j + d_j x_j$. Write a factored disjunctive model.

7.55. The classical assignment problem minimizes $\sum_i c_{ix_i}$ subject to the constraint $\mathrm{alldiff}(x_1, \ldots, x_m)$, where each $x_i \in \{1, \ldots, n\}$ and $m \leq n$. Show that the standard 0-1 model

$$\min_{y_{ij} \in \{0,1\}} \left\{ \sum_{ij} c_{ij} y_{ij} \,\middle|\, \sum_j y_{ij} = 1, \text{ all } i; \sum_i y_{ij} \leq 1, \text{ all } j \right\}$$

can be obtained by disjunctive modeling. Since the coefficient matrix is totally unimodular, $y_{ij} \in \{0,1\}$ can be replaced by $0 \leq y_{ij} \leq 1$. *Hint:* There are two sets of decisions. For each i, write a disjunction expressing the choice of value for x_i, using 0-1 variables y_{ij}. For each j, write a disjunction expressing the choice of which x_i will have value j (if any), using 0-1 variables $\bar{y}_{ji}$. Add the logical constraint that $y_{ij} = \bar{y}_{ji}$ and simplify.

7.56. The quadratic assignment problem minimizes $\sum_{ij} c_{ij} d_{x_i x_j}$ subject to $\mathrm{alldiff}(x_1, \ldots, x_m)$. It can be viewed as assigning facilities i to locations x_i to minimize the cost of traffic between all pairs of facilities, where c_{ij} is the cost per unit distance of traffic between i and j, and $d_{k\ell}$ is the distance from location k to location ℓ. A quadratic 0-1 model can be written

$$\min_{y_{ik} \in \{0,1\}} \left\{ \sum_{ijk\ell} c_{ij} d_{k\ell} y_{ik} y_{j\ell} \,\middle|\, \sum_k y_{ik} = 1, \text{ all } i; \sum_i y_{ik} \leq 1, \text{ all } k \right\}$$

Use disjunctive reasoning similar to that of the previous exercise to obtain a linear 0-1 model. *Hint:* For each pair i, j ($i \neq j$), formulate a disjunction, and similarly for each pair k, ℓ ($k \neq \ell$).

7.57. In a disjunctive scheduling problem, each of n jobs j has a duration p_j, release time r_j, and deadline d_j. Only one job can run at a time. The schedule can be viewed as a sequence of events k, each of which is the start of some job. Thus, one can write a disjunction of n alternatives for each job j, corresponding to the n events that might be the start of j. Let t_k be the time at which event k occurs, and write a factored disjunctive model for the problem. Use assignment constraints to make sure that no two jobs are assigned to the same event.

7.58. In a cumulative scheduling problem, each of n jobs j has a duration p_j, release time r_j, deadline d_j, and rate of resource consumption c_j. Jobs can run simultaneously as long as the total rate of resource consumption at any time is at most C. The schedule can be viewed as a sequence of $2n$ events, each of which is the start or end of some job. So, for each job j, one can write a disjunction of alternatives corresponding to the pairs k, k' of events ($k < k'$) that might represent the start and end of job j. Let t_k be the time at which event k occurs, and let z_k be the total rate of resource consumption immediately after event k occurs. Thus, when event k is the start of job j, $z_k = z_{k-1} + c_j$, and similarly when event k is the end of a job. Write a factored disjunctive model, using assignment constraints to make sure that no two jobs involve the same event.

7.7 Propositional Logic

The Boolean values *true* and *false* have no structure, aside from the fact that there are two of them. However, if they are interpreted as the numbers 1 and 0, the rich structure of the unit hypercube becomes available for building continuous relaxations of logical propositions.

The most straightforward way to relax a logical formula is to convert it to a set of logical clauses (Section 6.4.1), since each logical clause can be written as a 0-1 linear inequality. However, relaxations of the individual clauses may provide a weak relaxation for the clause set as a whole. Three strategies can address this problem:

- develop tight relaxations for elementary logical formulas other than clauses
- process the clause set with some version of the resolution algorithm to obtain a tighter relaxation (this is effect generates cutting planes)
- derive separating cuts for the clause set.

7.7.1 Common Logical Formulas

Certain logical expressions tend to occur frequently, and it is useful to build a catalog of convex hull relaxations for them.

Logical clauses are perhaps the most basic logical expression, because any logical formula can be converted to clausal form, and a logical clause has an obvious 0-1 representation. For example, the clause $x_1 \vee x_2 \vee \neg x_3$ can be represented as the 0-1 linear inequality $x_1 + x_2 + (1 - x_3) \geq 1$, which is relaxed by replacing $x_j \in \{0, 1\}$ with $0 \leq x_j \leq 1$. It is convenient to call this the *clausal inequality* that corresponds to the clause $x_1 \vee x_2 \vee \neg x_3$.

More generally, a logical clause C may be written

$$\bigvee_{j \in P} x_j \vee \bigvee_{j \in N} \neg x_j \tag{7.128}$$

It is relaxed by the corresponding clausal inequality and bounds on x_j:

$$\sum_{j \in P} x_j + \sum_{j \in N} (1 - x_j) \geq 1, \quad 0 \leq x_j \leq 1 \text{ for } j \in P \cup N \tag{7.129}$$

It is convenient to denote the clausal inequality for C by linear(C).

Theorem 7.26. *The system* (7.129) *is a convex hull relaxation of the logical clause* (7.128).

Proof. Suppose, without loss of generality, that all the literals in (7.128) are positive, so that $P = \{1, \ldots, n\}$ and $N = \emptyset$ (negated variables can be complemented). Let S be the polyhedron described by (7.129). Since the integral solutions of (7.129) are the feasible solutions of (7.128), it suffices to show that S is an integral polyhedron. Since the right-hand sides of (7.129) are integral, for this it suffices by Theorem 7.11 to show that the coefficient matrix for (7.129) is totally unimodular. The matrix has the form

$$\begin{bmatrix} e \\ -I \end{bmatrix}$$

where e is a row of ones and I is the identity matrix. This matrix satisfies the condition of Corollary 7.13 and is therefore totally unimodular. $\square$

Table 7.1 displays convex hull relaxations of several common logical formulas. Formula 1 is covered by Theorem 7.26. Formula 2 is equivalent to clause $\neg x_1 \vee x_2$ and therefore has the convex hull relaxation $(1 - x_1) + x_2 \geq 1$ (plus bounds). Formula 3 is also a clause. The relaxation given for Formula 4 describes the convex hull because it is totally unimodular (left as an exercise). Formula 5 is the result of complementing every variable in Formula 4, and so its convex hull relaxation can be derived from that of Formula 4. The relaxation of Formula 7, however, is not totally unimodular, and a different sort of proof is required.

Theorem 7.27. *A convex hull relaxation of the formula*

$$\bigvee_{j \in J_1} x_j \equiv \bigvee_{j \in J_2} x_j \tag{7.130}$$

is given by

$$x_i \leq \sum_{j \in J_2} x_j, \quad \text{all } i \in J_1 \qquad (a)$$

$$x_i \leq \sum_{j \in J_1} x_j, \quad \text{all } i \in J_2 \qquad (b) \tag{7.131}$$

$$0 \leq x_j \leq 1, \quad \text{all } j \in J_1 \cup J_2 \qquad (c)$$

Table 7.1 Convex hull relaxations of some common propositional formulas. The index sets J_1, J_2 are disjoint. The bounds $0 \leq x_j \leq 1$ are understood to be part of the relaxation.

Formula	Convex hull relaxation
1. $\bigvee_{j \in J} x_j$	$\sum_{j \in J} x_j \geq 1$
2. $x_1 \rightarrow x_2$	$x_1 \leq x_2$
3. $\bigwedge_{j \in J_1} x_j \rightarrow \bigvee_{j \in J_2} x_j$	$\sum_{j \in J_1} x_j \leq \sum_{j \in J_2} x_j + |J_1| - 1$
4. $\bigvee_{j \in J_1} x_j \rightarrow \bigvee_{j \in J_2} x_j$	$x_i \leq \sum_{j \in J_2} x_j, \text{ all } i \in J_1$
5. $\bigwedge_{j \in J_1} x_j \rightarrow \bigwedge_{j \in J_2} x_j$	$\sum_{j \in J_1} x_j \leq x_i + |J_1| - 1, \text{ all } i \in J_2$
6. $x_1 \equiv x_2$	$x_1 = x_2$
7. $\bigvee_{j \in J_1} x_j \equiv \bigvee_{j \in J_2} x_j$	$x_i \leq \sum_{j \in J_2} x_j, \text{ all } i \in J_1$
	$x_i \leq \sum_{j \in J_1} x_j, \text{ all } i \in J_2$
8. $\bigwedge_{j \in J_1} x_j \equiv \bigwedge_{j \in J_2} x_j$	$\sum_{j \in J_2} x_j \leq x_i + |J_2| - 1, \text{ all } i \in J_1$
	$\sum_{j \in J_1} x_j \leq x_i + |J_1| - 1, \text{ all } i \in J_2$
9. $\bigwedge_{j \in J_1} x_j \equiv \bigvee_{j \in J_2} x_j$	$\sum_{j \in J_1} x_j \leq \sum_{j \in J_2} x_j + |J_1| - 1$
	$x_i \leq \sum_{j \in J_2} x_j, \text{ all } i \in J_1$
10. $(x_1 \equiv x_2) \equiv x_3$	$x_1 + x_2 + x_3 \geq 1$
	$x_2 + x_3 \leq x_1 + 1$
	$x_1 + x_3 \leq x_2 + 1$
	$x_1 + x_2 \leq x_3 + 1$
11. $(x_1 \vee x_2) \wedge$ $(x_1 \vee x_3) \wedge$ $(x_2 \vee x_3)$	$x_1 + x_2 + x_3 \geq 2$
12. $(x_1 \vee x_2 \vee x_3) \wedge$ $(x_1 \vee x_4) \wedge$ $(x_2 \vee x_4) \wedge$ $(x_3 \vee x_4)$	$x_1 + x_2 + x_3 + 2x_4 \geq 3$

Proof. It suffices to show that all vertices of the polyhedron described by (7.131) are integral. The vertices can be enumerated by solving all linearly independent subsets S of $|J_1| + |J_2|$ of the following equations:

$$
\begin{aligned}
x_i &= \sum_{j \in J_2} x_j, \quad \text{all } i \in J_1 \ (i) \\
x_i &= \sum_{j \in J_1} x_j, \quad \text{all } i \in J_2 \ (ii) \\
x_j &= 0, \quad \text{all } j \in J_1 \cup J_2 \ (iii) \\
x_j &= 1, \quad \text{all } j \in J_1 \qquad (iv) \\
x_j &= 1, \quad \text{all } j \in J_2 \qquad (v)
\end{aligned}
\tag{7.132}
$$

Case I. S consists of the equations (i) and (ii). The coefficient matrix for these equations has the form

$$
\begin{bmatrix} -I & E \\ E & -I \end{bmatrix}
\tag{7.133}
$$

where E is a matrix of all ones, and it is clearly nonsingular. Since the right-hand side is zero, the solution is $x = 0$, which is integral.

Case II. S contains at least one of the equations (iv), but none of the equations (v). Since $x_j = 1$ for some $j \in J_1$, the constraints (7.131b) are redundant, and the equations (ii) need not be considered for S. The coefficient matrix of S is therefore some subset of rows of

$$
A = \begin{bmatrix} -I & E \\ I & 0 \\ 0 & I \\ I & 0 \end{bmatrix}
$$

But A is totally unimodular, due to Theorem 7.12. Given any subset $\bar{J}$ of columns of A, the desired partition of $\bar{J}$ can be obtained by placing the last $\lceil |J_2 \cap \bar{J}|/2 \rceil$ columns in one set and the remaining columns in the other. Thus, any solution obtained is integral.

Case III. S contains at least one of the equations (v), but none of the equations (iv). This is analogous to case II.

Case IV. S contains at least one of the equations (iv) and at least one of the equations (v). Since $x_j = 1$ for some $j \in J_1$ and for some

$j \in J_2$, constraints (7.131a) and (7.131b) are redundant, and equations (i) and (ii) need not be considered for S. Any independent set of the remaining equations has an integral solution.

Case V. S contains at least one of the equations (iii), but none of the equations (iv) and (v). In this case, one can eliminate the variables set to zero, and the resulting system of equations again has a coefficient matrix of the form (7.133). Since the right-hand side is zero, the solution is $x = 0$ and therefore integral. $\square$

Because formula 8 is the result of complementing all variables in formula 7, its convex hull relaxation can be obtained from that of formula 7. A special case of formula 8, namely $(x_1 \wedge x_2) \equiv x_3$, occurs frequently in mixed-integer modeling because it expresses such relations as $x_1 x_2 = x_3$. Another valid MILP model for this relation is

$$x_3 \geq x_1 + x_2 - 1$$
$$x_1 + x_2 \geq 2x_3$$

This is not a convex hull formulation, however, because the nonintegral point $x = (1, 0, \frac{1}{2})$, for instance, is a vertex of the polyhedron it defines but does not lie in the convex hull. As for formula 9, the proof of its convex hull formulation is similar to that for formula 7.

Every inequality listed in Table 7.1 defines a facet of the associated convex hull; that is, an $(n-1)$-dimensional face of the convex hull. The inequality $\sum_j x_j \geq 1$, for example, defines a facet of the convex hull of $\bigvee_j x_j$ because n affinely independent[3] points in the convex hull lie on the hyperplane defined by $\sum_j x_j = 1$. In particular, the affinely independent points $e^1, \ldots, e^n$ lie on the hyperplane, where e^j is the jth unit vector.

The facet-defining inequality for formula 11 is the smallest example (in terms of the number of variables) of a nonclausal facet-defining inequality for a set of 0-1 points. The inequality for formula 12 is the smallest example of a facet-defining inequality with a coefficient not in $\{0, 1, -1\}$.

---

[3] Points $a^1, \ldots, a^n$ are affinely independent when $a^2 - a^1, \ldots, a^n - a^1$ are linearly independent.

7.7.2 Resolution as a Tightening Technique

Any formula of propositional logic can be written as a clause set S, which can, in turn, be given a factored relaxation. This relaxation, denoted linear(S), contains the inequality linear(C) corresponding to each clause C in S, along with bounds $0 \leq x_j \leq 1$. Although this relaxation tends to be weak, it can generally be tightened by first applying some form of resolution (Section 6.4.2) to S. If S' is the result of applying resolution to S, then linear(S') is generally tighter than linear(S).

Resolution can clearly tighten a relaxation. The clause set S consisting of $x_1 \lor x_2$ and $x_1 \lor \neg x_2$, for example, has the feasible set $\{(1,0),(1,1)\}$. Linear(S) consists of

$$x_1 + x_2 \geq 1, \quad x_1 + (1 - x_2) \geq 1, \quad 0 \leq x_j \leq 1, \ j = 1,2 \qquad (7.134)$$

This is not a convex hull relaxation, because $(x_1, x_2) = (\frac{1}{2}, \frac{1}{2})$ satisfies linear(S) but does not belong to the convex hull of the feasible set, which is the line segment connecting the two feasible points. However, applying the resolution algorithm to S yields the single clause x_1, and linear($\{x_1\}$) is

$$x_1 \geq 1, \quad 0 \leq x_j \leq 1, \ j = 1,2$$

This is a convex hull relaxation and therefore tighter than linear(S).

A resolvent always tightens the factored relaxation of its parents because it is a rank 1 Chvátal–Gomory cut (Section 7.2.1). To put this more precisely,

Theorem 7.28. *Given two clauses C, D with resolvent R, linear(R) is a rank 1 Chvátal–Gomory cut for linear($\{C, D\}$).*

For example, clause (c) below is the resolvent R of (a) and (b):

$$
\begin{array}{ll}
x_1 \lor x_2 \lor x_3 & (a) \\
\neg x_1 \lor x_2 \quad\quad \lor \neg x_4 & (b) \\
\quad\quad x_2 \lor x_3 \lor \neg x_4 & (c)
\end{array}
$$

Linear(R) is $x_2 + x_3 + (1 - x_4) \geq 1$, which can be obtained by computing the weighted sum

$$
\begin{aligned}
x_1 + x_2 + x_3 &\geq 1 \; (\tfrac{1}{2}) \\
-x_1 + x_2 \quad - x_4 &\geq -1 \; (\tfrac{1}{2}) \\
x_3 &\geq 0 \; (\tfrac{1}{2}) \\
- x_4 &\geq -1 \; (\tfrac{1}{2}) \\
\hline
x_2 + x_3 - x_4 &\geq -\tfrac{1}{2}
\end{aligned}
$$

using the weights indicated on the right (note that the constants on the left-hand side of the clausal inequalities are moved to the right). By rounding up the $-\tfrac{1}{2}$ on the right-hand side of the weighted sum, one obtains the rank 1 Chvátal–Gomory cut $x_2 + x_3 - x_4 \geq 0$, which is linear(R).

In general, one can obtain linear(R) for the resolvent R of clauses C and D by taking a nonnegative linear combination of linear(C), linear(D), bounds $x_j \geq 0$ for all literals x_j that occur in C or D but not both, and bounds $-x_j \geq -1$ for all $\neg x_j$ that occur in C or D but not both. Weight $\tfrac{1}{2}$ is assigned to each of these, and the right-hand side of the result is rounded up.

Each iteration of the resolution algorithm generates additional cuts. The algorithm does not in general result in a convex hull relaxation, however. For example, the clause set S consisting of $x_1 \vee x_2$, $x_1 \vee x_3$, $x_2 \vee x_3$ is unchanged by the resolution algorithm. Yet the point $(x_1, x_2, x_3) = (\tfrac{1}{2}, \tfrac{1}{2}, \tfrac{1}{2})$ satisfies linear(S):

$$
x_1 + x_2 \geq 1, \quad x_1 + x_3 \geq 1, \quad x_2 + x_3 \geq 1, \quad 0 \leq x_j \leq 1, \; j = 1, 2, 3
$$

and does not lie within the convex hull of the feasible set

$$
\{(0, 1, 1), (1, 0, 1), (1, 1, 0), (1, 1, 1)\}
$$

Although resolution does not generally yield a convex hull relaxation, it can nonetheless tighten the relaxation considerably. When the clause set is large, the full resolution algorithm may consume too much time and it may be advantageous to use an incomplete form of resolution.

One incomplete form, unit resolution (Section 6.4.3), is very fast but unhelpful. It is easy to see that if R is the resolvent of a unit clause U with another clause C, then linear(R) is simply the sum of linear(U) and linear(C). Linear($\{U, C, R\}$) therefore describes the same polyhedron as linear($\{U, C\}$). This means that applying the unit resolution algorithm to a clause set S has no effect on the tightness of linear(S).

Input resolution, which lies between full resolution and unit resolution in strength, is a more attractive alternative. The input resolution algorithm is similar to the full resolution algorithm, except that at least one parent of each resolvent must belong to the original clause set. Input resolution therefore requires less computation than full resolution, but tightens the factored relaxation somewhat.

Some indication of how much input resolution tightens the relaxation is provided by an interesting fact that will be established in Section 7.7.4: input resolution generates all rank 1 Chvátal–Gomory cuts that have clausal form. That is, applying the input resolution algorithm to a clause set S generates precisely the clauses C such that $\text{linear}(C)$ is a rank 1 cut for $\text{linear}(S)$. So applying input resolution to a clause set before relaxing it is tantamount is relaxing it first and then generating all rank 1 clausal cuts.

The input resolution algorithm goes as follows. Let S_0 be the original clause set S. In each iteration $k \geq 1$, let S_k contain every clause R such that (a) R is the resolvent of a clause in S_{k-1} with a clause in S_0, and (b) R is absorbed by no clause in $\bigcup_{i=0}^{k-1} S_i$. The algorithm terminates at step m if S_m is empty, at which point all the clauses in $S' = \bigcup_{i=0}^{m-1} S_i$ are implied by S. One may then delete from S' all clauses that are absorbed by other clauses in S'.

Consider, for example, the clause set $S = S_0$:

$$
\begin{aligned}
\neg x_1 \lor \;\; & x_2 \lor \;\; x_3 \lor x_4 \\
& x_2 \lor \neg x_3 \\
x_1 \quad\quad & \lor \;\; x_3 \lor x_4 \\
\neg x_2 \quad\quad\quad & \lor x_5
\end{aligned}
\tag{7.135}
$$

S_1 contains all resolvents of clauses in S_0:

$$
\begin{aligned}
\neg x_1 \lor x_2 \quad\quad & \lor x_4 \\
x_2 \lor x_3 \lor x_4 & \\
\neg x_1 \quad\quad & \lor x_3 \lor x_4 \lor x_5 \\
x_1 \lor x_2 \quad\quad & \lor x_4 \\
\neg x_3 \lor \quad & \lor x_5
\end{aligned}
\tag{7.136}
$$

S_2 contains all resolvents of clauses in S_1 with those in S_0:

$$\begin{aligned}
\neg x_1 \qquad &\lor x_4 \lor x_5 \\
x_2 \quad &\lor x_4 \\
x_3 \lor x_4 &\lor x_5 \\
\neg x_1 \lor x_2 \quad &\lor x_4 \lor x_5 \\
x_1 \qquad &\lor x_4 \lor x_5
\end{aligned} \qquad (7.137)$$

S_3 contains the single clause

$$x_4 \lor x_5 \qquad (7.138)$$

which is the only resolvent of a clause in S_2 with one in S_0 that is not absorbed by a clause previously generated. Now S' consists of all the clauses (7.135)–(7.138). Most of these clauses are absorbed by others, reducing S' to the set of clauses on the left:

$$\begin{aligned}
x_1 \lor x_3 \lor x_4 \qquad & x_1 + x_2 + x_4 \geq 1 \\
x_2 \lor \neg x_3 \qquad & x_2 + (1 - x_3) \geq 1 \\
x_2 \lor x_4 \qquad & x_2 + x_4 \geq 1 \\
\neg x_3 \lor x_5 \qquad & (1 - x_3) + x_5 \geq 1 \\
x_4 \lor x_5 \qquad & x_4 + x_5 \geq 1 \\
& 0 \leq x_j \leq 1, \quad j = 1, \ldots, 5
\end{aligned} \qquad (7.139)$$

The inequalities on the right comprise linear(S'), which is a tighter relaxation of S than linear(S).

7.7.3 Refutation by Linear Relaxation

Before establishing the connection between input resolution and rank 1 cuts, it is useful to investigate the ability of input and unit resolution to prove infeasibility. This inquiry is also interesting in its own right. It is easy to show that input resolution proves infeasibility precisely when the linear relaxation is infeasible, and similarly for unit resolution. It follows that input resolution and unit resolution have the same power to detect infeasibility.

Let a *unit refutation* of clause set S be a unit resolution proof that S is infeasible, and similarly for an *input refutation*.

Theorem 7.29. *Clause set S has a unit refutation if and only if linear(S) is infeasible.*

Proof. First suppose that S has a unit refutation. It was observed in Section 7.7.2 that adding unit resolvents to S has no effect on the feasible set of linear(S). Because one of the unit resolvents added during the unit refutation is the empty clause, which corresponds to the inequality $0 \geq 1$, linear(S) is infeasible.

For the converse, suppose linear(S) has no unit refutation. Apply the unit resolution algorithm to S to obtain the clause set S'. Make each unit clause in S' true by setting its variable to 0 or 1 as required. Because the remaining variables x_j occur only in clauses C with two or more literals, setting each $x_j = \frac{1}{2}$ satisfies linear(C). Linear(S') is therefore feasible. Since linear(S) has the same feasible set, it is likewise feasible. $\square$

The converse of the following theorem is also true, but its proof must await the next section.

Theorem 7.30. *Clause set S has an input refutation if linear(S) is infeasible.*

Proof. It suffices to show that if S has no input refutation, then linear(S) is feasible. Let S' be the result of applying the input resolution algorithm to S. If the unit clause x_j occurs in S' but the unit clause $\neg x_j$ does not, then set $x_j = 1$. Similarly, if the unit clause $\neg x_j$ occurs in S' but x_j does not, set $x_j = 0$. We know that no unit clause $\neg x_j$ for which x_j is set to 1 occurs in S, because otherwise it would have resolved with x_j to produce an input refutation. Similarly, no unit clause x_j for which x_j is set to 0 occurs in S. Now setting all remaining variables in S to $\frac{1}{2}$ satisfies linear(S). $\square$

7.7.4 Input Resolution and Rank 1 Cuts

Recall that a rank 1 Chvátal–Gomory cut for a 0-1 system $Ax \geq b$ is the result of rounding up the coefficients and right-hand side of some surrogate inequality of $Ax \geq b$. Section 7.7.2 relied on the following theorem:

Theorem 7.31. *The input resolution algorithm applied to a clause set S generates precisely the clauses C for which linear(C) is a rank 1 Chvátal–Gomory cut for linear(S).*

It is convenient to prove the theorem in two parts, one showing that input resolution generates nothing but rank 1 clausal cuts, and the other showing that it generates all of them.

Theorem 7.32. *If the input resolution algorithm applied to S generates clause C, then linear(C) is a rank l cut for linear(S).*

The idea of the proof can be seen in an example. The clause $x_4 \lor x_5$ was derived from the clause set S in (7.135) by the following series of input resolutions:

$$
\begin{array}{ll}
                        & \neg x_1 \lor x_2 \lor x_3 \lor x_4 \\
x_1 \lor x_3 \lor x_4   & x_2 \lor x_3 \lor x_4 \\
x_2 \lor \neg x_3   & x_2 \lor x_4 \\
\neg x_2 \lor x_5   & x_4 \lor x_5
\end{array}
\tag{7.140}
$$

Each clause on the right (after the first) is the resolvent of the clause above and the clause to the left. Since the clause to the left is always a premise in S, these are input resolutions (the first clause on the right also belongs to S). To show that the inequality $x_4 + x_5 \geq 1$ corresponding to $x_4 \lor x_5$ is a rank 1 cut, construct a surrogate inequality as follows. First sum the inequalities that correspond to the parents of the first resolvent to obtain $x_2 + 2x_3 + 2x_4 \geq 1$ as shown below, where the inequalities on the left correspond to the premises on the left in (7.140):

$$
\begin{array}{ll}
                              & -x_1 + x_2 + x_3 + x_4 \geq 0 \\
(1)\ x_1 + x_3 + x_4 \geq 1   & x_2 + 2x_3 + 2x_4 \geq 0 \\
(2)\ x_2 - x_3 \geq 0   & 3x_2 + 2x_4 \geq 1 \\
(3)\ -x_2 + x_5 \geq 0   & 2x_4 + 3x_5 \geq 1
\end{array}
\tag{7.141}
$$

The next step is to take a weighted sum of $x_2 + 2x_3 + 2x_4 \geq 1$ with the premise $x_2 - x_3 \geq 0$ by giving the latter a weight of 2 (shown in parentheses), to cancel out x_3. Do the same for the rest of the premises as shown. The resulting weights provide a linear combination of premises that eliminates variables x_1, x_2, and x_3 and yields the weighted sum $2x_4 + 3x_5 \geq 1$:

$$\begin{array}{llll} -x_1 + x_2 + x_3 + & x_4 & & \geq 0 \ (1) \\ x_1 & + x_3 + & x_4 & \geq 1 \ (1) \\ & x_2 - x_3 & & \geq 0 \ (2) \\ & - x_2 & + & x_5 \geq 0 \ (3) \\ \hline & & 2x_4 + 3x_5 \geq 1 & \end{array} \qquad (7.142)$$

$$\begin{array}{ll} x_4 & \geq 0 \ (5) \\ & x_5 \geq 0 \ (4) \\ \hline 7x_4 + 7x_5 \geq 1 & \end{array}$$

By adding to $2x_4 + 3x_5 \geq 1$ multiples of the bounds $x_4 \geq 0$ and $x_5 \geq 0$ as shown, dividing the sum by 7 (the sum of the premise weights), and rounding up the $\frac{1}{7}$ on the right, one obtains the desired $x_4 + x_5 \geq 1$, which is therefore a rank 1 cut for linear(S).

For the proof of Theorem 7.32, it is convenient to write a clausal inequality in the form $ax \geq 1 + n(a)$, where $n(a)$ is the sum of the negative components of a.

Proof of Theorem 7.32. Let $P_0, \ldots, P_m$ be the premises used in the input resolution proof of C (repeated use of a premise is possible). Let linear(C) be $bx \geq 1 + n(b)$, and let linear(P_i) be $a^i x \geq 1 + n(a^i)$. The goal is to show that $bx \geq 1 + n(b)$ is a rank 1 cut. The ith step of the input resolution proof is

$$\begin{array}{cc} & b^{i-1} x \geq 1 + n(b^{i-1}) \\ a^i x \geq 1 + n(a^i) & b^i x \geq 1 + n(b^i) \end{array}$$

Now let $u^i x \geq v_i$ be the weighted sum of the inequalities above and to the left, as shown below, using weights 1 and w_i, respectively:

$$\begin{array}{cc} & u^{i-1} x \geq v_{i-1} \\ a^i x \geq 1 + n(a^i) & u^i x \geq v_i \end{array}$$

where $u^0 x \geq v_0$ is $a^0 x \geq 1 + n(a^0)$. Let sgn$(\alpha) = 1$ if $\alpha > 0$, -1 if $\alpha < 0$, and 0 if $\alpha = 0$. First show inductively that

$$\begin{array}{ll} b^i_j = \text{sgn}(u^i_j) \text{ for all } j & (a) \\ v_i = 1 + n(u^i) & (b) \end{array} \qquad (7.143)$$

which is trivially true for $i = 1$. Assume then that (7.143) holds for $i - 1$. Let x_k be the variable on which resolution takes place, so that

$a_k^i = -b_k^{i-1}$, and a_j^i and b_j^{i-1} do not have opposite signs for $j \neq k$. The weight w_i needed to make $a_k x_k$ and $u_k^{i-1} x_k$ cancel is a positive number (namely, $|u_k^{i-1}|$) because the induction hypothesis and $a_k^i = -b_k^{i-1}$ imply that a_k^i and u_k^{i-1} have opposite signs. Also, since a_j^i and b_j^{i-1} do not have opposite signs for $j \neq k$, the induction hypothesis implies that a_j^i and u_j^{i-1} do not have opposite signs for $j \neq k$, and (a) follows. To show (b), note that

$$v_i = w_i(1 + n(a^i)) + v_{i-1} = w_i(1 + n(a^i)) + 1 + n(u^{i-1})$$
$$= w_i + n(w_i a^i + u^{i-1}) - w_i + 1 = 1 + n(u^i)$$

where the first equality is by definition of v_i, the second is by the induction hypothesis, and the third is due to the fact that one negative component cancels in the sum $w_i a^i + u^{i-1}$.

To complete the proof, it suffices to show that $b^m x \geq 1 + n(b^m)$ can be obtained by rounding a nonnegative linear combination of the inequality $u^m x \geq 1 + n(u^m)$ and bounds. Let $W = w_1 + \cdots + w_m$ and add to $u^m x \geq 1 + n(u^m)$ the inequalities $(W - u_j^m)x_j \geq 0$ for each j for which $u_j^m > 0$ and $(-W - u_j^m)x_j \geq -W - u_j^m$ for each j for which $u_j^m < 0$. Due to (7.143), this yields the inequality

$$Wb^m x \geq 1 + n(u^m) - \sum_{u_j^m < 0}(W + u_j^m) = 1 + Wn(b^m) \qquad (7.144)$$

Dividing by W and rounding up on the right yields the desired inequality $b^m x \geq 1 + n(b^m)$. $\square$

At this point, the work of the previous section can be completed.

Corollary 7.33 *Clause set S has an input refutation if and only if* linear(S) *is infeasible.*

Proof. From Theorem 7.30, if S has no input refutation, then linear(S) is feasible. Conversely, if S has an input refutation, then by the proof of Theorem 7.32, $0 \geq \alpha$ is a surrogate of linear(S) for some positive α. This implies linear(S) is infeasible. $\square$

This and Theorem 7.29 imply:

Corollary 7.34 *A clause set has a unit refutation if and only if it has an input refutation.*

This does not say that the unit resolution and input resolution algorithms derive the same clauses. It says that they detect infeasibility in the same instances.

Before moving to the converse of Theorem 7.32, it is necessary to prove a lemma. A variable x_j is *monotone* for C in a clause set S if it always occurs with the same sign as in C.

Lemma 7.35 *If linear(C) is a rank 1 cut for linear(S), then it is a rank 1 cut for linear(S') for some subset S' of S in which every variable of C is monotone for C.*

Proof. Suppose, without loss of generality, that C contains only positive literals, and let linear(C) be $bx \geq 1 + n(b)$. Since $bx \geq 1 + n(b)$ is a rank 1 cut, it is the result of rounding up some surrogate $ux \geq v$ of linear(S). Suppose that some variable x_k in C is negated in an inequality $ax \geq 1 + n(a)$ that has weight, say, w in the nonnegative linear combination that yields $ux \geq v$. Then one can remove $ax \geq 1 + n(a)$ from the linear combination and compensate by adding inequalities, as follows:

$$\text{For } x_j \text{ in } C, \text{ add } \begin{cases} 2wx_j \geq 0 & \text{if } a_j = 1 \\ wx_j & \text{if } a_j = 0 \\ \text{nothing} & \text{if } a_j = -1 \end{cases}$$

$$\text{For } x_j \text{ not in } C, \text{ add } \begin{cases} wx_j \geq 0 & \text{if } a_j = 1 \\ \text{nothing} & \text{if } a_j = 0 \\ -wx_j \geq -w & \text{if } a_j = -1 \end{cases}$$

Let $u'x \geq v'$ be the resulting linear combination. If x_j is not in C, then $u'_j = u_j = 0$, and otherwise $u'_j = u_j + w$. Also, if s is the number of variables in C that are negated in $ax \geq 1 + n(a)$, one can check that $v' = v + (s - 1)w$. Thus, $u'x \geq v'$ can be written

$$\frac{1+w}{w} u \geq v + (s-1)w$$

Multiplying this inequality by $w/(1 + w)$, one obtains

$$ux \geq \frac{w}{1+w}v + \frac{s-1}{w} \geq \frac{w}{1+w}v$$

where the second inequality is due to $s \geq 0$. This, after rounding, yields $bx \geq 1 + n(b)$. $\square$

The converse of Theorem 7.32 can now be proved.

Theorem 7.36. *If linear(C) is a rank 1 cut for linear(S), then the input resolution algorithm applied to S generates clause C.*

Proof. Suppose, without loss of generality, that C contains only positive literals, and let linear(C) be $bx \geq 1 + n(b)$. By Lemma 7.35, linear(C) is a rank 1 cut for linear(S'), where S' is some subset of C in which all the variables of C are monotone for C. Then there is a surrogate $ux \geq v$ of linear(S') that rounds up to $bx \geq 1 + n(b)$, which means, in particular, that $u_j \leq b_j$ for all j and $v > 0$. For each x_j not in C, add $(b_j - u_j)x_j \geq 0$ to $ux \geq v$, and let the result be $u'x \geq v$. Thus, $u'x \geq v'$ is a surrogate of linear(S) such that $u'_j = 0$ for all j not in C. Let clause set S' be the result of removing the variables in C from all the clauses in S. Then the same set of multipliers that produced $u'x \geq v$ yields a surrogate $0 \geq v'$ of linear(S'). Because all the literals removed from S are positive, due to the monotonicity of the variables in C, the right-hand sides in linear(S') are the same as in linear(S). Thus, $v' \geq v > 0$, which means that linear(S') is infeasible. By Theorem 7.30, there is an input refutation of S'. Since all the variables in C occur positively in S, one can restore these variables to the premises in the input refutation and obtain an input resolution proof of C. $\square$

7.7.5 Separating Resolvents

An alternative to generating all resolvents, or all input resolvents, is to generate only separating resolvents. There are efficient algorithms for doing so. Generating separating resolvents is usually much faster than generating all resolvents, but one may be obliged to solve several linear relaxations before accumulating enough resolvents to obtain a tight relaxation.

Given a clause set S and a solution $\bar{x}$ of linear(S), a *separating resolvent* of S is a resolvent R such that $\bar{x}$ violates linear(R). One way to identify separating resolvents is to identify clauses of S that are potential parents of a separating resolvent. Then, when searching for separating resolvents, it is necessary only to examine pairs of the identified clauses.

There is a simple method for identifying potential parents of a separating resolvent. For a given clause C in S and a solution $\bar{x}$ of linear(S), let

$$\bar{x}_j^C = \begin{cases} \bar{x}_j & \text{if literal } x_j \text{ is in } C \\ 1 - \bar{x}_j & \text{if } \neg x_j \text{ is in } C \\ 0 & \text{otherwise} \end{cases}$$

Then a resolvent R of clauses in S is separating if and only if $\sum_j \bar{x}_j^R < 1$. Now consider either parent C of R. All the literals of C occur in R except the variable x_k on which the resolution takes place. Thus, R can be separating only if

$$\sum_{j \neq k} \bar{x}_j^C < 1 \qquad (7.145)$$

which implies $\sum_j \bar{x}_j^C < 2$. Furthermore, $\sum_j \bar{x}_j^C \geq 1$ because $\bar{x}$ satisfies linear(C). Supposing that x_k occurs positively in C, this inequality and (7.145) imply

$$\bar{x}_k = \sum_j \bar{x}_j^C - \sum_{j \neq k} \bar{x}_j^C > 0$$

Since $\neg x_k$ occurs in R's other parent, similar reasoning shows that $1 - \bar{x}_k > 0$. Thus, one need only examine pairs of clauses that respectively contain literals $x_k, \neg x_k$ with $0 < \bar{x}_k < 1$. The following has been shown:

Theorem 7.37. *If $\bar{x}$ is a solution of linear(S) for clause set S, then a clause C in S can be a parent of a separating resolvent of S only if $\sum_j \bar{x}_j^C < 2$. Furthermore, a separating resolvent can be obtained from C only by resolving on a variable x_k for which $\bar{x}_k$ is fractional.*

Consider for example the clause set S:

$$
\begin{array}{lcccc}
x_1 \vee x_2 \vee & x_3 & & & C_1 \\
\neg x_1 & \vee & x_3 \vee \neg x_4 & & C_2 \\
\neg x_2 \vee & x_3 \vee & x_4 & & C_3 \\
x_1 \vee x_2 \vee \neg x_3 & & & & C_4 \\
x_2 \vee & x_3 \vee \neg x_4 & & & C_5 \\
0 \quad \tfrac{1}{2} \quad \tfrac{1}{2} \quad 0 & & & & \bar{x}
\end{array}
$$

Suppose for the sake of the example that the objective is to minimize the number of true variables while satisfying S. A solution of linear(S)

Table 7.2 Values used to screen out clauses that cannot be parents of a separating resolvent.

Clause C_i	$\sum_j \bar{x}_j^{C_i}$
C_1	1
C_2	$2\frac{1}{2}$
C_3	1
C_4	1
C_5	2

that minimizes $\sum_j x_j$ is $(\bar{x}_1, \bar{x}_2, \bar{x}_3, \bar{x}_4) = (0, \frac{1}{2}, \frac{1}{2}, 0)$, shown along the bottom. The values of $\sum_j \bar{x}_j^{C_i}$ for $i = 1, \ldots, 5$ are given in Table 7.2. These values indicate that only clauses C_1, C_3, and C_4 need be considered as potential parents of a separating resolvent. Furthermore, since only $\bar{x}_2$ and $\bar{x}_3$ are nonintegral, it suffices to consider pairs of clauses that resolve on x_2 and x_3. Two pairs (C_1, C_3 and C_1, C_4) satisfy these conditions. They respectively yield the resolvents $x_1 \vee x_3 \vee x_4$ and $x_1 \vee x_2$. As it happens, both are separating. If these two clauses are added to S, then a solution of linear(S) that minimizes $\sum_j x_j$ is $(\bar{x}_1, \bar{x}_2, \bar{x}_3, \bar{x}_4) = (1, 0, 0, 0)$, which is integral and therefore an optimal solution of the original problem.

Exercises

7.59. Show that the relaxation given for formula 4 in Table 7.1 is a convex hull relaxation by showing that its coefficient matrix is totally unimodular.

7.60. Show that the coefficient matrix for the relaxation of formula 7 in Table 7.1 is not totally unimodular.

7.61. Use the information in Table 7.1 to write a convex hull relaxation for $(x_1 \rightarrow x_2) \rightarrow (x_3 \rightarrow x_4)$.

7.62. Write a convex hull relaxation for $\bigwedge_{j \in J_1} x_j \vee \bigwedge_{j \in J_2} \neg x_j$, and prove it is a convex hull relaxation.

7.63. Consider the clause set

$$x_1 \vee x_2$$
$$x_1 \vee \neg x_2 \vee x_3$$
$$x_1 \vee \neg x_2 \vee \neg x_3$$

Show that input resolution yields the same result as full resolution but requires more steps. Also show, using this example, that input resolution can infer fewer clauses if absorbed clauses are deleted from the original clause set before the algorithm is completed. Show that this is not true for full resolution.

7.64. Consider the clause set S, consisting of

$$x_1 \vee x_2 \vee x_3$$
$$x_1 \vee x_2 \vee \neg x_3$$
$$x_1 \vee \neg x_2 \vee x_3$$
$$x_1 \vee \neg x_2 \vee \neg x_3$$

Let S' be the result of applying input resolution. Show that S' does not contain all prime implications of S (which means input resolution is weaker than full resolution), but $\mathrm{linear}(S')$ is a tighter relaxation than $\mathrm{linear}(S)$.

7.65. Show that although input resolution has the same refutation power as unit resolution, it can infer clauses that unit resolution cannot.

7.66. The proof of Theorem 7.30 contains a method for constructing a solution of $\mathrm{linear}(S)$ when input resolution does not infer a contradiction from S. Use this method when S is

$$x_1 \vee x_2$$
$$\neg x_1 \vee x_3$$
$$\neg x_2 \vee x_3$$
$$x_1 \vee x_2 \vee \neg x_3$$

after applying input resolution to S. One can also find an input refutation for S when $\mathrm{linear}(S)$ is infeasible, but this is more complicated.

7.67. A unit clause C was derived in the input resolution proof of Exercise 7.66. Use the construction in the proof of Theorem 7.31 to show that $\mathrm{linear}(C)$ is a rank 1 Chvátal–Gomory cut for $\mathrm{linear}(S)$.

7.68. In the proof of Theorem 7.31, verify that (7.144) is a nonnegative linear combination of $u^m x \geq 1 + n(u^m)$ and bounds.

7.69. Input resolution derives the clause $x_1 \vee x_3$ from the clause set S, consisting of

$$x_1 \vee x_2 \qquad (a)$$
$$\neg x_1 \vee x_2 \vee x_3 \qquad (b)$$
$$x_1 \vee \neg x_2 \vee x_3 \qquad (c)$$

by resolving (a) and (b), and resolving the resulting clause with (c). Using the mechanism of Theorem 7.31, construct a set of multipliers to show that $\mathrm{linear}(x_1 \vee x_3)$ is a rank 1 cut for $\mathrm{linear}(S)$. Now use the procedure in the proof of Lemma 7.35 to show that $\mathrm{linear}(x_1 \vee x_3)$ is a rank 1 cut for a monotone subset of clauses, namely (a) and (c).

7.70. If S is the clause set

$$
\begin{array}{llllll}
x_1 \vee & x_2 \vee & x_3 & & \vee \; x_5 \vee \neg x_6 \\
x_1 \vee & x_2 & & \vee x_4 \\
x_1 & & \vee \; x_3 \vee x_4 & & \vee \neg x_6 \\
& x_2 \vee & x_3 \vee x_4 & & \vee \neg x_6 \\
\neg x_1 \vee & x_2 \vee & x_3 & & \vee \neg x_5 \\
& \neg x_2 \vee & x_3 \vee x_4 \vee & x_5 \\
& & \neg x_3 \vee x_4 & & \vee \; x_6
\end{array}
$$

then linear(S) has a basic solution $x = (\frac{1}{3}, \frac{1}{3}, \frac{1}{3}, \frac{1}{3}, 0, 1)$. Identify which clauses can be parents of separating resolvents. Derive their resolvents, and identify which one(s) are separating.

7.8 The Element Constraint

There are several continuous relaxations for the element constraint, most of them based on the relaxations for disjunctive constraints presented in Section 7.4. These relaxations are particularly useful when the variables have fairly tight bounds. There are also effective relaxations for the indexed linear element constraint. Section 6.7 discusses modeling applications for the various forms of the element constraint.

The goal is to derive a continuous relaxation for element$(y, z \,|\, a)$ in terms of z, for element(y, x, z) in terms of z and $x = (x_1, \ldots, x_m)$, and for the indexed linear constraint element$(y, x, z \,|\, a)$ in terms of x and z. Thus, if S is the feasible set of element$(y, z \,|\, a)$, then a relaxation should describe a set that contains the projection of S onto z, and onto x and z for the other element constraints. A convex hull relaxation in each case describes the convex hull of the corresponding projection.

A *vector-valued* generalization of the element constraint is important for some modeling applications. It typically has the form

$$
\text{element}(y, z \,|\, (a^1, \ldots, a^m))
$$

where z and each a^i are tuples. There is also a vector-valued indexed linear constraint

$$
\text{element}(y, x, z \,|\, (a^1, \ldots, a^m))
$$

where x is a scalar variable. It sets z equal to the yth tuple in the list $xa^1, \ldots, xa^m$. Both of these have useful continuous relaxations.

7.8.1 Convex Hull Relaxations

The simplest element constraint, $\text{element}(y, z \,|\, a)$, has an obvious convex hull relaxation:

$$\min_{k \in D_y}\{a_k\} \leq z \leq \max_{k \in D_y}\{a_k\} \tag{7.146}$$

Thus, a convex hull relaxation for $\text{element}(y, (2, 4, 5), z)$ is $2 \leq z \leq 5$, if $D_y = \{1, 2, 3\}$.

A useful convex hull relaxation can be derived for $\text{element}(y, x, z)$ if there are lower and/or upper bounds $L \leq x \leq U$ on the variables. The first step is to observe that $\text{element}(y, x, z)$ and these bounds imply the disjunction

$$\bigvee_{k \in D_y} \begin{pmatrix} z = x_k \\ L \leq x \leq U \end{pmatrix} \tag{7.147}$$

where $x = (x_1, \ldots, x_m)$. Theorem 7.14 can now be applied.

Theorem 7.38. *If $L_i \leq x_i \leq U_i$ for all $i \in D_y$, then the following is a convex hull relaxation of $\text{element}(y, (x_1, \ldots, x_m), z)$:*

$$z = \sum_{k \in D_y} x_k^k, \quad \sum_{k \in D_y} y_k = 1, \quad x_i = \sum_{k \in D_y} x_i^k, \text{ all } i \in D_y$$

$$Ly_k \leq x^k \leq Uy_k, \quad y_k \geq 0, \quad \text{all } k \in D_y \tag{7.148}$$

For example, a convex hull relaxation of

$$\text{element}(y, (x_1, x_2), z), \quad 1 \leq x_1 \leq 4, \quad 3 \leq x_2 \leq 5 \tag{7.149}$$

is

$$z = x_1^1 + x_2^2, \quad x_1 = x_1^1 + x_1^2, \quad x_2 = x_2^1 + x_2^2$$
$$y \leq x_1^1 \leq 4y, \quad y \leq x_2^1 \leq 4y$$
$$3(1 - y) \leq x_1^2 \leq 5(1 - y), \quad 3(1 - y) \leq x_2^2 \leq 5(1 - y)$$
$$0 \leq y \leq 1$$

The projection onto the (z, x) space is

$$\tfrac{2}{3}x_1 + \tfrac{1}{3} \leq z \leq 5, \quad \tfrac{4}{3}x_1 + x_2 - \tfrac{16}{3} \leq z \leq \tfrac{2}{3}x_1 + x_2 + \tfrac{2}{3}$$
$$1 \leq x_1 \leq 4, \quad 3 \leq x_2 \leq 5$$

In general, the projection is quite complex and is not computed explicitly in this fashion.

When every variable x_i has the same bounds, the convex hull relaxation simplifies considerably.

Theorem 7.39. *Suppose that $L_0 \leq x_i \leq U_0$ for all $i \in D_y$. Then a convex hull relaxation of $element(y, (x_1, \ldots, x_m), z)$ is given by*

$$\sum_{i \in D_y} x_i - (|D_y| - 1)U_0 \leq z \leq \sum_{i \in D_y} x_i - (|D_y| - 1)L_0 \qquad (7.150)$$

and the bounds $L_0 \leq z \leq U_0$, $L_0 \leq x_i \leq U_0$ for $i \in D_y$.

Proof. Without loss of generality, the origin can be moved to $x = (L_0, \ldots, L_0)$ and $z = L_0$, and (7.150) becomes

$$\sum_{i \in D_y} x_i - (|D_y| - 1)U_0 \leq z \leq \sum_{i \in D_y} x_i \qquad (7.151)$$

It suffices to show that any point $(\bar{z}, \bar{x})$ that satisfies (7.151) and bounds $0 \leq x_i \leq U_0$, $0 \leq z \leq U_0$ is a convex combination of points that satisfy the disjunction (7.147). Due to the second inequality in (7.151), one can write $\bar{z} = \alpha \sum_i \bar{x}_i$ for $\alpha \in [0, 1]$. For convenience, suppose $D_y = \{1, \ldots, m\}$. It will be shown that $(\bar{z}, \bar{x})$ is the following convex combination:

$$\begin{bmatrix} \bar{z} \\ \bar{x}_1 \\ \bar{x}_2 \\ \bar{x}_3 \\ \vdots \\ \bar{x}_m \end{bmatrix} = \frac{\alpha \bar{x}_1}{U_0} \begin{bmatrix} U_0 \\ U_0 \\ b_2 \\ b_3 \\ \vdots \\ b_m \end{bmatrix} + \frac{\alpha \bar{x}_2}{U_0} \begin{bmatrix} U_0 \\ b_1 \\ U_0 \\ b_3 \\ \vdots \\ b_m \end{bmatrix} + \cdots + \frac{\alpha \bar{x}_m}{U_0} \begin{bmatrix} U_0 \\ b_1 \\ b_2 \\ b_3 \\ \vdots \\ U_0 \end{bmatrix}$$

$$+ \frac{1}{m}\left(1 - \frac{\alpha}{U_0}\sum_i \bar{x}_i\right)\left(\begin{bmatrix} 0 \\ 0 \\ c_2 \\ c_3 \\ \vdots \\ c_m \end{bmatrix} + \begin{bmatrix} 0 \\ c_1 \\ 0 \\ c_3 \\ \vdots \\ c_m \end{bmatrix} + \cdots + \begin{bmatrix} 0 \\ c_1 \\ c_2 \\ c_3 \\ \vdots \\ 0 \end{bmatrix}\right)$$

where b_i, c_i will be defined shortly. Note that each vector

$$(U_0, b_1, \ldots, b_{i-1}, U_0, b_{i+1}, \ldots, b_m)$$

and each vector

$$(0, c_1, \ldots, c_{i-1}, 0, c_{i+1}, \ldots, c_m)$$

satisfy the disjunct $z = x_i$. It must also be shown that

(i) $0 \leq b_i \leq U_0$ for all i

(ii) $0 \leq c_i \leq U_0$ for all i

(iii) $(\bar{z}, \bar{x})$ is equal to the linear combination shown

(iv) the linear combination is a convex combination

For (iv), it is enough to show that the coefficients are nonnegative, because they obviously sum to one. But this follows from

$$0 \leq \frac{\alpha}{U_0} \sum_i \bar{x}_i \leq 1$$

where the first inequality is due to $\bar{x}_i \geq 0$ and the second to the fact that

$$\alpha \sum_i \bar{x}_i = \bar{z} \leq U_0 \tag{7.152}$$

Now, for each i, there are two cases.

Case I. $\bar{x}_i \leq \bar{z}$. In this case, set

$$b_i = U_0 \left(\frac{1 - \alpha}{\alpha} \right) \left(\frac{\bar{x}_i}{\sum_i \bar{x}_i - \bar{x}_i} \right)$$

Obviously, $b_i \geq 0$. Also, $b_i \leq U_0$, due to the fact that

$$b_i = U_0 \frac{\bar{x}_i - \alpha \bar{x}_i}{\alpha \sum_i \bar{x}_i - \alpha \bar{x}_i} \leq U_0 \frac{\bar{z} - \alpha \bar{x}_i}{\bar{z} - \alpha \bar{x}_i} = U_0$$

where the inequality is due to the case hypothesis and $\bar{z} = \alpha \sum_i \bar{x}_i$. This establishes (i). One can now set $c_i = 0$, so that (ii) holds, and direct calculation verifies (iii).

Case II. $\bar{x}_i > \bar{z}$. In this case, set $b_i = U_0$ so that (i) obviously holds. Also set

$$c_i = \frac{\bar{x}_i - \alpha \sum_j \bar{x}_j}{1 - \frac{\alpha}{U_0} \sum_j \bar{x}_j}$$

Here, (iii) is easily verified. It remains to show (ii). The fact that $c_i \geq 0$ follows from the case hypothesis and (7.152). To see that $c_i \leq U_0$, write

$$c_i \leq \frac{U_0 - \sum_j \bar{x}_j + (m-1)U_0}{1 - \frac{1}{U_0}\sum_j \bar{x}_j + (m-1)} = U_0$$

where the inequality is due to $\bar{x}_i \leq U_0$, the first inequality in (7.151), and (7.152). This completes the proof. $\square$

When the variables x_i have different bounds L_i, U_i, the inequality (7.150) remains valid if L_0 and U_0 are set to the most extreme bounds.

Corollary 7.40 *If $L_i \leq x_i \leq U_i$ for all $i \in D_y$, the following is a relaxation of element$(y, (x_1, \ldots, x_m), z)$:*

$$\sum_{i \in D_y} x_i - (|D_y| - 1)U_0 \leq z \leq \sum_{i \in D_y} x_i - (|D_y| - 1)L_0 \tag{7.153}$$

$$L_0 \leq z \leq U_0, \quad L_0 \leq x_i \leq U_0, \ i \in D_y$$

where $L_0 = \min_{i \in D_y}\{L_i\}$ and $U_0 = \max_{i \in D_y}\{U_i\}$.

For example, a valid relaxation can be obtained for (7.149) by setting $L_0 = 1$, $U_0 = 5$:

$$x_1 + x_2 - 5 \leq z \leq x_1 + x_2 - 1, \quad 1 \leq z \leq 5, \quad 1 \leq x_i \leq 5, \ i = 1, 2$$

A convex hull relaxation is easy to write for the indexed linear element constraint, because the relaxation contains only two variables.

Corollary 7.41 *If $L \leq x \leq U$, then the following is a convex hull relaxation of element$(y, x, z \,|\, (a_1, \ldots, a_m))$:*

$$z \geq \frac{U_{\min} - L_{\min}}{U - L}x + \frac{UL_{\min} - LU_{\min}}{U - L}$$

$$z \leq \frac{U_{\max} - L_{\max}}{U - L}x + \frac{UL_{\max} - LU_{\max}}{U - L}$$

$$L \leq x \leq U$$

where

$$L_{\min} = \min_{i \in D_y}\{a_i L\}, \quad U_{\min} = \min_{i \in D_y}\{a_i U\}$$

$$L_{\max} = \max_{i \in D_y}\{a_i L\}, \quad U_{\max} = \max_{i \in D_y}\{a_i U\}$$

If $L = 0$, the relaxation simplifies to

$$\min_{i \in D_y}\{a_i\}x \leq z \leq \max_{i \in D_y}\{a_i\}x, \quad 0 \leq x \leq U$$

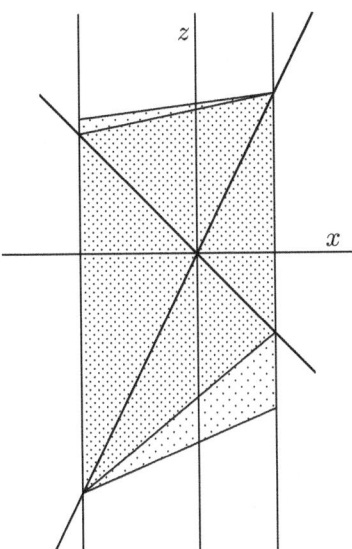

Fig. 7.18 Convex hull relaxation (darker shading) and big-M relaxation (entire shaded area) for an indexed linear element constraint.

For the constraint element$(y, x, z \,|\, (-1, 2))$ with $x \in [-3, 2]$,

$$(L_{\min}, U_{\min}) = (-6, -2) \quad \text{and} \quad (L_{\max}, U_{\max}) = (3, 4)$$

So, the convex hull relaxation is

$$\tfrac{4}{5}x - \tfrac{18}{5} \le z \le \tfrac{1}{5}x + \tfrac{18}{5}, \quad -3 \le x \le 2$$

as illustrated in Fig. 7.18.

7.8.2 Big-M Relaxations

Big-M relaxations of the element constraint are generally not as tight as a convex hull relaxation, but they involve fewer variables and constraints. A relaxation can be formed by applying Theorem 7.17 to the disjunction

$$\bigvee_{k \in D_y} (-z + x_k = 0) \tag{7.154}$$

Using (7.82),

$$M'_k = U_k^{\max} - L_k, \quad M''_k = U_k - L_k^{\min}$$

where

$$U_k^{\max} = \max_{\ell \neq k}\{U_k\}, \quad L_k^{\min} = \min_{\ell \neq k}\{L_\ell\}$$

A direct application of Theorem 7.17 yields the following corollary.

Corollary 7.42 *The big-M relaxation of element(y, x, z) with bounds $L \leq x \leq U$ is equivalent to the following:*

$$\begin{aligned}
& z \geq x_k - \left(U_k - L_k^{\min}\right)(1 - \alpha_k), \ \text{all } k \in D_y \\
& z \leq x_k + \left(U_k^{\max} - L_k\right)(1 - \alpha_k), \ \text{all } k \in D_y \\
& \sum_{k \in D_y} \alpha_k = 1, \quad \alpha_k \geq 0, \ \text{all } k \in D_y
\end{aligned} \tag{7.155}$$

When all the lower bounds L_j are equal to L_0, and all upper bounds U_i are equal to U_0, the relaxation (7.155) is dominated by the convex hull relaxation (7.153). When the lower bounds or upper bounds differ, however, it may be advantageous to use both relaxations.

The relaxation (7.153) for example (7.149) was found in the previous section to be

$$x_1 + x_2 - 5 \leq z \leq x_1 + x_2 - 1 \tag{7.156}$$

plus bounds. The big-M relaxation (7.155) for this example, when projected onto x_1 and x_2, is

$$\begin{aligned}
& 4x_1 + x_2 - 4 \leq z \leq -x_1 + x_2 + 1 \\
& x_1 - 1 \leq z \leq x_1 + 4 \\
& x_2 - 4 \leq z \leq x_2 + 1
\end{aligned}$$

plus bounds. Neither relaxation is redundant of the other.

There is no point in using a big-M relaxation for the indexed linear constraint element$(y, (a_1x, \ldots, a_mx)z)$, because the convex hull relaxation is already quite simple. A big-M relaxation for the vector form of the constraint element$(y, (A_1x, \ldots, A_mx), z)$ could be desirable, however, and is represented in the next section.

7.8.3 Vector-Valued Element

The vector-valued element constraint

$$\text{element}(y, z \,|\, (a^1, \ldots, a^m)) \tag{7.157}$$

can be very useful because its continuous relaxation is much tighter than the result of relaxing each component constraint individually.

This can be seen in an example. Suppose that the production level of a shop is a_1^y, and the corresponding cost is a_2^y, where $y \in \{1, 2, 3, 4\}$. The possible production levels are 10, 20, 40, and 90 and the corresponding costs are 100, 150, 200, and 250, respectively. Perhaps there are various constraints on y and other problem variables, but the constraints of interest here are

$$\text{element}(y, z_1 \mid (10, 20, 40, 90))$$
$$\text{element}(y, z_2 \mid (100, 150, 200, 250)) \tag{7.158}$$

which define z_1 to be the production level and z_2 to be the cost. The individual convex hull relaxations of the two element constraints are $10 \leq z_1 \leq 90$ and $100 \leq z_2 \leq 150$. However, a vector-valued element constraint that combines the two

$$\text{element}\left(y, \begin{bmatrix} z_1 \\ z_2 \end{bmatrix} \mid \left(\begin{bmatrix} 10 \\ 100 \end{bmatrix}, \begin{bmatrix} 20 \\ 150 \end{bmatrix}, \begin{bmatrix} 40 \\ 200 \end{bmatrix}, \begin{bmatrix} 90 \\ 250 \end{bmatrix} \right) \right)$$

has a much tighter convex hull relaxation (Fig. 7.19).

In general, one can obtain a convex hull relaxation for (7.157) by computing a convex hull description of the points $a^1, \ldots, a^m$ in z-space.

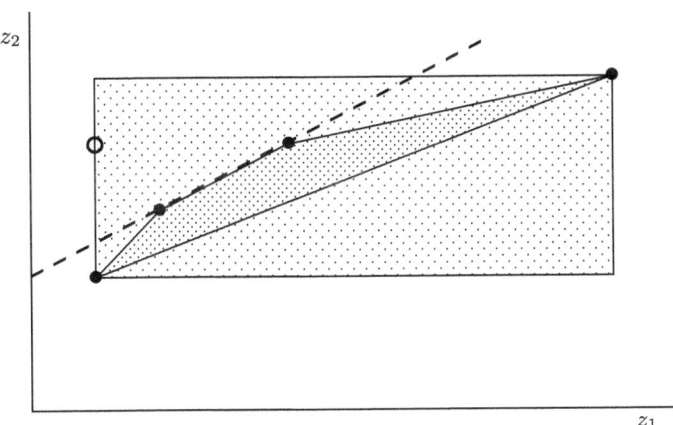

Fig. 7.19 Convex hull relaxation of a vector-valued element constraint (darker shading) and of two element constraints considered separately (entire shaded area). The dashed line is an optimal separating cut for the point represented by the small circle.

This is practical when the dimensionality of z is 2 or 3, using various algorithms developed for computational geometry.

For higher dimensionality, it may be useful to find a separating cut $dz \geq \delta$ that is maximally violated by the solution $\bar{z}$ of the current relaxation. That is, d and δ are chosen so that $\bar{z}$ maximally violates $dx \geq \delta$, and $a^1, \ldots, a^m$ satisfy it. This may be done by solving the linear optimization problem

$$
\begin{array}{c}
\max \ \delta - d\bar{z} \\
\|d\| \leq 1, \quad da^i \geq \delta, \ i = 1, \ldots, m
\end{array}
\tag{7.159}
$$

where d, δ are unrestricted in sign and $\| \cdot \|$ is a convenient norm. If the L_∞ norm ($\max_j \{|d_j|\}$) is used, the constraint $\|d\| \leq 1$ becomes $-1 \leq d_j \leq 1$ for all j.

If $\bar{z} = (10, 200)$ in the example of Fig. 7.19, the optimal separating cut using the L_∞ norm is $z_1 - 0.4z_2 \geq -40$, as shown in the figure.

There is also a vector-valued form of the indexed linear element constraint:

$$
\text{element}(y, x, z \,|\, (a^1, \ldots, a^m))
\tag{7.160}
$$

It can be used to implement indexed linear expressions of the form xa^y, where each a^k is a vector and x a scalar variable, by replacing xa^y with z and adding the constraint (7.160). This type of expression occurs, for instance, in the product configuration problem of Section 2.7.

Because (7.160) implies the disjunction

$$
\bigvee_{k \in D_y} \left(\begin{array}{c} z = xa^k \\ L \leq x \leq U \end{array} \right)
$$

Theorem 7.14 can be applied to obtain a convex hull relaxation for (7.160).

Corollary 7.43 If $L \leq x \leq U$, $\text{element}(y, x \,|\, (a^1, \ldots, a^m))$ has the convex hull relaxation

$$
z = \sum_{k \in D_y} a^k x_k, \quad x = \sum_{k \in D_y} x_k, \quad \sum_{k \in D_y} y_k = 1
\tag{7.161}
$$

$$
Ly_k \leq x_k \leq Uy_k, \quad y_k \geq 0, \ \text{all } k \in D_y
$$

A big-M relaxation can be written for (7.160), as follows. Suppose, as before, that $L \leq x \leq U$. Since z will be equated with xa^k for some k, it can be assumed that $\bar{L} \leq z \leq \bar{U}$, where

$$\bar{L}_j = \min_{k \in D_y} \left\{ \min\{a_j^k L, a_j^k U\} \right\}, \quad \text{all } j$$

$$\bar{U}_j = \max_{k \in D_y} \left\{ \max\{a_j^k L, a_j^k U\} \right\}, \quad \text{all } j$$

(7.162)

Using (7.65), the big-Ms are

$$M^k = \begin{bmatrix} \bar{U} - L^k \\ U^k - \bar{L} \end{bmatrix}$$

where

$$L^k = \min\{0, a^k\}U + \max\{0, a^k\}L$$
$$U^k = \min\{0, a^k\}L + \max\{0, a^k\}U$$

and the minimum and maximum are taken componentwise. So, the big-M relaxation (7.66) becomes

$$xa^k - (U^k - \bar{L})(1 - y_k) \le z \le xa^k + (\bar{U} - L^k)(1 - y_k)$$
$$L \le x \le U, \quad \sum_{k \in D_y} y_k = 1, \quad y_k \ge 0, \quad \text{all } k \in D_y$$

Section 7.8.1 derived a convex hull relaxation for the constraint element$(y, x \mid (-1, 2))$ with $x \in [-3, 2]$. Although a big-M relaxation would not in practice be used for this constraint, it is interesting to contrast it with the convex hull relaxation. The projection of the big-M relaxation onto z, x is

$$\tfrac{8}{19}x - \tfrac{90}{19} \le z \le \tfrac{1}{8}x + \tfrac{15}{4}, \quad -3 \le x \le 2$$

This is illustrated along with the convex hull relaxation in Fig. 7.18.

Exercises

7.71. Use Theorem 7.38 to write a convex hull relaxation for the constraint element$(y, (x_1, x_2), z)$, where $y \in \{1, 2\}$, $x_1 \in [-1, 2]$, and $x_2 \in [0, 3]$.

7.72. Prove Theorem 7.38 as a corollary of Theorem 7.14.

7.73. Use Theorem 7.39 to write a convex hull relaxation for the constraint element$(y, (x_1, x_2), z)$ where $y \in \{1, 2\}$ and $x_1, x_2 \in [-1, 3]$.

7.74. The point $(z, x_1, x_2) = (3, 2, 2)$ is not feasible for element$(y, (x_1, x_2), z)$ in Exercise 7.73 but belongs to the convex hull of the feasible set. Use the mechanism in the proof of Theorem 7.39 to construct a convex combination of feasible points that yields $(3, 2, 2)$. Note that the origin must be shifted before applying the proof.

7.75. Use Corollary 7.40 to write an alternate relaxation for the element constraint in Exercise 7.71.

7.76. Use Corollary 7.41 to write a convex hull relaxation for the indexed linear constraint element$(y, x, z, \mid (-2, 3))$ with $x \in [-5, 6]$.

7.77. Prove Corollary 7.41.

7.78. Use Corollary 7.42 to write a big-M relaxation for the element constraint in Exercise 7.71. Compare it with the convex hull relaxation, as well as the relaxation obtained for the same constraint in Exercise 7.75.

7.79. Prove Corollary 7.42.

7.80. Write the linear programming problem whose solution obtains an optimal separating cut for the point $(10, 200)$ in Fig. 7.19, using the L_∞ norm. Verify that the cut $z_1 - 0.4z_2 \geq -40$ is optimal.

7.81. Prove Corollary 7.43.

7.9 The All-Different Constraint

As noted in Section 6.8, the constraint

$$\text{alldiff}(x_1, \ldots, x_n)$$

appears in many constraint programming models, not only because many modeling situations call for it, but also because there are fast and effective domain filtering algorithms for the constraint. Continuous relaxations can be created for the constraint as well. One type of relaxation uses only the original variables $x_1, \ldots, x_n$, while another is based on an integer programming model of the constraint and introduces 0-1 variables.

Systems of all-different constraints frequently occur in applications, such as scheduling, parallel processing, register allocation, and

Latin square (quasigroup completion) problems—including the popular sudoku puzzles. The famous graph coloring problem is a multiple-alldiff problem in which x_i denotes the color assigned to vertex i of a graph. Because adjacent vertices cannot receive the same color, the problem is formulated by imposing an alldiff constriant for every clique of vertices.

Alldiff systems, too, can be given relaxations either in the original variables x_j or in 0-1 variables. In fact, valid inequalities in the original variables can be easily converted to valid inequalities for the 0-1 formulation, often resulting in a stronger relaxation than would otherwise be available. This suggests a general method for obtaining better cuts for 0-1 models: derive valid inequalities for a CP representation of the problem, and then map these inequalities into 0-1 space. The resulting inequalities may be stronger than known cuts for the 0-1 model.

7.9.1 Convex Hull Relaxation

The constraint alldiff$(x_1, \ldots, x_n)$ can be given a continuous relaxation in the original variables $x_1, \ldots, x_n$, if they take numerical values. If all the variable domains are the same, it is a convex hull relaxation. Unfortunately, the relaxation contains exponentially many constraints, but there is a simple way to generate separating cuts.

Suppose first that every variable has the same domain. A continuous relaxation of the alldiff constraint can be based on the fact that the sum of any k variables must be at least as large as the k smallest domain elements.

Theorem 7.44. *If the domain D_{x_j} is $\{v_1, \ldots, v_m\}$ for $j = 1, \ldots, n$, where $v_1 < \cdots < v_m$ and $m \geq n$, then the following is a convex hull relaxation of alldiff$(x_1, \ldots, x_n)$:*

$$\sum_{j=1}^{|J|} v_j \leq \sum_{j \in J} x_j \leq \sum_{j=m-|J|+1}^{m} v_j, \quad all \ J \subset \{1, \ldots, n\} \qquad (7.163)$$

Proof. Let S be the set of feasible points for the alldiff constraint. Clearly all points in S satisfy (7.163). It remains to show that every point satisfying (7.163) is a convex combination of points in S.

Rather than show this directly, it is convenient to use a dual approach. Since the convex hull of S is the intersection of all half planes containing S, it suffices to show that any half plane containing S contains all points satisfying (7.163). That is, it suffices to show that any valid inequality $ax \geq b$ for the alldiff is implied by (7.163), or equivalently, is dominated by a surrogate (nonnegative linear combination) of (7.163).

The first step is to prove the theorem when $a \geq 0$ or $a \leq 0$. Index the variables so that $a_1 \geq \cdots \geq a_n$, and consider a linear combination of the following inequalities from (7.163):

$$\sum_{j=1}^{i} x_j \geq \sum_{j=1}^{i} v_j, \quad i = 1, \ldots, n-1 \ (a_i - a_{i+1} \text{ if } a \geq 0, \ 0 \text{ otherwise})$$

$$\sum_{j=1}^{n} x_j \geq \sum_{j=1}^{n} v_j \qquad\qquad (a_n \text{ if } a \geq 0, \ 0 \text{ otherwise})$$

$$-\sum_{j=i}^{n} x_j \geq -\sum_{j=m-n+i}^{m} v_j, \quad i = 2, \ldots, n \ (a_{i-1} - a_i \text{ if } a \leq 0, \ 0 \text{ otherwise})$$

$$-\sum_{j=1}^{n} x_j \geq -\sum_{j=m-n+1}^{m} v_j \qquad (-a_1 \text{ if } a \leq 0, \ 0 \text{ otherwise})$$

where each inequality has the nonnegative multiplier shown on the right. The result of the linear combination is $ax \geq a\bar{v}$ where

$$\bar{v} = \begin{cases} (v_1, \ldots, v_n) & \text{if } a \geq 0 \\ (v_{m-n+1}, \ldots, v_m) & \text{if } a \leq 0 \end{cases}$$

But since $x = \bar{v}$ is a feasible solution of the alldiff, the validity of $ax \geq b$ implies $a\bar{v} \geq b$, which means that $ax \geq b$ is dominated by the surrogate $ax \geq a\bar{v}$ of (7.163).

Now, take an arbitrary valid inequality $ax \geq b$ in which the components of a may have any sign. Index the variables so that $a_1 \geq \cdots \geq a_n$, and suppose that $a_j \geq 0$ for $j = 1, \ldots, k$ and $a_j < 0$ for $j = k+1, \ldots, n$. Since $v_1 < \cdots < v_m$, it is clear that $ax \geq b$ cannot be valid unless $b \leq av^*$, where $v^* = (v_1, \ldots, v_k, v_{m-n+k+1}, \ldots, v_m)$. Also,

$$\sum_{j=1}^{k} a_j x_j \geq \sum_{j=1}^{k} a_j v_j \tag{7.164}$$

is valid because $a_1 \geq \cdots \geq a_k \geq 0$ and $v_1 < \cdots < v_m$, and

$$\sum_{j=k+1}^{n}(-a_j)x_j \le \sum_{j=m-n+k+1}^{m}(-a_j)v_j$$

is valid because $0 \le -a_{k+1} \le \cdots \le -a_n$ and $v_1 < \cdots < v_m$. The last inequality can be written

$$\sum_{j=k+1}^{n} a_j x_j \ge \sum_{j=m-n+k+1}^{m} a_j v_j \qquad (7.165)$$

The sum of (7.164) and (7.165) is $ax \ge av^*$, which dominates $ax \ge b$ because $b \le av^*$. But since (7.164) and (7.165) are valid and have coefficients that are all nonnegative or all nonpositive, they are surrogates of (7.163). The same is therefore true of $ax \ge b$, and the theorem follows. $\square$

For example, consider the constraint

$$\text{alldiff}(x_1, x_2, x_3), \quad x_j \in \{1, 5, 8\}, \ j = 1, 2, 3$$

The convex hull relaxation is

$$
\begin{aligned}
& x_1, x_2, x_3 \ge 1 && (a) \\
& x_1 + x_2 \ge 6, \quad x_1 + x_3 \ge 6, \quad x_2 + x_3 \ge 6 && (b) \\
& x_1 + x_2 + x_3 \ge 14 && (c) \\
& x_1 + x_2 + x_3 \le 14 && (d) \\
& x_1 + x_2 \le 13, \quad x_1 + x_3 \le 13, \quad x_2 + x_3 \le 13 && (e) \\
& x_1, x_2, x_3 \le 8 && (f)
\end{aligned}
$$

Since in this case each domain contains exactly n elements, constraints (e) and (f) are redundant, and the relaxation simplifies to

$$
\begin{aligned}
& x_1, x_2, x_3 \ge 1 && (a) \\
& x_1 + x_2 \ge 6, \quad x_1 + x_3 \ge 6, \quad x_2 + x_3 \ge 6 \ (b) \\
& x_1 + x_2 + x_3 = 14 && (c)
\end{aligned}
\qquad (7.166)
$$

Corollary 7.45 *If $m = n$ in Theorem 7.44, then the following is a convex hull relaxation of $\text{alldiff}(x_1, \ldots, x_n)$:*

$$
\sum_{j \in J} x_j \ge \sum_{j=1}^{|J|} v_j, \quad \text{all } J \subset \{v_1, \ldots, v_n\} \text{ with } |J| < n \ (a)
$$

$$
\sum_{j=1}^{n} x_j = \sum_{j=1}^{n} v_j \qquad (b)
$$

$$(7.167)$$

The relaxation in Theorem 7.44 contains exponentially many constraints because there are exponentially subsets J. However, one can begin by using only the constraints

$$\sum_{j=1}^{n} v_j \leq \sum_{j=1}^{n} x_j \leq \sum_{j=m-n+1}^{m} v_j \tag{7.168}$$

and bounds on the variables, and then generate separating cuts as needed. Let $\bar{x}$ be the solution of the current relaxation of the problem, and renumber the variables so that $\bar{x}_1 \leq \cdots \leq \bar{x}_n$. Then, for $i = 2, \ldots, n-1$, one can generate the cut

$$\sum_{j=1}^{i} x_j \geq \sum_{j=1}^{i} v_j \tag{7.169}$$

whenever

$$\sum_{j=1}^{i} \bar{x}_j < \sum_{j=1}^{i} v_j$$

Also, for each $i = n-1, \ldots, 2$, generate the cut

$$\sum_{j=i}^{n} x_j \leq \sum_{j=m-n+i}^{m} v_j$$

whenever

$$\sum_{j=i}^{n} \bar{x}_j > \sum_{j=m-n+i}^{m} v_j$$

There is no separating cut if $\bar{x}$ lies within the convex hull of the alldiff feasible set. If $m = n$, one can start with (7.167b) and bounds, and generate the cut (7.169) for $i = 1, \ldots, n-1$ whenever

$$\sum_{j=1}^{i} \bar{x}_j < \sum_{j=1}^{i} v_j$$

For example, suppose one wishes to solve the problem

$$
\begin{aligned}
&\min 2x_1 + 3x_2 + 4x_3 \\
&x_1 + 2x_2 + 3x_3 \geq 32, \quad \text{alldiff}(x_1, x_2, x_3) \\
&x_j \in \{1, 5, 8\}, \ j = 1, 2, 3
\end{aligned}
\tag{7.170}
$$

If one replaces the alldiff and domains in (7.170) with (7.166), the solution of the resulting relaxation is $\bar{x} = (1, 8, 5)$. Since this is feasible for the alldiff and domains, no branching is necessary and the problem is solved at the root node. Alternatively, once can use only (7.166c) and bounds $1 \leq x_j \leq 8$ in the relaxation, and add separating cuts as needed. In this case, the solution is again $\bar{x} = (1, 8, 5)$ and there is no need for a separating cut.[4] One can check that no separating cut exists by noting that $\bar{x}_1 \geq 1$ and $\bar{x}_1 + \bar{x}_3 \geq 6$.

A similar relaxation can be written when the variable domains are arbitrary finite sets, although it is not in general a convex hull relaxation. One can write valid inequalities of the form

$$L(J) \leq \sum_{j \in J} x_j \leq U(J), \quad \text{all } J \subset \{1, \ldots, n\} \tag{7.171}$$

where

$$L(J) = \min \left\{ \sum_{j \in J} x_j \,\middle|\, \text{alldiff}(x_j \mid j \in J),\ x_j \in D_{x_j} \text{ for all } j \in J \right\}$$

$$U(J) = \max \left\{ \sum_{j \in J} x_j \,\middle|\, \text{alldiff}(x_j \mid j \in J),\ x_j \in D_{x_j} \text{ for all } j \in J \right\}$$

Theorem 7.46. *If $L(J), U(J)$ are defined as above, then (7.171) is a relaxation of alldiff$(x_1, \ldots, x_n)$.*

One can compute $L(J)$ and $U(J)$ for a given J by solving a minimum and a maximum cost network flow problem (7.58), for which fast specialized algorithms are known. The network contains a node i for each $i \in J$, representing variable x_i, and a node v for each $v \in D_J = \bigcup_{j \in J} D_{x_j}$. An arc with cost $c_{iv} = v$ runs from each i to each $v \in D_{x_i}$, and an arc with capacity $U_{vt} = 1$ runs from each $v \in D_J$ to a sink node t. All other costs are zero, and all other capacities infinite. Also, $s_i = 1$ for $i \in J$, $s_v = 0$ for $v \in D_J$, and $s_t = -|J|$. A flow of $y_{iv} = 1$ represents assigning value v to x_i. $L(J)$ is the cost of a minimum-cost flow, and $U(J)$ the cost of a maximum cost flow.

For example, consider the problem (7.170) where the domains are

$$x_1 \in \{1, 4, 6\}, \quad x_2, x_3 \in \{5, 6\} \tag{7.172}$$

---

[4] One could have predicted that the same solution would result, because in this small example the constraints (b) are implied by (c) and the bounds.

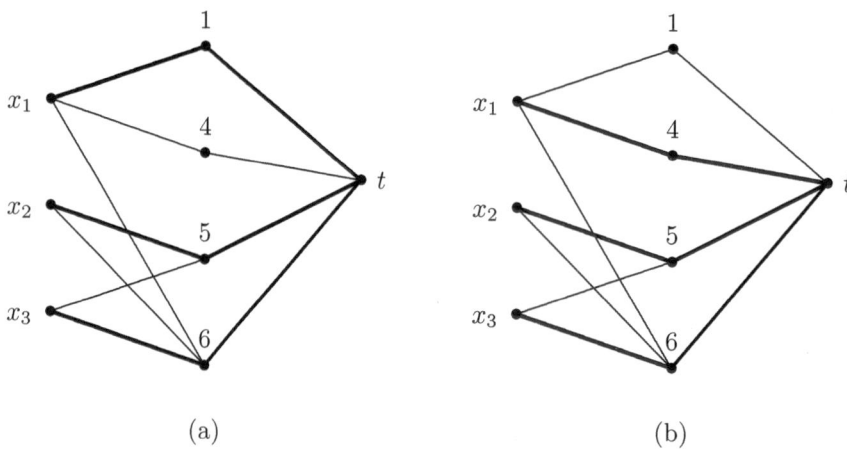

Fig. 7.20 Network flow model for a relaxation of alldiff with unequal variable domains. Heavy lines show a minimum-cost flow in diagram (a) and a maximum cost flow in (b).

The relaxation (7.171) is

$$12 \leq x_1 + x_2 + x_3 \leq 15$$
$$6 \leq x_1 + x_2 \leq 11$$
$$6 \leq x_1 + x_3 \leq 11$$
$$11 \leq x_2 + x_3 \leq 11 \tag{7.173}$$
$$1 \leq x_1 \leq 6$$
$$5 \leq x_2 \leq 6$$
$$5 \leq x_3 \leq 6$$

The minimum-cost network flow problems for obtaining the two constraints on the top line are illustrated in Fig. 7.20. The optimal solution of (7.170) with relaxation (7.173) replacing the alldiff and domains is $\bar{x} = (4, 5, 6)$. Since this is feasible for the alldiff and domains, the problem is solved at the root node.

7.9.2 Convex Hull Mixed-Integer Formulation

The alldiff constraint can obviously be written as an assignment model in 0-1 variables. This model is totally unimodular and therefore provides a convex hull relaxation in the 0-1 space. Furthermore, if the original variables x_i are numerical, a relaxation can be written that contains both the original variables and 0-1 variables. This is likewise a convex hull formulation, in the sense that the projection of its feasible set onto the x_i-space is the convex hull of the feasible set.

Let $\{v_1, \ldots, v_m\}$ be the union of the variable domains D_{x_i}. Define binary variables y_{ij} that take the value 1 when $x_i = v_j$. The assignment formulation is

$$
\begin{align}
\sum_{j=1}^{m} y_{ij} = 1, \quad i = 1, \ldots, n \qquad &(a) \\
\sum_{i=1}^{n} y_{ij} \leq 1, \quad j = 1, \ldots, m \qquad &(b) \qquad (7.174) \\
y_{ij} = 0 \text{ all } i, j \text{ with } j \notin D_{x_i} \quad &(c) \\
y_{ij} \in \{0,1\}, \quad \text{all } i, j
\end{align}
$$

The continuous relaxation is obtained by replacing $y_{ij} \in \{0,1\}$ with $0 \leq y_{ij} \leq 1$. Because the constraint set is totally unimodular, this is a convex hull relaxation.

If the domain values $v_1, \ldots, v_m$ are numerical, the following channeling constraints can be added to the assignment model:

$$
x_i = \sum_{j=1}^{m} v_j y_{ij}, \quad i = 1, \ldots, n \qquad (7.175)
$$

Theorem 7.47. *The continuous relaxation of* (7.174)–(7.175) *projected onto* $x_1, \ldots, x_n$ *is a convex hull relaxation of* alldiff$(x_1, \ldots, x_n)$ *and* $x_i \in D_{x_i}$ *for all* i.

Proof. The constraints (7.174)–(7.175) are clearly valid. It remains to show that the projection onto x of every feasible solution of the constraints is a convex combination of feasible solutions of the alldiff and $x_i \in D_{x_i}$. So let $(\bar{x}, \bar{y})$ be a feasible solution of (7.174)–(7.175). Note first that the coefficient matrix of constraints (7.174) is totally unimodular, using Theorem 7.12. So all vertices of the polyhedron defined by (7.174) and $y_{ij} \geq 0$ are 0-1 points, which means that $\bar{y}$ is a convex combination $\sum_k \alpha_k y^k$ of 0-1 points y^k satisfying (7.174). Now since $\bar{x}_i = \sum_j v_j \bar{y}_{ij}$,

$$
\bar{x}_i = \sum_j v_i \left(\sum_k \alpha_k y_{ij}^k \right) = \sum_k \alpha_k \left(\sum_j v_j y_{ij}^k \right) = \sum_k \alpha_k x_i^k
$$

where $x_i^k = \sum_j v_j y_{ij}^k$. But because $\sum_j y_{ij}^k = 1$, $y_{ik}^k \in \{0,1\}$, and $y_{ij}^k = 0$ when $v_j \notin D_{x_i}$, it follows that $x_i^k \in D_{x_i}$. Also since $\sum_i y_{ij}^k \leq 1$ for all j and k, one can infer alldiff$(x_1^k, \ldots, x_n^k)$ for each k. Thus, $\bar{x}$ is a convex combination of points x^k that are feasible for the alldiff and $x_i \in D_{x_i}$. $\square$

For example, alldiff(x_1, x_2, x_3) with domains (7.172) has the convex hull relaxation

$$
\begin{aligned}
x_1 &= y_{11} + 4y_{12} + 6y_{14} & y_{11} + y_{12} + y_{14} &= 1 \\
x_2 &= 5y_{23} + 6y_{24} & y_{23} + y_{24} &= 1 \\
x_3 &= 5y_{33} + 6y_{34} & y_{33} + y_{34} &= 1 \\
& & y_{23} + y_{33} &\leq 1 \\
& & y_{14} + y_{24} + y_{34} &\leq 1 \\
& & y_{ij} &\geq 0, \text{ all } i, j
\end{aligned}
\tag{7.176}
$$

If this relaxation replaces the alldiff and domains in (7.170), the optimal solution is $\bar{x} = (4, 5, 6)$, with $\bar{y}_{12} = \bar{y}_{23} = \bar{y}_{34} = 1$ and all other $\bar{y}_{ij} = 0$. Since this is feasible for the alldiff and domains, the problem is solved without branching. Incidentally, the projection of (7.176) onto x simplifies in this case to $x_2 + x_3 = 11$ with bounds $x_1 \in [1, 4]$ and $x_2, x_3 \in [5, 6]$. In general, the projection is quite complex and is not computed.

7.9.3 Relaxing the Objective Function

A relaxation is useful for obtaining LP bounds only if the objective function is linear and the variables have numerical values. However, the variables $x_1, \ldots, x_n$ that occur in an alldiff constraint may not represent numerical values, and if they do, the objective function may be nonlinear. Assignment problems, for example, often call for the nonlinear objective function $\sum_i c_{ix_i}$.

Some problems, such as (7.170), already have numerical variables and linear objective functions. In other cases, one can substitute numerical for nonnumeric values and linearize the objective. For example, to minimize the number of colors in a graph coloring problem, one can denote the colors with distinct numbers rather than color names and minimize z subject to $z \geq x_i$ for all i.

Even when the objective function is nonlinear, it may become linear when the relaxation is reformulated in 0-1 variables. As noted earlier, a 0-1 model of alldiff constraints can be obtained by using numeric domain elements and replacing each x_i by $\sum_j j y_{ij}$, where binary variable $y_{ij} = 1$ when $x_i = j$ (vertex i is assigned color j). Each constraint alldiff$(x_1, \ldots, x_n)$ becomes a set of assignment constraints (7.174). The nonlinear objective function $\sum_i c_{ix_i}$ can now be replaced with the linear function $\sum_{ij} c_{ij} y_{ij}$, and a linear relaxation results. In addition, valid linear inequalities derived in x_i-space can be transformed to valid linear 0-1 inequalities by substituting $\sum_j j y_{ij}$ for x_i.

The facet-defining inequalities obtained in the previous section are not useful in a 0-1 model of a single alldiff contraint, because the 0-1 model is totally unimodular. Mapping these inequalities into 0-1 space only produces cuts that are redundant of the assignment constraints. However, the situation changes for a *system* of alldiff constraints. The 0-1 model is no longer totally unimodular, and facets for the alldiff system map into valid 0-1 cuts that may be stronger than previously known cuts for the 0-1 graph coloring problem. This idea is explored in the next section.

7.9.4 Alldiff Systems

An alldiff system can be written

$$\text{multipleAlldiff}\{X_1, \ldots, X_q\} \tag{7.177}$$

where $X_i \subset X = \{x_1, \ldots, x_n\}$ for $i = 1, \ldots, q$. It is assumed, without loss of generality, that no X_i contains another. The alldiff system is equivalent to a vertex coloring problem on a graph whose vertices correspond to $x_1, \ldots, x_n$ and whose maximal cliques correspond to $X_1, \ldots, X_q$. Variable $x_i = j$ when vertex i receives color j. If the domain of each x_i is $\{0, 1, \ldots, d - 1\}$ (or some other set of at least n numbers), the number of colors can be minimized by minimizing z subject to (7.177) and $z \geq x_i$ for all i.

The convex hull of an alldiff system is not fully understood, but certain facts are known. For example, when the sets X_k have an certain *inclusion property*, the facets of the convex hull are all facets of individual alldiff constraints in (7.177). Let J_k be the index set of variables in X_k, and T_k be the index set of variables in X_k that occur in some other set $X_{k'}$. Then the alldiffs have the inclusion property if the index sets can be ordered $J_1, \ldots, J_n$ so that $T_1 \supset \cdots \supset T_q$. For example, four alldiffs with

$$
\begin{aligned}
X_1 &= \{\, x_1, x_2, x_3, x_4, && \} \\
X_2 &= \{\, x_1, x_2, x_3, && x_5, && \} \\
X_3 &= \{\, x_1, x_2, && x_6, x_7, && \} \\
X_4 &= \{\, x_1, && x_8, x_9 \,\}
\end{aligned}
$$

have the inclusion property because $T_1 \supset T_2 \supset T_3 \supset T_4$.

Additional facets are known for some types of systems that lack the inclusion property, and they can strengthen a relaxation. One such system is an *odd cycle* in which each X_k intersects only X_{k-1} and X_{k+1} for $k = 1, \ldots, q$ and q is odd, where X_0 is identified with X_q and X_{q+1} with X_1. A cycle of five alldiffs is illustrated in Fig. 7.21. Suppose that $S_1, \ldots, S_q \subset X$ satisfy

$$
S_k \subset J_k \cap J_{k+1}, \ |S_k| = s > 0, \ k = 1, \ldots, q,
$$

where J_{q+1} is identified with J_1. Then if $S = S_1 \cup \cdots \cup S_q$, it can be shown that the following is a facet-defining inequality for the odd cycle:

$$
\sum_{i \in S} x_i \geq r(s) \tag{7.178}
$$

where

$$
r(s) = \frac{q-1}{2}\frac{(L-1)(L-2)}{2} + (L-1)\left(sq - \frac{q-1}{2}(L-1) \right)
$$

and

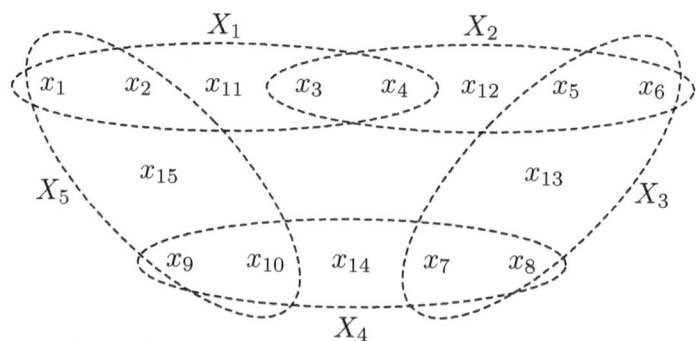

Fig. 7.21 A system of five alldiff constraints that form an odd cycle.

$$L = \left\lceil \frac{2sq}{q-1} \right\rceil$$

For example, the system of Fig. 7.21 with domain values $0, \ldots, 5$ has the facet-defining inequality

$$x_1 + \cdots + x_{10} \geq 20$$

when $s = 2$. When $s = 1$, $r(s)$ simplifies to $(q-1)/2 + 2$.

Additional cuts may be available for a particular objective function. If the problem is to minimize the number of colors, then as noted earlier, the objective is to minimize z subject to $z \geq x_i$ for all i. For the simple case in which $s = 1$, there is a valid cut:

$$z \geq \frac{1}{q} \left(\sum_i x_i + \frac{q-1}{2} + 2 \right) \tag{7.179}$$

Using this cut along with (7.178) results in a tighter relaxation bound when the problem consists solely of an odd cycle. That is, when the variables are allowed to take continuous values, the minimum of z subject to (7.178) and (7.179) is larger than subject to (7.178) alone.

Valid linear inequalities for an alldiff system can be mapped into a 0-1 model to strengthen its relaxation. Again let 0-1 variable $y_{ij} = 1$ when vertex i receives color j. Let binary variable $v_j = 1$ when color j is used. The problem of minimizing the number of colors can be written

$$\min \sum_j v_j$$

$$\sum_j y_{ij} = 1, \quad \text{all } i \qquad (a)$$

$$v_j \geq \sum_{x_i \in X_k} y_{ij}, \quad \text{all } j, k \qquad (b) \tag{7.180}$$

$$y_{jk} \in \{0, 1\}, \quad \text{all } j, k$$

Constraint (a) requires that every vertex receive a color, while (b) enforces the alldiff constraints and determines whether color j is used. The valid inequalities (7.178) and (7.179) can be included in the model by replacing each x_i with $\sum_j j y_{ij}$ to obtain

$$\sum_{i \in S} \sum_j j y_{ij} \geq r(s) \qquad (a)$$

$$\sum_j v_j \geq \frac{1}{q} \left(\sum_{i \in S} \sum_j j y_{ij} + \frac{q-1}{2} + 2 \right) \qquad (b) \qquad (7.181)$$

These cuts can be compared with standard odd hole cuts for the constraints (7.180c). Inequality (a) can improve on odd hole cuts when $s \geq 2$. When $s = 1$, the combination of (a) and (b) is stronger than odd hole cuts in a coloring problem.

To make this more precise, suppose G is the graph to be colored, and H is the index set of vertices in an odd hole of G. Then

$$\sum_{i \in H} y_{ij} \leq \frac{q-1}{2}$$

is an odd hole cut for any color j (see Theorem 7.3). Odd hole cuts can be obtained for (7.180c) by letting H contain one i from each S_k. These cuts, along with the assigment constraints, imply (a) in (7.181) when $s = 1$, but not when $s \geq 2$. Furthermore, when cuts (a) and (b) are added to (7.180) for $s = 1$, the bound obtained from the continuous relaxation is tighter than when the odd hole cut is added.

Separation is straightforward for the odd cycle cuts (7.178). Let $\bar{y}$ be the solution of the current relaxation of the 0-1 model, and set $\bar{x}_i = \sum_j j \bar{y}_{ij}$ for all j. Then if there is a separating cut of the form (7.178), the following polynomial-time algorithm finds one. Index the variables x_i so that $S_k = \{\ell_k, \ldots, u_k\}$ and $\bar{x}_{\ell_k} \leq \cdots \leq \bar{x}_{u_k}$ for each k. For $s = 1, \ldots, \min_i\{|S_i|\}$ do the following: if

$$\sum_{k=1}^{q} \sum_{i=\ell_k}^{\ell_k+s-1} \bar{x}_i < r(s)$$

then generate the cut

$$\sum_{k=1}^{q} \sum_{i=\ell_k}^{\ell_k+s-1} x_i \geq r(s)$$

and map it into 0-1 space.

Exercises

7.82. Write a convex hull relaxation for alldiff(x_1, x_2, x_3) when each $x_j \in \{1, 3, 5, 6\}$.

7.83. Minimize $2x_1 + 3x_2 + 4x_3$ subject to $x_1 + 2x_2 + 3x_3 \geq 14$ and the alldiff constraint of Exercise 7.82 by replacing the alldiff with its convex hull relaxation. Branching should be unnecessary. In many cases, however, a good deal of branching is required, because the convex hull relaxation of alldiff is quite weak. The relaxation is least effective when the objective function and other constraints tend to push all the x_js near the middle of their possible range.

7.84. Solve the problem of Exercise 7.83 but with a partial relaxation of the alldiff that contains (7.163) for $J = \{1, 2, 3\}$ only, along with bounds $1 \leq x_j \leq 6$ for $j = 1, 2, 3$. Generate a separating alldiff cut for the resulting solution and re-solve the relaxation with this cut. This should yield a feasible solution.

7.85. The proof of Theorem 7.44 simplifies considerably when $m = n$ (as in Corollary 7.45). Write this simpler proof.

7.86. Use Theorem 7.46 to write a relaxation for alldiff(x_1, x_2, x_3) with $x_1 \in \{1, 3, 5\}$, $x_2 \in \{1, 6\}$, and $x_3 \in \{3, 5, 6\}$. Solve the appropriate minimum and maximum cost flow problems to obtain the right-hand sides.

7.87. Minimize $2x_1 + 3x_2 + 4x_3$ subject to $x_1 + 2x_2 + 3x_3 \geq 14$ and the domains of Exercise 7.86 by replacing the alldiff with the relaxation obtained in that exercise. Is the resulting solution feasible?

7.88. Formulate the convex hull MILP relaxation of alldiff(x_1, x_2, x_3) using the domains of Exercise 7.86.

7.89. Minimize $2x_1 + 3x_2 + 4x_3$ subject to $x_1 + 2x_2 + 3x_3 \geq 14$ and the domains of Exercise 7.86 by replacing the alldiff with the MILP-based relaxation of Exercise 7.88. Is the resulting solution feasible?

7.90. Verify that the constraint matrix of (b)–(c) in (7.174) is totally unimodular as claimed in the proof of Theorem 7.47.

7.91. Show that when an alldiff system consists of a single odd cycle and $s = 1$, the continuous relaxation that minimizes z subject to (7.178), (7.179) and $z \geq x_i$ for all i has no integrality gap.

7.92. Suppose again that an alldiff system consists of a single odd cycle and that $s = 1$. Show that the odd cycle cut (7.181a) is redundant of an odd hole inequality and assignment constraints in the continuous relaxation of a 0-1 model of the alldiff system.

7.93. In Exercise 7.92, show by example that (7.181a) is not necessarily redundant when $s = 2$.

7.94. Show that inequality (7.179) is valid when $s = 1$.

7.95. Show that inequality (7.178) is valid.

7.10 The Cardinality Constraint

The cardinality constraint generalizes the alldiff constraint and has convex hull relaxations analogous to those given for alldiff in the previous section.

Recall that the constraint

$$\text{cardinality}(x \mid v, \ell, u)$$

requires each value v_j to occur at least ℓ_j times, and at most u_j times, among the variables $x = (x_1, \ldots, x_n)$. For purposes of relaxation, it is assumed that the v_js are numbers. It is also assumed that the domain of each x_i is a subset of $\{v_1, \ldots, v_m\}$. This incurs no loss of generality because any domain element that does not occur in v can be placed in v with upper and lower bounds of 0 and n.

7.10.1 Convex Hull Relaxation

The cardinality constraint has a continuous relaxation in the original variables $x_1, \ldots, x_n$ that is similar to that given for alldiff in Theorem 7.44. Again, it is a convex hull relaxation if all variables have the same domain.

Any sum $\sum_{j \in J} x_j$ must be at least as large the sum of $|J|$ values from $\{v_1, \ldots, v_m\}$, selecting the smallest value as many times as possible, then selecting the second smallest value as many times as possible, and so forth. Similarly, $\sum_{j \in J} x_j$ must be at least as small as the largest feasible sum of $|J|$ values from $\{v_1, \ldots, v_m\}$. Thus, if $v_1 < \cdots < v_m$, one can state

$$\sum_{i=1}^{m} p(|J|, i)v_i \leq \sum_{j \in J} x_j \leq \sum_{i=1}^{m} q(|J|, i)v_i, \quad \text{all } J \subset \{1, \ldots, n\} \quad (7.182)$$

where $p(k, i)$ is the largest number of times one can select v_i when minimizing a sum of k x_is, and $q(k, i)$ the largest number of times one can select v_i when maximizing a sum of k x_is. Thus,

$$p(k, i) = \min \left\{ p_i, k - \sum_{j=1}^{i-1} p(k, j) \right\}, \quad q(k, i) = \min \left\{ q_i, k - \sum_{j=i+1}^{m} q(k, j) \right\}$$

where

$$p_i = \min \left\{ u_i, \ n - \sum_{j=1}^{i-1} p_j - \sum_{j=i+1}^{m} \ell_j \right\}, \quad i = 1, \ldots, m$$

$$q_i = \min \left\{ u_i, \ n - \sum_{j=i+1}^{m} q_j - \sum_{j=1}^{i-1} \ell_j \right\}, \quad i = m, \ldots, 1$$

For example, consider the constraint

$$\text{cardinality}(\{x_1, \ldots, x_5\} \mid (20, 30, 60), (1, 2, 1), (3, 3, 1))$$

Here, $(p_1, p_2, p_3) = (2, 2, 1)$ and $(q_1, q_2, q_3) = (1, 3, 1)$. The inequalities (7.182) are

$$1 \cdot 20 \le x_j \le 1 \cdot 60, \quad \text{all } j$$

$$40 = 2 \cdot 20 \le \sum_{j \in J} x_j \le 1 \cdot 60 + 1 \cdot 30 = 90, \quad \text{all } J \text{ with } |J| = 2$$

$$70 = 2 \cdot 20 + 1 \cdot 30 \le \sum_{j \in J} x_j \le 1 \cdot 60 + 2 \cdot 30 = 120, \quad \text{all } J \text{ with } |J| = 3$$

$$100 = 2 \cdot 20 + 2 \cdot 30 \le \sum_{j \in J} x_j \le 1 \cdot 60 + 3 \cdot 30 = 150, \quad \text{all } J \text{ with } |J| = 4$$

$$160 = 2 \cdot 20 + 2 \cdot 30 + 1 \cdot 60 \le \sum_{j=1}^{5} x_j \le 1 \cdot 60 + 3 \cdot 30 + 1 \cdot 20 = 170$$

where $J \subset \{1, \ldots, 5\}$. This is, in fact, a convex hull relaxation.

Theorem 7.48. *If $x = (x_1, \ldots, x_n)$, $v = (v_1, \ldots, v_m)$, $v_1 < \cdots < v_m$, and $D_{x_i} = \{v_1, \ldots, v_m\}$ for $i = 1, \ldots, n$, then (7.182) is a convex hull relaxation of $\text{cardinality}(X \mid v, \ell, u)$.*

The proof is very similar to the proof of Theorem 7.44, and is left as an exercise.

The separation algorithm is similar to that given for alldiff in Section 7.9.1. Constraints

$$\sum_{i=1}^{m} p_i v_i \leq \sum_{j=1}^{n} x_j \leq \sum_{i=1}^{m} q_i v_i$$

and variable bounds can be included in the initial relaxation. Let $\bar{x}$ be the solution of the relaxation, and renumber the variables so that $\bar{x}_1 \leq \cdots \leq \bar{x}_n$. Then, for each $i = 2, \ldots, n-1$, one can generate the cut

$$\sum_{j=1}^{i} x_j \geq \sum_{j=1}^{m} p(i,j) v_j$$

whenever

$$\sum_{j=1}^{i} \bar{x}_j < \sum_{j=1}^{m} p(i,j) v_j$$

Also for each $i = n-1, \ldots, 2$ one can generate the cut

$$\sum_{j=i}^{n} x_j \leq \sum_{j=1}^{m} q(i,j) v_j$$

whenever

$$\sum_{j=i}^{n} \bar{x}_j > \sum_{j=1}^{m} q(i,j) v_j$$

In the above example, suppose $(\bar{x}_1, \ldots, \bar{x}_5) = (20, 20, 20, 30, 70)$. Then one can generate the separating cuts

$$x_1 + x_2 + x_3 \geq 70, \quad x_1 + x_2 + x_3 + x_4 \geq 100, \quad x_4 + x_5 \leq 90$$

7.10.2 Convex Hull Mixed-Integer Formulation

The constraint cardinality$(x \mid v, \ell, u)$ can be given a mixed-integer formulation very similar to that given in Section 7.9.2 for alldiff. Its continuous relaxation is a convex hull relaxation for arbitrary variable domains, provided the domain elements are numbers.

Let $\{v_1, \ldots, v_m\}$ be the union of the variable domains D_{x_i}. Define binary variables y_{ij} that take the value 1 when $x_i = v_j$. The cardinality constraint can be formulated

$$x_i = \sum_{j=1}^{m} v_j y_{ij}, \quad i = 1, \ldots, n$$

$$\sum_{j=1}^{m} y_{ij} = 1, \quad i = 1, \ldots, n$$

$$\ell_j \leq \sum_{i=1}^{n} y_{ij} \leq u_j, \quad j = 1, \ldots, m \tag{7.183}$$

$$y_{ij} = 0 \text{ all } i, j \text{ with } j \notin D_{x_i}$$

$$y_{ij} \in \{0, 1\}, \quad \text{all } i, j$$

The continuous relaxation of (7.183) is formed by replacing $y_{ij} \in \{0, 1\}$ with $y_{ij} \geq 0$.

Theorem 7.49. *Let $x = (x_1, \ldots, x_n)$ and $v = (v_1, \ldots, v_m)$. The continuous relaxation of (7.183) projected onto $x_1, \ldots, x_n$ is a convex hull relaxation of cardinality$(X \mid v, \ell, u)$ and $x_i \in D_{x_i}$ for all i.*

The proof is almost the same as the proof of Theorem 7.47.

Exercises

7.96. Write a convex hull relaxation for

$$\text{cardinality}((x_1, \ldots, x_5) \mid (2, 5, 6, 7), (2, 1, 1, 1), (3, 3, 2, 2))$$

where each $x_j \in \{2, 5, 6, 7\}$. Note that the lower and upper bounds on $\sum_{j=1}^{5} x_j$ are equal.

7.97. Suppose that $(x_1, \ldots, x_5) = (2, 2, 4, 7, 7)$, which is infeasible for the cardinality constraint in Exercise 7.96. Identify the separating cuts. One can also branch on x_3 by setting $x_3 \leq 2$ and $x_3 \geq 5$.

7.98. Prove Theorem 7.48.

7.99. Identify a family of valid cuts for the cardinality constraint when the domains differ, following the pattern of Theorem 7.46. Formulate minimum and maximum flow problems that can be used to obtain the right-hand sides.

7.100. Write a convex hull MILP relaxation for the cardinality constraint in Exercise 7.96.

7.101. Prove Theorem 7.49.

7.11 The Circuit Constraint

Recall that the circuit constraint

$$\text{circuit}(x_1, \ldots, x_n) \tag{7.184}$$

requires that $x_1, \ldots, x_n, x_1$ describe a permutation, where x_i is the item that follows item i in the permutation.

Because the circuit constraint describes the feasible set of the traveling salesman problem (Section 6.13.1), relaxations developed for the traveling salesman problem can serve as relaxations for circuit. These relaxations have been developed almost entirely for a particular 0-1 model of the problem.

The traveling salesman problem (TSP) minimizes $\sum_i c_{ix_i}$ subject to (7.184), where each D_{x_i} is a subset of $\{1, \ldots, n\}$. The problem can be viewed as defined on a directed graph G with a vertex for each x_i and an edge (i, j) when $j \in D_{x_i}$. Every feasible solution corresponds to a *tour*, or a Hamiltonian cycle in G. The objective is to find a minimum-length tour, where c_{ij} is the length of edge (i, j).

The TSP is said to be *symmetric* when $c_{ij} = c_{ji}$ and $j \in D_{y_i}$ if and only if $i \in D_{x_i}$ for all i, j. Otherwise, it is *asymmetric*. The symmetric problem is normally associated with an undirected graph, on which the objective is to find a minimum-cost undirected tour. The symmetric and asymmetric problems receive somewhat different, albeit related, analyzes in the literature. The asymmetric problem is discussed here, because it subsumes the symmetric problem as a special case.

A relaxation for the TSP can be developed in terms of the original variables, or within a 0-1 model of the problem. The focus here is on a particular 0-1 model, because it has been intensively studied. However, facet-defining inqualities can be also written in terms of the original variables x_i, and if desired, mapped into a 0-1 model by substituting $\sum_j j y_{ij}$ for each x_i. This approach can yield valid inequalities other than those normally used for the 0-1 model.

7.11.1 0-1 Programming Model

The most widely studied integer programming model of the asymmetric TSP is the following. Let the 0-1 variable y_{ij} (for $i \neq j$) take the value 1 when $x_i = j$; that is, when edge (i, j) is part of the selected tour. Then, the objective is to minimize $\sum_{ij} c_{ij} y_{ij}$ subject to

$$\sum_j y_{ij} = \sum_j y_{ji} = 1, \quad \text{all } i \qquad (a)$$

$$\sum_{(i,j) \in \delta(S)} y_{(i,j)} \geq 1, \quad \text{all } S \subset \{1, \ldots, n\} \text{ with } 2 \leq |S| \leq n - 2 \qquad (b)$$

$$y_{ij} \in \{0, 1\}, \quad \text{all } i, j$$

$$(7.185)$$

Here, $\delta(S)$ is the set of edges (i, j) of G for which $i \in S$ and $j \notin S$. If $j \notin D_{x_i}$, the variable y_{ij} is omitted from (7.185). Constraints (a) are vertex-degree constraints that require every vertex to have one incoming edge and one outgoing edge in the tour. The *subtour-elimination constraints* (b) exclude Hamiltonian cycles on all proper subsets S of $\{1, \ldots, n\}$ by requiring that at least one edge connect a vertex in S to one outside S.

The 0-1 variables y_{ij} are related to the original variables x_i via the channeling constraints

$$x_i = \sum_j j y_{ij}, \quad \text{all } i \qquad (7.186)$$

7.11.2 Continuous Relaxations

The simplest relaxation of (7.185) is the *assignment relaxation*, which consists of the vertex degree constraints (a) and $y_{ij} \geq 0$. The assignment relaxation can be strengthened by adding subtour-elimination inequalities (b). It is impractical to add all of these, however, because there are exponentially many. The usual practice is to add separating inequalities as needed. One can also exclude subtours on two vertices *a priori* by adding all constraints of the form $y_{ij} + y_{ji} \leq 1$.

Separating inequalities can be found as follows. If $\bar{y}$ is a solution of the current continuous relaxation, let the *capacity* of edge (i, j) be $\bar{y}_{ij}$.

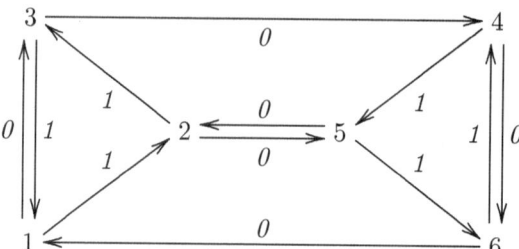

Fig. 7.22 Graph for a circuit constraint. An optimal solution of the assignment relaxation is indicated by the numbers on the edges.

Select a proper subset S of the vertices for which the total capacity of edges leaving S (i.e., the *outflow capacity* of S) is a minimum. The subtour-elimination constraint (b) corresponding to S is a separating cut if this minimum outflow capacity is less than 1.

Consider for example the graph in Fig. 7.22, which corresponds to the constraint $\text{circuit}(x_1, \ldots, x_6)$ with domains $D_{x_1} = \{2, 3\}$, $D_{x_2} = \{3, 5\}$, $D_{x_3} = \{1, 4\}$, $D_{x_4} = \{5, 6\}$, $D_{x_5} = \{2, 6\}$, and $D_{x_6} = \{1\}$. The assignment relaxation is

$$
\begin{aligned}
&y_{12} + y_{13} = 1, \quad y_{23} + y_{25} = 1, \quad y_{31} + y_{34} = 1 \\
&y_{45} + y_{46} = 1, \quad y_{52} + y_{56} = 1, \quad y_{64} + y_{61} = 1 \\
&y_{31} = 1, \quad y_{12} + y_{32} + y_{62} = 1, \quad y_{13} + y_{23} + y_{63} = 1 \\
&y_{54} = 1, \quad y_{25} = y_{45} + y_{65} = 1, \quad y_{26} + y_{46} + y_{56} = 1 \\
&y_{13} + y_{31} \leq 1, \quad y_{25} + y_{52} \leq 1, \quad y_{46} + y_{64} \leq 1 \\
&x_{ij} \geq 0, \quad \text{all } i, j
\end{aligned}
\tag{7.187}
$$

If the costs are $c_{34} = c_{25} = c_{52} = c_{61} = 1$ and $c_{ij} = 0$ for other i, j, then the optimal solution of the relaxation, shown in Fig. 7.22, defines two subtours. There are two minimum-capacity vertex sets, namely $S = \{1, 2, 3\}$ and $S = \{4, 5, 6\}$, both of which have outflow capacity of zero. They correspond to the violated subtour-elimination inequalities

$$
y_{34} + y_{25} \geq 1, \quad y_{52} + y_{61} \geq 1
\tag{7.188}
$$

When these are added to the relaxation (7.187), the resulting solution is $y_{ij} = \frac{1}{2}$ for all i, j. This satisfies all subtour-elimination inequalities. Yet, it is infeasible because it is nonintegral, and one must either generate more cutting planes or branch.

There are fast algorithms for finding an S with a minimum-outflow capacity (i.e., for finding a *minimum-capacity cut*) [198, 385], but in

---

Let $S = S_0$, $S_{\min} = S_0$, $C_{\min} = \sum_{(i,j)\in\delta(S_0)} \bar{y}_{ij}$, $C = C_{\min}$.

For all $i \notin S$ let $b_i = \sum_{j\in S} \bar{y}_{ji}$.

While $S \neq \{1, \ldots, n\}$ repeat:

    Select a vertex $i \notin S$ that maximizes b_i.

    Let $S = S \cup \{i\}$, $C = C + 2 - 2b_i$.

    For all $j \notin S$, let $b_j = b_j + \bar{y}_{ij}$.

    If $C < C_{\min}$ then let $C_{\min} = C$ and $S_{\min} = S$.

Fig. 7.23 Max back heuristic for finding a minimum capacity vertex set containing S_0. The quantity $C_{\min}$ records the minimum outflow capacity found so far. At the termination of the heuristic, a subtour-elimination inequality is generated for $S_{\min}$ if $C_{\min} < 1$.

practice a simple heuristic is often used. Known in this context as the *max back* heuristic (Fig. 7.23), it adds vertices one at a time to an initial set S (perhaps a single vertex) and keeps track of the resulting outflow capacity. The process continues until S contains all the vertices. The set with the smallest outflow capacity is selected to generate a subtour-elimination inequality, provided its outflow capacity is strictly less than 1. The vertex i added to S in each iteration is one with the largest *max back value* $b_i = \sum_{j\in S} \bar{y}_{ji}$, on the theory that S will have smaller outflow capacity if large-capacity edges are brought within S. The procedure can be restarted at different vertices until one or perhaps several separating inequalities have been found.

In the example of Fig. 7.22, one might start the max back heuristic with $S = \{1\}$, which has outflow capacity 1. Vertex 2 is added next, and then vertex 3, at which point the outflow capacity is 0. Since subsequent sets cannot have smaller outflow capacity, $S_{\min} = \{1, 2, 3\}$ is selected and the subtour-elimination inequality $y_{25} + y_{34} \geq 1$ is generated. If further separating inequalities are desired, one might restart the max back heuristic at a vertex outside $S_{\min}$.

7.11.3 Comb Inequalities

Several classes of cutting planes have been developed to strengthen a continuous relaxation of (7.185). By far the most widely used are *comb*

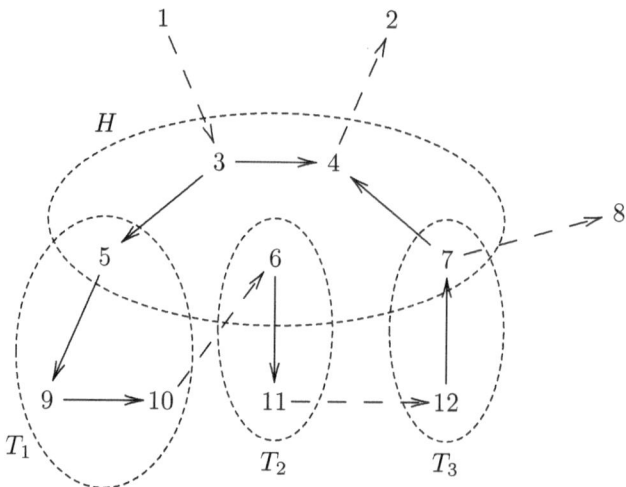

Fig. 7.24 Example of a comb with handle H and teeth T_1, T_2, T_3. The solid arrows indicate edges of the comb, and the dashed arrows show other edges of the graph.

inequalities. Suppose H is a subset of vertices of G, and $T_1, \ldots, T_m$ are pairwise disjoint sets of vertices (where m is odd) such that $H \cap T_k$ and $T_k \setminus H$ are nonempty for each k. The subgraph of G induced by the vertices of H and the T_ks is a *comb* with *handle* H and *teeth* T_k. The associated comb inequality is

$$\sum_{(i,j)\in\delta(H)} y_{ij} + \sum_{k=1}^{m} \sum_{(i,j)\in\delta(T_k)} y_{ij} \geq \tfrac{1}{2}(3m+1) \qquad (7.189)$$

For example, the comb shown in Fig. 7.24 corresponds to the inequality

$$(y_{42} + y_{59} + y_{6,11} + y_{78}) + (y_{10,6} + y_{11,12} + y_{74} + y_{78}) \geq 5$$

Theorem 7.50. *Any comb with handle H and teeth T_k for $k = 1, \ldots, m$ gives rise to a valid inequality* (7.189).

Proof. Define $T_k^o = T_k \setminus H$ and $T_k' = T_k \cap H$. Also, for a vertex set S let $y(S)$ abbreviate $\sum_{(i,j)\in\delta(S)} y_{ij}$. The following will be shown:

$$y(H) + \sum_k y(T_k) \geq \frac{1}{2}\left[\sum_k y(T_k^o) + \sum_k y(T_k') + \sum_k y(T_k)\right] \geq \frac{3}{2}m$$
$$(7.190)$$

Because $y(H) + \sum_k y(T_k)$ is integral, and m is odd, the right-hand side of (7.190) can be rounded up to $\frac{1}{2}(3m+1)$ and the theorem follows. To show (7.190), note that the second inequality follows from the fact that each term of the middle expression must be at least one, because any tour must contain at least one edge leaving a given vertex set. The first inequality of (7.190) can be established by a bookkeeping argument. Let O be the set of vertices outside the comb, let $H^o = H \setminus \bigcup_k T_k$, and let $y(S_1, S_2)$ abbreviate $\sum_{i \in S_1, j \in S_2} y_{ij}$. Then the first inequality of (7.190) can be verified by writing each expression in terms of $y(H^o, O)$, $y(T_k', T_k^o)$, $y(T_k', O)$, $y(H^o, T_k')$, $y(T_k', T_\ell')$, $y(T_k', O)$, and $y(T_k^o, O)$. $\square$

Several separation heuristics have been developed for comb inequalities. One goes as follows. Define a graph to be *2-connected* if the removal of any one vertex does not disconnect it. Given a solution $\bar{y}$ of the current relaxation, let the *support graph* $\bar{G}$ be the subgraph of G whose edges (i, j) correspond to fractional $\bar{y}_{ij}$s. Select a 2-connected component of $\bar{G}$ and let it be the handle H. Let $\{i, j\}$ be a tooth T_k if $\bar{y}_{ij} = 1$ and exactly one of i, j is in H. If $\bar{y}$ violates the corresponding comb inequality (7.190), then the comb inequality is separating. Otherwise a variation of the max back heuristic can be used. Start adding more teeth to the comb in the following way, keeping track of how the comb inequality (7.190) changes each time a tooth is added. Select a vertex of H that belongs to another 2-connected component of $\bar{G}$, and let this connected component be a tooth. Note that all teeth added in this way will be pairwise disjoint. Continue until the comb inequality becomes violated, or until no further teeth can be added.

As an example, consider again the graph of Fig. 7.22, only this time with different costs: let $c_{34} = c_{61} = 0$ and $c_{ij} = 1$ for all other i, j. The optimal solution of the relaxation puts $y_{34} = y_{61} = 1$, $y_{31} = y_{64} = 0$, and $y_{ij} = \frac{1}{2}$ for all other i, j. The resulting support graph $\bar{G}$ is shown in Fig. 7.25. There are three 2-connected components of $\bar{G}$, one on vertices $\{1, 2, 3\}$, one on $\{4, 5, 6\}$, and one on $\{2, 5\}$. Using the heuristic just described, one can let the first component be the handle $H = \{1, 2, 3\}$ and include two teeth $T_1 = \{3, 4\}$ and $T_2 = \{1, 6\}$. The corresponding comb inequality is not violated by $\bar{y}$ (in fact it is not actually a comb inequality, because there are an even number of teeth). Applying the max back heuristic, the 2-connected component containing vertex 2 can be added as a third tooth $T_3 = \{2,5\}$. The corresponding comb inequality

$$(y_{34} + y_{25}) + (y_{31} + y_{45} + y_{46} + y_{23} + y_{56} + y_{12} + y_{13} + y_{64}) \geq 5 \quad (7.191)$$

is violated by $\bar{y}$, and there is no need to add more teeth. The separating inequality (7.191) is added to the relaxation (7.187), which now has solution $\bar{y}_{12} = \bar{y}_{23} = \bar{y}_{34} = \bar{y}_{45} = \bar{y}_{56} = \bar{y}_{61} = 1$ and $\bar{y}_{ij} = 0$ for other i, j. This defines a tour and is therefore an optimal solution.

Because comb and subtour-elimination inequalities do not completely describe the convex hull, they may not furnish a separating cut. This is illustrated by the relaxation (7.187)–(7.188) obtained in Section 7.11.2, whose solution is $\bar{y} = \frac{1}{2}$ for all i, j. This solution satisfies all comb and subtour-elimination inequalities. To find an integral solution, one must branch or identify another family of cutting planes.

Exercises

7.102. Write the assignment relaxation for the problem of minimizing $\sum_i c_i x_i$ subject to $\text{circuit}(x_1, \ldots, x_6)$ and draw the associated graph, where $c_{24} = c_{61} = 1$ and all other $c_{ij} = 0$. The domains are $x_1 \in \{2\}$, $x_2 \in \{3,4\}$, $x_3 \in \{1,4\}$, $x_4 \in \{5\}$, and $x_6 \in \{1,3,4\}$. Solve the relaxation and note that the solution is infeasible.

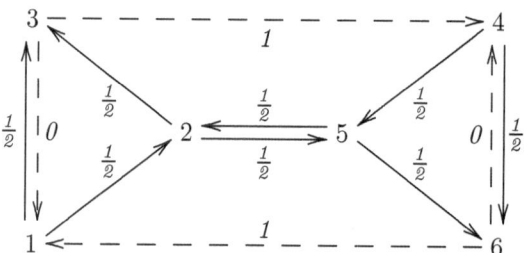

Fig. 7.25 A solution of the assignment relaxation that satisfies all subtour-elimination constraints. The edges of the support graph are shown with solid arrows.

7.103. Use the max back heuristic to identify one or more separating subtour-elimination inequalities for the solution obtained in the previous exercise. Re-solve the relaxation with these cuts added and obtain an optimal tour.

7.104. Given a graph with directed edges $(1,2)$, $(1,3)$, $(2,3)$, $(3,1)$, $(3,4)$, $(3,6)$, $(4,5)$, $(5,6)$, $(6,7)$, $(7,1)$, and $(7,4)$, write one or more comb inequalities and verify that they are satisfied by the only tour in the graph.

7.105. Verify the first inequality in (7.190).

7.12 Disjunctive Scheduling

The constraint noOverlap$(s \, | \, p)$ introduced in Section 6.14 requires that jobs be scheduled sequentially so as not to overlap in time. Here $s = (s_1, \ldots, s_n)$ is a vector of start time variables for jobs $1, \ldots, n$, and the parameter $p = (p_1, \ldots, p_n)$ is a vector of processing times. Each job j also has a release time r_j and deadline d_j. The time windows are implicit in the initial domain $[r_j, d_j - p_j]$ of each variable s_j.

The noOverlap constraint can be relaxed in at least three ways. One is to create convex hull or big-M relaxations for individual disjunctions implied by the constraint. The convex hull relaxations add a large number of auxiliary variables, while the big-M relaxations use only the original variables but are considerably weaker. Another approach is to use the continuous relaxation of a mixed-integer model of the constraint. Such models can be formulated for both discrete and continuous time domains. A third approach is to write a family of valid inequalities that are obtained by solving a certain optimization problem in advance. The second and third relaxation methods will also prove useful for cumulative scheduling.

None of the relaxations presented here for disjunctive and cumulative scheduling are particularly tight. This is probably because not even the convex hull relaxations, were they known, would be tight. It is therefore important to use relaxations in conjunction with the sophisticated filtering methods that have been developed for these constraints, some of which are described in Sections 6.14 and 6.15.

Relaxations of disjunctive scheduling subproblems are important in logic-based Benders methods, as Section 2.8 indicates. The relaxations are expressed in terms of master problem variables rather than the start time variables of the scheduling subproblems. They will be discussed in connection with cumulative scheduling in Section 7.13. Relaxations for the disjunctive scheduling problem are a special case of those for the cumulative scheduling problem, and there is no advantage in examining this case separately.

7.12.1 Disjunctive Relaxations

The most straightforward way to relax $\text{noOverlap}(s \,|\, p)$ is to relax each disjunction

$$(s_i + p_i \le s_j) \vee (s_j + p_j \le s_i) \tag{7.192}$$

individually—thus creating a factored relaxation (Section 7.6.3). The disjunctions can be given either a convex hull or a big-M relaxation. Unfortunately, both tend to result in a weak relaxation for the scheduling problem, except perhaps when the time windows are small.

The convex hull relaxation introduces a large number of auxiliary variables. Applying Theorem 7.14, the convex hull relaxation of each disjunction (7.192) is

$$
\begin{array}{ll}
-s'_i + s'_j \ge p_i y & s''_i - s''_j \ge p_i(1-y) \\
r_i y \le s'_i \le \ell_i y & r_i(1-y) \le s''_i \le \ell_i(1-y) \\
r_j y \le s'_j \le \ell_j y & r_j(1-y) \le s''_j \le \ell_j(1-y) \\
s_i = s'_i + s''_i, \;\; s_j = s'_j + s''_j, \;\; 0 \le y \le 1
\end{array}
\tag{7.193}
$$

where $\ell_i = d_i - p_i$ and where s'_i, s''_i, s'_j, s''_j and y are new variables. Because a separate copy of the new variables must be made for each disjunction, the relaxation of $\text{disjunctive}(s \,|\, p)$ becomes

$$
\left.
\begin{array}{ll}
-s'_{ij} + s'_{ji} \ge p_i y_{ij} & s''_{ij} - s''_{ji} \ge p_i(1-y_{ij}) \\
r_i y_{ij} \le s'_{ij} \le \ell_i y_{ij} & r_i(1-y_{ij}) \le s''_{ij} \le \ell_i(1-y_{ij}) \\
r_j y_{ij} \le s'_{ij} \le \ell_j y_{ij} & r_j(1-y_{ij}) \le s''_{ji} \le \ell_j(1-y_{ij})
\end{array}
\right\}
\begin{array}{l}
\text{all } i,j \\
\text{with} \\
i < j
\end{array}
$$
$$
s_i = s'_{ij} + s''_{ij}, \;\; \text{all } i,j \text{ with } i \ne j
$$
$$
0 \le y_{ij} \le 1, \;\; \text{all } i,j \text{ with } i < j
\tag{7.194}
$$

As an example, consider the scheduling problem illustrated in Fig. 7.26. The problem is to minimize the sum of the completion times

$$(s_1 + p_1) + (s_2 + p_2) + (s_3 + p_3) = s_1 + s_2 + s_3 + 6 \tag{7.195}$$

subject to disjunctive$(s \mid p)$ and the time windows shown in the figure. The optimal solution, which has value 13, is also shown in the figure.

In this instance, there are three disjunctions

$$
\begin{aligned}
(s_1 + p_1 \le s_2) \lor (s_2 + p_2 \le s_1) \\
(s_1 + p_1 \le s_3) \lor (s_3 + p_3 \le s_1) \\
(s_2 + p_2 \le s_3) \lor (s_3 + p_3 \le s_2)
\end{aligned}
\tag{7.196}
$$

Figure 7.27 shows the feasible set of the relaxation (7.193) of each disjunction, projected onto the variables s_j. Although each disjunction receives a convex hull relaxation, the relaxation (7.194) as a whole does not describe the convex hull of the feasible set. This is evident in the fact that the relaxed problem of minimizing (7.195) subject to (7.194) has a nonintegral optimal solution $(y_{12}, y_{13}, y_{23}) = (0, \frac{1}{2}, 1)$. Also, the optimal value 11 of the relaxation is less than the optimal value 13 of the original problem, although it provides a reasonably good lower bound in this small example.

One can simplify matters considerably by using a big-M relaxation of each disjunction rather than a convex hull relaxation. Due to Theorem 7.15, each disjunction (7.192) has the relaxation

$$
\begin{aligned}
\left(\frac{1}{d_j - r_i} - \frac{1}{d_i - r_j} \right) s_i + \left(\frac{1}{d_i - r_j} - \frac{1}{d_j - r_i} \right) s_j \\
\ge \frac{p_i}{d_i - r_j} + \frac{p_j}{d_j - r_i} - 1
\end{aligned}
\tag{7.197}
$$

along with the time window bounds. By writing this inequality for each pair i, j with $i < j$, and the bounds $r_j \le s_j \le d_j - p_j$ for each j, one can obtain a continuous relaxation without adding any variables.

Unfortunately, this relaxation tends to be weak. In the example of Fig. 7.26, it simplifies to

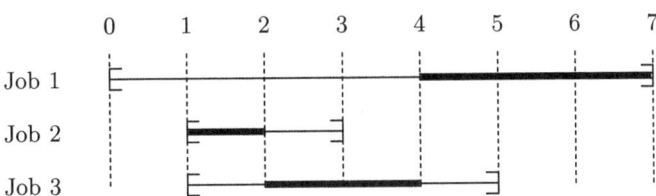

Fig. 7.26 Time windows (horizontal lines) for a 3-job disjunctive scheduling problem. The heavy lines show the solution that minimizes the sum of completion times. Note that the processing times are $(p_1, p_2, p_3) = (3, 1, 2)$.

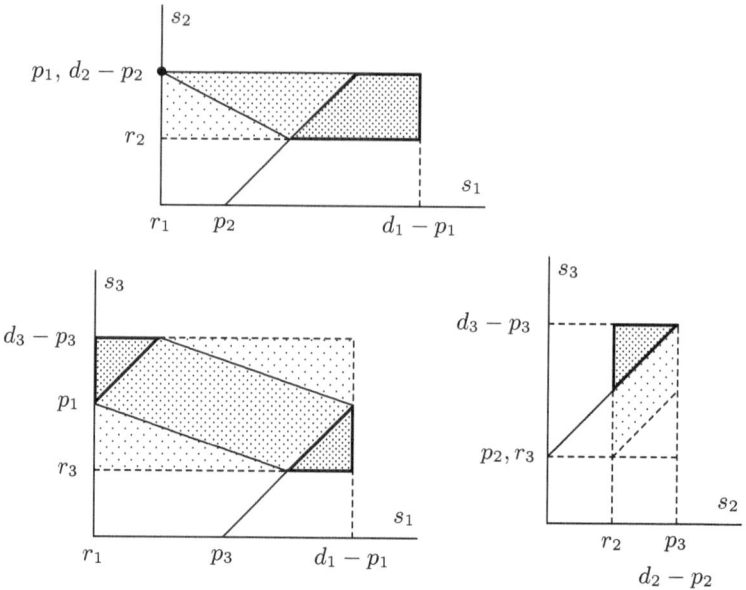

Fig. 7.27 Feasible sets (dark shading) of three disjunctions associated with the scheduling problem of Fig. 7.26. The dark and medium shading represent the feasible set of the convex hull relaxation of each disjunction, and the entire shaded area represents the feasible set of the big-M relaxation.

$$s_1 - s_2 \geq 2, \quad s_3 \geq s_2$$
$$0 \leq s_1 \leq 4, \quad 1 \leq s_2 \leq 2, \quad 1 \leq s_3 \leq 3 \tag{7.198}$$

As the figure illustrates, the relaxation is only slightly tighter than the time window bounds alone. In fact, when the relaxation is solved by minimizing (7.195) subject to (7.198), the solution is the same as would be obtained by using only the bounds; namely, $(s_1, s_2, s_3) = (0, 1, 1)$, with an optimal value of 8.

7.12.2 Mixed-Integer Relaxations

An alternative to using factored disjunctive relaxations is to use continuous relaxations of time-indexed or event-indexed MILP models. The time-indexed models may be practical when time can be discretized into a reasonably small number of equal units. The event-indexed models use continuous time variables but are more complicated and tend to have weaker relaxations. Yet, they may be more practical when the application requires a time horizon and granularity that result in many time increments.

The time-indexed model is the easiest to formulate. Let the 0-1 variable x_{jt} be 1 when job i starts at discrete time t. The variable appears for a particular pair i, t only when job j can start at time t without violating its time window; that is, when $r_j \leq t < d_j - p_j$. Note that, due to the discreteness of time, the job must finish strictly before its deadline. An MILP model can be written

$$\sum_j \sum_{t' \in T_{jt}} x_{jt'} \leq 1, \quad \text{all } t \quad (a)$$

$$\sum_t x_{jt} = 1, \quad \text{all } j \quad (b)$$

(7.199)

where each $x_{jt} \in \{0, 1\}$ and $T_{jt} = \{t' \mid t - p_j < t' \leq t\}$ is the set of times job j could have started if it is running at time t. Constraint (a) ensures that at most one job is running at any one time. Constraint (b) requires each job to be assigned exactly one start time. As usual, the continuous relaxation is formed by replacing $x_{jt} \in \{0, 1\}$ with $0 \leq x_{jt} \leq 1$. A variation on this model introduces an inventory variable z_t that is equal to 1 when a job is running at time t:

$$z_t = z_{t-1} + \sum_j x_{jt} - \sum_j x_{j,t-p_j} \quad \text{all } t$$

$$\sum_t x_{jt} = 1, \quad \text{all } j$$

(7.200)

$$z_t \leq 1, \quad \text{all } t$$

where each $x_{jt} \in \{0, 1\}$. If the first time is $t = 0$, the initial inventory at time $t = -1$ is $z_{-1} = 0$. The models (7.199) and (7.200) have equivalent relaxations, because (7.199) can be obtained from (7.200) by substitution of variables.

The example of Fig. 7.26 has the time-indexed relaxation

$$x_{10} + x_{11} + x_{21} + x_{31} \leq 1$$
$$x_{10} + x_{11} + x_{12} + x_{22} + x_{31} + x_{32} \leq 1$$
$$x_{11} + x_{12} + x_{13} + x_{32} + x_{33} \leq 1$$
$$x_{12} + x_{13} + x_{14} + x_{33} \leq 1$$
$$\sum_{t=0}^{4} x_{1t} = \sum_{t=1}^{2} x_{2t} = \sum_{t=1}^{3} x_{3t} = 1,$$
$$0 \leq x_{jt} \leq 1, \quad \text{all } j, t$$

Three redundant constraints are omitted. The objective function can be written $\sum_{jt}(t+p_j)x_{jt} = \sum_{jt} tx_{jt} + \sum_j p_j$. This relaxation is stronger than either of the disjunctive relaxations solved in the previous section, because its optimal value is 13, the same as the optimal value of the original problem. In fact, the optimal solution of the relaxation is the solution shown in Fig. 7.26.

When there are a large number of discrete times, it may be advantageous to use an event-indexed model in which time is continuous. In such a model, there are n events, each of which is the start of a job, and the 0-1 variable x_{jk} is 1 if event k is the start of job j. The continuous variable t_k is the time at which event k occurs. The model for disjunctive$(s\,|\,p)$ is

$$\sum_k x_{jk} = 1, \quad \text{all } j \qquad\qquad (a)$$

$$\sum_j x_{jk} = 1, \quad \text{all } k \qquad\qquad (b)$$

$$t_{k+1} - t_k \geq \sum_j p_j x_{jk}, \quad \text{all } k < n \quad (c)$$

$$t_k \geq \sum_j x_{jk} r_j, \quad \text{all } k \qquad\qquad (d) \qquad (7.201)$$

$$t_k \leq \sum_j (d_i - p_j)x_{jk} \quad \text{all } k \qquad (e)$$

$$x_{jk} \in \{0,1\}, \quad \text{all } j, k$$

Constraints (a) and (b) ensure that each job is assigned to one event and vice-versa. Constraint (c) prevents jobs from overlapping. Constraints (d) and (e) enforce release times and deadlines. In the small example of Fig. 7.26, the continuous relaxation of (7.201) yields the optimal solution.

Another MILP model for the disjunctive scheduling problem is suggested in Exercise 7.57 of Section 7.6.

7.12.3 A Class of Valid Inequalities

One way to relax a constraint is to generate valid inequalities of the form $f(x) \geq \alpha$, where $f(x)$ is some function of the constraint variables x. Here, α is the minimum of $f(x)$ subject to a relaxation R of the constraint that is chosen to make α easy to compute in advance. One can also maximize $f(x)$ subject to R and obtain a valid inequality $f(x) \leq \alpha$. Valid inequalities of this sort can then be used to form or enrich a relaxation of the constraint—perhaps a continuous relaxation. This can be advantageous when the resulting relaxation R' is tighter than R, or when the objective function of interest can be easily optimized subject to R' but not subject to R.

A simple way to apply this technique to disjunctive scheduling is to let $f(s)$ be a weighted sum $\sum_{j \in J} a_j s_j$ of some subset of start times. Let the relaxation R be the original disjunctive constraint, but with time windows relaxed so that every release time is $r_J = \min_{j \in J}\{r_j\}$ and every deadline is $d_J = \max_{j \in J}\{d_j\}$. Let $J = \{j_1, \ldots, j_k\}$ where the jobs are indexed so that $p_{j_1} \leq \cdots \leq p_{j_k}$. Now, if $a_{j_1} \geq \cdots \geq a_{j_k} \geq 0$, the problem of minimizing $f(s)$ subject to R can be solved by a greedy method: let job j_1 start at r_J, job j_2 immediately after it at $r_J + p_{j_1}$, and so forth. Then job j_i starts at

$$s_i^{\min} = r_J + \sum_{\ell=1}^{i-1} p_{j_\ell} \tag{7.202}$$

This solution clearly minimizes the weighted sum of finish times $\sum_{j \in J} a_j(s_j + p_j)$, and therefore minimizes the weighted sum of start times $\sum_{j \in J} a_j s_j$. One can also maximize $f(s)$ by scheduling the jobs in nonincreasing order of processing time, with the last job finishing at time d_J and job j_i starting at

$$s_i^{\max} = d_J - \sum_{\ell=1}^{i} p_{j_\ell} \tag{7.203}$$

So, one obtains the following result:

Theorem 7.51. *If $J = \{j_1, \ldots, j_k\} \subset \{1, \ldots, n\}$, $p_{j_1} \leq \cdots \leq p_{j_k}$, and $a_{j_1} \geq \cdots \geq a_{j_k} \geq 0$, then*

$$\sum_{j \in J} a_j s_j \geq \sum_{i=1}^{k} a_{j_i} s_i^{\min} \qquad \sum_{j \in J} a_j s_j \leq \sum_{i=1}^{k} a_{j_i} s_i^{\max} \tag{7.204}$$

are valid inequalities for disjunctive$(s \mid p)$, where $s_i^{\min}$ and $s_i^{\max}$ are given by (7.205) and (7.203), respectively. In particular (letting each $a_j = 1$),

$$\sum_{j \in J} s_j \geq kr_J + \sum_{i=1}^{k-1}(k-i)p_{j_i}$$

$$\sum_{j \in J} s_j \leq kd_J - \sum_{i=1}^{k}(k-i+1)p_{j_i}$$
(7.205)

are valid inequalities.

Let $J(t_1, t_2)$ be the set of jobs with time windows in the interval $[t_1, t_2]$; that is,

$$J(t_1, t_2) = \{j \mid t_1 \leq r_j, \ d_j \leq t_2\}$$

It is clear that all valid inequalities (7.204)–(7.205) are dominated by those corresponding to sets $J = J(r_j, d_k)$ for which $r_j < d_k$. In the example of Fig. 7.26, these sets are $J(0,3) = J(1,3) = \{2\}$, $J(0,5) = J(1,5) = J(1,7) = \{2,3\}$, and $J(0,7) = \{1,2,3\}$. The corresponding valid inequalities (7.205), omitting the singleton $\{2\}$, are

$$3 \leq s_2 + s_3 \leq 6$$
$$4 \leq s_1 + s_2 + s_3 \leq 11$$
(7.206)

The two upper bounds are redundant of the time window bounds $r_j \leq s_j \leq d_j - p_j$, but the two lower bounds are not.

One way to use the inequalities (7.206) is to add them to the time window bounds and big-M relaxation (7.198) to obtain a continuous relaxation R' in the original variables s. In this particular case, one can easily minimize the desired objective function (7.195) subject to R, by using the greedy method, but it is still advantageous to use R', because it is tighter than R and easy to solve as a linear programming problem. The minimum of (7.195) subject to R is 10, but its minimum subject to R' is 12.

Since the strength of the inequalities (7.204) tends to rise rapidly with the size of the set J, in larger problems it is reasonable to select a few large sets $J(r_j, d_k)$ to generate inequalities. It is not hard to design selection heuristics that yield strong inequalities. Such heuristics should recognize, however, that an inequality corresponding to $[r_k, d_k]$ does not necessarily dominate one corresponding to a proper subset of $[r_j, d_k]$.

Exercises

7.106. Verify that (7.193) is a convex hull relaxation of (7.192).

7.107. Consider the problem of minimizing $s_1 + s_2$ subject to the constraint disjunctive$((s_1, s_2) \,|\, (2, 2))$, with domains $s_1 \in [0, 2]$ and $s_2 \in [1, 3]$. Draw a graph of the problem and identify the feasible set. Write a disjunctive relaxation using the convex hull relaxation of each disjunction in the model. Since there is only one disjunction in this problem, this yields a convex hull relaxation for the entire problem. Given that the optimal solution is obvious upon inspection, indicate what must be the optimal value of each variable in the relaxation without actually computing a solution of the relaxation.

7.108. Show that (7.197) is a valid big-M relaxation of the disjunction (7.192).

7.109. Write the relaxation (7.197) for the problem of Exercise 7.107. Note that it and the bounds define a convex hull relaxation, although this is not true in general.

7.110. Write a discrete-time MILP model for the problem of Exercise 7.107.

7.111. Formulate a continuous-time MILP model for the problem of Exercise 7.107. In this case, solution of its continuous relaxation is feasible in the original problem.

7.112. Show that if $a_1 \geq \cdots \geq a_n$, $p_1 \leq \cdots \leq p_n$, and each $x_j \geq 0$, the minimum of $\sum_j a_j x_j$ subject to noOverlap$(x \,|\, p)$ is the greedy solution $x_j = \sum_{i=1}^{j-1} p_i$.

7.113. Verify the validity of cuts (7.205) in Theorem 7.51.

7.114. What are the cuts (7.205) for the example of Exercise 7.107?

7.115. It is possible to obtain cuts by applying Theorem 7.51 for a given J with $a \neq (1, \ldots, 1)$ that are not redundant of the cuts obtained with $a = (1, \ldots, 1)$?

7.116. Valid cuts are obtained above by defining a relaxation of the constraint noOverlap$(s \,|\, p)$ that can by solved by a greedy algorithm. In particular, the constraint is relaxed by replacing each time window $[r_j, d_j]$ with $[r_J, d_J]$. Other relaxations can be solved in a greedy fashion—for example a relaxation obtained by changing each time window to $[r_j, d_J]$ and each p_j to $\min_{j \in J}\{p_j\}$. Show how to obtain cuts from this relaxation. The resulting cuts could be useful when the processing times are about the same, but the release times differ substantially. Symmetric cuts can, of course, be obtained by setting each time window to $[r_J, d_j]$.

7.13 Cumulative Scheduling

Cumulative scheduling requires that jobs be scheduled so that their total rate of resource consumption at any one time never exceeds a given limit. Section 6.15 presents several filtering methods for the most popular cumulative scheduling constraint. The task of the present section is to derive some continuous relaxations for it.

Two of the three relaxation methods presented for disjunctive scheduling in Section 7.12 can be extended to cumulative scheduling. Continuous relaxations of MILP models can be written for the constraint and valid inequalities can be derived—in this case inequalities based on energetic reasoning.

Recall that in cumulative$(s \mid p, c, C)$, variables $s = (s_1, \ldots, s_n)$ represent the start time of the jobs. The parameter $p = (p_1, \ldots, p_n)$ contains the processing time p_j of each job j, and $c = (c_1, \ldots, c_n)$ contains the rate c_j of resource consumption for each job j. The constraint requires that the total rate of resource consumption of the jobs running at any time t never exceed C:

$$\sum_{\substack{j \\ s_j \le t \le s_j + p_j}} c_j \le C, \quad \text{for all times } t$$

There is also a time window $[r_j, d_j]$ for each job j, which is reflected in the initial domain $[r_j, d_j - p_j]$ of s_j.

7.13.1 Mixed-Integer Models

As in the case of disjunctive scheduling, one can formulate time-indexed and event-indexed models for cumulative scheduling. The time-indexed model is a straightforward generalization of the disjunctive scheduling model (7.199). Again the 0-1 variable x_{jt} is 1 when job i starts at discrete time t. The variable appears for a particular pair i, t only when $r_j \le t < d_j - p_j$. The model is

$$
\begin{aligned}
\sum_{j} \sum_{t' \in T_{jt}} c_j x_{jt'} &\le C, \quad \text{all } t \quad (a) \\
\sum_{t} x_{jt} &= 1, \quad \text{all } j \qquad (b) \\
x_{jt} &\in \{0, 1\}, \quad \text{all } j, t \qquad (c)
\end{aligned}
\tag{7.207}
$$

where $T_{jt} = \{t' \mid t - p_j < t' \le t\}$. Constraint (a) ensures that the total rate of resource consumption is at most C at any one time. Constraint (b) requires each job to be assigned exactly one start time. The continuous relaxation is formed by replacing (c) with $0 \le x_{jt} \le 1$.

A variation on the model uses an inventory variable z_t that keeps track of the resource consumption at each time t:

$$z_t = z_{t-1} + \sum_j c_j x_{jt} - \sum_j c_j x_{j,t-p_j} \quad \text{all } t$$

$$\sum_t x_{jt} = 1, \quad \text{all } j \qquad\qquad (7.208)$$

$$z_t \le C, \quad \text{all } t$$

$$x_{jt} \in \{0, 1\}, \quad \text{all } j, t$$

The initial inventory at time $t = -1$ is $z_{-1} = 0$. The models (7.199) and (7.200) have equivalent continuous relaxations.

A small cumulative scheduling problem appears in Fig. 7.28. If the objective is to minimize the sum of the finish times, the optimal value is 11. The time-indexed relaxation (7.207) for this problem instance is

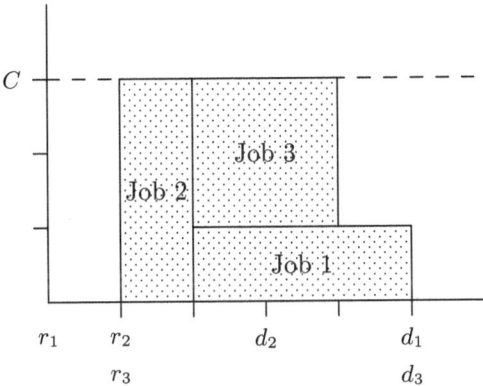

Fig. 7.28 Optimal solution of a small cumulative scheduling problem in which the objective is to minimize the sum of completion times. The horizontal axis represents time and the vertical axis represents resource consumption.

$$x_{10} + x_{11} + 3x_{21} + 2x_{31} \leq 3$$
$$x_{10} + x_{11} + x_{12} + 3x_{22} + 2(x_{31} + x_{32}) \leq 3$$
$$x_{11} + x_{12} + 2(x_{32} + x_{33}) \leq 3 \qquad\qquad (7.209)$$
$$x_{10} + x_{11} + x_{12} = x_{21} + x_{22} = x_{31} + x_{32} + x_{33} = 1$$
$$x_{jt} \geq 0, \quad \text{all } j, t$$

where two redundant capacity constraints are omitted. The minimum value of the objective function $\sum_{jt} t x_{jt} + \sum_j p_j$ subject to (7.209) is $9\frac{5}{6}$, which is a lower bound on the optimal value of the original problem.

When there are a large number of discrete times, it may be advantageous to use one of two event-indexed models that employ continuous time variables. In these models, there are $2n$ events, each of which can be the start of a job or the finish of a job. In one model, the 0-1 variable $x_{jkk'}$ is 1 if event k is the start of job j and event k' is the finish of job j. The continuous variable t_k is the time at which event k occurs. The inventory variable z_k keeps track of how much resource is being consumed when event k occurs; obviously one wants $z_k \leq C$. The model for cumulative$(s \mid p, c, C)$ is

$$z_k = z_{k-1} + \sum_j \sum_{k'>k} c_j x_{jkk'} - \sum_j \sum_{k'<k} c_j x_{jk'k}, \quad \text{all } k \qquad (a)$$

$$z_0 = 0, \quad 0 \leq z_k \leq C, \quad \text{all } k \qquad (b)$$

$$\sum_k \sum_{k'>k} x_{jkk'} = 1, \quad \text{all } j \qquad (c)$$

$$\sum_j \sum_{k'>k} x_{jkk'} + \sum_j \sum_{k'<k} x_{jk'k} = 1, \quad \text{all } k \qquad (d)$$

$$t_{k'} - t_k = \sum_j p_j x_{jkk'}, \quad \text{all } k, k' \text{ with } k < k' \qquad (e) \qquad (7.210)$$

$$t_k \geq \sum_j \sum_{k'>k} x_{jkk'} r_j, \quad \text{all } k \qquad (f)$$

$$t_k \leq \sum_j \sum_{k'>k} x_{jkk'}(d_j - p_j) + \sum_j \sum_{k'<k} x_{jk'k} d_j, \quad \text{all } k \qquad (g)$$

$$x_{jkk'} \in \{0,1\}, \text{all } j, k, k'$$

where $d_{\max} = \max_j\{d_j\}$ and M is a large number (e.g., the total length of the time horizon). Constraint (a) keeps track of how much resource is being consumed, and (b) imposes the upper limit. Constraint (c) makes sure that each job starts once and finishes once. Constraint (d) requires each event to be associated with the start or finish of exactly one job. Constraint (e) defines the time lapse between the start and finish events for a job. Constraint (f) enforces release times by requiring that any start event be scheduled no early than the release time of the associated job. If event k is a finish event, the constraint is innocuous because the right-hand side is zero. Constraint (g) enforces deadlines by requiring that a start event allow time for the deadline and processing time, and a finish event occur no later than the deadline. The latter is redundant, but writing the constraint in this fashions ensures that the right-hand side never vanishes (because every event is a start or finish).

If one wishes to relate the event times t_k to the start times s_j, the following constraints can be added

$$s_j = \sum_k \sum_{k'>k} t_k x_{jkk'}, \quad \text{all } j$$

However, this constraint may be unnecessary in a particular context. For instance, if the objective is to minimize latest finish time, one can minimize y and add the constraint $y \geq t_k$ for all k, and the start time variables s_j are unnecessary.

The objective for the example of Fig. 7.28, which is to minimize the sum of finish times, can also be written in terms of the event times t_k without using the start time variables s_k. If $\sum_S$ is the sum of start times and $\sum_F$ the sum of finish times, then

$$\sum_F = \sum_S + \sum_j p_j$$

So,

$$2\sum_F = \left(\sum_S + \sum_F\right) + \sum_j p_j = \sum_j t_j + \sum_j p_j$$

which implies

$$\sum_F = \tfrac{1}{2}\sum_j t_j + \tfrac{1}{2}\sum_j p_j$$

Thus, one can minimize this quantity subject to the relaxation (7.210).

Model (7.201) is quite large due to the triply indexed variables $x_{jkk'}$. An alternative is to use separate variables for start-events and finish-events, which reduces the triple index to a double index at the cost of producing a weaker relaxation. Let the 0-1 variable x_{jk} be 1 when event k is the start of job j, and $y_{jk} = 1$ when event k is the finish of job j. The new continuous variable f_j is the finish time of job j:

$$z_k = z_{k-1} + \sum_j c_j x_{jk} - \sum_j c_j y_{jk}, \quad \text{all } k \qquad (a)$$

$$z_0 = 0, \quad z_k \leq C, \quad \text{all } k \qquad (b)$$

$$\sum_k x_{jk} = 1, \quad \sum_k y_{jk} = 1, \quad \text{all } j \qquad (c)$$

$$\sum_j x_{jk} + y_{jk} = 1, \quad \text{all } k \qquad (d)$$

$$t_{k-1} \leq t_k, \quad x_{jk} \leq \sum_{k'>k} y_{jk'}, \quad \text{all } k > 1, \text{ all } j \qquad (e) \qquad (7.211)$$

$$t_k \geq \sum_j r_j x_{jk}, \quad \text{all } k \qquad (f)$$

$$t_k + p_j x_{jk} - d_{\max}(1 - x_{jk}) \leq f_j$$
$$\leq t_k + d_{\max}(1 - y_{jk}), \quad \text{all } j, k \qquad (g)$$

$$f_j \leq d_j, \quad \text{all } j \qquad (h)$$

$$x_{jk}, y_{jk} \in \{0, 1\}$$

where $d_{\max} = \max_j \{d_j\}$. Constraints (a) and (b) perform the same function as before. Constraints (c) and (d) require that each job start once and end once, but not as the same event. Constraints (e) are redundant but may tighten the relaxation. One constraint requires the events to occur in chronological order, and one requires a job's start-event to have a smaller index than its finish-event. Constraint (f) observes the release times. The new element is constraint (g). The first inequality defines the finish time f_j of each job by forcing it to occur no earlier than p_j time units after the start time. The second inequality forces the time associated with the finish-event to be no earlier than the finish time. Finally, constraint (h) enforces the deadlines.

Again, the objective function that measures the sum of finish times is $\frac{1}{2} \sum_j s_j + \frac{1}{2} \sum_j p_j$. Another mixed-integer model for the cumulative scheduling problem is suggested in Exercise 7.58 of Section 7.6.

7.13.2 A Class of Valid Inequalities

As in the case of disjunctive scheduling, one can develop a class of valid inequalities $\sum_{j \in J} a_j s_j \geq \alpha$ for cumulative scheduling by letting α be the minimum of $\sum_{j \in J} a_j s_j$ subject to a relaxation R of the constraint.

In this case, the relaxation R not only relaxes the time windows to $[r_J, d_J]$ for every job j, but it also allows any job j to be replaced by any job with the same energy consumption $p_j c_j$ and with resource consumption rate at most C. Let the jobs be indexed so that $J = \{j_1, \ldots, j_k\}$ and $p_{j_1} c_{j_1} \leq \cdots \leq p_{j_k} c_{j_k}$. Then, one feasible solution of R is a *compressed* schedule that replaces each job j with a job having duration $p_j c_j / C$ and resource consumption rate C, and schedules jobs $j_1, \ldots, j_k$ consecutively with job j_1 starting at r_J. Then each job j_i in the compressed schedule begins at

$$s_i^{\min} = r_J + \frac{1}{C} \sum_{\ell=1}^{i-1} p_{j_\ell} c_{j_\ell} \tag{7.212}$$

and finishes at $s_i^{\min} + p_{j_i} c_{j_i} / C$. The compressed schedule for the problem of Fig. 7.28 is illustrated in Fig. 7.29.

It will be shown that if $a_{j_1} \geq \cdots \geq a_{j_k} \geq 0$ and f_j is the finish time of job j, the compressed schedule minimizes

$$\sum_{j \in J} a_j f_j = g(f)$$

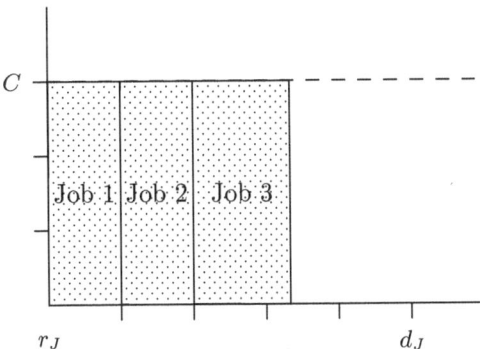

Fig. 7.29 Compressed schedule for the cumulative scheduling problem of Fig. 7.28. The order of jobs 1 and 2 could be reversed.

subject to R. One can also maximize $g(f)$ by scheduling each job j_i at

$$s_i^{\max} = d_J - \frac{1}{C} \sum_{\ell=1}^{i} p_{j_\ell} c_{j_\ell} \qquad (7.213)$$

This leads to the following theorem.

Theorem 7.52. *If $J = \{j_1, \ldots, j_k\} \subset \{1, \ldots, n\}$, $p_{j_1}c_{j_1} \leq \cdots \leq$*
$p_{j_k}c_{j_k}$, and $a_{j_1} \geq \cdots \geq a_{j_k} \geq 0$, then

$$\sum_{j \in J} a_j s_j \geq \sum_{i=1}^{k} a_{j_i} s_i^{\min} - \sum_{j \in J} a_j p_j \left(1 - \frac{c_j}{C}\right) \qquad (7.214)$$

and

$$\sum_{j \in J} a_j s_j \leq \sum_{i=1}^{k} a_{j_i} s_i^{\max} \qquad (7.215)$$

are valid inequalities for cumulative($s \mid p, c, C$), where $s_i^{\min}$ and $s_i^{\max}$
are given by (7.212) and (7.213). In particular (letting each $a_j = 1$),

$$\sum_{j \in J} s_j \geq k r_J + \frac{1}{C} \sum_{i=1}^{k} (k - i + 1) p_{j_i} c_{j_i} - \sum_{i=1}^{k} p_{j_i}$$

$$\sum_{j \in J} s_j \leq k d_J - \frac{1}{C} \sum_{i=1}^{k} (k - i + 1) p_{j_i} c_{j_i} \qquad (7.216)$$

are valid inequalities.

Proof. Let $\hat{R}$ be a restriction of R in which each job j has duration $p_j c_j / C$ and resource consumption rate C. Then it is clear that the compressed schedule $\hat{f}$ minimizes $g(f)$ subject to $\hat{R}$. It will be shown that $g(\hat{f})$ is less than or equal to the minimum of $g(f)$ subject to R. Since R is a relaxation of $\hat{R}$, it follows that $\hat{f}$ minimizes $g(f)$ subject to R, which implies that

$$\sum_{j \in J} a_j f_j \geq \sum_{i=1}^{k} a_{j_i} \left(s_i^{\min} + \frac{p_{j_i} c_{j_i}}{C}\right)$$

is a valid inequality. Since $f_j = s_j - p_j$ in the original problem, inequality (7.214) follows. Inequality (7.215) is derived in a similar fashion.

It remains to show that $g(\hat{f}) \leq g(\bar{f})$ for any feasible solution $\bar{f}$ of R. This will be shown by induction on the number k of jobs. For $k = 1$, it suffices to note that if a job j finishes at $\bar{f}_j$, then because the job starts no earlier than r_J and has duration of at least $p_j c_j / C$, the compressed schedule $\hat{f}_j = r_J + p_j c_j / C$ results in $\hat{f}_j \leq \bar{f}_j$. This implies $g(\hat{f}_j) \leq g(\bar{f}_j)$.

Now suppose that the claim is true for $k-1$ jobs and show it is true for k jobs. Take any feasible schedule $\bar{f}$ of R, and index the jobs so that job k is the last job to finish. By the induction hypothesis, there is a compressed schedule $(\hat{f}_1, \ldots, \hat{f}_{k-1})$ on jobs $1, \ldots, k-1$ for which

$$g(\hat{f}_1, \ldots, \hat{f}_{k-1}) \leq g(\bar{f}_1, \ldots, \bar{f}_{k-1}) \tag{7.217}$$

It must be shown that the compressed schedule $\hat{f}$ for jobs $1, \ldots, k$ is feasible and satisfies $g(\hat{f}) \leq g(\bar{f})$. Note that $\hat{f}_k \leq \bar{f}_k$, because no schedule for k jobs can finish all the jobs sooner than the compressed schedule does. Thus $\hat{f}_k \leq d_J$, since $\bar{f}_k \leq d_J$, and $\hat{f}$ is therefore feasible. Also, $\hat{f}_k \leq \bar{f}_k$ and (7.217) imply $g(\hat{f}) \leq g(\bar{f})$. $\square$

As in the case of disjunctive scheduling, one need only consider inequalities corresponding to sets $J = J(r_j, d_k)$ with $r_j < d_k$. As noted in Section 7.12.3, these sets for the example of Fig. 7.28 are $\{1, 2, 3\}$ and $\{2, 3\}$ (ignoring the singleton $\{2\}$). The corresponding inequalities (7.216) are

$$\tfrac{1}{3} \leq s_1 + s_2 + s_3 \leq 8\tfrac{2}{3}$$
$$2\tfrac{1}{3} \leq s_2 + s_3 \leq 6\tfrac{2}{3}$$

Two of the bounds ($8\tfrac{2}{3}$ and $2\tfrac{1}{3}$) are nonredundant of the time window bounds.

The inequalities (7.214)–(7.215) grow rapidly in strength as the size of J increases. It is therefore advisable to design a heuristic that selects a few large sets J to generate inequalities.

7.13.3 Relaxation of Benders Subproblems

The success of Benders methods for planning and scheduling can depend critically on the inclusion of a subproblem relaxation in the master problem. Since the subproblem is a disjunctive or cumulative scheduling problem, it is essential to develop relaxations for this type of problem. This type of relaxation differs from those discussed above, because it must be expressed in terms of the master problem variables, which assign jobs to facilities, rather than variables that specify start times. The relaxation should also take the form of linear inequalities, because the master problem is frequently modeled as an integer linear programming problem. The relaxations presented here are designed for cumulative scheduling, but are readily specialized to disjunctive scheduling.

The Benders approach to planning and scheduling is described in Section 2.8, which presents a problem in which several jobs are assigned to facilities and scheduled on them. Sections 6.14.3 and 6.15.5 discuss in more detail how Benders cuts may be generated for the master problem.

Time-Window Relaxation

The simplest relaxation is based on the fact that the total running time of any subset of jobs assigned to the same facility must fit within the earliest release time and latest deadline for those jobs. As illustrated in the example of Section 2.8, a relaxation of this sort can contain many redundant inequalities. The following analysis enables one to identify the redundant inequalities and delete them.

To review the notation, p_{ij} is the processing time of job j on facility i, and c_{ij} is the rate of resource consumption. The maximum rate of resource consumption on facility i is C_i. The master problem contains 0-1 variables x_{ij} that are equal to 1 when job j is assigned to facility i. The subproblem contains variables s_j that represent the start time of job j. The task is to write a linear relaxation of the time window constraints on s_j in terms of the variables x_{ij}. Such a relaxation is useful whenever there are hard constraints on the release time and finish time of the jobs, such as in the minimum-cost problem (6.108).

Recall that the energy consumed by job j on machine i is $p_{ij}c_{ij}$, and $p_{ij}c_{ij}/C_i$ is the task interval of the job. Let $J(t_1, t_2)$ be the set of jobs j whose time windows lie in the interval $[t_1, t_2]$, so that

$$J(t_1, t_2) = \{j \mid [r_j, d_j] \subset [t_1, t_2]\}$$

The total task interval of jobs in $J(t_1, t_2)$ assigned to a given machine i must fit into the time interval $[t_1, t_2]$:

$$\frac{1}{C_i} \sum_{j \in J_i(t_1, t_2)} p_{ij}c_{ij}x_{ij} \leq t_2 - t_1 \qquad (7.218)$$

It is convenient to refer to the inequality (7.218) as $R_i(t_1, t_2)$. Let $r_1 < \cdots < r_p$ be the distinct release times and $d_1 < \cdots < d_q$ the distinct deadlines. Then, the following is a valid linear relaxation for facility i:

$R_i(r_j, d_k)$, all $j \in \{1, \ldots, p\}$ and $k \in \{1, \ldots, q\}$ with $r_j < d_k$

This relaxation can be added to the master problem for each facility i. A relaxation for disjunctive scheduling is obtained by setting $C_i = c_{ij} = 1$ for all i, j.

Redundant inequalities in the relaxation are identified as follows. Let

$$T_i(t_1, t_2) = \frac{1}{C_i} \sum_{j \in J(t_1, t_2)} p_{ij} c_{ij} - t_2 + t_1$$

measure the *tightness* of $R_i(t_1, t_2)$. The following is easily verified.

Lemma 7.53 $R_i(t_1, t_2)$ *dominates* $R_i(u_1, u_2)$ *when* $[t_1, t_2] \subset [u_1, u_2]$ *and* $T_i(t_1, t_2) \geq T_i(u_1, u_2)$.

The algorithm of Fig. 7.30 now allows one to delete redundant inequalities. It has $O(n^3)$ complexity in the worst case, where n is the number of jobs, because it is possible that none of the inequalities are eliminated. This occurs, for example, when each job j has release time $j - 1$ and deadline j, and $p_{ij} = 2$.

Relaxation for Minimizing Makespan

The minimum makespan problem is formulated in (6.113). If M is the makespan of a schedule on facility i, then for any time t the total task interval of the jobs in $J(t, \infty)$ must fit in the interval $[t, M]$. This leads to a lower bound on the makespan:

Let $\mathcal{R}_i = \emptyset$.
For $j = 1, \ldots, p$:
    Set $k' = 0$.
    For $k = 1, \ldots, q$:
        If $d_k > r_j$ and $T_i(r_j, d_k) > T_i(r_j, d_{k'})$ then
            Remove from $\mathcal{R}_i$ all $R_i(r_{j'}, d_k)$ for which $j' < j$ and
            $T_i(r_j, d_k) \geq T_i(r_{j'}, d_k)$.
            Add $R_i(r_j, d_k)$ to $\mathcal{R}_i$ and set $k' = k$.

Fig. 7.30 Algorithm for generating an inequality set $\mathcal{R}_i$ that relaxes the time window constraints for facility i, where $r_1 < \cdots < r_p$ are the distinct release times and $d_1 < \cdots < d_q$ are the distinct deadlines. By convention, $d_0 = -\infty$ and $T_i(r_j, -\infty) = 0$.

> Let $\mathcal{R}_i = \emptyset$ and set $k = q + 1$.
> For $j = p, \ldots, 1$:
> If $T_i(r_j) > T_i(r_k)$ then add $R_i(r_j)$ to $\mathcal{R}_i$ and set $k = j$.

Fig. 7.31 Algorithm for generating a relaxation $\mathcal{R}_i$ for minimizing makespan, where $r_1 < \cdots < r_p$ are the distinct release times. By convention, $r_{q+1} = \infty$ and $T_i(\infty) = 0$.

$$M \geq t + \frac{1}{C_i} \sum_{j \in J(t,\infty)} p_{ij} c_{ij} x_{ij}$$

Let this be inequality $R_i(t)$. Then if $r_1 < \cdots < r_p$ are the distinct release times, the following relaxation can be added to 0-1 formulation of the master problem (6.114) for each facility i:

$$R_i(r_j), \text{ all } j \in \{1, \ldots, p\}$$

A relaxation of this sort is illustrated in Section 2.8.

Let the tightness of $R_i(t)$ be

$$T_i(t) = t + \frac{1}{C_i} \sum_{j \in J(t,\infty)} p_{ij} c_{ij}$$

Then, $R_i(t)$ dominates $R_i(u)$ if $t \leq u$ and $T_i(t) \geq T_i(u)$. An $O(n)$ algorithm for eliminating redundant inequalities appears in Fig. 7.31.

Relaxation for Minimizing Late Jobs

The problem of minimizing the number of late jobs is formulated in (6.119). A linear lower bound for the number of late jobs can be derived much as for the makespan and included in the 0-1 formulation of the master problem.

As before, the total running time of the jobs in $J(t_1, t_2)$ assigned to facility i is at least their total task interval

$$\frac{1}{C_i} \sum_{\ell \in J(t_1,t_2)} c_{i\ell} p_{i\ell} x_{i\ell}$$

If this quantity is greater than $t_2 - t_1$, then at least one job is late. In fact the number L_i of late jobs on facility i can be bounded as follows:

$$L_i \geq \frac{\dfrac{1}{C_i} \displaystyle\sum_{\ell \in J(t_1,t_2)} c_{i\ell} p_{i\ell} x_{i\ell} - t_2 + t_1}{\displaystyle\max_{\ell \in J(t_1,t_2)} \{p_{i\ell}\}}$$

If this inequality is denoted $\bar{R}_i(t_1, t_2)$, then one can add to the master problem the following bound on the total number L of late jobs. Again, let $r_1, \ldots, r_p$ be the distinct release times and $d_1, \ldots, d_q$ be the distinct deadlines.

$$L \geq \sum_i L_i$$

$\bar{R}_i(r_j, d_k)$, all $j \in \{1, \ldots, p\}$ and $k \in \{1, \ldots, q\}$ with $r_j < d_k$, all i

$L_i \geq 0$, all i

$$(7.219)$$

As an example, suppose that four jobs are to be assigned to, and scheduled on, two facilities. The problem data appear in Table 7.3. There is one distinct release time $r_1 = 0$ and four distinct deadlines $(d_1, d_2, d_3, d_4) = (2, 3, 4, 5)$. The relaxation (7.219) is

$$
\begin{aligned}
&L \geq L_1 + L_2 \\
&L_1 \geq x_{11} - 1, \quad L_1 \geq \tfrac{1}{2}x_{11} + \tfrac{2}{3}x_{12} - \tfrac{3}{4}, \\
&\quad L_1 \geq \tfrac{2}{5}x_{11} + \tfrac{8}{15}x_{12} + \tfrac{1}{3}x_{13} - \tfrac{4}{5}, \\
&\quad L_1 \geq \tfrac{1}{3}x_{11} + \tfrac{4}{9}x_{12} + \tfrac{5}{18}x_{13} + \tfrac{1}{3}x_{14} - \tfrac{5}{6} \\
&L_2 \geq x_{21} - \tfrac{1}{2}, \quad L_2 \geq \tfrac{4}{5}x_{11} + \tfrac{2}{3}x_{22} - \tfrac{3}{5}, \\
&\quad L_2 \geq \tfrac{2}{3}x_{21} + \tfrac{5}{9}x_{22} + \tfrac{1}{3}x_{23} - \tfrac{2}{3}, \\
&\quad L_2 \geq \tfrac{2}{3}x_{21} + \tfrac{5}{9}x_{22} + \tfrac{1}{3}x_{23} + \tfrac{5}{6}x_{24} - \tfrac{5}{6} \\
&L_1, L_2 \geq 0
\end{aligned}
\tag{7.220}
$$

Table 7.3 Data for a planning and scheduling problem instance with four jobs and two facilities. The release times r_j are all zero, and each facility i has capacity $C_i = 3$.

		Facility 1			Facility 2		
j	d_j	p_{1j}	c_{1j}	$p_{1j}c_{1j}$	p_{2j}	c_{2j}	$p_{2j}c_{2j}$
1	2	2	3	6	4	3	12
2	3	4	2	8	5	2	10
3	4	5	1	5	6	1	6
4	5	6	3	18	5	3	15

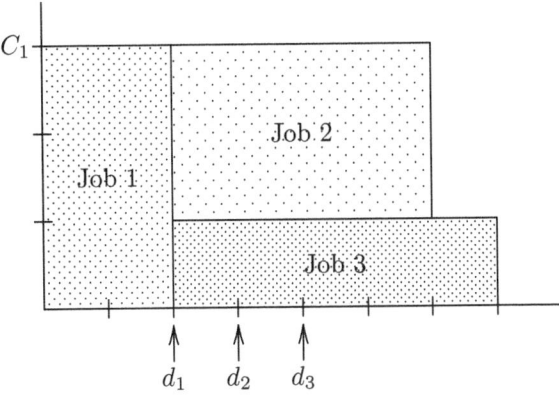

Fig. 7.32 Schedule for jobs 1–3 on facility 1 that minimizes the number of late jobs (2) and minimizes total tardiness (6).

For instance, if jobs 1, 2 and 3 are assigned to facility 1 (this occurs when $x_{11} = x_{12} = x_{13} = 1$ and $x_{14} = 0$), the bounds in (7.220) for facility 1 corresponding to d_1, d_2, d_3 become $L_1 \geq 0$, $L_1 \geq \frac{5}{12}$, and $L_1 \geq \frac{7}{15}$, respectively (the bound for d_4 is redundant). This implies that at least $\lceil \frac{7}{15} \rceil = 1$ job is late on facility 1. In fact, two jobs must be late, as in the optimal schedule of Fig. 7.32.

Relaxation for Minimizing Total Tardiness

The problem of minimizing total tardiness has almost the same formulation (6.119) as that for minimizing the number of late jobs. The differences are that the objective function is $\sum_j T_j$ (where T_j is the tardiness of job j) and the conditional constraint is dropped.

Two types of bounds can be developed for the total tardiness. The first and simpler bound is similar to that obtained for minimizing the number of late jobs. It is based on the following lemma. Let α^+ denote $\max\{0, \alpha\}$.

Lemma 7.54 *Consider a minimum total tardiness problem in which jobs $j = 1, \ldots, n$ with time windows $[r_j, d_j]$ are scheduled on a single facility i, where $\min_j \{r_j\} = 0$. The total tardiness incurred by any feasible solution is bounded below by*

$$\left(\frac{1}{C_i} \sum_{j \in J(0, d_k)} p_{ij} c_{ij} - d_k \right)^+$$

for each $k = 1, \ldots, n$.

Proof. For any k, the last scheduled job in the set $J(0, d_k)$ can finish no earlier than time $t = \frac{1}{C_i} \sum_{j \in J(0,d_k)} p_{ij} c_{ij}$. Since the last job has due date no later than d_k, its tardiness is no less than $(t - d_k)^+$. Thus, total tardiness is no less than $(t - d_k)^+$. $\square$

In the example of Table 7.3, suppose again that jobs 1–3 are assigned to facility 1. The bounds of Lemma 7.54 are

$$
\begin{aligned}
\text{for } J(0, d_1) = \{1\}: & \quad \left(\tfrac{1}{3}(6) - 2\right)^+ = 0 \\
\text{for } J(0, d_2) = \{1, 2\}: & \quad \left(\tfrac{1}{3}(6 + 8) - 3\right)^+ = 1\tfrac{2}{3} \\
\text{for } J(0, d_3) = \{1, 2, 3\}: & \quad \left(\tfrac{1}{3}(6 + 8 + 5) - 4\right)^+ = 2\tfrac{1}{3}
\end{aligned}
$$

Since the data are integral, the minimum tardiness is at least $\lceil 2\tfrac{1}{3} \rceil = 3$; it is 6 in the optimal schedule of Fig. 7.32.

Lemma 7.54 gives rise to the following bounds, which can be added to a master problem in which T represents the total tardiness:

$$
T \geq \sum_i T_i^L
$$

$$
T_i^L \geq \frac{1}{C_i} \sum_{j \in J(0,d_k)} p_{ij} c_{ij} x_{ij} - d_k, \quad \text{all } i, k \tag{7.221}
$$

$$
T_i^L \geq 0, \quad \text{all } i
$$

In the example, the relaxation (7.221) becomes

$$
\begin{aligned}
T \geq {} & T_1^L + T_2^L \\
T_1^L \geq {} & 2x_{11} - 2, \quad T_1^L \geq 2x_{11} + \tfrac{8}{3}x_{12} - 3, \\
T_1^L \geq {} & 2x_{11} + \tfrac{8}{3}x_{12} + \tfrac{5}{3}x_{13} - 4, \\
T_1^L \geq {} & 2x_{11} + \tfrac{8}{3}x_{12} + \tfrac{5}{3}x_{13} + 2x_{14} - 5 \\
T_2^L \geq {} & 4x_{21} - 2, \quad T_2^L \geq 4x_{11} + \tfrac{10}{3}x_{22} - 3, \\
T_2^L \geq {} & 4x_{21} + \tfrac{10}{3}x_{22} + 2x_{23} - 4, \\
L_2 \geq {} & 4x_{21} + \tfrac{10}{3}x_{22} + 2x_{23} + 5x_{24} - 5 \\
T_1^L, T_2^L \geq {} & 0
\end{aligned} \tag{7.222}
$$

A second type of bound can be developed on the basis of the following lemma. For each facility i, let π_i be a permutation of $\{1, \ldots, n\}$ that orders the jobs by increasing energy on facility i; that is, $p_{i\pi_i(1)}c_{i\pi_i(1)} \leq \cdots \leq p_{\pi_i(n)}c_{\pi_i(n)}$. One can obtain a lower bound on the tardiness incurred by the job with the kth latest deadline by supposing that it finishes no sooner than the task interval of the k jobs with the smallest energies. More precisely:

Lemma 7.55 *Consider a minimum total tardiness problem in which jobs $1, \ldots, n$ with time windows $[r_j, d_j]$ are scheduled on a single facility i. Assume $\min_j\{r_j\} = 0$ and index the jobs so that $d_1 \leq \cdots \leq d_n$. Then the total tardiness T of any feasible solution is bounded below by $\underline{T} = \sum_{k=1}^{n} \underline{T}_k$, where*

$$\underline{T}_k = \left(\frac{1}{C_i} \sum_{j=1}^{k} p_{i\pi_i(j)}c_{i\pi_i(j)} - d_k \right)^+, \quad k = 1, \ldots, n$$

Proof. Consider any feasible solution of the one-facility minimum tardiness problem, in which jobs $1, \ldots, n$ are respectively scheduled at times $t_1, \ldots, t_n$. The minimum tardiness is

$$T^* = \sum_{k=1}^{n}(t_k + p_{ik} - d_k)^+ \tag{7.223}$$

Let $\sigma_0(1), \ldots, \sigma_0(n)$ be the order in which jobs are scheduled in this solution, so that $t_{\sigma_0(1)} \leq \cdots \leq t_{\sigma_0(n)}$. For an arbitrary permutation σ of $\{1, \ldots, n\}$, let

$$\underline{T}_k(\sigma) = \left(\frac{1}{C_i} \sum_{j=1}^{k} p_{i\pi_i(j)}c_{i\pi_i(j)} - d_{\sigma(k)} \right)^+ \tag{7.224}$$

and $\underline{T}(\sigma) = \sum_{k=1}^{n} \underline{T}_k(\sigma)$.

It will first be shown that $T^* \geq \underline{T}(\sigma_0)$. Since σ_0 is a permutation, one can write (7.223) as

$$T^* = \sum_{k=1}^{n} \left(t_{\sigma_0(k)} + p_{i\sigma_0(k)} - d_{\sigma_0(k)} \right)^+$$

Observe that

$$T^* \geq \sum_{k=1}^{n} \left(\frac{1}{C_i} \sum_{i=1}^{k} p_{i\sigma_0(j)}c_{i\sigma_0(j)} - d_{\sigma_0(k)} \right)^+$$

$$\geq \sum_{k=1}^{n} \left(\frac{1}{C_i} \sum_{j=1}^{k} p_{i\pi_0(j)}c_{\pi_0(j)} - d_{\sigma_0(k)} \right)^+ = \underline{T}(\sigma_0)$$

where the first inequality is based on the energy required by jobs, and the second inequality is due to the definition of π_i.

Now, perform a bubble sort on the integers $\sigma_0(1), \ldots, \sigma_0(n)$ to put them in increasing order, and let $\sigma_0, \ldots, \sigma_P$ be the resulting series of permutations. Thus $(\sigma_P(1), \ldots, \sigma_P(n)) = (1, \ldots, n)$, and σ_{p+1} is obtained from σ_p by swapping two adjacent terms $\sigma_p(k)$ and $\sigma_p(k+1)$, where $\sigma_p(k) > \sigma_p(k+1)$. This means σ_p and σ_{p+1} are the same, except that $\sigma_{p+1}(k) = \sigma_p(k+1)$ and $\sigma_{p+1}(k+1) = \sigma_p(k)$. Because $T^* \geq \underline{T}(\sigma_0)$ and $\underline{T}(\sigma_P) = \underline{T}$, to prove the theorem it suffices to show $\underline{T}(\sigma_0) \geq \cdots \geq \underline{T}(\sigma_P)$.

For any two adjacent permutations σ_p, σ_{p+1}, it can be shown as follows that $\underline{T}(\sigma_p) \geq \underline{T}(\sigma_{p+1})$. One can observe that

$$\underline{T}(\sigma_p) = \sum_{j=1}^{k-1} \underline{T}_j(\sigma_p) + \underline{T}_k(\sigma_p) + \underline{T}_{k+1}(\sigma_p) + \sum_{j=k+2}^{n} \underline{T}_j(\sigma_p)$$

$$\underline{T}(\sigma_{p+1}) = \sum_{j=1}^{k-1} \underline{T}_j(\sigma_p) + \underline{T}_k(\sigma_{p+1}) + \underline{T}_{k+1}(\sigma_{p+1}) + \sum_{j=k+2}^{n} \underline{T}_j(\sigma_p)$$

$$(7.225)$$

Using (7.224), note that $\underline{T}_k(\sigma_p) = (a - B)^+$, $\underline{T}_{k+1}(\sigma_p) = (A - b)^+$, $\underline{T}_k(\sigma_{p+1}) = (a - b)^+$, and $\underline{T}_{k+1}(\sigma_{p+1}) = (A - B)^+$ if one sets

$$a = \frac{1}{C_i} \sum_{j=1}^{k} p_{i\pi_i(j)} c_{i\pi_i(j)}, \quad A = \frac{1}{C_i} \sum_{j=1}^{k+1} p_{i\pi_i(j)} c_{i\pi_i(j)}$$

$$b = d_{\sigma_p(k+1)}, \quad B = d_{\sigma_p(k)}$$

Clearly, $a \leq A$. Also $b \leq B$ because $\sigma_p(k) > \sigma_p(k+1)$ and $d_1 \leq \cdots \leq d_n$. From (7.225), one has

$$\underline{T}(\sigma_p) - \underline{T}(\sigma_{p+1}) = (a - B)^+ + (A - b)^+ - (a - b)^+ - (A - B)^+ \quad (7.226)$$

It is straightforward to check that this quantity is always nonnegative when $a \leq A$ and $b \leq B$. The theorem follows. $\square$

In the example, suppose again that jobs 1–3 are assigned to facility 1. The permutation π_1 is $(\pi_1(1), \pi_1(2), \pi_1(3)) = (3, 1, 2)$. The lower bound $\underline{T}$ of Lemma 7.55 is $\underline{T}_1 + \underline{T}_2 + \underline{T}_3$, where

$$\begin{aligned} \underline{T}_1 &= \left(\tfrac{1}{3}(5) - 2 \right)^+ = 0 \\ \underline{T}_2 &= \left(\tfrac{1}{3}(5 + 6) - 3 \right)^+ = \tfrac{2}{3} \\ \underline{T}_3 &= \left(\tfrac{1}{3}(5 + 6 + 8) - 4 \right)^+ = 2\tfrac{1}{3} \end{aligned} \qquad (7.227)$$

The bound is $\underline{T} = 3$, which in this case is slightly stronger than the bound of $2\frac{1}{3}$ obtained from Lemma 7.54.

The bound of Lemma 7.55 can be written in terms of the variables x_{ik}:

$$\sum_{k=1}^{n} \underline{T}'_{ik} x_{ik}$$

where

$$\underline{T}'_{ik} \geq \frac{1}{C_i} \sum_{j=1}^{k} p_{i\pi_i(j)} c_{i\pi_i(j)} x_{i\pi_i(j)} - d_k, \quad k = 1, \ldots, n$$

and $\underline{T}'_{ik} \geq 0$. One can linearize the bound by writing it as

$$\sum_{k=1}^{n} \underline{T}_{ik} \tag{7.228}$$

where

$$\underline{T}_{ik} \geq \frac{1}{C_i} \sum_{j=1}^{k} p_{i\pi_i(j)} c_{i\pi_i(j)} x_{i\pi_i(j)} - d_k - (1 - x_{ik}) U_{ik}, \quad k = 1, \ldots, n \tag{7.229}$$

and $\underline{T}_{ik} \geq 0$. The big-$M$ term U_{ik} is given by

$$U_{ik} = \frac{1}{C_i} \sum_{j=1}^{k} p_{i\pi_i(j)} c_{i\pi_i(j)} - d_k$$

Note that, although U_{ik} can be negative, the right-hand side of (7.229) is never positive when $x_{ik} = 0$. Finally, to obtain a bound on total tardiness, sum (7.228) over all facilities and write

$$T \geq \sum_{i=1}^{m} \sum_{k=1}^{n} \underline{T}_{ik} \tag{7.230}$$

One can now add to the master problem a bound consisting of (7.221), (7.230), and (7.229) for $i = 1, \ldots, m$. The bound is valid only when jobs are indexed so that $d_1 \leq \cdots \leq d_n$. In the example, the bound consists of

$$T \geq \underline{T}_{11} + \underline{T}_{12} + \underline{T}_{13} + \underline{T}_{21} + \underline{T}_{22} + \underline{T}_{23}$$
$$\underline{T}_{11} \geq \tfrac{5}{3}x_{11} - 2 + \tfrac{1}{3}(1 - x_{11})$$
$$\underline{T}_{12} \geq \tfrac{5}{3}x_{11} + 2x_{12} - 3 - \tfrac{2}{3}(1 - x_{12})$$
$$\underline{T}_{13} \geq \tfrac{5}{3}x_{11} + 2x_{12} + \tfrac{8}{3}x_{13} - 4 - \tfrac{7}{3}(1 - x_{13})$$
$$\underline{T}_{14} \geq \tfrac{5}{3}x_{11} + 2x_{12} + \tfrac{8}{3}x_{13} + 6x_{14} - 5 - \tfrac{22}{3}(1 - x_{14})$$
$$\underline{T}_{21} \geq 2x_{21} - 2 \qquad\qquad\qquad\qquad\qquad\qquad (7.231)$$
$$\underline{T}_{22} \geq 2x_{21} + \tfrac{10}{3}x_{12} - 3 - \tfrac{7}{3}(1 - x_{22})$$
$$\underline{T}_{23} \geq 2x_{21} + \tfrac{10}{3}x_{12} + 4x_{13} - 4 - \tfrac{16}{3}(1 - x_{23})$$
$$\underline{T}_{24} \geq 2x_{21} + \tfrac{10}{3}x_{12} + 4x_{13} + 5x_{24} - 5 - \tfrac{28}{3}(1 - x_{24})$$
$$\underline{T}_{ik} \geq 0$$

When the master problem contains the bounds (7.227) and (7.231), its solution assigns jobs 1–3 to facility 1 and job 4 to facility 2.

Exercises

7.117. Consider the constraint cumulative$(s \,|\, p, c, C)$ with $p = (2, 1, 1, 2)$, $c = (1, 2, 3, 2)$, and $C = 3$. The release times are $r = (0, 0, 0, 1)$ and the deadlines $d = (5, 5, 5, 5)$. Write all the valid, nonredundant inequalities given by Theorem 7.52 with $a = (1, 1, 1, 1)$.

7.118. The problem is to minimize $\sum_j s_j$ subject to the cumulative constraint and time windows in Exercise 7.117. What is the optimal solution and its optimal value? What is the minimum value of $\sum_j s_j$ subject to the relaxation obtained in Exercise 7.117? The relaxation should include the bounds on s_j.

7.119. Show how the valid inequalities (7.214) and (7.215) in Theorem 7.52 are derived.

7.120. Verify that (7.216) follows from (7.214)–(7.215).

7.121. Theorem 7.52 obtains valid inequalities from a relaxation of the cumulative constraint that can be solved by a greedy algorithm. In particular, the relaxation replaces each time window $[r_j, d_j]$ with $[r_J, d_J]$ and allows p_j, c_j for each job to be replaced by any values p'_j, c'_j such that $p'_j c'_j = p_j c_j$. Another relaxation that can be solved in a greedy fashion replaces each $[r_j, d_j]$ with $[r_J, d_J]$ and allows p_j, c_j to be replaced by any values p'_j, c'_j for which $p'_j c'_j \geq \min_i\{p_i c_i\}$. Derive valid cuts from this relaxation. These cuts could be useful when the energies $p_j c_j$ of the jobs are about the same but the release times differ considerably. Symmetrical cuts can be derived by replacing $[r_j, d_j]$ with $[r_J, d_j]$.

7.122. Use Lemma 7.53 to identify the nonredundant inequalities (2.28) for machine B in the example problem of Section 2.8.

7.123. Prove Lemma 7.53.

7.124. Suppose that four jobs are to be assigned to machines and scheduled so as to minimize the number of late jobs. The jobs have release times $r = (0, 0, 0, 1)$ and deadlines $d = (2, 2, 2, 2)$. On machine 1, the job durations are $(p_{11}, \ldots, p_{14}) = (2, 1, 1, 2)$, and the rates of resource consumption are $(c_{11}, \ldots, c_{14}) = (1, 2, 3, 2)$. The resource capacity of machine 1 is $C_1 = 3$. Write two nonredundant valid bounds on the minimum number of late jobs on machine 1 in terms of 0-1 variables x_{1j}, where $x_{1j} = 1$ when job j is assigned to machine 1. If all four jobs are assigned to machine 1, what is the lower bound on the number of late jobs? What is the actual minimum number of late jobs?

7.125. Suppose that all four jobs are assigned to machine 1 in Exercise 7.124. Use Lemmas 7.54 and 7.55 to obtain lower bounds on the minimum total tardiness. What is the actual minimum?

7.126. Show that inequality (7.226) in the proof of Lemma 7.55 is valid.

7.127. Show that the right-hand side of (7.229) is nonpositive when $x_{ik} = 0$.

7.128. Exhibit an example in which Lemma 7.54 provides a tighter bound than Lemma 7.55.

7.14 Bibliographic Notes

Section 7.1. The convex hull of a piecewise linear function is used by [380, 381, 413] in an integrated problem-solving context. Fast methods for computing the convex hull in two dimensions can be found in [460] (pages 351–352) and [399].

Section 7.2. Theorem 7.1, which states that all valid inequalities for a 0-1 system are Chvátal–Gomory cuts, is due to Chvátal [133]. The cuts are named for Gomory as well due to his pioneering work in cutting-plane theory [235, 237] and the fact that Gomory cuts are a special case of Chvátal–Gomory cuts (Theorem 7.6).

Cover inequalities for knapsack constraints originate in [30, 254, 384, 506]. The sequential lifting procedure presented here for cover inequalities is developed in [30, 254, 506] and summarized in [346]. The dual dynamic programming recursion for computing $h^*(t)$ is described in Section II.6.1 of [371] (Proposition 1.6). The approach to sequence-independent lifting described here (Theorem 7.2) is based on [248, 249, 506].

The odd cycle inequalities (Theorem 7.3) are shown in [383] to be facet-defining when the odd cycle is an odd hole. A polynomial-time separation algorithm can be based on Lemma 9.1.11 of [246]. The clique inequalities [215] are shown in [383] to be facet-defining when the clique is maximal. Separation algorithms are discussed in [108]. See the entry for the set-packing constraint in Chapter 8.

Section 7.3. The completeness proof of Chvátal–Gomory cuts for general integer inequality systems (Theorem 7.5) appears in [133]. Gomory cuts originate in [235, 237]. Mixed-integer rounding cuts (Theorems 7.7–7.9) are due to [347].

The equivalence of integrality and total unimodularity stated in Theorem 7.11 is proved in [271], and a simpler proof (essentially the one presented here) is provided in [493]. The necessary and sufficient condition for total unimodularity (Theorem 7.12) is proved in [225]. After these early results, much of the research on integral polyhedra has explored the properties of *balanced* and *ideal matrices*. An excellent survey of this work is [144].

Much of the recent work on cutting planes has focused on split cuts, intersection cuts for corner polyhedra, and related ideas that have roots in [28, 239, 240]. Most general-purpose cuts used in MILP solvers are split cuts, including Gomory cuts (Section 7.3.2), mixed-integer rounding cuts (Section 7.3.3), and lift-and-project cuts [34]. Split cuts are, in turn, a special case of intersection cuts [28]. There is recent interest in intersection cuts derived from two or more rows of the simplex tableau (e.g., [14]), although these ideas have not reached practical application. Some of this work is surveyed in [145].

A related line of research generalizes mixed-integer rounding (MIR) cuts to *n-step MIR cuts* [148, 159, 315, 316]. These cuts are derived in a multi-step procedure in which the mixed-integer cut of Fig. 7.8, which serves as the basis for MIR cuts, is replaced with a broader family of *base facets*.

Section 7.4. The convex hull relaxation of a disjunction of linear systems (Theorem 7.14) is due to [29, 31]. The tightened big-M model using (7.70) is described in [295], which gives a necessary and sufficient condition for its integrality. This article also points out that the solution of the big-M relaxation of a disjunction may be fractional, even when it satisfies the disjunction. The simplified big-M relaxation for a disjunction of individual linear inequalities (Theorem 7.15) originates with [50]. The procedure for tightening a big-M cut to a supporting cut (Theorem 7.16) is given in [295]. Disjunctive cuts, characterized by Theorem 7.18 [31], play a role in several areas of cutting-plane theory. The perspective given here (Theorem 7.19) appears in [295]. Disjunctive models and cutting planes are surveyed and further developed in [435, 437].

Section 7.5. Theorem 7.20, which provides the basis for the convex hull relaxation of a disjunction of nonlinear systems, can be found in the convex analysis literature, e.g. [269]. The convex hull formulation (Theorem 7.21) is due to [336, 465]. Two methods are described in [436] for dealing with numerical instability that may result from a small ϵ.

Section 7.6. The original MILP representability result was proved in [302] for unions of polyhedra, along with a similar result for unions of convex nonlinear systems. The MILP result was extended in [292] to unions of mixed-integer polyhedra (Theorem 7.23). It is observed in [502] that the original representability theorem is valid for big-M as well as convex hull modeling, and this likewise is extended in [292] to disjunctions of mixed-integer polyhedra (Theorem 7.25). Factored disjunctive models form a hierarchy of models, described in [32, 437], that move successively closer to a product model and have successively tighter relaxations. General discussions of mixed-integer modeling can be found in [292, 293, 500, 501, 502]. The package delivery problem is taken from [480].

Section 7.7. The connection between resolution and cutting planes (e.g., Theorem 7.28) is observed in [146, 273, 498]. The equivalence of unit refutation and linear programming on clause sets (Theorem 7.29) is remarked in [97]. The fact that input resolution generates the clauses that are rank 1 cuts (Theorems 7.31, 7.32, 7.36) is proved in [274]. The equivalence of input and unit refutation (Corollary 7.34), which follows from Theorem 7.31, is given a purely logic-based proof in [128]. The characterization of separating resolvents (Theorem 7.37) appears in [294], which states an algorithm that identifies all of the separating resolvents.

Section 7.8. The convex hull relaxation of the element constraint (Theorem 7.38) is described in [279]. The simplified relaxation that results when all variables have the same bounds (Theorem 7.39) is also proved in [279].

Section 7.9. Theorem 7.44 describes the convex hull of an alldiff constraint in which the n variables have the same domain. The special case in which the domain size is n (i.e., Corollary 7.45) is proved in [279, 503]. The polyhedral result for alldiff systems with the inclusion property is proved in [17]. The valid inequalities for odd cycles, along with the separation algorithm and other classes of valid inequalities, are derived in [80].

Section 7.11. The popular 0-1 formulation of the traveling salesman problem given here is due to [158]. Comb inequalities were first described in a restricted form by [134]. The larger family of inequalities described here appears in [247]. A number of heuristics have been proposed for solving the separation problem for the subtour-elimination and comb inequalities. The max back heuristic described here is adapted from a similar heuristic for the symmetric problem described in [368, 369, 370]. Comprehensive discussions of polyhedral analysis of the traveling salesman problem include [36, 368]. Solution techniques for the problem have been intensively studied. One survey of this voluminous literature may be found in [307]. Facet-defining inequalities in terms of the original variables have received little attention but are studied in [221].

Section 7.13. The mixed-integer model (7.208) for cumulative scheduling is similar to a model for the scheduling of chemical plants proposed in [123], and model (7.211) is suggested by a model for a related problem in [484]. The time window relaxation for the Benders scheduling subproblem and Lemma 7.53 appear in [283]. The relaxations for minimizing the number of late jobs and for minimizing total tardiness (Lemmas 7.54 and 7.55) are from [285].

Chapter 8
Dictionary of Constraints

This book advocates modeling with metaconstraints, which represent structured groups of more elementary constraints. As explained in Chapter 1, this allows modelers to write succinct formulations and to communicate the structure of the problem to the solver. Ideally, modelers would have at hand a menu of constraints, organized by problem domain, or in other ways that guide them to the right choice of constraints. This chapter provides a sampling of some metaconstraints that might appear on that menu.

There is no pretense of comprehensiveness. The list contains most of the better-known global constraints from the constraint programming world, and some of the major problem classes in the mathematical programming tradition. Generally, a global constraint is included only if filtering algorithms are discussed in the open literature. Some of the mathematical programming structures appear in the guise of global constraints, such as *flow* (capacitated network flows), *path* (shortest path), and *network design* (fixed-charge network flows). Others appear under global constraints because they are formulated in order to relax the constraint; thus, the assignment problem appears under *alldiff*, the traveling salesman problem under *circuit*, and the vehicle routing problem under *cycle*. Several of the constraints are known under multiple names. There is no attempt here to dictate a standard name but only to choose a suggestive name that is reasonably well known. Some constraints have variations and extensions that are mentioned within the entries (and listed in the Index).

For brevity, the dictionary refers to a filter that achieves domain consistency as a *complete* filter. *Boolean* variables are variables whose two values can be identified with *true* and *false* in some natural way.

When a constraint is written in the form $constraintName(x, y \mid a, b)$, the symbols before the vertical bar are variables (or tuples/sets of variables), and those after the bar are parameters. This distinction is important, because achieving domain consistency becomes harder when parameters become variables whose domains must be filtered.

When two or more instances of the same metaconstraint appear in a model, they may for convenience be grouped together under that name. For example, the two constraints alldiff(x) and alldiff(y) may be written

$$\text{Alldiff:} \begin{cases} (x) \\ (y) \end{cases}$$

The two constraints are nonetheless filtered and relaxed separately. To filter or relax them jointly, one must use a constraint designed for this purpose (the multipleAlldiff constraint). If there are two or more instances of a metaconstraint that is specified as a set of elementary constraints, they should be combined into a single metaconstraint. Thus, if there are two *linear integer* metaconstraints, the inequalities they list should be merged to form a single integer linear metaconstraint (or if they are not, the modeling system should merge them automatically).

Examples of modeling with many of these constraints appear in Chapter 2 and elsewhere in the book. These passages are referenced below, as are numerous sources from the literature.

An exhaustive catalog of global constraints used in the constraint programming community is provided in [59]. Briefer surveys of global constraints can be found in [423, 487].

0-1 linear

A *0-1 linear* constraint consists of a system of linear inequalities or equations in 0-1 variables.

Notation. A 0-1 linear constraint can be written

$$\text{01Linear} \, (Ax \geq b)$$

where A is an $m \times n$ matrix and x is a tuple $(x_1, \ldots, x_n)$ of variables with domain $\{0, 1\}$. A system of equations $Ax = b$ is a special case because it can be written $Ax \geq b$, $-Ax \geq -b$.

Usage.   0-1 linear inequalities are a subset of mixed-integer linear inequalities, which provide a versatile modeling language. Modeling principles and examples are presented in Section 7.6 as well as [500, 501].

Inference.   Inferences may be drawn from each individual 0-1 inequality, or from the entire set considered jointly. Section 6.5 shows how to check for domination between individual knapsack inequalities, Section 6.5.2 shows how to derive implied logical clauses from a single inequality, and Section 6.5.3 shows how to derive implied cardinality clauses [44, 273]. 0-1 resolution [273, 275] is a complete inference method for sets of 0-1 linear inequalities considered jointly (Section 6.5.4). A weaker form of 0-1 resolution achieves k-completeness (Section 6.5.5), and a still weaker form achieves strong k-consistency (Section 6.5.6). Inference duality provides the basis for deducing Benders cuts [279, 296] (Sections 6.6.1, 6.6.2). One can use bounds propagation (Section 6.2.1) and dual multipliers for the continuous relaxation (Section 6.2.3) to fix variables.

Relaxation.   A continuous relaxation can be obtained by replacing each $x_j \in \{0, 1\}$ with $0 \leq x_j \leq 1$. The relaxation of a single knapsack inequality can be strengthened by the addition of 0-1 knapsack cuts, also known as *cover inequalities* when generated for knapsack packing constraints (Section 7.2.2). They can often be strengthened by lifting techniques [30, 154] (Sections 7.2.3–7.2.4). Knapsack cuts and lifting are discussed in [346, 371, 510] and generalized in [195, 509]. Cutting planes can also be inferred from the system of inequalities considered jointly. They can all, in principle, be generated by Chvátal's procedure [133] (Section 7.2.1). General separating cuts, known as Gomory cuts [235, 237], can be derived as well (Section 7.3.2). Other general cuts include lift-and-project cuts [34]. There are a number of specialized cuts that presuppose special structure in the problem, some of which are discussed in [346, 371, 510].

Related constraints.   There is also an *integer linear* constraint, in which the variables take general integer values.

All-different

The *all-different* (*alldiff*) constraint requires a set of variables to take distinct values.

Notation.   The constraint is written

$$\text{alldiff}(X)$$

where X is a tuple $\{x_1, \ldots, x_n\}$ of variables with finite domains. It requires that each x_i take a distinct value. The constraint can also be formulated with a cost variable w:

$$\text{alldiff}(X, w \mid c) \tag{8.1}$$

where c is a matrix whose elements c_{ij} indicate the cost of assigning value j to x_i. Constraint (8.1) enforces alldiff(X) and $w = \sum_i c_{ix_i}$. This has been called the minimum weight alldiff [443].

Usage.   The constraint is used in a wide variety of discrete models, particularly where assignment, scheduling, and allocation are involved. The classical assignment problem minimizes w subject to (8.1), as discussed in Section 7.9.3. The alldiff constraint is illustrated in Sections 2.5 (employee scheduling), 6.8 (job assignment), and 6.13.1 (contrast with *circuit*, traveling salesman).

Inference.   A complete filter for alldiff(X) based on maximum cardinality matching [150, 416] is described in Section 6.8.2. A complete filter for bounds consistency is based on finding Hall intervals [401, 341] or on the convexity of the graph [353], and the latter is presented in Section 6.8.3.

Relaxation.   A convex hull relaxation in the original variables for alldiff with n-element numerical domains appears in [279, 503]. It is generalized to numerical domains of arbitrary but equal cardinality in Section 7.9.1. A convex hull relaxation for arbitrary domains of any cardinality can be written by adding $O(n^2)$ new variables (Section 7.9.2).

Related constraints.   The *k-diff* constraint [417] requires the variables in X to take at least k different values. A set-valued alldiff constraint, alldiff($S_1, \ldots, S_n$), requires the sets $S_1, \ldots, S_n$ to be distinct and has been offered by commercial solvers. The *multiple alldiff* constraint defines a system of alldiff constraints with some variables in common, allowing the individual alldiffs to be jointly filtered. Relaxations of k-alldiff are discussed in Section 7.9.4 and [15, 16, 17, 80, 324]. The *alldiff matrix* (or *Latin square*) constraint requires that each row and each column of a matrix contain all different entries. It is a special case of the cardinality matrix constraint, for which a filtering algorithm appears in [424].

Among

The *among* constraint bounds the number of variables that take one of a given set of values.

Notation.   Several forms of the constraint are described in [61], one of which is

$$\text{among}(X \mid V, \ell, u) \tag{8.2}$$

where X is a set $\{x_1, \ldots, x_n\}$ of variables, V a set of values, and ℓ and u are nonnegative integers. It requires that at least ℓ and at most u of the variables take values in the set V. Another form is

$$\text{among}(X \mid V, \ell, u, s) \tag{8.3}$$

in which s is a positive integer. It requires that in any sequence of s consecutive variables, at least ℓ and at most u variables take a value in v.

Usage. Constraint (8.2) can be used in recourse-constrained sequencing models. A sequence of product types are to be processed on an assembly line, and the product types in V require a certain resource. One can specify that at most u products use the resource during a given time period. Here x_i is the product type assigned to slot i of the sequence, and $x_1, \ldots, x_n$ represent the slots that are prodcessed during the time period. Such constraints may be imposed for several resources and stations. Constraint (8.3) can prevent uneven usage of a resource over time.

Inference. Filtering to obtain domain consistency is straightforward and described in Section 6.10.1.

Related constraints. The *cardinality* constraint is similar to (8.2), except that it counts occurrences of a specific value rather than a set of values. The *sequence* constraint imposes a series of overlapping among constraints.

Bin Packing

The (one-dimensional) *bin-packing* constraint require that a given set of items be packed into bins so that the weight in no bin exceeds a given limit.

Notation. The constraint is

$$\text{bin-packing}(x \mid w, u, k) \tag{8.4}$$

where x is an n-tuple of variables x_i with domain $\{1, \ldots, k\}$ indicating which bin holds item i, w an n-tuple of item weights, u the bin capacity (same for every bin), and k the number of bins. The constraint requires that

$$\sum_{\substack{i \\ x_i = j}} w_i \leq u, \quad j = 1, \ldots, k$$

The number k of bins is treated as a variable in

$$\text{bin-packing}(x, k \mid w, u) \tag{8.5}$$

The classical bin-packing problem minimizes k subject to (8.5). One can also treat u as a variable

$$\text{bin-packing}(x, u \mid w, k) \tag{8.6}$$

and minimize the maximum load u subject to (8.6).

Usage.   Many resource-allocation problems have a bin-packing component. The problem has received attention in the operations research community at least since 1974 [304].

Inference.   Achieving domain consistency is NP-hard. Incomplete filtering algorithms for (8.4) are discussed in [364, 365, 366, 367, 438, 444, 453].

Relaxation.   MILP-based and other relaxations are discussed in [165, 348, 349]. There are many approximation algorithms for the problem [141].

Related constraints.   Constraint (8.4) is a special case of the *cumulative scheduling* constraint with all processing times equal to one. See also the *0-1 linear* constraint.

Cardinality

The *cardinality* (*distribute, gcc, generalized cardinality*) constraint bounds the number of variables that take each of a given set of values.

Notation.   The constraint is written

$$\text{cardinality}(X \mid v, \ell, u) \tag{8.7}$$

where x is a set $\{x_1, \ldots, x_n\}$ of variables, v an m-tuple of values, and ℓ, u are m-tuples of nonnegative integers. The constraint requires, for $j = 1, \ldots, m$, that at least ℓ_j, and at most u_j, of the variables take the value v_j. Another form of the constraint treats the number of occurrences as a variable:

$$\text{cardinality}(x, y \mid v) \tag{8.8}$$

where $y = (y_1, \ldots, y_m)$. This constraint says that exactly y_j variables must take the value v_j. The number of occurrences of v_j can be bounded by bounding y_j.

The constraint can also be formulated with a cost variable w:

$$\text{cardinality}(X, w \mid v, \ell, u, c) \tag{8.9}$$

where c is a matrix whose elements c_{ij} indicate the cost of assigning value j to x_i. Constraint (8.9) enforces cardinality$(X \mid v, \ell, u)$ and $w = \sum_i c_{ix_i}$.

Usage.   This is a highly generic constraint that often finds application when there are bounds on cardinalities. It is illustrated in Section 2.5 (employee scheduling).

Inference.   The simplest complete filtering algorithm for (8.7) [418] is based on a network flow model and is presented in Sections 6.9.2 and 6.9.3. Some improved algorithms appear in [403], which shows that achieving domain consistency for (8.8) is NP-hard. Bounds consistency algorithms for (8.7) generalize those for alldiff. One [404] finds Hall intervals, and another [311] exploits convexity of the graph. The algorithm in [311] also computes bounds consistency for (8.8). The cardinality constraint with costs (where w is treated as a bound on cost rather than a variable) is given filtering algorithms in [202, 203] based on reduced costs, and filtering algorithms that achieve domain consistency in [418, 420].

Relaxation. A convex hull relaxation in the original variables for cardinality with numerical domains of equal cardinality is presented in Section 7.10.1. A convex hull relaxation for arbitrary domains can be obtained by adding $\mathcal{O}(n^2)$ 0-1 variables (Section 7.10.2).

Related constraints. The *cardinality matrix* constraint requires that each of a given set of values occur a certain number of times in each row and column of a matrix. In [424], the number of occurrences is treated as a variable. The constraint can be written

$$\text{cardinalityMatrix}\,(X, Y, Z \mid v)$$

where X is an $m \times n$ matrix of variables x_{ij}, Y an $m \times p$ matrix of variables y_{ik}, Z an $n \times p$ matrix of variables z_{jk}, and v a tuple of values $(v_1, \ldots, v_p)$. The constraint requires that y_{ik} variables x_{ij} in row i of X take value v_k and z_{jk} variables in column j take value v_k for all i, j, k. The *alldiff matrix* (or *Latin square*) constraint is a special case. Filtering algorithms that achieve domain consistency are presented in [320, 424].

Cardinality Clause

A *cardinality clause* specifies that at least a certain number of Boolean variables must be true.

Notation. The constraint can be written

$$\sum_{j=1}^{n} x_j \geq k$$

where $x_1, \ldots, x_n$ are Boolean variables and k is a nonnegative integer. It requires that at least k of the variables be true. If desired, the constraint can be generalized to

$$\sum_{j=1}^{n} L_j \geq k$$

where each L_j is a literal x_j or $\neg x_j$, to require that at least k literals be true. This constraint can also be written as a 0-1 inequality

$$ax \geq k + (a) \tag{8.10}$$

where each $a_j \in \{0, 1, -1\}$ and $n(a)$ is the sum of the negative components of a. Setting x_j to 1 or 0 corresponds to making x_j true or false, respectively.

Usage. Cardinality clauses allow cardinality conditions to be expressed in a quasi-logical language for which inference is much easier than general 0-1 inequalities.

Inference. It is easy to check when a 0-1 inequality implies a cardinality clause, or when one cardinality clause implies another [273] (Section 6.5.3). There is a method for generating all cardinality clauses implied by a 0-1 linear system [44].

Relaxation. The inequality (8.10), together with $x_j \in [0, 1]$ for all j, is a convex hull representation of the cardinality clause it represents.

Related constraints. The cardinality clause becomes a logical clause when $k = 1$ and is generalized by a *cardinality conditional*.

Cardinality Conditional

A *cardinality conditional* specifies that, if at least k variables in one set of Boolean variables are true, then at least ℓ of another set must be true.

Notation. The constraint can be written

$$\left(\sum_{i=1}^{m} x_i \geq k \right) \rightarrow \left(\sum_{j=1}^{n} y_j \geq \ell \right)$$

where $x_1, \ldots, x_n$ and $y_1, \ldots, y_m$ are Boolean variables and k, ℓ are integers. It states that if at least k of the variables x_i are true, then at least ℓ of the variables y_j must be true. If desired, one can permit negated variables as in a cardinality clause.

Usage. A cardinality conditional allows one to express a conditional of this kind conveniently without using the general *conditional* constraint and thereby losing the special structure. A large number of 0-1 inequalities may be required to capture the meaning.

Relaxation. A convex hull relaxation is given in [513]. This result is generalized in [33].

Related constraints. The cardinality conditional becomes a *cardinality clause* when $k = 0$.

Change

The *change* constraint counts the number of times a given type of change occurs in a sequence of variables.

Notation.   The constraint is

$$\text{change}(x, k \mid \text{rel})$$

where $x = (x_1, \ldots, x_n)$, k is an integer-valued variable, and *rel* is a binary relation such as $=$, $\neq$, $\leq$, $\geq$, $<$, or $>$. The constraint requires that k be the number of times two consecutive variables x_i, x_{i+1} satisfy x_i rel x_{i+1}.

Usage.   The constraint can be used in employee scheduling and other timetabling problems to constrain the number of times a change in shift or work assignment occurs.

Inference.   Filtering algorithms are discussed in [53].

Related constraints.   The *stretch* constraint can limit the type of changes that occur.

Circuit

The *circuit* constraint describes a Hamiltonian cycle on a directed graph.

Notation.   The constraint is

$$\text{circuit}(x)$$

where $x = (x_1, \ldots, x_n)$ is a tuple of variables whose domains are subsets of $\{1, \ldots, n\}$. The constraint requires that $y_1, \ldots, y_n$ be a permutation of $1, \ldots, n$, where each $y_{i+1} = x_{y_i}$ and y_{n+1} is identified with y_1. The constraint can be viewed as describing a Hamiltonian cycle on a directed graph G that contains an edge (i, j) if and only if j belongs to the domain of x_i. An edge (i, j) of G is selected when $x_i = j$, and *circuit* requires that the selected edges form a Hamiltonian cycle. One can add a cost variable w to obtain the constraint

$$\text{circuit}(x, w \mid c) \tag{8.11}$$

where c is a matrix of cost coefficients c_{ij} that indicate the cost incurred when j immediately follows i in a Hamiltonian cycle. Constraint (8.11) enforces circuit(x) and $w = \sum_i c_{ix_i}$.

Usage.   Section 6.13.1 discusses modeling with *circuit*. The traveling salesman problem minimizes w subject to (8.11) (Section 6.1.2).

Inference.   Achieving domain consistency for circuit(x) is NP-hard. Section 6.13.2 describes two elementary incomplete filtering methods, which are related to those in [122, 457]. Sections 6.13.3 and 6.13.4 strengthen the filter by analyzing a separator graph [220].

Relaxation.   The circuit constraint can be relaxed by adding 0-1 variables and writing valid inequalities for the traveling salesman problem, as described in Sections 7.11.1–7.11.3. Valid inequalities, facet-defining inequalities, and separating cuts for this problem have been intensively studied; see [36, 307, 368] for surveys. Facet-defining inequalities in terms of the original variables are studied in [221].

Related constraints.   The *cycle* constraint [61] generalizes *circuit* by specifying the number of subtours allowed.

Clique

A *clique* constraint requires that a given graph contain a clique of a specified size.

Notation.   One way to write a clique constraint is

$$\text{clique}(x, k \mid G) \qquad\qquad (8.12)$$

where x is an n-tuple of Boolean variables, k is the number of true x_js, and G is an undirected graph. The constraint requires that the set V of vertices i for which $x_i = true$ induce a clique of G (i.e, every pair of vertices in V is connected by an edge of G). The *maximum clique* problem maximizes k subject to (8.12).

Usage.   The maximum clique problem has received much attention and has seen applications in coding theory, fault diagnosis, pattern recognition, cutting-plane generation, and other areas; see [105] for a survey.

Inference.   Achieving domain consistency is NP-hard. Incomplete filtering algorithms are presented in [189, 421].

Relaxation.   MILP-based and other relaxations are described in [35, 37].

Related constraints.   See the *set-packing* constraint.

Conditional

A *conditional* constraint states conditions under which a given constraint set is enforced.

Notation.   A conditional constraint has the form

$$\mathcal{D} \Rightarrow \mathcal{C}$$

where $\mathcal{D}$ and $\mathcal{C}$ are constraint sets. It states that the constraints in $\mathcal{C}$ must hold if the constraints in $\mathcal{D}$ hold. If the conditional is to be implemented in practice, the set $\mathcal{D}$ must be limited to constraints for which the solver can easily check whether *all* solutions belonging to the current domains satisfy $\mathcal{D}$. One option is to restrict $\mathcal{D}$ to domain constraints (constraints of the form $x_j \in D$) or logical propositions in which the atoms are domain constraints, such as $(x_1 \in D_1 \wedge x_2 \in D_2) \rightarrow x_3 \notin D_3$. A conditional constraint should be distinguished from a logical conditional $F \rightarrow G$, which is a formula of propositional logic (Section 6.4) and can be used in a *logic* constraint.

Usage.   Conditional constraints are a very convenient device but can be overused (see below).

Inference.   The solver posts the constraints in $\mathcal{C}$ whenever the search process reduces domains to the point that $\mathcal{D}$ is satisfied by any set of values belonging to the current domains. Logical conditionals should not be written as conditional constraints, because this prevents the application of inference algorithms designed especially for propositional logic.

Relaxation.   Conditionals should not be used when a more specialized meta-constraint is available. For example, a set of conditionals in which the consequent $\mathcal{C}$ is a linear system can often be re-expressed as one or more linear disjunctions that exploit the problem structure for relaxation purposes. This is illustrated by the production planning problem of Section 2.4.

Related constraints.   See the discussion of *linear disjunctions*.

Cumulative Scheduling

A *cumulative scheduling* constraint requires that jobs be scheduled so that the total rate of resource consumption at no time exceeds a given limit.

Notation.   The constraint is written

$$\text{cumulative}(s \mid p, c, C)$$

where $s = (s_1, \ldots, s_n)$ is a tuple of real-valued[1] variables s_j representing the start time of job j. The parameter $p = (p_1, \ldots, p_n)$ is a tuple of processing times for each job, $c = (c_1, \ldots, c_n)$ resource consumption rates,

---

[1] These variables are traditionally integer-valued in constraint programming systems, but this is not necessary in an integrated solver.

and C a limit on total resource consumption rate at any one time. The constraint requires that the total rate of resource consumption of the jobs underway at any time t be at most C:

$$\sum_{\substack{j \\ s_j \le t \le s_j + p_j}} c_j \le C, \quad \text{for all times } t$$

The current domains $[L_j, U_j]$ of the start time variables s_j impose release times L_j and deadlines $U_j + p_j$.

Variations of cumulative scheduling allow *preemptive* scheduling (in which one job can interrupt another), multiple resources, and resource limits that are variable over time.

Usage. The cumulative scheduling constraint is widely used for scheduling tasks subject to one or more resource constraints. It is illustrated in Sections 6.15 and 7.13.

Inference. Achieving bounds consistency is NP-hard. Incomplete filtering methods include timetabling methods [210, 387, 462], edge finding [121, 376, 377] (Section 6.15.1), extended edge finding [375, 39] (Section 6.15.2), not-first/not-last rules [376, 377] (Section 6.15.3), and energetic reasoning [186, 187] (Section 6.15.4). These and other methods are described in [39]. Generation of logic-based Benders cuts [279, 284, 285] is described in Section 6.15.5.

Relaxation. The *cumulative* constraint can be given MILP relaxations (Section 7.13.1) or relaxations in the original variables (Section 7.13.2). When it appears as the subproblem of a Benders formulation, there are relaxations for minimizing cost and makespan [284, 291], as well as number of late jobs and total tardiness [285, 289, 291] (Section 7.13.3).

Related constraints. When c is a tuple of ones and $C = 1$, the cumulative scheduling constraint becomes a *disjunctive scheduling* constraint.

Cutset

A *cutset* constraint requires that a set of vertices cut all directed cycles in a graph.

Notation. The cutset constraint is

$$\text{cutset}(x, k \,|\, G) \tag{8.13}$$

where x is an n-tuple of Boolean variables, k is the number of true x_js, and G is a directed graph. The constraint requires that the set V' of vertices

i for which $x_i = true$ be a cutset; that is, if V is the set of vertices in G, then $V \setminus V'$ induces a subgraph of G that contains no cycles. The *minimum cutset* problem minimizes k subject to (8.13).

Usage. The minimum cutset problem has applications in deadlock breaking, program verification, and Bayesian inference [429, 452, 514].

Inference. Filtering algorithms are presented in [188].

Cycle

The *cycle* constraint [61] specifies the number of subtours that must cover a directed graph.

Notation. The constraint is
$$\text{cycle}(y, x)$$
where y is an integer-valued variable and x a tuple $(x_1, \ldots, x_n)$ of variables with domains in $\{1, \ldots, n\}$. Let G be a graph with vertices $1, \ldots, n$, in which (i, j) is an edge if and only if $j \in D_{x_i}$. The variables x select an edge (i, j) when $x_i = j$. The constraint requires that x select edges that form exactly y directed cycles, such that every vertex belongs to exactly one cycle.

Usage. The cycle constraint can be used in vehicle routing problems where x_i is the city a vehicle visits immediately after city i. The constraint requires that exactly y vehicles cover all the cities.

Inference. The elementary filtering methods for *circuit* in Section 6.13.2 can be adapted to *cycle*.

Relaxation. The cycle constraint can be given an integer programming model for vehicle routing, which can in turn be relaxed by dropping integrality. Formulations include the two-index, three-index, set-partitioning, and multicommodity flow models; several are surveyed in [147, 199, 478].

Related constraints. The cycle constraint becomes *circuit* when y is fixed to 1.

Diffn

The *diffn* constraint arranges a given set of multidimensional boxes in n-space so that they do not overlap.

Notation. The basic diffn constraint [6, 61] is

$$\text{diffn}((x^1, \Delta x^1), \ldots, (x^m, \Delta x^m))$$

where each x^i is an n-tuple of variables indicating the coordinates of one corner of the ith box to be placed, and Δx^i is an n-tuple of variables indicating the size of the box along each dimension. The constraint requires that for every pair of boxes i, i', there is at least one dimension j such that $x^{i'}_j \geq x^i_j + \Delta x^i_j$ or $x^i_j \geq x^{i'}_j + \Delta x^{i'}_j$. Bounds on the corner-positioning and size of the boxes are implicit in the initial variable domains. More elaborate versions of the constraint allow one to bound the volume of each box, to bound the area within which each box may lie (which is slightly different than bounding x^i and Δx^i), and to bound the distance between boxes. The constraint has been generalized to convex polytopes [63, 426].

Usage. The constraint is used for space or space-time packing. The two-dimensional version is used for resource-constrained scheduling, in which each box represents a job whose horizontal dimension is the processing time and vertical dimension is the rate of resource consumption.

Inference. A sweep algorithm for filtering for the two-dimensional case appears in [54]. Propagation for the general case is discussed in [63] and generalized to convex polytopes in [63, 426].

Related constraints. Recent work on general non-overlapping constraints in space or space-time appears in [1, 57].

Disjunctive Scheduling

A *disjunctive scheduling* constraint requires that jobs be scheduled sequentially without overlapping.

Notation. A disjunctive scheduling constraint has the form

$$\text{noOverlap} (s \mid p)$$

where $s = (s_1, \ldots, s_n)$ is a tuple of real-valued[2] variables s_j indicating the start time of job j. The parameter $p = (p_1, \ldots, p_n)$ consists of processing times for each job. The constraint enforces

$$(s_i + p_i \leq s_j) \vee (s_j + p_j \leq s_i)$$

for all i, j with $i \neq j$. The current domains $[L_j, U_j]$ of the start time variables s_j impose release times L_j and deadlines $U_j + p_j$. A variation of the constraint allows for *preemptive scheduling*, in which one job may interrupt another.

---

[2] These variables are traditionally integer-valued in constraint programming systems, but this is not necessary in an integrated solver.

Usage. Disjunctive scheduling constraints are widely used for scheduling tasks sequentially (i.e., under a *unary resource* constraint). It is illustrated in Section 2.8 (machine scheduling).

Inference. Achieving bounds consistency is NP-hard. Incomplete filtering methods are based on timetabling, edge finding, and not-first/not-last rules. Timetabling filters appear in [210, 387, 462]. Edge finding originated in [117]. An $O(n^2)$ edge finding algorithm (where n is the number of jobs) based on the Jackson preemptive schedule (Section 6.14.1) is given in [118]. Another $O(n^2)$ algorithm appears in [39, 375, 378]. An algorithm that achieves $O(n \log n)$ complexity with complex data structures is given in [119], and an $O(n^3)$ algorithm that allows incremental updates in [121]. Extensions that take setup times into account are presented in [112, 206]. Propagation algorithms for not-first/not-last rules appear in [38, 180, 477] (Section 6.14.2). A comprehensive treatment of scheduling is [39]. Generation of logic-based Benders cuts [279, 284, 285, 300] is described in Section 6.14.3.

Relaxation. The disjunctive scheduling constraint can be given disjunctive relaxations (Section 7.12.1) or MILP relaxations (Section 7.12.2). When it appears as the subproblem of a Benders formulation, there are relaxations for minimizing cost and makespan [284, 291] as well as number of late jobs and total tardiness [285, 289, 291] (Section 7.13.3).

Related constraints. The *cumulative scheduling* constraint allows jobs to be scheduled in parallel, subject to a resource limit.

Element

The *element* constraint selects a specified value from a list and assigns it to a variable.

Notation. The simplest form of the constraint is written

$$\text{element}\,(y, z \mid a) \qquad (8.14)$$

where y is a variable representing a positive integer, z is a variable, and a is a tuple $(a_1, \ldots, a_m)$ of values. The constraint requires z to take the yth value in the tuple. Another form is

$$\text{element}\,(y, x, z) \qquad (8.15)$$

where x is a tuple $(x_1, \ldots, x_m)$ of variables. The constraint requires z to take the same value as the yth variable in the tuple. The constraint becomes multidimensional when the index variable y is a p-tuple, in which case a and x become p-dimensional arrays.

A *vector-valued element* constraint

$$\text{element}\left(y, z \mid (a^1, \dots, a^m)\right) \qquad (8.16)$$

requires the tuple $z = (z_1, \dots, z_m)$ of variables to take the values in the tuple a^y.

Usage.   The element constraint is useful for implementing *variable indices* (*variable subscripts*). An expression of the form a_y, in which y is a variable index, is processed by replacing it with z and adding constraint (8.14). An expression of the form x_y, where x is a variable, can be similarly processed with constraint (8.15). When y is a tuple representing multiple indices, a multidimensional element is used.

Illustrations of *element* appear in Sections 2.5 (employee scheduling) and 2.5 (product configuration). An example of the vector-valued constraint (8.16) is given in Section 2.4 (production planning).

Inference.   Filtering for domain and bounds consistency for (8.14) and (8.15) is straightforward. Domain consistency is discussed in Section 6.7.1 and bounds consistency in Section 6.7.2. Filtering for the multidimensional constraint is the same as for the one-dimensional constraint.

Relaxation.   Relaxation of (8.14) is trivial (Section 7.8.1). Relaxation of (8.15) is based on relaxations for a disjunction of linear systems. Section 7.8.1 describes the convex hull relaxation and Section 7.8.2 a big-M relaxation. Relaxation of the multidimensional constraint is the same as for the one-dimensional constraint. Relaxation of the vector-valued constraint is discussed in Sections 2.7 and 7.8.3.

Related constraints.   The indexed linear element constraint

$$\text{element}\left(y, (A_1 x, \dots, A_m x), z\right)$$

is used in the *indexed linear* metaconstraint and is discussed in connection with it. The constraint also has a vector-valued form. The *sum* constraint implements index sets with variable indices.

Flow

The *flow* constraint requires flow conservation on a capacitated network and computes variable costs.

Notation.   One way to notate the constraint is

$$\text{flow}\left(x, z \mid N, s, A, \ell, u, c\right) \qquad (8.17)$$

where x is a tuple of real-valued flow variables x_{ij}, and z is a real-valued variable representing cost. N is a tuple of nodes, and s is a tuple of net supply values corresponding to the nodes in N. A is a tuple of directed arcs, and ℓ, u, c are tuples of lower flow bounds, upper flow bounds, and unit costs corresponding to the arcs in A. The constraint requires that

$$z = \sum_{(i,j) \in A} c_{ij} x_{ij}$$

$$\sum_{(i,j) \in A} x_{ij} - \sum_{(j,i) \in A} x_{ji} = s_i, \text{ all } i \in N$$

$$\ell_{ij} \le x_{ij} \le u_{ij}, \text{ all } (i,j) \in A$$

The classical minimum-cost network flow problem minimizes z subject to (8.17). A slight generalization of the model allows *gains*, which specify that the flow leaving (i, j) is α_{ij} times the flow entering it. There is a multicommodity version of the problem, but it does not enjoy a very fast specialized solution algorithm as does the single-commodity problem.

Usage. The capacitated network flow model has countless applications [8, 9, 45]. It is also used to filter several constraints, such as *cardinality* (Sections 6.9.2 and 6.9.3) and *circuit* (Section 6.13.4). For other constraint programming applications, see [103].

Related constraints. The *network design* constraint selects the arcs that will belong to the network and charges a fixed cost for each.

Indexed Linear

The *indexed linear* constraint is used to implement linear inequalities in which the coefficients are indexed by variables.

Notation. The constraint has the form

$$\text{indexedLinear} \begin{cases} \sum_{i \in I} z^i + \sum_{i \in I'} A_i x_i \ge b & (a) \\ \\ \text{element } \left(y_i, x_i, z^i \mid A_i\right), \quad i \in I & (b) \end{cases} \tag{8.18}$$

where each x_i is a real-valued variable, each z^i is a tuple of real-valued variables, each y_i is a variable representing a positive integer, and each A_i is a tuple of the same length as z^i. Each element constraint i requires that z^i be equal to $A_{iy_i} x_i$. This form of the element constraint is an *indexed linear element* constraint.

Usage. The indexed linear constraint is used to implement linear inequality systems of the form

$$\sum_{i \in I} A_{iy_i} x_i + \sum_{i \in I'} A_i x_i \geq b \tag{8.19}$$

It is more efficient to define a single constraint of this structure than simply to replace each term $A_{iy_i} x_i$ in (8.19) with z^i and write the element constraint (8.18b) for each. A single constraint allows the solver to generate cover inequalities for the knapsack constraint (8.18a) on the basis of information about the coefficients in the element constraints, as explained in Section 2.7. Normally, an indexed linear constraint would not occur explicitly in a model but would be generated automatically by the modeling system when a constraint of the form (8.19) is present. A model that uses the constraint is presented in Section 2.7 (product configuration).

Inference.   Filtering is described in Sections 6.7.1 and 6.7.2.

Relaxation.   Relaxation methods appear in [474] and Sections 7.8.1–7.8.2.

Related constraints.   Other forms of the *element* constraint are listed under that entry.

Integer Linear

An *integer linear* constraint consists of a system of linear inequalities or equations in integer-valued variables.

Notation.   An integer linear system can be written

$$Ax \geq b$$

where A is an $m \times n$ matrix and x is a tuple $(x_1, \ldots, x_n)$ of variables whose domain is the set of integers. A system of equations $Ax = b$ is a special case because it can be written $Ax \geq b$, $-Ax \geq -b$.

Usage.   Integer linear inequalities are a subset of mixed-integer linear inequalities, which provide a versatile modeling language. Modeling principles and examples are presented in Section 7.6 as well as [500, 501].

Inference.   Bounds propagation for integer linear inequalities is the same as for linear inequalities (Section 6.2.1), except that bounds can be rounded to the appropriate integer value. Dual multipliers for the continuous relaxation (Section 6.2.3) can be used to fix variables, again with rounding. Reduced-cost variable fixing is a special case. Inference duality provides the basis for deducing Benders cuts [279, 296] (Sections 6.6.1, 6.6.2).

Relaxation.   A continuous relaxation can be obtained by dropping the integrality requirement. The relaxation of an individual inequality can be strengthened by the addition of integer knapsack cuts [26] (Section 2.3).

They, in turn, can often be strengthened by lifting techniques [25, 249]. Cutting planes can also be inferred from the system of inequalities considered jointly. All valid cuts can, in principle, be generated by Chvátal's procedure [133] (Section 7.3.1). General separating cuts, known as Gomory cuts [235, 237], can be derived as well (Section 7.3.2). For a general discussion of cutting planes, see [214, 346, 371, 510]. Another approach to relaxation is based on group theory and Gröbner bases [238, 239, 467], surveyed in [2, 472]. Recent work on cutting planes includes investigation of *split cuts, intersection cuts* for corner polyhedra, and related ideas that have roots in [28, 239, 240]. Some of this work is surveyed in [145].

Related constraints.   There is also a *0-1 linear* constraint, in which the variables take 0-1 values.

Knapsack

A knapsack constraint enforces a bound on the value of an integer knapsack problem, which is an integer programming problem with one inequality constraint. An important special case is the 0-1 knapsack problem.

Notation.   A knapsack constraint can be written

$$\text{knapsack}\,(x \mid a, u, c, \ell)$$

where $x = (x_1, \ldots, x_n)$ is a tuple of integer (or 0-1) variables, $a = (a_1, \ldots, a_n)$ a tuple of weights, u a maximum weight, $c = (c_1, \ldots, c_n)$ a tuple of profits, and ℓ a minimum profit. The values in c and a are normally assumed to be integers, with little loss of generality. The constraint requires x be a solution of the knapsack problem with profit at least ℓ. That is, x must satisfy $ax \leq u$ and $cx \geq \ell$.

Usage.   The problem can be viewed as selecting x_i items of type i, weight a_i, and unit profit contribution c_i to put in a knapsack, so that the total weight is at most u and the total profit at least ℓ. There are many applications of this idea.

Inference.   Achieving domain consistency is NP-hard. An incomplete filtering algorithm is proposed in [191] and refined in [310, 444, 446].

Relaxation.   The knapsack problem is an integer programming problem and can be relaxed accordingly (see the entries on 0-1 and integer linear constraints). In particular, one can generate knapsack cuts from the single inequality constraint. 0-1 knapsack cuts, also known as *cover inequalities*, are discussed in Section 7.2.2. They can often be strengthened by lifting techniques [30, 154] (Sections 7.2.3 and 7.2.4). Knapsack cuts and lifting are discussed in [346, 371, 510] and generalized in [195, 509]. General integer knapsack cuts [26] are briefly discussed in Section 2.3.3. They, too, can often be strengthened by lifting techniques [25, 249].

Related constraints. There is also a *0-1 linear* constraint, in which the variables take 0-1 values.

Lex Greater

The *lex greater* constraint requires that one tuple of variables be lexicographically greater than another.

Notation. The constraint is

$$x >_{\text{lex}} y$$

where x, y are n-tuples of variables whose domains are subsets of a totally ordered set. The constraint requires that x be lexicographically greater than y; that is, $x \geq_{\text{lex}} y$, and $x_n > y_n$ if $x_i = y_i$ for $i = 1, \ldots, n-1$. The relation $x \geq_{\text{lex}} y$ holds if and only if $x_1 \geq y_1$ and, for $i = 2, \ldots, n$, $x_i \geq y_i$ whenever $x_j = y_j$ for all $j < i$.

Usage. The constraint is used for symmetry breaking in matrices of variables [201] as, for example, in sports scheduling [268], and for multicriteria optimization.

Inference. Complete filtering algorithms are given in [213, 319].

Related constraints. LexGreater is one of several lex-ordering constraints [213].

Linear Disjunction

linear disjunction is a disjunction of linear systems.

Notation. The constraint can be written

$$\text{linearDisjunction:} \bigvee_{k \in K} \begin{bmatrix} y_k \\ A^k x \geq b^k \end{bmatrix} \tag{8.20}$$

where x is a tuple $(x_1, \ldots, x_n)$ of real-valued variables and each y_k is a Boolean variable. The constraint enforces

$$\begin{array}{ll} \bigvee_{k \in K} y_k & (a) \\ y_k \rightarrow A^k x \geq b^k, \quad \text{all } k \in K & (b) \end{array} \tag{8.21}$$

Usage. It is common for linear systems to be enforced under certain conditions y_k, as expressed by conditionals like those in (8.21b). When a disjunction like (8.21a) is known to hold, one can obtain a much stronger relaxation by writing a single linear disjunction (8.20), rather than the individual constraints in (8.21). Disjunctions such as (8.21a) can be derived by computing the prime implications of the logical constraints in the problem. Section 2.4 illustrates the idea with a production planning example, and Section 7.6.3 discusses disjunctive modeling in general. See also [32, 292, 293, 437].

Relaxation. The constraint (8.20) implies

$$\bigvee_{\substack{k \\ 1 \in D_{y_k}}} A^k x \geq b^k$$

which can be given a convex hull relaxation [31] (Section 7.4.1) or a big-M relaxation (Section 7.4.2). There are specialized relaxations for the case where each disjunct is a single inequality [50] (Section 7.4.3) or a single equation (Section 7.4.4). Cutting planes can also be generated [31, 34, 295, 435, 437] (Section 7.4.5).

Related constraints. Certain *nonlinear disjunctions* can be given convex hull relaxations.

Logic

A *logic* constraint consists of logical clauses. The conjunction of the clauses is a propositional formula in conjunctive normal form. The constraint can be extended to allow other types of logical propositions as well.

Notation. A logic constraint has the form

$$\text{logic:} \bigvee_{j \in J_i} x_j \vee \bigvee_{j \in \bar{J}_i} \neg x_j, \quad i \in I$$

where each x_j is a Boolean variable in the set $\{x_1, \ldots, x_n\}$. Each $i \in I$ corresponds to a logical *clause* that contains the literals x_j for $j \in J_i$ and $\neg x_j$ (not x_j) for $j \in \bar{J}_i$. The clause requires that at least one of the literals be true.

It may be convenient to allow all formulas of propositional logic, and not just logical clauses, to appear in a logic constraint, because they can be automatically converted to conjunctive normal form, as described in Section 6.4.1.

Usage. The logic constraint expresses logical relations between variables, generally Boolean variables that take the values F, T or $0, 1$. The constraint is illustrated in Section 2.5 (employee staffing).

Inference. The resolution method is a complete inference method for general clause sets and therefore achieves domain consistency (Sections 6.4.2 and 6.4.4). Unit resolution does the same for Horn sets (Section 6.4.3). k-resolution achieves strong k-consistency on general clause sets (Section 6.4.5). Inference duality permits the deduction of nogoods (conflict clauses) from a clause set (Section 5.2.4). Parallel resolution [279] makes the nogood set easy to satisfy (Sections 5.2.7, 6.4.6).

Relaxation. Convex hull relaxations of common logical formulas [290] appear in Section 7.7.1. General clause sets can be relaxed by dropping the integrality constraints from a 0-1 formulation that is strengthened by resolvents (Section 7.7.2). In particular, input resolution generates all clausal rank 1 cuts [274] (Section 7.7.4). Section 7.7.5 describes how to generate separating resolvents [294].

Related constraints. Variables that appear in a logic constraint are often related to other constraints by means of *conditional, linear disjunction,* or *nonlinear disjunction* constraints. For example, to enforce at least one of the constraints C_1, C_2, C_3, one can write $x_1 \lor x_2 \lor x_3$ and the conditionals $x_i \rightarrow C_i$ for $i = 1, 2, 3$. If each C_i is a linear system, the conditionals should be replaced by the constraint

$$\text{linearDisjunction:} \quad \begin{bmatrix} x_1 \\ C_1 \end{bmatrix} \lor \begin{bmatrix} x_2 \\ C_2 \end{bmatrix} \lor \begin{bmatrix} x_3 \\ C_3 \end{bmatrix}$$

and similarly if the C_is are nonlinear systems.

Logical clauses can be generalized to include disjunctions of the form $\bigvee_{j \in J} (x_j \in D_j)$, where x_j is any finite domain variable. A complete multivalent resolution method for such clauses is described in [279, 295].

Lot Sizing

Lot sizing constraints can be defined to capture multiperiod production and inventory problems with setup costs.

Notation. A basic single-product lot-sizing constraint can be written

$$\text{lotSizing}\,(x, s, y, z \mid d, u, c, f, h) \tag{8.22}$$

Here $x = (x_1, \dots, x_n)$ are real-valued production variables, $s = (s_1, \dots, s_n)$ are real-valued stock level variables, and $y = (y_1, \dots, y_n)$ are 0-1 variables

y_i indicating whether product i is manufactured in period i. Also variable z indicates total production and holding costs. The problem data consist of unit variable costs $c = (c_1, \ldots, c_n)$ of production, fixed costs $f = (f_1, \ldots, f_n)$ of production, and unit holding costs $h = (h_1, \ldots, h_n)$ for each peripod. There are also demands $d = (d_1, \ldots, d_n)$ and maximum production levels $u = (u_1, \ldots, u_n)$ for each period. Constraint (8.22) requires that

$$z = \sum_i (p_i x_i + h_i s_i + f_i y + i)$$

$$s_{i-1} + x_i = d_i + s_i, \quad i = 1, \ldots, n \qquad (8.23)$$

$$x_i \leq u_i y_i, \quad i = 1, \ldots, n$$

$$x_i, s_i \geq 0 \text{ and } y_i \in \{0, 1\} \quad i = 1, \ldots, n$$

A generalization of the problem accounts for startup costs that are assessed when $y_i = 0$ and $y_{i+1} = 1$. A further generalization allows multiple product types and assesses changeover costs when there is switch from one product in period i to another product in period $i + 1$.

Usage.   The multiperiod lotsizing problem is a fundamental problem of operations management.

Relaxation.   The model (8.23) can be viewed as a special case of the network design problem, and a number of cutting-plane families have been developed on that basis [346]. In particular, there classes of cutting planes for the uncapacitated case ($u_i = \infty$), the constant capacity case (all u_is equal), and the variable capacity case (arbitrary u_is). MILP formulations and cutting planes for startup and changeover costs are discussed in [151, 346, 511].

Related constraints.   There are additional variations of the lot sizing problem for which MILP models and/or cutting planes have been developed (e.g., [27, 71]).

Mixed Integer

A *mixed-integer* constraint consists of a system of linear inequalities or equations in which at least some of the variables must take integer values.

Notation.   An MILP constraint can be written

$$\text{MILP:} \quad Ax + By \geq b$$

where A and B are matrices, x is a tuple of integer-valued variables, and y a tuple of real-valued variables.

Usage.   Mixed-integer linear inequalities provide a versatile modeling language. Modeling principles and examples are presented in Section 7.6 as well as [292, 293, 500, 501, 502]. Many constraints can be relaxed by formulating them as an MILP constraint and taking its continuous relaxation.

Inference. Bounds propagation (Section 6.2.1) and dual multipliers for the continuous relaxation (Section 6.2.3) can be used to reduce domains.

Relaxation. A continuous relaxation can be obtained by dropping the integrality requirement. The relaxation can be strengthened by the addition of Gomory's mixed-integer cuts [236], mixed-integer rounding cuts [347] (Sections 7.3.3 and 7.3.4), and lift-and-project cuts [34], among others. For a general discussion of cutting planes, see [214, 346, 371, 510]. Recent work on cutting planes includes investigation of *split cuts*, *intersection cuts* for corner polyhedra, and related ideas that have roots in [28, 239, 240]. Some of this work is surveyed in [145]. In addition, mixed-integer rounding (MIR) cuts have been generalized to *n-step MIR cuts* [148, 159, 315, 316].

Related constraints. See the *integer linear* and *0-1 linear* constraints for additional inference and relaxation techniques.

Min-n

The *min-n* constraint selects the rth smallest value taken by a set of variables.

Notation. The constraint
$$\text{min-}n(X, v \mid r)$$
requires that v be the rth smallest distinct value taken by the finite domain variables $X = \{x_1, \ldots, x_n\}$, or the largest value if there are fewer than r distinct values.

Usage. The constraint has been used to require that a task not start until at least r other tasks have started.

Inference. A complete filtering algorithm is given in [52].

Related constraints. There is also a max-n constraint. Min-n is closely related to *nvalues*.

Network Design

The *network design (fixed-charge network flow)* constraint requires that capacitated arcs be selected and flows placed on them to satisfy flow conservation constraints. It also computes variable and fixed costs.

Notation.   The constraint can be written

$$\text{network-design}\,(x, y, z \mid N, s, A, u, c, f) \qquad (8.24)$$

where x is a tuple of real-valued flow variables x_{ij}, y a tuple of corresponding integer variables y_{ij}, and z a real-valued variable representing cost. N a tuple of nodes, and s a tuple of net supply values corresponding to the nodes in N. A is a tuple of directed arcs, and u, c, and f are tuples of flow capacities, unit variable costs, and fixed costs corresponding to the arcs in A. The variables y_{ij} can take integral values greater than one to allow for the possibility that multiple arcs of a given capacity may connect nodes i and j, as when one lays two or more telecommunication cables. Constraint (8.24) requires that

$$z = \sum_{(i,j) \in A} c_{ij} x_{ij} + \sum_{(i,j) \in A} f_{ij} y_{ij}$$

$$\sum_{(i,j) \in A} x_{ij} - \sum_{(j,i) \in A} x_{ji} = s_i, \text{ all } i \in N \qquad (8.25)$$

$$0 \le x_{ij} \le u_{ij} y_{ij}, \text{ all } (i,j) \in A$$

$$y_{ij} \ge 0 \text{ and } y_{ij} \text{ integral, all } i,j$$

The fixed-charge network flow problem minimizes z subject to (8.24). There is a multicommodity version of the problem that allows one to require that messages from one node travel to a certain other node.

Usage.   The model (8.25) and variations of it are widely used for the design of telecommunication and other networks. Special cases of the problem include certain lot sizing problems [40] and the capacitated facility location problem [182].

Relaxation.   The model (8.25) can be relaxed by dropping the integrality requirement. Several types of cutting planes, known as *flow cuts*, have been developed for the model and its variations [346].

Related constraints.   The *flow* constraint is a special case in which the choice of arcs is fixed.

Nonlinear Disjunction

A *nonlinear disjunction* is a disjunction of convex, bounded nonlinear systems.

Notation.   The constraint is written

$$\text{nonlinearDisjunction:} \ \bigvee_{k \in K} g^k(x) \le 0 \qquad (8.26)$$

where each $g^k(x)$ is a tuple of functions $g_i^k(x)$, and x a tuple of real-valued variables. It is assumed that $\ell \leq x \leq u$, and $g^k(x)$ is bounded when $\ell \leq x \leq u$. It is further assumed that each $g_i^k(x)$ is a convex function on $[\ell, u]$. The constraint enforces

$$\bigvee_{k \in K} y_k$$

$$y_k \rightarrow \left(g^k(x) \leq 0\right), \quad \text{all } k \in K$$

Usage. Nonlinear disjunctions are used in a manner analogous to *linear disjunctions*.

Relaxation. The constraint (8.26) implies

$$\bigvee_{\substack{k \\ 1 \in D_{y_k}}} g^k(x) \leq 0$$

which can be given a convex hull relaxation [336, 465, 436] (Section 7.5.1) or a big-M relaxation (Section 7.5.2).

Related constraints. There is a special constraint for *linear disjunctions*.

Nvalues

The *nvalues* constraint bounds the number of distinct values taken by a set of variables.

Notation   The simplest form of the constraint is

$$\text{nvalues}(x \mid \ell, u) \tag{8.27}$$

where x is a tuple $(x_1, \ldots, x_n)$ of variables with finite domains, and ℓ and u are nonnegative integers. The constraint requires that the variables take at least ℓ and at most u different values. In other versions of the constraint, the lower bound is a variable:

$$\text{nvalues}(x, \ell \mid u) \tag{8.28}$$

or the upper bound is a variable:

$$\text{nvalues}(x, u \mid \ell) \tag{8.29}$$

Usage. The constraint is illustrated in Section 2.5 (employee scheduling).

Inference. As Section 6.9.4 notes, a network flow model provides complete filtering for (8.27). A more elaborate model [395] provides the same for (8.28). Achieving domain consistency for (8.29) is NP-hard [91]. An incomplete filter is presented in [85], and a less thorough one in [52].

Related constraints. *Alldiff* is a special case of *nvalues* in which $\ell = u = n$. A *weighted nvalues* constraint is presented in [60], which also describes a filter. There are a number of other variations, such as ninterval (counts values as distinct only when they lie in different intervals) and nclass (counts values as distinct only when they lie in different sets) [52].

Path

The *path* constraint finds a simple path in a graph having at most a given length. The constraint may have no length bound, and a constraint with a length bound has been called the *shorter-path* constraint.

Notation. The constraint can be written

$$\text{path}(x, w \mid G, c, s, t) \tag{8.30}$$

where x is a tuple of Boolean variables corresponding to the vertices of a directed or undirected graph G, w is an integer, G is a directed or undirected graph on n vertices, c contains an integral edge length c_{ij} for every edge (i, j) of G, and s, t are vertices of G. It is assumed there are no cycles in G of negative length. The constraint requires that G contain a simple path (no repeated vertices) of length at most w from s to t, where x_i is true if and only if vertex i lies on the path. The *shortest-path problem* on G minimizes w subject to (8.30) when the domain of every x_j is {true, false}.

Usage. Shortest-path models have countless applications. An application of a specially structured path constraint is presented in Section 5.1.7.

Inference. When the domain of every x_j contains *false*, the minimum value of w subject to (8.30) can be computed in polynomial time by a shortest-path algorithm on an induced subgraph of G. Shortest-path algorithms, some of which compute shortest paths between all pairs of vertices, are explained in [8, 45, 334] and exhaustively surveyed in [173]. Finding a shortest simple path that contains certain vertices is NP-hard, however, as is achieving domain consistency for (8.30). Incomplete filtering methods are presented in [445, 449]. If the graph is acyclic, complete filtering can be accomplished in polynomial time [190].

Related constraints. The *path-partitioning* constraint [67] partitions the vertex set of a directed graph to define a specified number of disjoint paths, each of which must terminate in a given subset of vertices.

Piecewise Linear

The *piecewise linear* constraint sets a variable equal to a piecewise linear function of one variable.

Notation.   The constraint can be written

$$\text{piecewiseLinear}(x, z \mid L, U, c, d)$$

where x and z are real-valued variables. $L = (L_1, \ldots, L_m)$ and $U = (U_1, \ldots, U_m)$ define the intervals $[L_k, U_k]$ on which the piecewise linear function is defined. Adjacent intervals should intersect in at most a point. On each interval, the graph of the function is a line segment connecting (L_k, c_k) with (U_k, d_k). The constraint sets z equal to

$$\frac{U_k - x}{U_k - L_k} c_k + \frac{x - L_k}{U_k - L_k} d_k \ \text{ for } x \in [L_k, U_k], \ \ k = 1, \ldots, m$$

Usage.   A piecewise linear constraint (possibly with gaps) is a versatile modeling device. Piecewise linear functions commonly occur in practice, and moreover they can approximate many nonlinear functions that would be hard to relax directly. A separable nonlinear function $\sum_j h_j(x_j)$ can be approximated to a high degree of accuracy by replacing each nonlinear function $h_j(x)$ with a variable z_j that is set equal to a piecewise linear function having sufficiently many linear pieces. This is illustrated in Section 7.1, and applications are described in [380, 381, 413].

Inference.   The piecewise linear function can interact with variable bounds to tighten the latter, as described in Section 7.1 and [413].

Relaxation.   A convex hull relaxation can be generated in linear time without using additional variables. Fast methods for computing the convex hull in two dimensions can be found in [460] (pages 351–352) and [399].

Range

The *range* constraint identifies the set of values taken by specified variables.

Notation.   The range constraint [86] can be written

$$\text{range}(X, S, T) \tag{8.31}$$

where $X = \{x_1, \ldots, x_n\}$ is a set of variables, variable S a subset of the indices $\{1, \ldots, n\}$, and variable T a set of values. The constraint requires that T be the set of values taken by the variables in X indexed by S. That is, $T = \{v(x_i) \mid i \in S\}$, where $v(x_i)$ is the value of x_i.

Usage. The range and roots constraints have been proposed as a pair of constraints that, together with set inclusion and set cardinality, can represent a number of counting-type constraints [85, 90].

Inference. Filtering algorithms appear in [87].

Regular

The *regular* constraint restricts possible sequences of events or states by specifying state transitions in the form of a deterministic (or nondeterministic) finite automaton.

Notation. The regular constraint [390, 392] (Section 6.12.1) can be written

$$\text{regular}\,(x \mid A) \tag{8.32}$$

where $x = (x_1, \ldots, x_n)$ is a tuple of variables. Also $A = (S, D, t, \alpha, F)$ is a deterministic finite automaton in which S is a set of states, D a set of controls, t a partial transition function, α the initial state, and F a set of accepting states. The transition function specifies the result $t(s, x)$ of applying control x in state s. It may not be defined for all pairs (s, x), in which case control x cannot be applied in state s. It is assumed that $t(s, x) \neq t(s, x')$ when $x \neq x'$. The automaton *accepts* a string $x_1, \ldots, x_n$ when $s_1 = \alpha$, $s_{n+1} \in F$, and $s_{i+1} = t(x_i, x_i)$ for $i = 1, \ldots, n$, and the set of strings accepted by such an automaton is a regular language. The regular constraint (8.32) is satisfied when A accepts string x.

There is a cyclic version of the constraint [406] (Section 6.12.5) :

$$\text{regularCycle}\,(x \mid A) \tag{8.33}$$

where $x = (x_1, \ldots, x_{n+1})$. It requires that automaton A accept the string $x_1, \ldots, x_n, x_1$.

The regular constraint can also be defined in terms of a nondeterministic finite automaton [55, 56, 406] (Section 6.12.4) :

$$\text{regular}\,(x \mid N)$$

It is satisfied when the automaton N accepts string $x = (x_1, \ldots, x_n)$.

Usage. The regular constraint can be used in sequencing and scheduling problems that impose restrictions on the sequence of events. For example, the constraint may require that employees have a a day off after working a certain number of days, or that a machine not manufacture products A, B, and C consecutively. However, it is not always obvious how to model possible state transitions as a deterministic finite automaton. The cyclic version of the constraint is used for weekly, monthly, and other repeating schedules. The nondeterministic form can be useful when a deterministic automaton requires a large state space to capture the constraint (Section 6.12.4).

Inference. The constraint is naturally formulated with a dynamic programming model, for which complete filtering is straightforward[390, 392, 479] (Section 6.12.2) . Domain consistency can also be achieved (with the same time complexity) by a decomposition strategy that filters the individual state transition constraints and propagates them in a forward and backward pass [55, 56, 405, 406] (Section 6.12.3) . The cyclic regular constraint can be filtered by coverting it to a normal regular constraint on a larger automaton [406] (Section 6.12.5), or by propagating a dynamic programming model with an additional state variable (Section 6.12.5). The regular constraint based on a nondeterministic automaton can be filtered in the same fashion as a normal regular constraint [55, 56, 406] (Section 6.12.4).

Related constraints. The *stretch* constraint (Section 6.11) is a special case of the regular constraint that provides more efficient filtering when the problem has this particular form. A *dynamic programming* constraint (Section 6.12.6) can be filtered in exactly the same way as the regular constraint but provides more flexibility in modeling. A regular language constraint can be extended to other types of languages in the Chomsky hierarchy [447].

Roots

The *roots* constraint identifies the subset of variables that take values in a given set.

Notation. The roots constraint [86] can be written

$$\text{roots}(X, T, S) \qquad (8.34)$$

where $X = \{x_1, \ldots, x_n\}$ is a set of variables, variable T a set of values, and variable S a subset of the indices $\{1, \ldots, n\}$. The constraint requires that S be the index set of variables that take values in T. That is, $S = \{i \mid v(x_i) \in T\}$, where $v(x_i)$ is the value of x_i.

Usage. The range and roots constraint have been proposed as a pair of constraints that, together with set inclusion and set cardinality, can represent a number of counting-type constraints [85, 90].

Inference. A decomposition approach to filtering is described in [88]. Although complete filtering is NP-hard, the paper states that its linear-time filter is complete "in many situations met in practice."

Same

The *same* constraint requires that two sets of equally many variables take the same multiset of values.

Notation. The constraint is

$$\text{same}(X, Y)$$

where X and Y are sets of n variables having finite domains. The constraint requires that the multiset of values taken by $x_1, \ldots, x_n$ equal the multiset of values taken by $y_1, \ldots, y_n$.

Usage. The following example appears in [66]. The organization Doctors without Borders wishes to pair doctors and nurses for emergency missions. Each x_i is the date on which doctor i departs, and the domain of x_i consists of the dates the doctor is available to depart. Each y_i has the same meaning for nurses. In another example [58], x_i is the shift one person works on day i, and y_i is the same for another person. The same constraint is enforced for all pairs of persons to ensure fairness, in that all work the same multiset of shifts.

Inference. A complete filter based on network flows (but a different model than used for *alldiff* and *cardinality*) appears in [66].

Related constraints. The constraint usedby(X, Y) allows X to contain more variables than Y and requires that the multiset of values used by Y be contained in the multiset used by X [65]. The sort(x, y) constraint [98, 353] is slightly stronger than *same*—it requires that variables $y = (y_1, \ldots, y_n)$ be the result of sorting variables $x = (x_1, \ldots, x_n)$.

Sequence

The *sequence* constraint imposes *among* constraints on successive subsequences of q consecutive variables. The *generalized sequence* constraints imposes along constraints on arbitrary subsequences of consecutive variables.

Notation. The sequence constraint is

$$\text{sequence}\,(x \mid q, V, \ell, u)$$

where $x = (x_1, \ldots, x_n)$ is a tuple of variables having finite domains, q a positive integer, V a set of values, and ℓ, u nonnegative integers. The constraint imposes an among constraint on each subsequence of q consecutive variables and is therefore equivalent to

$$\text{among}\,(X_j \mid V, \ell, u)\,, \quad j = 1, \ldots, n - q + 1$$

where each $X_j = \{x_j, \ldots, x_{j+q-1}\}$. The generalized sequence constraint is

$$\text{genSequence}\,(x \mid \mathcal{X}, V, \ell, u)$$

where $\mathcal{X} = (X_1, \ldots, X_m)$, each X_i is a subset of consecutive variables occurring in x, $\ell = (\ell_1, \ldots, \ell_m)$, and $u = (u_1, \ldots, u_m)$. The constraint imposes

$$\text{among}\,(X_i \mid V, \ell_i, u_i)\,, \quad i = 1, \ldots, m$$

Usage. The constraint naturally applies to assembly line sequencing, where a limited number of product types requiring a certain resource can be scheduled in each subsequence of q products. Here V is the set of product types that require the resource, and x_i is the ith product type in the sequence. A car sequencing example is given in Section 6.10.2. The constraint can also be used for employee scheduling problems in which the number of days off or number of night shifts in any 7-day period must be controlled.

Inference. Incomplete filtering algorithms for sequence appear in [53] and [425]. The first of several complete, polynomial-time filters appeared in [488] and is described in Section 6.10.3. Complete filters based in alternate encodings of the constraint are presented in [110]. Complete filtering based on a network flow model is introduced in [344] and described in Section 6.10.4. The filter in [488] achieves domain consistency for genSequence as well. The flow-based filter in [344] may achieve domain consistency for genSequence, depending on the structure of the problem, as discussed in Section 6.10.4.

Related constraints. The *global sequencing constraint* [425] combines a sequence constraint with a cardinality constraint, because they frequently occur together in assembly line sequencing problems. An incomplete filter appears in [425]. Computational results for this and other multiple-among filters appear in [489]. A *multiple sequence* constraint is proposed in [110], which presents a filter that is based on filtering for the regular constraint. It is assumed that the sets V of values for each constraint are pairwise disjoint. The *sliding sum* constraint is like the sequence constraint except that the sum of variables in each subsequence is controlled. A complete flow-based filter appears in [344].

Set Covering

Given a collection of sets, the *set-covering* constraint selects at most k sets that have the same union as the collection.

Notation.   The constraint can be written

$$\text{setCovering}(x, k \mid S_1, \ldots, S_n) \tag{8.35}$$

where x is an n-tuple of Boolean variables x_j that are true when set S_j is selected, and k is a positive integer. The constraint requires that at most k sets be selected whose union is $\bigcup_{j=1}^{n} S_j$. A 0-1 representation of the constraint is also natural:

$$Ax \geq e, \quad \sum_{j=1}^{n} x_j \leq k, \quad x_j \in \{0,1\}, \ j = 1, \ldots, n \tag{8.36}$$

Here, e is an m-tuple of ones, and A is an $m \times n$ 0-1 matrix in which $A_{ij} = 1$ when $i \in S_j$. The *set-covering problem* is to minimize k subject to (8.35), or more generally to minimize $\sum_j c_j x_j$.

Usage.   Set-covering models arise in a wide variety of contexts [496]. For example, one may wish to buy the fewest possible CDs that contain all of one's favorite songs.

Inference.   Achieving domain consistency is NP-hard, because the set-covering problem is NP-hard. The upper bound on k can be tightened by solving the set-covering problem approximately. The problem has a long history of algorithmic development. One survey of algorithms is [389].

Relaxation.   A continuous relaxation can be obtained by dropping the integrality constraints from the 0-1 model (8.36) and adding valid cuts [374, 434], which are surveyed in [108].

Related constraints.   The *set-packing* constraint is the analogous constraint for $\leq$ inequalities.

Set Packing

Given a collection of sets, the *set-packing* constraint selects at least k sets that are pairwise disjoint.

Notation.   The constraint can be written

$$\text{setPacking}(x, k \mid S_1, \ldots, S_n) \tag{8.37}$$

where x is an n-tuple of Boolean variables x_j that are true when set S_j is selected, and k is a positive integer. The constraint requires that at least k pairwise disjoint sets be selected. A 0-1 representation of the constraint is also natural:

$$Ax \leq e, \quad \sum_{j=1}^{n} x_j \geq k, \quad x_j \in \{0,1\}, \ j = 1, \ldots, n \tag{8.38}$$

Here, e is an m-tuple of ones, and A is an $m \times n$ 0-1 matrix in which $A_{ij} = 1$ when $i \in S_j$. The *set-packing problem* is to maximize k subject to (8.37), or more generally to maximize $\sum_j c_j x_j$.

Usage. Set-packing models arise in wide variety of contexts [496]. For example, in a combinatorial auction, each bid is for a bundle of items. The auctioneer may wish to select bids to maximize revenue, but with no two bids claiming the same item. Set-packing inequalities are generated as knapsack cuts in mixed-integer solvers and strengthened with cutting planes (Section 7.2.5).

Inference. Achieving domain consistency is NP-hard, because the set-packing problem is NP-hard. The lower bound on k can be tightened by solving the set-packing problem approximately. The problem has a long history of algorithmic development. One survey of algorithms is [389].

Relaxation. A continuous relaxation can be obtained by dropping the integrality constraints from the 0-1 model (8.38) and adding valid cuts. These include odd cycle inequalities [383] (Section 7.2.5), for which there is a polynomial-time separation algorithm (Lemma 9.1.11 of [246]); clique inequalities [215, 383] (Section 7.2.5), for which separation algorithms are surveyed in [108]; orthonormal representation inequalities, which include clique inequalities and can be separated in polynomial time [246]; and other families surveyed in [108]. General discussions of these cutting planes may be found in [214, 346]

Related constraints. The *set-covering* constraint is the analogous constraint for $\geq$ inequalities.

Soft Alldiff

The *soft alldiff* constraint limits the degree to which a set of variables fail to take all different values.

Notation. Two varieties of the *soft* alldiff constraint have been studied, both written

$$\text{alldiff}(x, k)$$

One requires k to be an upper bound on the number of pairs of variables that have the same value. The other requires k to be an upper bound on the minimum number of variables that must change values to satisfy $\text{alldiff}(x)$.

Usage. The constraints are designed for overconstrained problems in which the object is to obtain a solution that is close to feasibility.

Inference.   An incomplete filter for the first soft alldiff constraint appears in [395] and a complete one in [486]. A complete filter for the second constraint is given in [395].

Related constraints.   Soft alldiff is an extension of the *all-different* constraint. Soft constraints in general have received much recent attention in constraint programming. Introductions to the area can be found in [43, 355].

Sort

The *sort* constraint arranges contents of tuple in nondecreasing order.

Notation.   The sort constraint [98] is written

$$\text{sort}\,(x, y)$$

where $x = (x_1, \ldots, x_n)$ and $y = (y_1, \ldots, y_n)$ are tuples of variables on whose domains a complete order is defined. The constraint requires that y contain a sorted list of the values in x. That is, the constraint is satisfied if there is a permutation π of $1, \ldots, n$ such that $y_i = x_{\pi(i)}$ for $i = 1, \ldots, n$ and $y_i \leq y_{i+1}$ for $i = 1, \ldots, n-1$. Another form of the constraint [520] includes the permutation π as one of the arguments:

$$\text{sort}\,(x, \pi, y)$$

Usage.   The sort constraint has been used to solve job shop scheduling problems [520].

Inference.   A filter that achieves bounds consistency appears in [353]. Its running time is $\mathcal{O}(n)$ plus the time required to sort the endpoints of the variable domains.

Spanning Tree

The *spanning tree* constraint locates minimum spanning trees in a graph, or spanning trees having at most a given weight.

Notation.   The constraint is defined on graph-valued variables, which are similar to set-valued variables (Section 5.1.7). The value of a graph variable G is the pair (V, E) consisting of the vertex set V and edge set E. The domain of G is stored in the form of a lower bounding graph L_G and upper bounding graph U_G, indicating that any feasible value of G must be a subgraph of U_G and have L_G as a subgraph. Bounds consistency is defined in analogy with ordinary bounds consistency (Section 6.1.3).

A spanning tree of a graph G is a subgraph of G that is a tree and contains all the vertices of G. The spanning tree constraint has been studied in two forms. The *minimum spanning tree* constraint [176] can be written

$$\text{spanningTree}\,(G, T, w) \qquad (8.39)$$

where G is a graph, T a tree, and w a matrix of scalar variables w_{ij} representing the weight of edge (i, j). The constraint requires that T be a minimum-weight spanning tree of G, using weights w, where the weight of a tree is the sum of its edge weights. The *weight-bounded spanning tree* constraint [177]

$$\text{spanningTree}\,(G, T, w, w_{\max}) \qquad (8.40)$$

requires that T be a spanning tree of G with weight at most $w_{\max}$ using edge weights w. The two constraints are very similar, but require different approaches to propagation. In fact, bounds consistency can be achieved in polynomial time for (8.39), but is an NP-hard problem for (8.40).

Usage.   The constraint can be used in network design problems, particularly when it is important to know when a particular edge must be part of an optimal or acceptable solution.

Inference.   The bounds consistency problem was partially solved for (8.39) with fixed G by [22] and completely for the general form of (8.39) by [176]. It is shown in [177] that achieving bounds consistency is NP-hard for the general constraint (8.40), but that it can be done efficiently when the weights w are fixed, as well as when the vertex set of both G and T is fixed.

Related constraints.   The *tree* constraint of [62] partitions a directed graph into vertex-disjoint anti-arborescences (i.e., trees in which there is a directed path from every vertex to the root). The constraint has applications in circuit design, biology, and linguistics. The *resource-forest* constraint [64] and an undirected form of the tree constraint. Given a graph in which certain vertices are identified as resources, it identifies a forest (set of trees) that covers the graph and in which every tree contains a resource vertex. For example, the resource vertices might represent printers, and the remaining vertices computers that must be connected to a printer. The *proper-forest* constraint [64] identifies a forest that covers the graph and in which every tree contains at least two vertices. The vertices might represent computers that must be connected to at least one backup computer.

Spread

The *spread* constraint controls the mean and L_p norm of a set of variables. It was originally proposed [393] under the name "spread" for $p = 2$, in which case the constraint controls the mean and standard deviation. The constraint has been called the *deviation* constraint for $p = 1$ and the *balance* constraint for general p.

Notation. The constraint is written

$$\text{spread}(X, \mu, L \mid p)$$

where X is a set $\{x_1, \ldots, x_n\}$ of variables, μ and L are real numbers, and p is normally 0, 1, 2, or ∞. The constraint requires that the mean and L_p norm of $x_1, \ldots, x_n$ be equal to μ and L respectively. That is,

$$\mu = \frac{1}{n} \sum_i x_i, \quad L = \left(\sum_i |x_i - \mu|^p \right)^{\frac{1}{p}}$$

when $p \geq 1$, and $L = |\{i \mid x_i \neq \mu\}|$ when $p = 0$. Here the variable L is interpreted

$$L = \begin{cases} \text{number of } x_i\text{s different from the mean} & \text{if } p = 0 \\ n \text{ times mean absolute deviation} & \text{if } p = 1 \\ n \text{ times standard deviation} & \text{if } p = 2 \\ \text{maximum absolute deviation from the mean} & \text{if } p = \infty \end{cases}$$

Usage. The constraint is motivated by situations in which balance or fairness is important. It may be desirable to balance the workload of assembly line stations, the number of weekends or night shifts assigned to workers, or the number of customers or travel time assigned to vehicles in a routing problem.

Inference. An algorithm in [393] achieves bound consistency when $p = 2$. The algorithm simplifies when μ is treated as a constant [439]. The deviation constraint ($p = 1$) is filtered in [440].

Stretch

The *stretch* constraint was designed for scheduling workers in shifts. It specifies limits on how many consecutive days a worker can be assigned to each shift, and which shifts can follow another.

Notation. The constraint is

$$\text{stretch}(x \mid v, \ell, u, P)$$

where $x = (x_1, \ldots, x_n)$ is a tuple of variables with finite domains. Perhaps x_i denotes the shift that a given employee will work on day i. Also, v is an m-tuple of possible values of the variables, ℓ an m-tuple of lower bounds, and u an m-tuple of upper bounds. A stretch is a maximal sequence of consecutive variables that take the same value. Thus, $x_j, \ldots, x_k$ is a stretch if for some value v, $x_j, \ldots, x_k = v$, $x_{j-1} \neq v$ (or $j = 1$), and $x_{k+1} \neq v$ (or $k = n$). The stretch constraint requires that for each $j \in \{1, \ldots, m\}$, any stretch of value v_j in x have length at least ℓ_j and at most u_j. In addition, P is a set of *patterns*, which are pairs of values $(v_j, v_{j'})$. The constraint requires that when a stretch of value v_j immediately precedes a stretch of value $v_{j'}$, the pair $(v_j, v_{j'})$ must be in P.

There is also a cyclic version of the stretch constraint, *stretchCycle*, that recognizes stretches that continue from x_n to x_1. It can be used when every week must have the same schedule.

Usage.   Examples of scheduling with *stretch* appear in Section 2.5 and [391].

Inference.   A polynomial-time dynamic programming algorithm achieves domain consistency [264] (Section 6.11).

Related constraints.   *Stretch* is similar to the *pattern* constraint [109]. The *change* constraint limits the number of times a given type of shift change occurs. Stretch is a special case of the *regular* constraint.

Sum

The *sum* constraint computes a sum over an index set that depends on the value of a variable.

Notation.   The simplest form of the constraint is written

$$\text{sum}(y, z \mid a, S)$$

where y is a variable representing a positive integer, z is a variable, a is a tuple $(a_1, \ldots, a_m)$ of values, and S a tuple $(S_1, \ldots, S_n)$ of index sets, with each $S_i \subset \{1, \ldots, m\}$. The constraint requires z to be equal to $\sum_{j \in S_y} a_j$. Another form is

$$\text{sum}(y, x, z \mid S)$$

where x is a tuple $(x_1, \ldots, x_m)$ of variables. The constraint enforces $z = \sum_{j \in S_y} x_j$.

Usage.   The constraint implements sums over variable index sets, which frequently occur in modeling. An application to a production planning problem with sequence-dependent cumulative costs is described in [515].

Inference. A complete filter is described in [279].

Relaxation. A convex hull relaxation appears in [515].

Related constraints. The sum constraint can be viewed as an extension of *element*.

Symmetric Alldiff

Given a set of items, the *symmetric alldiff* constraint pairs each with another compatible item in the set.

Notation. One way to write the constraint is simply

$$\text{symAlldiff}(X)$$

where $X = \{x_1, \ldots, x_n\}$ is a set of variables with domains that are subsets of $\{1, \ldots, n\}$. The constraint requires alldiff(X) and $x_{x_i} = i$ for $i = 1, \ldots, n$. In [419], the constraint is written with a bijection that associates the domain elements with the variables.

Usage. The constraint can be used to pair items with compatible items. For example, x_i can denote a sports team that is to play team i. If team i plays team j ($x_i = j$), then team j plays team i ($x_j = i$), and no two teams play the same team—alldiff(x). The domain of x_i contains the teams that team i is allowed to play. Other applications include staffing problems in which people are assigned to work in pairs.

Inference. A complete filter and a faster, incomplete filter are presented in [419].

Related constraints. This is a restriction of *alldiff*.

Symmetric Cardinality

The *symmetric cardinality* constraint bounds the number of values assigned to each variable, as well as the number of variables to which each value is assigned.

Notation. The constraint can be written

$$\text{symCardinality}(\mathcal{X} \mid \bar{\ell}, \bar{u}, v, \ell, u)$$

where $\mathcal{X}$ is a set $\{X_1, \ldots, X_n\}$ of set-valued variables, $\bar{\ell}, \bar{u}$ are n-tuples of nonnegative integers, v an m-tuple of values, and ℓ, u are m-tuples of nonnegative integers. The constraint requires that (a) each X_i be assigned a set with cardinality at least $\bar{\ell}_i$ and at most $\bar{u}_i$, and (b) for $j = 1, \ldots, m$, at least ℓ_j and at most u_j of the variables be assigned a set containing v_j.

Usage.   The constraint is used to limit how many workers are assigned to each task, and how many tasks are assigned to each worker. The workers that are appropriate for each task i are indicated by the initial domain of X_i.

Inference.   A complete filter based on network flows is given in [320].

Related constraints.   This is an extension of *cardinality*.

Value Precedence

Given two values s and t, and a set of variables, the (integer) *value precedence* constraint requires that if a variable takes value t then a variable with a lower index takes value s.

Notation.   The constraint is

$$\mathrm{integerValuePrecedence}(x \mid s, t)$$

where x is a tuple of variables with integer domains and s, t are integers. The constraint requires that whenever a variable x_j takes value t, a variable x_i with $i < j$ takes value s.

Usage.   The constraint is used for symmetry breaking, in particular when two values are interchangeable in a set of variables [333].

Inference.   Filtering algorithms are presented in [333].

Related constraints.   There is an analogous constraint for set-valued variables [333].

References

1. M. Aagren, N. Beldiceanu, M. Carlsson, M. Sbihi, C. Truchet, and S. Zampelli. Six ways of integrating symmetries within non-overlapping constraints. In W.-J. van Hoeve and J. N. Hooker, editors, *Proceedings of the International Workshop on Integration of AI and OR Techniques in Constraint Programming for Combinatorial Optimization Problems (CPAIOR 2009)*, volume 5547 of *Lecture Notes in Computer Science*, pages 11–25, New York, 2009. Springer.

2. K. Aardal, R. Weismantel, and L. Wolsey. Non-standard approaches to integer programming. *Discrete Applied Mathematics*, 123:5–74, 2002.

3. T. Achterberg. Conflict analysis in mixed integer programming. *Discrete Optimization*, 4:4–20, 2007.

4. T. Achterberg. SCIP: Solving constraint integer programs. *Mathematical Programming Computation*, 1:1–41, 2008.

5. T. Achterberg, T. Berthold, T. Koch, and K. Wolter. Constraint integer programming: A new approach to integrate CP and MIP. In L. Perron and M. A. Trick, editors, *Proceedings of the International Workshop on Integration of Artificial Intelligence and Operations Research Techniques in Constraint Programming for Combinatorial Optimization Problems (CPAIOR 2008)*, volume 5015 of *Lecture Notes in Computer Science*, pages 6–20, New York, 2008. Springer.

6. A. Aggoun and N. Beldiceanu. Extending CHIP in order to solve complex scheduling and placement problems. *Mathematical and Computer Modelling*, 17:57–73, 1993.

7. R. K. Ahuja, M. Kodialam, A. K. Mishra, and J. B. Orlin. Computational investigations of maximum flow algorithms. *European Journal of Operational Research*, 97:509–542, 1997.

8. R. K. Ahuja, T. L. Magnanti, and J. B. Orlin. *Network Flows: Theory, Algorithms, and Applications.* Prentice-Hall, Englewood Cliffs, NJ, 1993.

9. R. K. Ahuja, T. L. Magnanti, and J. B. Orlin. Applications of network optimization. In M. O. Ball, T. L. Magnanti, C. L. Monma, and G. L. Nemhauser, editors, *Network Models*, Handbooks in Operations Research and Management Science, pages 1–84. Elsevier, Amsterdam, 1995.

10. S. B. Akers. Binary decision diagrams. *IEEE Transactions on Computers*, C-27:509–516, 1978.

11. H. Alt, N. Blum, K. Mehlhorn, and M. Paul. Computing maximum cardinality matching in time $O(n^{1.5}\sqrt{m/\log n})$. *Information Processing Letters*, 37:237–240, 1991.

12. H. R. Andersen. An introduction to binary decision diagrams. Lecture notes, available online, IT University of Copenhagen, 1997.

13. H. R. Andersen, T. Hadzic, J. N. Hooker, and P. Tiedemann. A constraint store based on multivalued decision diagrams. In C. Bessiere, editor, *Principles and Practice of Constraint Programming (CP 2007)*, volume 4741 of *Lecture Notes in Computer Science*, pages 118–132, New York, 2007. Springer.

14. K. Andersen, Q. Louveaux, R. Weismantel, and L. A. Wolsey. Cutting planes from two rows of a simplex tableau. In *Proceedings of the 12th International Conference on Integer Programming and Combinatorial Optimization (IPCO 2007)*, volume 4513 of *Lecture Notes in Computer Science*, pages 1–15, New York, 2007. Springer.

15. G. Appa, D. Magos, and I. Mourtos. Linear programming relaxations of multiple all-different predicates. In J. C. Régin and M. Rueher, editors, *Integration of AI and OR Techniques in Constraint Programming for Combinatorial Optimization Problems (CPAIOR 2004)*, volume 3011 of *Lecture Notes in Computer Science*, pages 364–369, New York, 2004. Springer.

16. G. Appa, D. Magos, and I. Mourtos. On the system of two all-different predicates. *Information Processing Letters*, 94:99–105, 2005.

17. G. Appa, D. Magos, and I. Mourtos. A polyhedral approach to the alldifferent system. *Mathematical Programming*, 124:1–52, 2010.

18. G. Appa, I. Mourtos, and D. Magos. Integrating constraint and integer programming for the orthogonal Latin squares problem. In P. Van Hentenryck, editor, *Principles and Practice of Constraint Programming (CP 2002)*, volume 2470 of *Lecture Notes in Computer Science*, pages 17–32, New York, 2002. Springer.

19. K. Apt and M. Wallace. *Constraint Logic Programming Using ECLiPSe*. Cambridge University Press, 2006.

20. S. Arnborg, D. G. Corneil, and A. Proskurowski. Complexity of finding embeddings in a *k*-tree. *SIAM Jorunal on Algebraic and Discrete Mathematics*, 8:277–284, 1987.

21. S. Arnborg and A. Proskurowski. Characterization and recognition of partial *k*-trees. *SIAM Jorunal on Algebraic and Discrete Mathematics*, 7:305–314, 1986.

22. I. Aron and P. Van Hentenryck. A constraint satisfaction approach to the robust spanning tree problem with interval data. In *Proceedings of the 18th Conference on Uncertainty in Artificial Intelligence (UAI 2002)*, pages 18–25. Morgan Kaufman, 2002.

23. I. Aron, J. N. Hooker, and T. H. Yunes. SIMPL: A system for integrating optimization techniques. In J. C. Régin and M. Rueher, editors, *Integration of AI and OR Techniques in Constraint Programming for Combinatorial Optimization Problems (CPAIOR 2004)*, volume 3011 of *Lecture Notes in Computer Science*, pages 21–36, New York, 2004. Springer.

24. S. Arora and B. Barak. *Computational Complexity: A Modern Approach*. Cambridge University Press, 2009.

25. A. Atamtürk. Sequence independent lifting for mixed-integer programming. *Operations Research*, 52:487–490, 2004.

26. A. Atamtürk. Cover and pack inequalities for (mixed) integer programming. *Annals of Operations Research*, 139:21–38, 2005.

27. A. Atamtürk and S. Küçükyavuz. Lot sizing with inventory bounds and fixed costs: Polyhedral study and computation. *Operations Research*, 53:711–730, 2005.

28. E. Balas. Intersection cuts: A new type of cutting planes for integer programming. *Mathematical Programming*, 19:19–39, 1971.

29. E. Balas. Disjunctive programming: Properties of the convex hull of feasible points. Techical report mssr-348, Carnegie Mellon University, 1974.

30. E. Balas. Facets of the knapsack polytope. *Mathematical Programming*, 8:146–164, 1975.

31. E. Balas. Disjunctive programming. *Annals of Discrete Mathematics*, 5:3–51, 1979.

32. E. Balas. Disjunctive programming and a hierarchy of relaxations for discrete optimization problems. *SIAM Journal on Algebraic and Discrete Methods*, 6:466–485, 1985.

33. E. Balas, A. Bockmayr, N. Pisaruk, and L. Wolsey. On unions and dominants of polytopes. *Mathematical Programming*, 99:223–239, 2004.

34. E. Balas, S. Ceria, and G. Cornuéjols. A lift-and-project cutting plane algorithm for mixed 0-1 programs. *Mathematical Programming*, 58:295–324, 1993.

35. E. Balas, S. Ceria, G. Cornuéjols, and G. Pataki. Polyhedral methods for the maximum clique problem. In D. S. Johnson and M. A. Trick, editors, *Cliques, Colorings and Satisfiability: 2nd DIMACS Implementation Challenge, 1993*, pages 11–28. American Mathematical Society, 1996.

36. E. Balas and M. Fischetti. Polyhedral theory for the asymmetric traveling salesman problem. In G. Gutin and A. P. Punnen, editors, *The Traveling Salesman Problem and Its Variations*, pages 117–168. Kluwer, Dordrecht, 2002.

37. E. Balas and C. S. Yu. Finding a maximum clique in an arbitrary graph. *SIAM Journal on Computing*, 14:1054–1068, 1986.

38. P. Baptiste and C. Le Pape. Edge-finding constraint propagation algorithms for disjunctive and cumulative scheduling. In *Proceedings of the Fifteenth Workshop of the U.K. Planning Special Interest Group*, Liverpool, U.K., 1996.

39. P. Baptiste, C. Le Pape, and W. Nuijten. *Constraint-Based Scheduling: Applying Constraint Programming to Scheduling Problems*. Kluwer, Dordrecht, 2001.

40. I. Barany, T. J. van Roy, and L. A. Wolsey. Strong formulations for multi-item capacitated lot-sizing. *Management Science*, 30:1255–1261, 1984.

41. C. Barnhart, E. L. Johnson, G. L. Nemhauser, M. W. P. Savelsbergh, and P. H. Vance. Branch-and-price: Column generation for solving huge integer programs. *Operations Research*, 46:316–329, 1998.

42. N. A. Barricelli. Numerical testing of evolution theories. Part II: Preliminary tests of performance, symbiogenesis and terrestrial life. *Acta Biotheoretica*, 16:99–126, 1963.

43. R. Barták. Modelling soft constraints: A survey. *Neural Network World*, 12:421–431, 2002.

44. P. Barth. *Logic-Based 0-1 Constraint Solving in Constraint Logic Programming*. Kluwer, Dordrecht, 1995.

45. M. S. Bazaraa, J. J. Jarvis, and H. D. Sherali. *Linear Programming and Network Flows*. John Wiley, New York, 3rd edition, 2004.

46. M. S. Bazaraa, J. J. Jarvis, and H. D. Sherali. *Linear Programming and Network Flows*. John Wiley, New York, 4th edition, 2009.

47. M. S. Bazaraa, H. D. Sherali, and C. M. Shetty. *Nonlinear Programming: Theory and Algorithms*. Wiley-Interscience, New York, 3rd edition, 2006.

48. E. M. L. Beale. An alternative method for linear programming. In *Proceeedings of the Cambridge Philosophical Society*, volume 50, pages 513–523, 1954.

49. P. Beame, H. Kautz, and A. Sabharwal. Understanding the power of clause learning. In *International Joint Conference on Artificial Intelligence (IJCAI 2003)*, 2003.

50. N. Beaumont. An algorithm for disjunctive programs. *European Journal of Operational Research*, 48:362–371, 1990.

51. C. Beeri, R. Fagin, D. Maier, and M. Yannakakis. On the desirability of acyclic database schemes. *Journal of the ACM*, 30:479–513, 1983.

52. N. Beldiceanu. Pruning for the *minimum* constraint family and for the *number of distinct values* constraint family. In T. Walsh, editor, *Principles and Practice of Constraint Programming (CP 2001)*, volume 2239 of *Lecture Notes in Computer Science*, pages 211–224, New York, 2001. Springer.

53. N. Beldiceanu and M. Carlsson. Revisiting the cardinality operator and introducing the cardinality-path constraint family. In P. Codognet, editor, *International Conference on Logic Programming (ICLP 2001)*, volume 2237 of *Lecture Notes in Computer Science*, pages 59–73, New York, 2001. Springer.

54. N. Beldiceanu and M. Carlsson. Sweep as a generic pruning technique applied to the non-overlapping rectangles constraints. In T. Walsh, editor, *Principles and Practice of Constraint Programming (CP 2001)*, volume 2239 of *Lecture Notes in Computer Science*, pages 377–391, New York, 2001. Springer.

55. N. Beldiceanu, M. Carlsson, R. Debruyne, and T. Petit. Reformulation of global constraints based on constraint checkers. *Constraints*, 10:339–362, 2005.

56. N. Beldiceanu, M. Carlsson, and T. Petit. Deriving filtering algoriths from constraint checkers. In M. Wallace, editor, *Principles and Practice of Constraint Programming (CP 2004)*, volume 3258 of *Lecture Notes in Computer Science*, pages 107–122, New York, 2004. Springer.

57. N. Beldiceanu, M. Carlsson, E. Poder, R. Sadek, and C. Truchet. A generic geometric constraint kernel in space and time for handling polymorphic k-dimensional objects. In C. Bessiere, editor, *Principles and*

Practice of Constraint Programming (CP 2007), volume 4741 of *Lecture Notes in Computer Science*, pages 180–194, New York, 2007. Springer.

58. N. Beldiceanu, M. Carlsson, and J.-X. Rampon. Global constraint catalog. SICS technical report T2005:08, Swedish Institute of Computer Science, 2005.

59. N. Beldiceanu, M. Carlsson, and J.-X. Rampon. Global constraint catalog. SICS technical report T2010:07, Swedish Institute of Computer Science, 2010.

60. N. Beldiceanu, M. Carlsson, and S. Thiel. Cost-filtering algorithms for the two sides of the *sum of weights of distinct values* constraint. SICS technical report, Swedish Institute of Computer Science, 2002.

61. N. Beldiceanu and E. Contejean. Introducing global constraints in CHIP. *Mathematical and Computer Modelling*, 12:97–123, 1994.

62. N. Beldiceanu, P. Flener, and X. Lorca. The *tree* constraint. In R. Barták and M. Milano, editors, *Integration of AI and OR Techniques in Constraint Programming for Combinatorial Optimization Problems (CPAIOR 2005)*, volume 3524 of *Lecture Notes in Computer Science*, pages 64–78, New York, 2005. Springer.

63. N. Beldiceanu, Q. Guo, and S. Thiel. Non-overlapping constraints between convex polytopes. In T. Walsh, editor, *Principles and Practice of Constraint Programming (CP 2001)*, volume 2239 of *Lecture Notes in Computer Science*, pages 392–407, New York, 2001. Springer.

64. N. Beldiceanu, I. Katriel, and X. Lorca. Undirected forest constraints. In J. C. Beck and B. M. Smith, editors, *Integration of AI and OR Techniques in Constraint Programming for Combinatorial Optimization Problems (CPAIOR 2006)*, volume 3990 of *Lecture Notes in Computer Science*, pages 29–43, New York, 2006. Springer.

65. N. Beldiceanu, I. Katriel, and S. Thiel. Filtering algorithms for the Same and UsedBy constraints. Research report mpi-i-2004-1-001, Max-Planck-Institut für Informatik, Saarbrücken, 2004.

66. N. Beldiceanu, I. Katriel, and S. Thiel. Filtering algorithms for the same constraint. In J. C. Régin and M. Rueher, editors, *Integration of AI and OR Techniques in Constraint Programming for Combinatorial Optimization Problems (CPAIOR 2004)*, volume 3011 of *Lecture Notes in Computer Science*, pages 65–79, New York, 2004. Springer.

67. N. Beldiceanu and X. Lorca. Necessary condition for path partitioning constraint. In E. Loute and L. Wolsey, editors, *Proceedings of the International Workshop on Integration of Artificial Intelligence and Operations Research Techniques in Constraint Programming for Combinatorial Optimization Problems (CPAIOR 2007)*, volume 4510 of *Lecture Notes in Computer Science*, pages 141–154, New York, 2007. Springer.

68. R. Bellman. The theory of dynamic programming. *Bulletin of the American Mathematical Society*, 60:503–516, 1954.

69. R. Bellman. *Dynamic Programming*. Princeton University Press, Princeton, NJ, 1957.

70. P. Belotti, J. Lee, L. Liberti, F. Margot, and A. Waechter. Branching and bound tightening techniques for non-convex MINLP. *Optimization Methods and Software*, 24:597–634, 2009.

71. G. Belvaux and L. A. Wolsey. Modelling practical lot-sizing problems as mixed-integer programs. *Management Science*, 47:993–1007, 2001.

72. P. Benchimol, J.-C. Régin, L.-M. Rousseau, M. Rueher, and W.-J. van Hoeve. Improving the Held and Karp approach with constraint programming. In A. Lodi, M. Milano, and P. Toth, editors, *Proceedings of the International Workshop on Integration of AI and OR Techniques in Constraint Programming for Combinatorial Optimization Problems (CPAIOR 2010)*, volume 6140 of *Lecture Notes in Computer Science*, pages 40–44, New York, 2010. Springer.

73. J. F. Benders. Partitioning procedures for solving mixed-variables programming problems. *Numerische Mathematik*, 4:238–252, 1962.

74. M. Benichou, J. M. Gautier, P. Girodet, G. Hentges, R. Ribiere, and O. Vincent. Experiments in mixed-integer linear programming. *Mathematical Programming*, 1:76–94, 1971.

75. L. Benini, D. Bertozzi, A. Guerri, and M. Milano. Allocation and scheduling for MPSOCs via decomposition and no-good generation. In *Principles and Practice of Constraint Programming (CP 2005)*, volume 3709 of *Lecture Notes in Computer Science*, pages 107–121, New York, 2005. Springer.

76. L. Benini, M. Lombardi, M. Mantovani, M. Milano, and M. Ruggiero. Multi-stage Benders decomposition for optimizing multicore architectures. In L. Perron and M. A. Trick, editors, *Proceedings of the International Workshop on Integration of AI and OR Techniques in Constraint Programming for Combinatorial Optimization Problems (CPAIOR 2008)*, volume 5015 of *Lecture Notes in Computer Science*, pages 36–50, New York, 2008. Springer.

77. T. Benoist, E. Gaudin, and B. Rottembourg. Constraint programming contribution to Benders decomposition: A case study. In P. Van Hentenryck, editor, *Principles and Practice of Constraint Programming (CP 2002)*, volume 2470 of *Lecture Notes in Computer Science*, pages 603–617, New York, 2002. Springer.

78. T. Benoist, F. Laburthe, and B. Rottembourg. Lagrange relaxation and constraint programming collaborative schemes for traveling tournament problems. In C. Gervet and M. Wallace, editors, *Proceedings of*

the *International Workshop on Integration of Artificial Intelligence and Operations Research Techniques in Constraint Programming for Combinatorial Optimization Problems (CPAIOR 2001)*, Ashford, U.K., 2001.

79. C. Berge. Two theorems in graph theory. *Proceedings of the National Academy of Sciences*, 43:842–844, 1957.

80. D. Bergman and J. Hooker. Polyhedral results for alldifferent systems. Technical report, Carnegie Mellon University, 2010.

81. D. Bergman, W.-J. van Hoeve, and J. N. Hooker. Manipulating MDD relaxations for combinatorial optimization. In T. Achterberg and J. C. Beck, editors, *Proceedings of the International Workshop on Integration of AI and OR Techniques in Constraint Programming for Combinatorial Optimization Problems (CPAIOR 2011)*, volume 6697 of *Lecture Notes in Computer Science*, pages 20–35, New York, 2011. Springer.

82. U. Bertele and F. Brioschi. *Nonserial Dynamic Programming*. Academic Press, New York, 1972.

83. T. Berthold, S. Heinz, M. E. Lübbecke, R. H. Möhring, and J. Schulz. A constraint integer programming approach for resource-constrained project scheduling. In A. Lodi, M. Milano, and P. Toth, editors, *Proceedings of the International Workshop on Integration of AI and OR Techniques in Constraint Programming for Combinatorial Optimization Problems (CPAIOR 2010)*, volume 6140 of *Lecture Notes in Computer Science*, pages 313–317, New York, 2010. Springer.

84. D. P. Bertsekas. *Dynamic Programming and Optimal Control*, volume 1 and 2. Athena Scientific, Nashua, NH, 3rd edition, 2007.

85. C. Bessiere, E. Hebrard, B. Hnich, Z. Kiziltan, and T. Walsh. Filtering algorithms for the nvalue constraint. In R. Barták and M. Milano, editors, *Integration of AI and OR Techniques in Constraint Programming for Combinatorial Optimization Problems (CPAIOR 2005)*, volume 3524 of *Lecture Notes in Computer Science*, pages 79–93, New York, 2005. Springer.

86. C. Bessiere, E. Hebrard, B. Hnich, Z. Kiziltan, and T. Walsh. The range and roots constraints: Specifying counting and occurrence problems. In *International Joint Conference on Artificial Intelligence (IJCAI 2003)*, pages 60–65, 2005.

87. C. Bessiere, E. Hebrard, B. Hnich, Z. Kiziltan, and T. Walsh. The range constraint: Algorithms and implementation. In J. C. Beck and B. M. Smith, editors, *Integration of AI and OR Techniques in Constraint Programming for Combinatorial Optimization Problems (CPAIOR 2006)*, volume 3990 of *Lecture Notes in Computer Science*, pages 59–73, New York, 2006. Springer.

88. C. Bessiere, E. Hebrard, B. Hnich, Z. Kiziltan, and T. Walsh. The ROOTS constraint. In C. Bessiere, editor, *Principles and Practice of Constraint Programming (CP 2007)*, volume 4741 of *Lecture Notes in Computer Science*, pages 75–90, New York, 2006. Springer.

89. C. Bessiere, E. Hebrard, B. Hnich, Z. Kiziltan, and T. Walsh. SLIDE: A useful special case of the CARDPATH constraint. In M. Ghallab, C.D. Spyropoulos, N. Fakotakis, and N. Avouris, editors, *European Conference on Artificial Intelligence (ECAI 2008)*, pages 475–479. IOS Press, 2008.

90. C. Bessiere, E. Hebrard, B. Hnich, Z. Kiziltan, and T. Walsh. Range and roots: Two common patterns for specifying and propagating counting and occurrence constraints. *Artificial Intelligence*, 174:1054–1078, 2009.

91. C. Bessiere, E. Hebrard, B. Hnich, and T. Walsh. The complexity of global constraints. In *National Conference on Artificial Intelligence (AAAI 2004)*, pages 112–117, 2004.

92. C. Bessiere, Z. Kiziltan, N. Narodytska, C.-G. Quimper, and T. Walsh. Decompositions of all different, global cardinality and related constraints. In *International Joint Conference on Artificial Intelligence (IJCAI 2009)*, pages 419–424, 2009.

93. R. E. Bixby, W. Cook, A. Cox, and E. K. Lee. Parallel mixed integer programming. Technical report CRPC-TR95554, Center for Research on Parallel Computation, 1995.

94. R. E. Bixby and W. Cunningham. Converting linear programs to network problems. *Mathematics of Operations Research*, 5:321–357, 1980.

95. C. E. Blair and R. G. Jeroslow. The value function of a mixed integer program: I. *Discrete Applied Mathematics*, 19:121–138, 1977.

96. C. E. Blair and R. G. Jeroslow. The value function of a mixed integer program. *Mathematical Programming*, 23:237–273, 1982.

97. C. E. Blair, R. G. Jeroslow, and J. K. Lowe. Some results and experiments in programming techniques for propositional logic. *Computers and Operations Research*, 13:633–645, 1988.

98. N. Bleuzen-Guernalec and A. Colmerauer. Narrowing a block of sortings in quadratic time. In G. Smolka, editor, *Principles and Practice of Constraint Programming (CP 1997)*, volume 1330 of *Lecture Notes in Computer Science*, pages 2–16, New York, 1997. Springer.

99. C. Bliek. Generalizing dynamic and partial order backtracking. In *National Conference on Artificial Intelligence (AAAI 1998)*, pages 319–325, Madison, WI, 1998.

100. C. Blum, J. Puchinger, G. Raidl, and A. Roli. Hybrid metaheuristics. In P. van Hentenryck and M. Milano, editors, *Hybrid Optimization: The Ten Years of CPAIOR*, pages 305–336. Springer, New York, 2011.

101. A. Bockmayr and T. Kasper. Branch-and-infer: A unifying framework for integer and finite domain constraint programming. *INFORMS Journal on Computing*, 10:287–300, 1998.

102. A. Bockmayr and N. Pisaruk. Detecting infeasibility and generating cuts for mixed integer programming using constraint progrmaming. In M. Gendreau, G. Pesant, and L.-M. Rousseau, editors, *Proceedings of the International Workshop on Integration of Artificial Intelligence and Operations Research Techniques in Constraint Programming for Combinatorial Optimization Problems (CPAIOR 2003)*, Montréal, 2003.

103. A. Bockmayr, N. Pisaruk, and A. Aggoun. Network flow problems in constraint programming. In T. Walsh, editor, *Principles and Practice of Constraint Programming (CP 2001)*, volume 2239 of *Lecture Notes in Computer Science*, pages 196–210, New York, 2001. Springer.

104. S. Bollapragada, O. Ghattas, and J. N. Hooker. Optimal design of truss structures by mixed logical and linear programming. *Operations Research*, 49:42–51, 2001.

105. I. M. Bomze, M. Budinich, P. M. Pardalos, and M. Pelillo. The maximum clique problem. In D.-Z. Du and P. M. Pardalos, editors, *Handbook of Combinatorial Optimization, Supplement Volume A*, pages 1–74. Kluwer, Dordrecht, 1999.

106. G. Boole. *Studies in Logic and Probability*, R. Rhees, editor. Open Court Publishing Company, La Salle, IL, 1952.

107. K. Booth and G. Lueker. Testing for the consecutive ones property, interval graphs and graph planarity using PQ-tree algorithms. *Journal of Computer and Systems Sciences*, 13:335–379, 1976.

108. R. Borndörfer. *Aspects of Set Packing, Partitioning, and Covering*. Shaker Verlag, Aachen, Germany, 1998.

109. S. Bourdais, P. Galinier, and G. Pesant. Hibiscus: A constraint programming application to staff scheduling in health care. In F. Rossi, editor, *Principles and Practice of Constraint Programming (CP 2003)*, volume 2833 of *Lecture Notes in Computer Science*, pages 153–167, New York, 2003. Springer.

110. S. Brand, N. Narodytska, C-G. Quimper, P. Stuckey, and T. Walsh. Encodings of the sequence constraint. In C. Bessiere, editor, *Principles and Practice of Constraint Programming (CP 2007)*, volume 4741 of *Lecture Notes in Computer Science*, pages 210–224, New York, 2007. Springer.

111. R. G. Brown, J. W. Chinneck, and G. M. Karam. Optimization with constraint programming systems. In R. Sharda et al., editor, *Impact of Recent Computer Advances on Operations Research*, volume 9 of *Publications in Operations Research Series*, pages 463–473, Williamsburg, VA, 1989. Elsevier.

112. P. Brucker and O. Thiele. A branch and bound method for the general-shop problem with sequence-dependent setup times. *OR Spektrum*, 18:145–161, 1996.

113. R. E. Bryant. Graph-based algorithms for boolean function manipulation. *IEEE Transactions on Computers*, C-35:677–691, 1986.

114. H. Cambazard, P.-E. Hladik, A.-M. Déplanche, N. Jussien, and Y. Trinquet. Decomposition and learning for a hard real time task allocation problem. In M. Wallace, editor, *Principles and Practice of Constraint Programming (CP 2004)*, volume 3258 of *Lecture Notes in Computer Science*, pages 153–167, New York, 2004. Springer.

115. H. Cambazard, E. O'Mahony, and B. O'Sullivan. Hybrid methods for the multileaf collimator sequencing problem. In A. Lodi, M. Milano, and P. Toth, editors, *Proceedings of the International Workshop on Integration of AI and OR Techniques in Constraint Programming for Combinatorial Optimization Problems (CPAIOR 2010)*, volume 6140 of *Lecture Notes in Computer Science*, pages 56–70, New York, 2010. Springer.

116. J. Carlier. One machine problem. *European Journal of Operational Research*, 11:42–47, 1982.

117. J. Carlier and E. Pinson. An algorithm for solving the job-shop problem. *Management Science*, 35:164–176, 1989.

118. J. Carlier and E. Pinson. A practical use of Jackson's preemptive schedule for solving the job shop problem. *Annals of Operations Research*, 26:269–287, 1990.

119. J. Carlier and E. Pinson. Adjustment of heads and tails for the job-shop problem. *European Journal of Operational Research*, 78:146–161, 1994.

120. M. Carlsson and N. Beldiceanu. From constraints to finite automata to filtering algorithms. In *European Symposium on Programming (ESOP 2004)*, pages 94–108, 2004.

121. Y. Caseau and F. Laburthe. Improved CLP scheduling with task intervals. In *Proceedings of the Eleventh International Conference on Logic Programming (ICLP 1994)*, pages 369–383, Cambridge, MA, 1994. MIT Press.

122. Y. Caseau and F. Laburthe. Solving small TSPs with constraints. In L. Naish, editor, *Proceedings, Fourteenth International Conference on Logic Programming (ICLP 1997)*, volume 2833, pages 316–330. The MIT Press, 1997.

123. P. M. Castro and I. E. Grossmann. An efficient MILP model for the short-term scheduling of single stage batch plants. Technical report, Departamento de Modelação e Simulação de Processos, INETI, Lisbon, 2006.

124. S. Ceria, C. Cordier, H. Marchand, and L. A. Wolsey. Cutting planes for integer programs with general integer variables. *Mathematical Programming*, 81:201214, 1998.

125. A. Chabrier. A cooperative CP and LP optimizer approach for the pairing generation problem. In *Proceedings of the International Workshop on Integration of Artificial Intelligence and Operations Research Techniques in Constraint Programming for Combinatorial Optimization Problems (CPAIOR 1999)*, Ferrara, Italy, 2000.

126. A. Chabrier. Heuristic branch-and-price-and-cut to solve a network design problem. In M. Gendreau, G. Pesant, and L.-M. Rousseau, editors, *Proceedings of the International Workshop on Integration of Artificial Intelligence and Operations Research Techniques in Constraint Programming for Combinatorial Optimization Problems (CPAIOR 2003)*, Montréal, 2003.

127. V. Chandru and J. N. Hooker. Detecting embedded Horn structure in propositional logic. *Information Processing Letters*, 42:109–111, 1992.

128. C. L. Chang. The unit proof and the input proof in theorem proving. *Journal of the ACM*, 14:698–707, 1970.

129. K. K. H. Cheung. A Benders approach for computing lower bounds for the mirrored traveling tournament problem. *Discrete Optimization*, 6:189–196, 2009.

130. D. Chhajed and T. J. Lowe. Solving structured multifacility location problems efficiently. *Transportation Science*, 28:104–115, 1994.

131. Y. Chu and Q. Xia. Generating Benders cuts for a class of integer programming problems. In J. C. Régin and M. Rueher, editors, *Integration of AI and OR Techniques in Constraint Programming for Combinatorial Optimization Problems (CPAIOR 2004)*, volume 3011 of *Lecture Notes in Computer Science*, pages 127–141, New York, 2004. Springer.

132. Y. Chu and Q. Xia. A hybrid algorithm for a class of resource-constrained scheduling problems. In R. Barták and M. Milano, editors, *Integration of AI and OR Techniques in Constraint Programming for Combinatorial Optimization Problems (CPAIOR 2005)*, volume 3524 of *Lecture Notes in Computer Science*, pages 110–124, New York, 2005. Springer.

133. V. Chvátal. Edmonds polytopes and a hierarchy of combinatorial problems. *Discrete Mathematics*, 4:305–337, 1973.

134. V. Chvátal. Edmonds polytopes and weakly hamiltonian graphs. *Mathematical Programming*, 5:29–40, 1973.

135. V. Chvátal. Tough graphs and hamiltonian circuits. *Discrete Mathematics*, 5:215–228, 1973.

136. V. Chvátal. *Linear Programming*. W. H. Freeman, New York, 1983.

137. V. Chvátal. Hamiltonian cycles. In E. L. Lawler, J. K. Lenstra, A. H. G. Rinooy Kan, and D. B. Shmoys, editors, *The Traveling Salesman Problem: A Guided Tour of Combinatorial Optimization*, pages 403–430. John Wiley, New York, 1985.

138. E. Coban and J. N. Hooker. Single-facility scheduling over long time horizons by logic-based Benders decomposition. In A. Lodi, M. Milano, and P. Toth, editors, *Proceedings of the International Workshop on Integration of AI and OR Techniques in Constraint Programming for Combinatorial Optimization Problems (CPAIOR 2010)*, volume 6140 of *Lecture Notes in Computer Science*, pages 87–91, New York, 2010. Springer.

139. E. Coban and J. N. Hooker. Single-facility scheduling by logic-based Benders decomposition. *Annals of Operations Research*, to appear.

140. G. Codato and M. Fischetti. Combinatorial Benders cuts for mixed-integer linear programming. *Operations Research*, 54:756–766, 2006.

141. E. G. Coffman, M. R. Garey, D. S., and Johnson. Approximation algorithms for bin-packing: A survey. In D. Hochbaum, editor, *Approximation Algorithms for NP-hard Problems*, pages 46–93. PWS Publishing Company, Boston, 1997.

142. Y. Colombani and S. Heipcke. Mosel: An extensible environment for modeling and programming solutions. In N. Jussien and F. Laburthe, editors, *Proceedings of the International Workshop on Integration of Artificial Intelligence and Operations Research Techniques in Constraint Programming for Combinatorial Optimization Problems (CPAIOR 2002)*, pages 277–290, Le Croisic, France, 2002.

143. Y. Colombani and S. Heipcke. Multiple models and parallel solving with Mosel. Xpress optimization suite white paper, FICO, 2004.

144. M. Conforti, G. Cornuéjols, and K. Vuskovic. Balanced matrices. *Discrete Mathematics*, 306:2411–2437, 2006.

145. M. Conforti, G. Cornuéjols, and G. Zambelli. Corner polyhedron and intersection cuts. *Surveys in Operations Research and Management Science*, to appear.

146. W. Cook, C. R. Coullard, and G. Turán. On the complexity of cutting plane proofs. *Discrete Applied Mathematics*, 18:25–38, 1987.

147. J.-F. Cordeau and G. Laporte. Modeling and optimization of vehicle routing and arc routing problems. In G. Appa, L. Pitsoulis, and H. P. Williams, editors, *Handbook on Modelling for Discrete Optimization*, pages 151–191. Springer, New York, 2006.

148. G. Cornuéjols and D. Vandenbussche. *k*-cuts: A variation of Gomory mixed integer cuts from the LP tableau. *INFORMS Journal on Computing*, 15:385–396, 2003.

149. A. I. Corréa, A. Langevin, and L. M. Rousseau. Dispatching and conflict-free routing of automated guided vehicles: A hybrid approach combining constraint programming and mixed integer programming. In J. C. Régin and M. Rueher, editors, *Integration of AI and OR Techniques in Constraint Programming for Combinatorial Optimization Problems (CPAIOR 2004)*, volume 3011 of *Lecture Notes in Computer Science*, pages 370–378, New York, 2004. Springer.

150. M.-C. Costa. Persistency in maximum cardinality bipartite matchings. *Operations Research Letters*, 15:143–149, 1994.

151. M. Costantino. A cutting plane approach to capacitated lot-sizing with start-up costs. *Mathematical Programming*, 75:353–376, 1996.

152. Y. Crama, P. Hansen, and B. Jaumard. The basic algorithm for pseudo-boolean programming revisited. *Discrete Applied Mathematics*, 29:171–185, 1990.

153. W. Cronholm and Farid Ajili. Strong cost-based filtering for Lagrange decomposition applied to network design. In M. Wallace, editor, *Principles and Practice of Constraint Programming (CP 2004)*, volume 3258 of *Lecture Notes in Computer Science*, pages 726–730, New York, 2004. Springer.

154. H. Crowder, E. Johnson, and M. W. Padberg. Solving large-scale zero-one linear programming problems. *Operations Research*, 31:803–834, 1983.

155. G. B. Dantzig. Maximization of a linear function of variables subject to linear inequalities. In T. C. Koopmans, editor, *Activity Ananlysis of Production and Allocation*, pages 339–347. John Wiley, New York, 1951.

156. G. B. Dantzig. *Linear Programming and Extensions*. Princeton University Press, Princeton, NJ, 1963.

157. G. B. Dantzig. Linear programming. In J. K. Lenstra, A. H. G. Rinnooy Kan, and A. Schrijver, editors, *History of Mathematical Programming: A Collection of Personal Reminiscences*, Handbooks in Operations Research and Management Science, pages 19–31. CWI, North-Holland, Amsterdam, 1991.

158. G. B. Dantzig, D. R. Fulkerson, and S. M. Johnson. Solution of a large scale traveling salesman problem. *Operations Research*, 2:393–410, 1954.

159. S. Dash and O. Günlük. Valid inequalities based on simple mixed-integer sets. *Mathematical Programming*, 105:29–53, 2006.

160. E. Davis. Constraint propagation with intervals labels. *Artificial Intelligence*, 32:281–331, 1987.

161. M. Davis, G. Logemann, and H. Putnam. A machine program for theorem proving. *Communications of the ACM*, 5:394–397, 1962.

162. M. Davis and H. Putnam. A computing procedure for quantification theory. *Journal of the ACM*, 7:201–215, 1960.

163. R. Davis. Diagnostic reasoning based on structure and behavior. *Journal of Artificial Intelligence*, 24:347–410, 1984.

164. M. Dawande and J. N. Hooker. Inference-based sensitivity analysis for mixed integer/linear programming. *Operations Research*, 48:623–634, 2000.

165. J. Valerio de Carvalho. Exact solution of bin-packing problems using column generation and branch-and-bound. *Annals of Operations Research*, 86:629–659, 1999.

166. S. de Givry and L. Jeannin. ToOLS: A library of partial and hybrid search methods. In M. Gendreau, G. Pesant, and L.-M. Rousseau, editors, *Proceedings of the International Workshop on Integration of Artificial Intelligence and Operations Research Techniques in Constraint Programming for Combinatorial Optimization Problems (CPAIOR 2003)*, pages 124–138, Montréal, 2003.

167. J. de Kleer and B. C. Williams. Diagnosing multiple faults. *Journal of Artificial Intelligence*, 32:97–130, 1987.

168. R. Dechter. Learning while searching in constraint-satisfaction-problems. In *AAAI Conference on Artificial Intelligence (AAAI 86)*, pages 178–185, 1986.

169. R. Dechter. Bucket elimination: A unifying framework for several probabilistic inference algorithms. In *Proceedings of the Twelfth Annual Conference on Uncertainty in Artificial Intelligence (UAI 96)*, pages 211–219, Portland, OR, 1996.

170. S. Demassey, C. Artiques, and P. Michelon. A hybrid constraint propagation-cutting plane procedure for the RCPSP. In N. Jussien and F. Laburthe, editors, *Proceedings of the International Workshop on Integration of Artificial Intelligence and Operations Research Techniques in Constraint Programming for Combinatorial Optimization Problems (CPAIOR 2002)*, Le Croisic, France, 2002.

171. S. Demassey, G. Pesant, and L.-M. Rousseau. Constraint-programming based column generation for employee timetabling. In R. Barták and M. Milano, editors, *Integration of AI and OR Techniques in Constraint Programming for Combinatorial Optimization Problems (CPAIOR 2005)*, volume 3524 of *Lecture Notes in Computer Science*, pages 140–154, New York, 2005. Springer.

172. E. V. Denardo. *Dynamic Programming: Models and Applications.* Dover, Mineola, NY, 2003.

173. N. Deo and C.-Y. Pang. Shortest-path algorithms: Taxonomy and annotation. *Networks*, 14:275–323, 1984.

174. K. Dhyani, S. Gualandi, and P. Cremonesi. A constraint programming approach to the service consolidation problem. In A. Lodi, M. Milano, and P. Toth, editors, *Proceedings of the International Workshop on Integration of AI and OR Techniques in Constraint Programming for Combinatorial Optimization Problems (CPAIOR 2010)*, volume 6140 of *Lecture Notes in Computer Science*, pages 97–101, New York, 2010. Springer.

175. B. Dilkina, C. P. Gomes, and A. Sabharwal. Tradeoffs in the complexity of backdoor detection. In C. Bessiere, editor, *Principles and Practice of Constraint Programming (CP 2007)*, volume 4741 of *Lecture Notes in Computer Science*, pages 256–270, New York, 2007. Springer.

176. G. Dooms and I. Katriel. The minimum spanning tree constraint. In F. Benhamou, editor, *Principles and Practice of Constraint Programming (CP 2006)*, volume 4204 of *Lecture Notes in Computer Science*, pages 152–166, New York, 2006. Springer.

177. G. Dooms and I. Katriel. The "not-too-heavy spanning tree" constraint. In C. Bessiere, editor, *Principles and Practice of Constraint Programming (CP 2007)*, volume 4741 of *Lecture Notes in Computer Science*, pages 59–70, New York, 2007. Springer.

178. M. Dorigo. Optimization, learning and natural algorithms. PhD thesis, Politecnico di Milano, 1992.

179. M. Dorigo and L.M. Gambardella. Ant colony system: A cooperative learning approach to the traveling salesman problem. *IEEE Transactions on Evolutionary Computation*, 1:53–66, 1997.

180. U. Dorndorf, E. Pesch, and T. Phan-Huy. Solving the open shop scheduling problem. *Journal of Scheduling*, 4:157–174, 2001.

181. W. F. Dowling and J. H. Gallier. Linear-time algorithms for testing the satisfiability of propositional Horn formulae. *Journal of Logic Programming*, 1:267–284, 1984.

182. Z. Drezner. *Facility Location: A Survey of Applications and Methods.* Springer, New York, 1995.

183. K. Easton, G. Nemhauser, and M. Trick. The traveling tournament problem description and benchmarks. In T. Walsh, editor, *Principles and Practice of Constraint Programming (CP 2001)*, volume 2239 of *Lecture Notes in Computer Science*, pages 580–584, New York, 2001. Springer.

184. K. Easton, G. Nemhauser, and M. Trick. Solving the traveling tournament problem: A combined integer programming and constraint programming approach. In *Proceedings of the International Conference on the Practice and Theory of Automated Timetabling (PATAT 2002)*, 2002.

185. K. Easton, G. Nemhauser, and M. Trick. CP based branch and price. In M. Milano, editor, *Constraint and Integer Programming: Toward a Unified Methodology*, pages 207–231. Kluwer, Dordrecht, 2004.

186. J. Erschler, P. Lopez, and P. Esquirol. Ordonnancement de tâches sous contraintes: Une approche énergétique. *RAIRO Automatique, Productique, Informatique Industrielle*, 26:453–481, 1992.

187. J. Erschler, P. Lopez, and C. Thuriot. Raisonnement temporel sous contraintes de ressource et problèmes d'ordonnancement. *Revue d'Intelligence Artificielle*, 5:7–32, 1991.

188. F. Fages and A. Lal. A global constraint for cutset problems. In M. Gendreau, G. Pesant, and L.-M. Rousseau, editors, *Proceedings of the International Workshop on Integration of Artificial Intelligence and Operations Research Techniques in Constraint Programming for Combinatorial Optimization Problems (CPAIOR 2003)*, Montréal, 2003.

189. T. Fahle. Cost based filtering vs. upper bounds for maximum clique. In N. Jussien and F. Laburthe, editors, *Proceedings of the International Workshop on Integration of Artificial Intelligence and Operations Research Techniques in Constraint Programming for Combinatorial Optimization Problems (CPAIOR 2002)*, pages 93–108, Le Croisic, France, 2002.

190. T. Fahle, U. Junker, S. E. Karish, N. Kohn, M. Sellmann, and B. Vaaben. Constraint programming based column generation for crew assignment. *Journal of Heuristics*, 8:59–81, 2002.

191. T. Fahle and M. Sellmann. Cost based filtering for the constrained knapsack problem. *Annals of Operations Research*, 115:73–93, 2002.

192. G. Farkas. A Fourier-féle mechanikai elv alkalmazásai (Hungarian) [On the applications of the mechanical principle of Fourier]. *Mathematikai és Természettudományi Értesítő*, 12:457–472, 1894.

193. G. Farkas. A Fourier-féle mechanikai elv alkalmazásának algebrai alapja (Hungarian) [The algebraic basis of the application of the mechanical principle of Fourier]. *Mathematikai és Természettudományi Értesítő*, 16:361–364, 1898.

194. M. M. Fazel-Zarandi and J. C. Beck. Solving a location-allocation problem with logic-based Benders decomposition. In I. P. Gent, editor, *Principles and Practice of Constraint Programming (CP 2009)*, volume 5732 of *Lecture Notes in Computer Science*, pages 344–351, New York, 2009. Springer.

195. C. E. Ferreira, A. Martin, and R. Weismantel. Solving multiple knapsack problems by cutting planes. *SIAM Journal on Optimization*, 6:858–877, 1996.

196. M. Fischetti, F. Glover, and A. Lodi. The feasibility pump. *Mathematical Programming*, 104:91–104, 2005.

197. M. Fischetti and A. Lodi. Repairing MIP infeasibility through local branching. *Computers and Operations Research*, 35:1436–1445, 2008.

198. M. Fischetti, A. Lodi, and P. Toth. Solving real-world ATSP instances by branch-and-cut. In M. Jünger, G. Reinelt, and G. Rinaldi, editors, *Combinatorial Optimization: Eureka, You Shrink! Papers Dedicated to Jack Edmonds*, volume 2570 of *Lecture Notes in Computer Science*, pages 64–77, New York, 2003. Springer.

199. M. Fisher. Vehicle routing. In M. O. Ball, T. L. Magnanti, C. L. Monma, and G. L. Nemhauser, editors, *Network Routing*, volume 8 of *Handbooks in Operations Research and Management*, pages 1–79. North-Holland, 1997.

200. M. L. Fisher. The Lagrangian relaxation method for solving integer programming problems. *Management Science*, 27:1–18, 1981.

201. P. Flener, A. M. Frisch, B. Hnich, Z. Kiziltan, I. Miguel, J. Pearson, and T. Walsh. Breaking row and column symmetries in matrix models. In P. Van Hentenryck, editor, *Principles and Practice of Constraint Programming (CP 2002)*, volume 2470 of *Lecture Notes in Computer Science*, pages 462–476, New York, 2002. Springer.

202. F. Focacci, A. Lodi, and M. Milano. Cost-based domain filtering. In J. Jaffar, editor, *Principles and Practice of Constraint Programming (CP 1999)*, volume 1713 of *Lecture Notes in Computer Science*, pages 189–203, New York, 1999. Springer.

203. F. Focacci, A. Lodi, and M. Milano. Integration of CP and OR methods for matching problems. In *Proceedings of the First International Workshop on Integration of Artificial Intelligence and Operations Research Techniques in Constraint Programming for Combinatorial Optimization Problems (CPAIOR 1999)*, Ferrara, Italy, 1999.

204. F. Focacci, A. Lodi, and M. Milano. Solving TSP with time windows with constraints. In *International Conference on Logic Programming (ICLP 1999)*, pages 515–529. MIT Press, 1999.

205. F. Focacci, A. Lodi, and M. Milano. Cutting planes in constraint programming: An hybrid approach. In R. Dechter, editor, *Principles and Practice of Constraint Programming (CP 2000)*, volume 1894 of *Lecture Notes in Computer Science*, pages 187–201, New York, 2000. Springer.

206. F. Focacci and W. P. M. Nuijten. A constraint propagation algorithm for scheduling with sequence dependent setup times. In U. Junker, S. E. Karisch, and S. Tschöke, editors, *Proceedings of the International Workshop on Integration of Artificial Intelligence and Operations Research Techniques in Constraint Programming for Combinatorial Optimization Problems (CPAIOR 2000)*, pages 53–55, Paderborn, Germany, 2000.

207. L. R. Ford and D. R. Fulkerson. Maximal flow through a network. *Canadian Journal of Mathematics*, 8:399–404, 1956.

208. L. R. Ford and D. R. Fulkerson. *Flows in Networks*. Princeton University Press, Princeton, NJ, 1962.

209. J. B. J. Fourier. [work reported in] Analyse des travaux de l'Académie Royale des Sciences, pendant l'année 1824, Partie mathématique. *Histoire de l'Académie Royale des Sciences de l'Institut de France*, 7:xlvii–lv, 1827.

210. B. R. Fox. Chronological and non-chronological scheduling. In *Proceedings of the First Annual Conference on Artificial Intelligence: Simulation and Planning in High Autonomy Systems*, Tucson, USA, 1990.

211. E. C. Freuder. Synthesizing constraint expressions. *Communications of the ACM*, 21:958–966, 1978.

212. E. C. Freuder. A sufficient condition for backtrack-free search. *Communications of the ACM*, 29:24–32, 1982.

213. A. Frisch, B. Hnich, Z. Kiziltan, I. Miguel, and T. Walsh. Global constraints for lexicographic orderings. In P. Van Hentenryck, editor, *Principles and Practice of Constraint Programming (CP 2002)*, volume 2470 of *Lecture Notes in Computer Science*, pages 93–108, New York, 2002. Springer.

214. A. Fügenschuh and A. Martin. Computational integer programming and cutting planes. In K. Aardal, G. L. Nemhauser, and R. Weismantel, editors, *Discrete Optimization*, Handbooks in Operations Research and Management Science, pages 69–121. Elsevier, Amsterdam, 2005.

215. D. R. Fulkerson. Blocking and anti-blocking pairs of polyhedra. *Mathematical Programming*, 1:168–194, 1971.

216. R. Garfinkel and G. L. Nemhauser. Optimal political districting by implicit enumeration techniques. *Management Science*, 16:B495–B508, 1970.

217. J. Gaschnig. Experimental studies of backtrack vs. waltz-type vs. new algorithms for satisficing-assignment problems. In *Proceedings, 2nd National Conference of the Canadian Society for Computational Studies of Intelligence*, pages 19–21, 1978.

218. J. M. Gautier and R. Ribiere. Experiments in mixed-integer linear programming using pseudo-costs. *Mathematical Programming*, 12:26–47, 1977.

219. T. Gellermann, M. Sellmann, and R. Wright. Shorter-path constraints for the resource constrained shortest path problem. In R. Barták and M. Milano, editors, *Integration of AI and OR Techniques in Constraint Programming for Combinatorial Optimization Problems (CPAIOR 2005)*, volume 3524 of *Lecture Notes in Computer Science*, pages 201–216, New York, 2005. Springer.

220. L. Genç-Kaya and J. N. Hooker. A filter for the circuit constraint. In F. Benhamou, editor, *Principles and Practice of Constraint Programming (CP 2006)*, volume 4204 of *Lecture Notes in Computer Science*, pages 706–710, New York, 2006. Springer.

221. L. Genç-Kaya and J. N. Hooker. The circuit polytope. Technical report, Carnegie Mellon University, 2011.

222. B. Gendron, H. Lebbah, and G. Pesant. Improving the cooperation between the master problem and the subproblem in constraint programming based column generation. In R. Barták and M. Milano, editors, *Integration of AI and OR Techniques in Constraint Programming for Combinatorial Optimization Problems (CPAIOR 2005)*, volume 3524 of *Lecture Notes in Computer Science*, pages 217–227, New York, 2005. Springer.

223. M. R. Genesereth. The use of design descriptions in automated diagnosis. *Journal of Artificial Intelligence*, 24:411–436, 1984.

224. A. M. Geoffrion. Generalized Benders decomposition. *Journal of Optimization Theory and Applications*, 10:237–260, 1972.

225. A. Ghouila-Houri. Caractérisation des matrices totalement unimodulaires. *Comptes rendus de l'Académie des Sciences de Paris*, 254:1192–1194, 1962.

226. M. L. Ginsberg. Dynamic backtracking. *Journal of Artificial Intelligence Research*, 1:25–46, 1993.

227. M. L. Ginsberg and D. A. McAllester. GSAT and dynamic backtracking. In *Principles and Practice of Constraint Programming (CP 1994)*, volume 874 of *Lecture Notes in Computer Science*, pages 216–225, New York, 1994. Springer.

228. F. Glover. A bound escalation method for the solution of integer linear programs. *Cahiers du Centre d'Etudes de Recherche Opérationelle*, 6:131–168, 1964.

229. F. Glover. Maximum matching in a convex bipartite graph. *Naval Research Logistics Quarterly*, 316:313–316, 1967.

230. F. Glover. Surrogate constraint duality in mathematical programming. *Operations Research*, 23:434–451, 1975.

231. F. Glover. Tabu search: Part I. *ORSA Journal on Computing*, 1:190–206, 1989.

232. F. Glover, D. Karney, and D. Klingman. Implementation and computational comparisons for primal, dual, and primal-dual computer codes for minimum cost network flow problems. *Networks*, 4:191–212, 1974.

233. O. Goldreich. *P, NP, and NP-Completeness: The Basics of Computational Complexity*. Cambridge University Press, 2003.

234. C. P. Gomes, H. Kautz, A. Sabharwal, and B. Selman. Satisfiability solver. In F. van Harmelen, V. Lifschitz, and B. Porter, editors, *Handbook of Knowledge Representation*, pages 89–134. Elsevier, Amsterdam, 2008.

235. R. E. Gomory. Outline of an algorithm for integer solutions to linear programs. *Bulletin of the American Mathematical Society*, 64:275–278, 1958.

236. R. E. Gomory. An algorithm for the mixed integer problem. RAND technical report, RAND Corporation, 1960.

237. R. E. Gomory. Solving linear programming problems in integers. In R. Bellman and M. Hall, editors, *Combinatorial Analysis*, volume 10 of *Symposia in Applied Mathematics*, pages 211–215. American Mathematical Society, 1960.

238. R. E. Gomory. On the relation between integer and noninteger solutions to linear programs. In *Proceedings of the National Academy of Sciences*, volume 53, pages 260–265, 1965.

239. R. E. Gomory. Some polyhedra related to combinatorial problems. *Linear Algebra and Its Applications*, 2:451–558, 1969.

240. R. E. Gomory and E. L. Johnson. Some continuous functions related to corner polyhedra I. *Mathematical Programming*, 3:23–85, 1972.

241. F. Granot and P. L. Hammer. On the use of boolean functions in 0-1 programming. *Methods of Operations Research*, 12:154–184, 1971.

242. F. Granot and P. L. Hammer. On the role of generalized covering problems. *Cahiers du Centre d'Études de Recherche Opérationnelle*, 17:277–289, 1975.

243. H. Greenberg. A branch-and-bound solution to the general scheduling problem. *Operations Research*, 8:353–361, 1968.

244. M. Grönkvist. Using constraint propagation to accelerate column generation in aircraft scheduling. In M. Gendreau, G. Pesant, and L.-M. Rousseau, editors, *Proceedings of the International Workshop on Integration of Artificial Intelligence and Operations Research Techniques in Constraint Programming for Combinatorial Optimization Problems (CPAIOR 2003)*, Montréal, 2003.

245. I. E. Grossmann, J. N. Hooker, R. Raman, and H. Yan. Logic cuts for processing networks with fixed charges. *Computers and Operations Research*, 21:265–279, 1994.

246. M. Grötschel, L. Lovász, and A. Schrijver. *Geometric Algorithms and Combinatorial Optimization*. Springer, New York, 1988.

247. M. Grötschel and M. W. Padberg. On the symmetric traveling salesman problem: I: Inequalities. *Mathematical Programming*, 16:265–280, 1979.

248. Z. Gu, G. L. Nemhauser, and M. W. P. Savelsbergh. Sequence independent lifting of cover inequalities. In *Proceedings of the 4th International Conference on Integer Programming and Combinatorial Optimization (IPCO 1995)*, volume 920 of *Lecture Notes in Computer Science*, pages 452–461, New York, 1995. Springer.

249. Z. Gu, G. L. Nemhauser, and M. W. P. Savelsbergh. Sequence independent lifting in mixed integer programming. *Journal of Combinatorial Optimization*, 4:109–129, 2000.

250. S. Gualandi. k-clustering minimum biclique completion via a hybrid CP and SDP approach. In W.-J. van Hoeve and J. N. Hooker, editors, *Proceedings of the International Workshop on Integration of AI and OR Techniques in Constraint Programming for Combinatorial Optimization Problems (CPAIOR 2009)*, volume 5547 of *Lecture Notes in Computer Science*, pages 87–101, New York, 2009. Springer.

251. N. El Hachemi, M. Gendreau, and L.-M. Rousseau. Hybrid LS/CP approach to solve the weekly log-truck scheduling problem. In W.-J. van Hoeve and J. N. Hooker, editors, *Proceedings of the International Workshop on Integration of AI and OR Techniques in Constraint Programming for Combinatorial Optimization Problems (CPAIOR 2009)*, volume 5547 of *Lecture Notes in Computer Science*, pages 319–320, New York, 2009. Springer.

252. T. Hadzic, J. N. Hooker, and P. Tiedemann. Propagating separable equalities in an MDD store. In L. Perron and M. A. Trick, editors, *Proceedings of the International Workshop on Integration of Artificial Intelligence and Operations Research Techniques in Constraint Programming for Combinatorial Optimization Problems (CPAIOR 2008)*, volume 5015 of *Lecture Notes in Computer Science*, pages 318–322, New York, 2008. Springer.

253. T. Hailperin. *Boole's Logic and Probability*. Studies in Logic and the Foundations of Mathematics **85**. North-Holland, Amsterdam, 2nd edition, 1986.

254. P. L. Hammer, E. L. Johnson, and U. N. Peled. Facets of regular 0-1 polytopes. *Mathematical Programming*, 8:179–206, 1975.

255. P. Hansen. The steepest ascent mildest descent heuristic for combinatorial programming. In presentation at *Congress on Numerical Methods in Combinatorial Optimization*, Capri, 1986.

256. R. M. Haralick and G. L. Elliot. Increasing tree search efficiency for constraint satisfaction problems. *Artificial Intelligence*, 14:263–313, 1980.

257. I. Harjunkoski and I. E. Grossmann. A decomposition approach for the scheduling of a steel plant production. *Computers and Chemical Engineering*, 25:1647–1660, 2001.

258. I. Harjunkoski and I. E. Grossmann. Decomposition techniques for multistage scheduling problems using mixed-integer and constraint programming methods. *Computers and Chemical Engineering*, 26:1533–1552, 2002.

259. W. D. Harvey and M. L. Ginsberg. Limited discrepancy search. In *Proceedings of the International Joint Conference on Artificial Intelligence (IJCAI 1995)*, pages 607–613, 1995.

260. E. Hebrard, E. O'Mahony, and B. O'Sullivan. Constraint programming and combinatorial optimisation in Numberjack. In A. Lodi, M. Milano, and P. Toth, editors, *Proceedings of the International Workshop on Integration of AI and OR Techniques in Constraint Programming for Combinatorial Optimization Problems (CPAIOR 2010)*, volume 6140 of *Lecture Notes in Computer Science*, pages 181–185, New York, 2010. Springer.

261. S. Heipcke. Hybrid MIP/CP solving with Xpress-Optimizer and Xpress-Kalis. Xpress white paper, FICO, 2005.

262. M. Held and R. M. Karp. The traveling-salesman problem and minimum spanning trees. *Operations Research*, 18:1138–1162, 1970.

263. M. Held and R. M. Karp. The traveling-salesman problem and minimum spanning trees: Part II. *Mathematical Programming*, 1:6–25, 1971.

264. L. Hellsten, G. Pesant, and P. van Beek. A domain consistency algorithm for the stretch constraint. In M. Wallace, editor, *Principles and Practice of Constraint Programming (CP 2004)*, volume 3258 of *Lecture Notes in Computer Science*, pages 290–304, New York, 2004. Springer.

265. P. Van Hentenryck. *Constraint Satisfaction in Logic Programming*. MIT Press, Cambridge, MA, 1989.

266. P. Van Hentenryck and J.-P.Carillon. Generality versus specificity: An experience with AI and OR techniques. In *Proceedings of the American Association for Artificial Intelligence (AAAI-88)*, 1988.

267. P. Van Hentenryck and L. Michel. *Constraint Based Local Search*. MIT Press, Cambridge, MA, 2005.

268. P. Van Hentenryck, L. Michel, L. Perron, and J.-C. Régin. Constraint programming in OPL. In *International Conference on Principles and Practice of Declarative Programming (PPDP 1999)*, Paris, 1999.

269. J. Hiriart-Urruty and C. Lemaréchal. *Convex Analysis and Minimization Algorithms*. Springer, New York, 1993.

270. S. Hoda, W.-J. van Hoeve, and J. N. Hooker. A systematic approach to MDD-based constraint programming. In D. Cohen, editor, *Principles and Practice of Constraint Programming (CP 2010)*, volume 6308 of *Lecture Notes in Computer Science*, pages 266–280, New York, 2010. Springer.

271. A. J. Hoffmann and J. B. Kruskal. Integral boundary points of convex polyhedra. In H. W. Kuhn and A. W. Tucker, editors, *Linear Inequalities and Related Systems*, pages 223–246. Princeton University Press, Princeton, NJ, 1956.

272. J. H. Holland. *Adaptation in Natural and Artificial Systems*. University of Michigan Press, Ann Arbor, 1975.

273. J. N. Hooker. Generalized resolution and cutting planes. *Annals of Operations Research*, 12:217–239, 1988.

274. J. N. Hooker. Input proofs and rank one cutting planes. *ORSA Journal on Computing*, 1:137–145, 1989.

275. J. N. Hooker. Generalized resolution for 0-1 linear inequalities. *Annals of Mathematics and Artificial Intelligence*, 6:271–286, 1992.

276. J. N. Hooker. Logic-based methods for optimization. In A. Borning, editor, *Principles and Practice of Constraint Programming (CP 2002)*, volume 874 of *Lecture Notes in Computer Science*, pages 336–349, New York, 1994. Springer.

277. J. N. Hooker. Inference duality as a basis for sensitivity analysis. In E. C. Freuder, editor, *Principles and Practice of Constraint Programming (CP 1996)*, volume 1118 of *Lecture Notes in Computer Science*, pages 224–236, New York, 1996. Springer.

278. J. N. Hooker. Constraint satisfaction methods for generating valid cuts. In D. L. Woodruff, editor, *Advances in Computational and Stochastic Optimization, Logic Programming and Heuristic Search*, pages 1–30. Kluwer, Dordrecht, 1997.

279. J. N. Hooker. *Logic-Based Methods for Optimization: Combining Optimization and Constraint Satisfaction*. John Wiley, New York, 2000.

280. J. N. Hooker. Integer programming duality. In C. A. Floudas and P. M. Pardalos, editors, *Encyclopedia of Optimization*, volume 2, pages 533–543. Kluwer, New York, 2001.

281. J. N. Hooker. Logic, optimization and constraint programming. *INFORMS Journal on Computing*, 14:295–321, 2002.

282. J. N. Hooker. A framework for integrating solution methods. In H. K. Bhargava and M. Ye, editors, *Computational Modeling and Problem Solving in the Networked World (Proceedings of ICS2003)*, pages 3–30. Kluwer, 2003.

283. J. N. Hooker. A hybrid method for planning and scheduling. In M. Wallace, editor, *Principles and Practice of Constraint Programming (CP 2004)*, volume 3258 of *Lecture Notes in Computer Science*, pages 305–316, New York, 2004. Springer.

284. J. N. Hooker. A hybrid method for planning and scheduling. *Constraints*, 10:385–401, 2005.

285. J. N. Hooker. Planning and scheduling to minimize tardiness. In *Principles and Practice of Constraint Programming (CP 2005)*, volume 3709 of *Lecture Notes in Computer Science*, pages 314–327, New York, 2005. Springer.

286. J. N. Hooker. A sesarch-infer-and-relax framework for integrating solution methods. In R. Barták and M. Milano, editors, *Integration of AI and OR Techniques in Constraint Programming for Combinatorial Optimization Problems (CPAIOR 2005)*, volume 3524 of *Lecture Notes in Computer Science*, pages 243–257, New York, 2005. Springer.

287. J. N. Hooker. Unifying local and exhaustive search. In L. Villaseñor and A. I. Martinez, editors, *Avances in la Ciencia de la Computación (ENC 2005)*, pages 237–243, Puebla, Mexico, 2005.

288. J. N. Hooker. Duality in optimization and constraint satisfaction. In J. C. Beck and B. M. Smith, editors, *Integration of AI and OR Techniques in Constraint Programming for Combinatorial Optimization*

Problems (CPAIOR 2006), volume 3990 of *Lecture Notes in Computer Science*, pages 3–15, New York, 2006. Springer.

289. J. N. Hooker. An integrated method for planning and scheduling to minimize tardiness. *Constraints*, 11:139–157, 2006.

290. J. N. Hooker. Logic-based modeling. In G. Appa, L. Pitsoulis, and H. P. Williams, editors, *Handbook on Modelling for Discrete Optimization*, pages 61–102. Springer, New York, 2006.

291. J. N. Hooker. Planning and scheduling by logic-based Benders decomposition. *Operations Research*, 55:588–602, 2007.

292. J. N. Hooker. A principled approach to mixed integer/linear problem formulation,. In J. W. Chinneck, B. Kristjansson, and M. Saltzman, editors, *Operations Research and Cyber-Infrastructure (ICS 2009)*, pages 79–100, New York, 2009. Springer.

293. J. N. Hooker. Hybrid modeling. In M. Milano and P. Van Hentenryck, editors, *Hybrid Optimization: The Ten Years of CPAIOR*, pages 11–62. Springer, New York, 2011.

294. J. N. Hooker and C. Fedjki. Branch-and-cut solution of inference problems in propositional logic. *Annals of Mathematics and Artificial Intelligence*, 1:123–139, 1990.

295. J. N. Hooker and M. A. Osorio. Mixed logical/linear programming. *Discrete Applied Mathematics*, 96–97:395–442, 1999.

296. J. N. Hooker and G. Ottosson. Logic-based Benders decomposition. *Mathematical Programming*, 96:33–60, 2003.

297. J. N. Hooker, G. Ottosson, E. S. Thorsteinsson, and H.-J. Kim. A scheme for unifying optimization and constraint satisfaction methods. *Knowledge Engineering Review*, 15:11–30, 2000.

298. J. N. Hooker and H. Yan. Logic circuit verification by Benders decomposition. In V. Saraswat and P. Van Hentenryck, editors, *Principles and Practice of Constraint Programming: The Newport Papers*, pages 267–288, Cambridge, MA, 1995. MIT Press.

299. J. E. Hopcroft and R. M. Karp. A $n^{5/2}$ algorithm for maximum matchings in bipartite graphs. *SIAM Journal on Computing*, 2:225–231, 1973.

300. V. Jain and I. E. Grossmann. Algorithms for hybrid MILP/CP models for a class of optimization problems. *INFORMS Journal on Computing*, 13:258–276, 2001.

301. R. G. Jeroslow. Cutting plane theory: Algebraic methods. *Discrete Mathematics*, 23:121–150, 1978.

302. R. G. Jeroslow. Representability in mixed integer programming, I: Characterization results. *Discrete Applied Mathematics*, 17:223–243, 1987.

303. F. John. Extremum problems with inequalities as subsidiary conditions. In K. O. Friedrichs, O. E. Neugebauer, and J. J. Stoker, editors, *Studies and Essays: Courant Anniversary Volume*, pages 187–204. Wiley-Interscience, New York, 1948.

304. D. S. Johnson. Fast algorithms for bin packing. *Journal of Computer and Systems Sciences*, 8:272–314, 1974.

305. E. L. Johnson. Programming in networks and graphs. Technical report ORC 65-1, Operations Research Center, University of California, Berkeley, 1965.

306. E. L. Johnson. Cyclic groups, cutting planes, and shortest paths. In T. C. Hu and S. Robinson, editors, *Mathematical Programming*, pages 185–211. Academic Press, 1973.

307. M. Jünger, G. Reinelt, and G. Rinaldi. The traveling salesman problem. In M. O. Ball, T. L. Magnanti, C. L. Monma, and G. L. Nemhauser, editors, *Network Models*, Handbooks in Operations Research and Management Science, pages 225–330. Elsevier, Amsterdam, 1995.

308. U. Junker, S. E. Karish, N. Kohl, B. Vaaben, T. Fahle, and M. Sellmann. A framework for constraint programming based column generation. In J. Jaffar, editor, *Principles and Practice of Constraint Programming (CP 1999)*, volume 1713 of *Lecture Notes in Computer Science*, pages 261–275, New York, 1999. Springer.

309. W. Karush. Minima of functions of several variables with inequalities as side conditions. Master's thesis, University of Chicago, 1939.

310. I. Katriel, M. Sellmann, E. Upfal, and P. Van Hentenryck. Propagating knapsack constraints in sublinear time. In *National Conference on Artificial Intelligence (AAAI 2007)*, pages 231–236, 2007.

311. I. Katriel and S. Thiel. Fast bound consistency for the global cardinality constraint. In F. Rossi, editor, *Principles and Practice of Constraint Programming (CP 2003)*, volume 2833 of *Lecture Notes in Computer Science*, pages 437–451, New York, 2003. Springer.

312. J. Kennedy and R. Eberhart. Particle swarm optimization. In *Proceedings of IEEE International Conference on Neural Networks*, pages 1942–1948, 1995.

313. M. O. Khemmoudj, H. Bennaceur, and A. Nagih. Combining arc consistency and dual Lagrangean relaxation for filtering CSPs. In R. Barták and M. Milano, editors, *Integration of AI and OR Techniques in Constraint Programming for Combinatorial Optimization Problems (CPAIOR 2005)*, volume 3524 of *Lecture Notes in Computer Science*, pages 258–272, New York, 2005. Springer.

314. M. Khichane, P. Albert, and C. Solnon. Strong combination of ant colony optimization with constraint programming optimization. In A. Lodi, M. Milano, and P. Toth, editors, *Proceedings of the International Workshop on Integration of AI and OR Techniques in Constraint Programming for Combinatorial Optimization Problems (CPAIOR 2010)*, volume 6140 of *Lecture Notes in Computer Science*, pages 232–246, New York, 2010. Springer.

315. K. Kianfar and Y. Fathi. Generalized mixed integer rounding inequalities: Facets for infinite group polyhedra. *Mathematical Programming*, 120:313–346, 2009.

316. K. Kianfar and Y. Fathi. Generating facets for finite master cyclic group polyhedra using n-step mixed integer rounding functions. *European Journal of Operational Research*, 207:105–109, 2010.

317. H.-J. Kim and J. N. Hooker. Solving fixed-charge network flow problems with a hybrid optimization and constraint programming approach. *Annals of Operations Research*, 115:95–124, 2002.

318. S. Kirkpatrick, C. D. Gelatt, and M. P. Vecchi. Optimization by simulated annealing. *Science*, 220:671–680, 1983.

319. Z. Kiziltan. Symmetry breaking ordering constraints. PhD thesis, Uppsala University, 2004.

320. W. Kocjan and P. Kreuger. Filtering methods for symmetric cardinality constraint. In J. C. Régin and M. Rueher, editors, *Integration of AI and OR Techniques in Constraint Programming for Combinatorial Optimization Problems (CPAIOR 2004)*, volume 3011 of *Lecture Notes in Computer Science*, pages 200–208, New York, 2004. Springer.

321. N. Kohl. Application of OR and CP techniques in a real world crew scheduling system. In *Proceedings of the International Workshop on Integration of Artificial Intelligence and Operations Research Techniques in Constraint Programming for Combinatorial Optimization Problems (CPAIOR 2000)*, Paderborn, Germany, 2000.

322. D. A. Kohler. Translation of a report by Fourier on his work on linear inequalities. *Opsearch*, 10:38–42, 1973 [original work 1827].

323. T. J. Koopmans. Optimum utilization of the transportation system. *Econometrica*, 17:3–4, 1949.

324. S. Kruk and S. Toma. Polytope of all-different predicate. *Congressus Numerantium*, 195:117–159, 2009.

325. H. W. Kuhn. The Hungarian method for the assignment problem. *Naval Research Logistics Quarterly*, 2:83–97, 1955.

326. H. W. Kuhn and A. W. Tucker. Nonlinear programming. In J. Neyman, editor, *Proceedings of the Second Berkeley Symposium on Mathematical Statistics and Probability*, pages 481–492, Berkeley, CA, 1951. Univ. of California Press.

327. P. Laborie. IBM ILOG CP optimizer for detailed scheduling illustrated on three problems. In W.-J. van Hoeve and J. N. Hooker, editors, *Proceedings of the International Workshop on Integration of AI and OR Techniques in Constraint Programming for Combinatorial Optimization Problems (CPAIOR 2009)*, volume 5547 of *Lecture Notes in Computer Science*, pages 148–162, New York, 2009. Springer.

328. P. Laborie, J. Rogerie, P. Vilím, and F. Wagner. ILOG CP optimizer: Detailed scheduling model and OPL formulation. Tech report 08-002, ILOG/IBM, 2008.

329. F. Laburthe and Y. Caseau. SALSA: A language for search algorithms. *Constraints*, 7:255–288, 2002.

330. M. Z. Lagerkvist and C. Schulte. Propagator groups. In I. P. Gent, editor, *Principles and Practice of Constraint Programming (CP 2009)*, volume 5732 of *Lecture Notes in Computer Science*, pages 524–538, New York, 2009. Springer.

331. J.-L. Laurière. A language and a program for stating and solving combinatorial problems. *Artificial Intelligence*, 10:29–127, 1978.

332. S. L. Lauritzen and D. J. Spiegelhalter. Local computations with probabilities on graphical structures and their application to expert systems. *Journal of the Royal Statistical Society B*, 50:157–224, 1988.

333. Y. C. Law and J. H. M. Lee. Global constraints for integer and set value precedence. In M. Wallace, editor, *Principles and Practice of Constraint Programming (CP 2004)*, volume 3258 of *Lecture Notes in Computer Science*, pages 362–376, New York, 2004. Springer.

334. E. L. Lawler. *Combinatorial Optimization: Networks and Matroids.* Holt, Rinehart and Winston, New York, 1976.

335. S. Lee and I. Grossmann. A global optimization algorithm for nonconvex generalized disjunctive programming and applications to process systems. *Computers and Chemical Engineering*, 25:1675–1697, 2001.

336. S. Lee and I. Grossmann. Generalized disjunctive programming: Nonlinear convex hull relaxation and algorithms. *Computational Optimization and Applications*, 26:83–100, 2003.

337. S. Lee and I. E. Grossmann. Global optimization of nonlinear generalized disjunctive programming with bilinear equality constraints: Applications to process networks. *Computers and Chemical Engineering*, 27:1557–1575, 2003.

338. C. E. Lemke. The dual method of solving the linear programming problem. *Naval Research Logistics Quarterly*, 1:36–47, 1954.

339. J. T. Linderoth and M. W. P. Savelsbergh. A computational study of search strategies for mixed integer programming. *INFORMS Journal on Computing*, 11:173–187, 1999.

340. J. Little and K. Darby-Dowman. The significance of constraint logic programming to operational research. In M. Lawrence and C. Wilsden, editors, *Operational Research Tutorial Papers (Invited tutorial paper to the Operational Research Society Conference, 1995)*, pages 20–45, 1995.

341. A. Lopez-Ortiz, C.-G. Quimper, J. Tromp, and P. van Beek. A fast and simple algorithm for bounds consistency of the alldifferent constraint. In *International Joint Conference on Artificial Intelligence (IJCAI 2003)*, pages 245–250, 2003.

342. D. G. Luenberger and Y. Ye. *Linear and Nonlinear Programming*. Springer, New York, 2008.

343. A. Mackworth. Consistency in networks of relations. *Artificial Intelligence*, 8:99–118, 1977.

344. M. J. Maher, N. Narodytska, C.-G. Quimper, and T. Walsh. Flow-based propagators for the SEQUENCE and related global constraints. In P. J. Stuckey, editor, *Principles and Practice of Constraint Programming (CP 2008)*, volume 5202 of *Lecture Notes in Computer Science*, pages 159–174, New York, 2008. Springer.

345. C. T. Maravelias and I. E. Grossmann. Using MILP and CP for the scheduling of batch chemical processes. In J. C. Régin and M. Rueher, editors, *Integration of AI and OR Techniques in Constraint Programming for Combinatorial Optimization Problems (CPAIOR 2004)*, volume 3011 of *Lecture Notes in Computer Science*, pages 1–20, New York, 2004. Springer.

346. H. Marchand, A. Martin, R. Weismantel, and L. Wolsey. Cutting planes in integer and mixed integer programming. *Discrete Applied Mathematics*, 123:397–446, 2002.

347. H. Marchand and L. A. Wolsey. Aggregation and mixed integer rounding to solve MIPs. *Operations Research*, 49:363–371, 2001.

348. S. Martello and P. Toth. *Knapsack Problems: Algorithms and Computer Implementations*. John Wiley, New York, 1990.

349. S. Martello and P. Toth. Lower bounds and reduction procedures for the bin packing problem. *Discrete Applied Mathematics*, 28:59–70, 1990.

350. A. Martin and R. Weismantel. Contribution to general mixed integer knapsack problems. Technical report SC 97-38, Konrad-Zuse-Zentrum für Informationstechnik Berlin, 1997.

351. D. A. McAllester. Partial order backtracking. Manuscript, AI Laboratory, MIT, Cambridge, MA, 1993.

352. G. P. McCormick. Computability of global solutions to factorable nonconvex programs: Part I: Convex underestimating problems. *Mathematical Programming*, 10:147–175, 1976.

353. K. Mehlhorn and S. Thiel. Faster algorithms for bound-consistency of the sortedness and the alldifferent constraint. In R. Dechter, editor, *Principles and Practice of Constraint Programming (CP 2000)*, volume 1894 of *Lecture Notes in Computer Science*, pages 306–319, New York, 2000. Springer.

354. A. Meisels and A. Schaerf. Modelling and solving employee timetabling problems. *Annals of Mathematics and Artificial Intelligence*, 239:41–59, 2002.

355. P. Meseguer, F. Rossi, and T. Schiex. Soft constraints. In F. Rossi, P. van Beek, and T. Walsh, editors, *Handbook of Constraint Programming*, pages 281–328. Elsevier, 2006.

356. N. Metropolis, A. W. Rosenbluth, M. N. Rosenbluth, A. H. Teller, and E. Teller. Equations of state calculations by fast computing machines. *Journal of Chemical Physics*, 21:1087–1092, 1953.

357. L. Michel and P. Van Hentenryck. Localizer: A modeling language for local search. *INFORMS Journal on Computing*, 11:1–14, 1999.

358. M. Milano and W. J. van Hoeve. Building negative reduced cost paths using constraint programming. In P. Van Hentenryck, editor, *Principles and Practice of Constraint Programming (CP 2002)*, volume 2470 of *Lecture Notes in Computer Science*, pages 1–16, New York, 2002. Springer.

359. R. Mohr and G. Masini. Good old discrete relaxation. In Y. Kodratoff, editor, *Proceedings of the 8th European Conference on Artificial Intelligence (ECAI 1988)*, pages 651–656. Pitman Publishers, 1988.

360. U. Montanari. Networks of constraints: Fundamental properties and applications to picture processing. *Information Science*, 7:95–132, 1974.

361. M. W. Moskewicz, C. F. Madigan, Y. Zhao, L. Zhang, and S. Malik. Chaff: Engineering an efficient SAT solver. In *Proceedings of the 38th Design Automation Conference (DAC 2001)*, pages 530–535, 2001.

362. T. S. Motzkin. Beiträge zur Theorie der linearen Ungelichungen. PhD thesis, University of Basel, 1936.

363. T. S. Motzkin. Contributions to the theory of inequalities [translation]. In D. Cantor, B. Gordon, and B. Rothschild, editors, *Theodore S. Motzkin: Selected Papers*, pages 1–80. Birkhäuser, Boston, 1983.

364. M. Müller-Hannemann, W. Stille, and K. Weihe. Evaluating the bin-packing constraint, Part I: Overview of the algorithmic approach. Technical report, Technische Universität Darmstadt, 2003.

365. M. Müller-Hannemann, W. Stille, and K. Weihe. Evaluating the bin-packing constraint, Part II: An adaptive rounding problem. Technical report, Technische Universität Darmstadt, 2003.

366. M. Müller-Hannemann, W. Stille, and K. Weihe. Evaluating the bin-packing constraint, Part III: Joint evaluation with concave constraints. Technical report, Technische Universität Darmstadt, 2003.

367. M. Müller-Hannemann, W. Stille, and K. Weihe. Patterns of usage for global constraints: A case study based on the bin-packing constraint. Technical report, Technische Universität Darmstadt, 2003.

368. D. Naddef. Polyhedral theory and branch-and-cut algorithms for the symmetric TSP. In G. Gutin and A. P. Punnen, editors, *The Traveling Salesman Problem and Its Variations*, pages 29–116. Kluwer, Dordrecht, 2002.

369. D. Naddef and S. Thienel. Efficient separation routines for the symmetric traveling salesman problem I: General tools and comb separation. *Mathematical Programming*, 92:237–255, 2002.

370. D. Naddef and S. Thienel. Efficient separation routines for the symmetric traveling salesman problem II: Separating multi handle inequalities. *Mathematical Programming*, 92:257–285, 2002.

371. G. L. Nemhauser and L. A. Wolsey. *Integer and Combinatorial Optimization*. John Wiley, New York, 1999.

372. A. Neumaier. Complete search in continuous global optimization and constraint satisfaction. In A. Iserles, editor, *Acta Numerica 2004*, pages 271–369. Cambridge University Press, 2004.

373. N. Nishimura, P. Ragde, and S Szeider. Detecting backdoor sets with respect to Horn and binary clauses. In *International Conference on Theory and Applications of Satisfiability Testing (SAT 2004)*, pages 96–103, 2004.

374. P. Nobili and A. Sassano. Facets and lifting procedures for the set covering polytope. *Mathematical Programming*, 45:111–137, 1989.

375. W. P. M. Nuijten. Time and resource constrained scheduling. PhD thesis, Eindhoven University of Technology, 1994.

376. W. P. M. Nuijten and E. H. L. Aarts. Constraint satisfaction for multiple capacitated job shop scheduling. In A. Cohn, editor, *Proceedings of the 11th European Conference on Artificial Intelligence (ECAI 1994)*, pages 635–639. John Wiley, 1994.

377. W. P. M. Nuijten and E. H. L. Aarts. A computational study of constraint satisfaction for multiple capacitated job shop scheduling. *European Journal of Operational Research*, 90:269–284, 1996.

378. W. P. M. Nuijten, E. H. L. Aarts, D. A. A. van Erp Taalman Kip, and K. M. van Hee. Randomized constraint satisfaction for job-shop scheduling. In *AAAI-SIGMAN Workshop on Knowledge-Based Production Planning, Scheduling and Control*, 1993.

379. M. Osorio and F. Glover. Logic cuts using surrogate constraint analysis in the multidimensional knapsack problem. In C. Gervet and M. Wallace, editors, *Proceedings of the International Workshop on Integration of Artificial Intelligence and Operations Research Techniques in Constraint Programming for Combinatorial Optimization Problems (CPAIOR 2001)*, Ashford, U.K., 2001.

380. G. Ottosson, E. Thorsteinsson, and J. N. Hooker. Mixed global constraints and inference in hybrid IP-CLP solvers. In *Proceedings of CP99 Post-Conference Workshop on Large-Scale Combinatorial Optimization and Constraints*, http://www.dash.co.uk/wscp99, pages 57–78, 1999.

381. G. Ottosson, E. Thorsteinsson, and J. N. Hooker. Mixed global constraints and inference in hybrid CLP-IP solvers. *Annals of Mathematics and Artificial Intelligence*, 34:271–290, 2002.

382. with contributions by I. Lustig, L. Michel, and J. F. Puget P. Van Hentenryck. *The OPL Optimization Programming Language*. MIT Press, Cambridge, MA, 1999.

383. M. Padberg. On the facial structure of set packing polyhedra. *Mathematical Programming*, 5:199–215, 1973.

384. M. Padberg. A note on zero-one programming. *Operations Research*, 23:833–837, 1975.

385. M. Padberg and G. Rinaldi. An efficient algorithm for the minimum capacity cut problem. *Mathematical Programming*, 47:19–36, 1990.

386. C. H. Papadimitriou and K. Steiglitz. *Combinatorial Optimization: Algorithms and Complexity*. Dover, 1998.

387. C. Le Pape. Implementation of resource constraints in ILOG SCHEDULE: A library for the development of constraint-based scheduling systems. *Intelligent Systems Engineering*, 3:55–66, 1994.

388. P. M. Pardalos and H. E. Romeijn, editors. *Handbook of Global Optimization*, volume 2. Springer, New York, 2002.

389. V. T. Paschos. A survey of approximately optimal solutions to some covering and packing problems. *ACM Computing Surveys*, 29:171–209, 1997.

390. G. Pesant. A regular language membership constraint for sequence of variables. In A. M. Frisch, editor, *Workshop on Modelling and Reformulating Constraint Satisfaction Problems*, pages 110–119, Kinsale, Ireland, 1993.

391. G. Pesant. A filtering algorithm for the stretch constraint. In T. Walsh, editor, *Principles and Practice of Constraint Programming (CP 2001)*, volume 2239 of *Lecture Notes in Computer Science*, pages 183–195, New York, 2001. Springer.

392. G. Pesant. A regular language membership constraint for finite sequences of variables. In M. Wallace, editor, *Principles and Practice of Constraint Programming (CP 2004)*, volume 3258 of *Lecture Notes in Computer Science*, pages 482–495, New York, 2004. Springer.

393. G. Pesant and J.-C. Régin. Spread: A balacing constraint based on statistics. In *Principles and Practice of Constraint Programming (CP 2005)*, volume 3709 of *Lecture Notes in Computer Science*, pages 460–474, New York, 2005. Springer.

394. B. Peterson and M. Trick. A Benders' approach to a transportation network design problem. In W.-J. van Hoeve and J. N. Hooker, editors, *Proceedings of the International Workshop on Integration of AI and OR Techniques in Constraint Programming for Combinatorial Optimization Problems (CPAIOR 2009)*, volume 5547 of *Lecture Notes in Computer Science*, pages 326–327, New York, 2009. Springer.

395. T. Petit, J. C. Régin, and C. Bessiere. Specific filtering algorithms for over-constrained problems. In T. Walsh, editor, *Principles and Practice of Constraint Programming (CP 2001)*, volume 2239 of *Lecture Notes in Computer Science*, pages 451–463, New York, 2001. Springer.

396. Q. D. Pham, Y. Deville, and P. Van Hentenryck. Constraint-based local search for constrained optimum paths problems. In A. Lodi, M. Milano, and P. Toth, editors, *Proceedings of the International Workshop on Integration of AI and OR Techniques in Constraint Programming for Combinatorial Optimization Problems (CPAIOR 2010)*, volume 6140 of *Lecture Notes in Computer Science*, pages 267–281, New York, 2010. Springer.

397. Y. Pochet and R. Weismantel. The sequential knapsack polytope. *SIAM Journal on Optimization*, 8:248–264, 1998.

398. Y. Pochet and L. A. Wolsey. Integer knapsack and flow covers with divisible coefficients: Polyhedra, optimization, and separation. *Discrete Applied Mathematics*, 59:57–74, 1995.

399. F. P. Preparata and S. J. Hong. Convex hulls of finite sets of points in two and three dimensions. *Communications of the ACM*, 20:87–93, 1977.

400. S. Prestwich. Exploiting relaxation in local search. In *First International Workshop on Local Search Techniques in Constraint Satisfaction (LSCS 2004)*, Toronto, 2004.

401. J.-F. Puget. A fast algorithm for the bound consistency of alldiff constraints. In *National Conference on Artificial Intelligence (AAAI 1998)*, pages 359–366. AAAI Press, 1990.

402. L. Quadrifoglio, M. M. Dessouky, and F. Ordóñez. Mobility allowance shuttle transit (MAST) services: MIP formulation and strengthening with logic constraints. In L. Perron and M. A. Trick, editors, *Proceedings of the International Workshop on Integration of Artificial Intelligence and Operations Research Techniques in Constraint Programming for Combinatorial Optimization Problems (CPAIOR 2008)*, volume 5015 of *Lecture Notes in Computer Science*, pages 387–391, New York, 2008. Springer.

403. C.-G. Quimper, A. López-Ortiz, P. van Beek, and A. Golynski. Improved algorithms for the global cardinality constraint. In M. Wallace, editor, *Principles and Practice of Constraint Programming (CP 2004)*, volume 3258 of *Lecture Notes in Computer Science*, pages 542–556, New York, 2004. Springer.

404. C.-G. Quimper, P. van Beek, A. López-Ortiz, A. Golynski, and S. B. Sadjad. An efficient bounds consistency algorithm for the global cardinality constraint. In F. Rossi, editor, *Principles and Practice of Constraint Programming (CP 2003)*, volume 2833 of *Lecture Notes in Computer Science*, pages 600–614, New York, 2003. Springer.

405. C.-G. Quimper and T. Walsh. Global grammar constraints. In F. Benhamou, editor, *Principles and Practice of Constraint Programming (CP 2006)*, volume 4204 of *Lecture Notes in Computer Science*, pages 751–755, New York, 2006. Springer.

406. C.-G. Quimper and T. Walsh. Decomposing global grammar constraints. In C. Bessiere, editor, *Principles and Practice of Constraint Programming (CP 2007)*, volume 4741 of *Lecture Notes in Computer Science*, pages 590–604, New York, 2007. Springer.

407. W. V. Quine. The problem of simplifying truth functions. *American Mathematical Monthly*, 59:521–531, 1952.

408. W. V. Quine. A way to simplify truth functions. *American Mathematical Monthly*, 62:627–631, 1955.

409. R. Raman and I. E. Grossmann. Modeling and computational techniques for logic based integer programming. *Computers and Chemical Engineering*, 20:563–578, 1994.

410. R. Rasmussen and M. A. Trick. A Benders approach to the constrained minimum break problem. *European Journal of Operational Research*, 177:198–213, 2007.

411. R. Rasmussen and M. A. Trick. A Benders approach to the constrained minimum break problem. *European Journal of Operational Research*, pages 198–213, 2007.

412. I. Rechenberg. *Evolutionsstrategie*. Holzmann-Froboog, Stuttgart, 1973.

413. P. Refalo. Tight cooperation and its application in piecewise linear optimization. In J. Jaffar, editor, *Principles and Practice of Constraint Programming (CP 1999)*, volume 1713 of *Lecture Notes in Computer Science*, pages 375–389, New York, 1999. Springer.

414. P. Refalo. Linear formulation of constraint programming models and hybrid solvers. In R. Dechter, editor, *Principles and Practice of Constraint Programming (CP 2000)*, volume 1894 of *Lecture Notes in Computer Science*, pages 369–383, New York, 2000. Springer.

415. P. Refalo. Impact-based search strategies for constraint programming. In M. Wallace, editor, *Principles and Practice of Constraint Programming (CP 2004)*, volume 3258 of *Lecture Notes in Computer Science*, pages 557–571, New York, 2004. Springer.

416. J.-C. Régin. A filtering algorithm for constraints of difference in CSP. In *National Conference on Artificial Intelligence (AAAI 1994)*, pages 362–367. AAAI Press, 1994.

417. J.-C. Régin. Développement d'outils algorithmiques pour l'intelligence artificielle: Application à la chimie organique. PhD thesis, Université de Montpellier II, 1995.

418. J.-C. Régin. Generalized arc consistency for *global cardinality* constraint. In *National Conference on Artificial Intelligence (AAAI 1996)*, pages 209–215. AAAI Press, 1996.

419. J.-C. Régin. The symmetric alldiff constraint. In T. Dean, editor, *Proceedings of the International Joint Conference on Artificial Intelligence (IJCAI 1999)*, volume 1, pages 420–425, Stockholm, 1996. Morgan Kaufmann.

420. J.-C. Régin. Arc consistency for global cardinality with costs. In J. Jaffar, editor, *Principles and Practice of Constraint Programming (CP 1999)*, volume 1713 of *Lecture Notes in Computer Science*, pages 390–404, New York, 1999. Springer.

421. J.-C. Régin. Using constraint propagation to solve the maximum clique problem. In F. Rossi, editor, *Principles and Practice of Constraint Programming (CP 2003)*, volume 2833 of *Lecture Notes in Computer Science*, pages 634–648, New York, 2003. Springer.

422. J.-C. Régin. Modeling problems in constraint programming. In *Tutorial presented at conference on Principles and Practice of Constraint Programming (CP 2004)*, Toronto, 2004.

423. J.-C. Régin. Global constraints: A survey. In P. van Hentenryck and M. Milano, editors, *Hybrid Optimization: The Ten Years of CPAIOR*, pages 63–134. Springer, New York, 2011.

424. J.-C. Régin and C. Gomes. The cardinality matrix constraint. In M. Wallace, editor, *Principles and Practice of Constraint Programming (CP 2004)*, volume 3258 of *Lecture Notes in Computer Science*, pages 572–587, New York, 2004. Springer.

425. J.-C. Régin and J.-F. Puget. A filtering algorithm for global sequencing constraints. In G. Smolka, editor, *Principles and Practice of Constraint Programming (CP 1997)*, volume 1330 of *Lecture Notes in Computer Science*, pages 32–46, New York, 1997. Springer.

426. C. Ribeiro and M. A. Carravilla. A global constraint for nesting problems. In J. C. Régin and M. Rueher, editors, *Integration of AI and OR Techniques in Constraint Programming for Combinatorial Optimization Problems (CPAIOR 2004)*, volume 3011 of *Lecture Notes in Computer Science*, pages 256–270, New York, 2004. Springer.

427. J. A. Robinson. A machine-oriented logic based on the resolution principle. *Journal of the ACM*, 12:23–41, 1965.

428. R. Rodošek, M. Wallace, and M. Hajian. A new approach to integrating mixed integer programming and constraint logic programming. *Annals of Operations Research*, 86:63–87, 1999.

429. B. K. Rosen. Robust linear algorithms for cutsets. *Journal of Algorithms*, 3:205–212, 1982.

430. L.-M. Rousseau. Stabilization issues for constraint programming based column generation. In J. C. Régin and M. Rueher, editors, *Integration of AI and OR Techniques in Constraint Programming for Combinatorial Optimization Problems (CPAIOR 2004)*, volume 3011 of *Lecture Notes in Computer Science*, pages 402–408. Springer, 2004.

431. L. M. Rousseau, M. Gendreau, and G. Pesant. Solving small VRPTWs with constraint programming based column generation. In N. Jussien and F. Laburthe, editors, *Proceedings of the International Workshop on Integration of Artificial Intelligence and Operations Research Techniques in Constraint Programming for Combinatorial Optimization Problems (CPAIOR 2002)*, Le Croisic, France, 2002.

432. R. Sadykov. A hybrid branch-and-cut algorithm for the one-machine scheduling problem. In J. C. Régin and M. Rueher, editors, *Integration of AI and OR Techniques in Constraint Programming for Combinatorial*

Optimization Problems (CPAIOR 2004), volume 3011 of *Lecture Notes in Computer Science*, pages 409–415. Springer, 2004.

433. T. Sandholm and R. Shields. Nogood learning for mixed integer programming. In *Workshop on Hybrid Methods and Branching Rules in Combinatorial Optimization*, 2006.

434. A. Sassano. On the facial structure of the set covering polytope. *Mathematical Programming*, 44:181–202, 1989.

435. N. W. Sawaya and I. E. Grossmann. A cutting plane method for solving linear generalized disjunctive programming problems. *Computers and Chemical Engineering*, 29:1891–1913, 2005.

436. N. W. Sawaya and I. E. Grossmann. Computational implementation of non-linear convex hull reformulation. *Computers and Chemical Engineering*, 31:856–866, 2007.

437. N. W. Sawaya and I. E. Grossmann. Reformulations, relaxations and cutting planes for linear generalized disjunctive programming. Research report, Department of Chemical Engineering, Carnegie Mellon University, 2008.

438. P. Schaus. Solving balancing and bin-packing problems with constraint programming. PhD thesis, Université catholique de Louvain de Louvain-la-Neuve, 2009.

439. P. Schaus, Y. Deville, P. Dupont, and J.-C. Régin. Simplification and extension of the spread constraint. In *Workshop on Constraint Propagation and Implementation*, Lecture Notes in Computer Science, pages 72–92, Nantes, France, 2006.

440. P. Schaus, Y. Deville, P. Dupont, and J.-C. Régin. The deviation constraint. In P. van Hentenryck and L. Wolsey, editors, *Integration of AI and OR Techniques in Constraint Programming for Combinatorial Optimization Problems (CPAIOR 2007)*, volume 4510 of *Lecture Notes in Computer Science*, pages 260–274, New York, 2007. Springer.

441. L. Schrage and L. Wolsey. Sensitivity analysis for branch and bound integer programming. *Operations Research*, 33:1008–1023, 1985.

442. A. Schrijver. *Theory of Linear and Integer Programming*. John Wiley, New York, 1986.

443. M. Sellmann. An arc-consistency algorithm for the minimum-weight all different constraint. In P. Van Hentenryck, editor, *Principles and Practice of Constraint Programming (CP 2002)*, volume 2470 of *Lecture Notes in Computer Science*, pages 744–749, New York, 2002. Springer.

444. M. Sellmann. Approximated consistency for knapsack constraints. In F. Rossi, editor, *Principles and Practice of Constraint Programming (CP 2003)*, volume 2833 of *Lecture Notes in Computer Science*, pages 679–693, New York, 2003. Springer.

445. M. Sellmann. Cost-based filtering for shorter path constraints. In F. Rossi, editor, *Principles and Practice of Constraint Programming (CP 2003)*, volume 2833 of *Lecture Notes in Computer Science*, pages 694–708, New York, 2003. Springer.

446. M. Sellmann. The practice of approximated consistency for knapsack constraints. In *National Conference on Artificial Intelligence (AAAI 2004)*, pages 179–184, 2004.

447. M. Sellmann. The theory of grammar constraints. In F. Benhamou, editor, *Principles and Practice of Constraint Programming (CP 2006)*, volume 4204 of *Lecture Notes in Computer Science*, pages 530–544, New York, 2006. Springer.

448. M. Sellmann and T. Fahle. Constraint programming based Lagrangian relaxation for a multimedia application. In C. Gervet and M. Wallace, editors, *Proceedings of the International Workshop on Integration of Artificial Intelligence and Operations Research Techniques in Constraint Programming for Combinatorial Optimization Problems (CPAIOR 2001)*, Ashford, U.K., 2001.

449. M. Sellmann, T. Gellermann, and R. Wright. Cost-based filtering for shorter path constraints. *Constraints*, 12:207–238, 2007.

450. M. Sellmann, K. Zervoudakis, P. Stamatopoulos, and T. Fahle. Crew assignment via constraint programming: Integrating column generation and heuristic tree search. *Annals of Operations Research*, 115:207–225, 2002.

451. G. Shafer, P. P. Shenoy, and K. Mellouli. Propagating belief functions in qualitative Markov trees. *International Journal of Approximate Reasoning*, 1:349–400, 1987.

452. A. Shamir. A linear time algorithm for finding minimum cutsets in reducible graphs. *SIAM Journal on Computing*, 8:645–655, 1979.

453. P. Shaw. A constraint for bin packing. In M. Wallace, editor, *Principles and Practice of Constraint Programming (CP 2004)*, volume 3258 of *Lecture Notes in Computer Science*, pages 648–662, New York, 2004. Springer.

454. P. Shaw. Constraint programming and local search hybrids. In P. van Hentenryck and M. Milano, editors, *Hybrid Optimization: The Ten Years of CPAIOR*, pages 271–304. Springer, New York, 2011.

455. P. P. Shenoy and G. Shafer. Propagating belief functions with local computation. *IEEE Expert*, 1:43–52, 1986.

456. Y. Shi and R. Eberhart. A modified particle swarm optimizer. In *Proceedings of IEEE International Conference on Neural Networks*, pages 69–73, 1998.

457. J. A. Shufelt and H. J. Berliner. Generating hamiltonian circuits without backtracking. *Theoretical Computer Science*, 132:347–375, 1994.

458. J. P. M. Silva and K. A. Sakallah. GRASP: A search algorithm for propositional satisfiability. *IEEE Transactions on Computers*, 48:506–521, 1999.

459. H. Simonis. Modelling in CP. In G. Pesant, editor, *Constraint Programming Tutorial, CPAIOR 2009*, Pittsburgh, USA, 2009.

460. S. S. Skiena. *The Algorithm Design Manual*. Springer, New York, 1997.

461. J. Skorin-Kapov and F. Granot. Nonlinear integer programming: Sensitivity analysis for branch and bound. *Operations Research Letters*, 6:269–274, 1987.

462. S. F. Smith. OPIS: A methodology and architecture for reactive scheduling. In M. Zweben and M. S. Fox, editors, *Intelligent Scheduling*, pages 29–66. Morgan Kaufmann, San Francisco, 1995.

463. V. Srinivasan and G. Thompson. Benefit–cost analysis of coding techniques for the primal transportation algorithm. *Journal of the ACM*, 20:194–213, 1973.

464. R. M. Stallman and G. J. Sussman. Forward reasoning and dependency-directed backtracking in a system for computer-aided circuit analysis. *Journal of Artificial Intelligence*, 9:135–196, 1977.

465. R. Stubbs and S. Mehrotra. A branch-and-cut method for 0-1 mixed convex programming. *Mathematical Programming*, 86:515–532, 1999.

466. P. J. Stuckey, M. G. de la Banda, M. Maher, K. Marriott, J. Slaney, Z. Somogyi, M. Wallace, and T. Walsh. The G12 project: Mapping solver independent models to efficient solutions. In P. van Beek, editor, *Principles and Practice of Constraint Programming (CP 2005)*, volume 3668 of *Lecture Notes in Computer Science*, pages 314–327, New York, 2005. Springer.

467. B. Sturmfels. *Gröbner Bases and Convex Polytopes*. American Mathematical Society, Providence, RI, 1995.

468. M. Tawarmalani and N. V. Sahinidis. *Convexification and Global Optimization in Continuous and Mixed-integer Nonlinear Programming: Theory, Algorithms, Software, and Applications*. Springer, New York, 2002.

469. M. Tawarmalani and N. V. Sahinidis. Global optimization of mixed-integer nonlinear programs: A theoretical and computational study. *Mathematical Programming*, 99:563–591, 2004.

470. M. Tawarmalani and N. V. Sahinidis. A polyhedral branch-and-cut approach to global optimization. *Mathematical Programming*, 103:225–249, 2005.

471. D. Terekhov, J. C. Beck, and K. N. Brown. Solving a stochastic queueing design and control problem with constraint programming. In *Proceedings of the 22nd National Conference on Artificial Intelligence (AAAI 2007)*, volume 1, pages 261–266. AAAI Press, 2007.

472. R. R. Thomas. The structure of group relaxations. In K. Aardal, G. L. Nemhauser, and R. Weismantel, editors, *Discrete Optimization*, Handbooks in Operations Research and Management Science, pages 123–170. Elsevier, Amsterdam, 2005.

473. E. Thorsteinsson. Branch and check: A hybrid framework integrating mixed integer programming and constraint logic programming. In T. Walsh, editor, *Principles and Practice of Constraint Programming (CP 2001)*, volume 2239 of *Lecture Notes in Computer Science*, pages 16–30, New York, 2001. Springer.

474. E. Thorsteinsson and G. Ottosson. Linear relaxations and reduced-cost based propagation of continuous variable subscripts. *Annals of Operations Research*, 115:15–29, 2001.

475. C. Timpe. Solving planning and scheduling problems with combined integer and constraint programming. *OR Spectrum*, 24:431–448, 2002.

476. J. Tind and L. A. Wolsey. An elementary survey of general duality theory in mathematical programming. *Mathematical Programming*, 21:241–261, 1981.

477. P. Torres and P. Lopez. On not-first/not-last conditions in disjunctive scheduling. *European Journal of Operational Research*, 127:332–343, 2000.

478. P. Toth and D. Vigo. Models, relaxations and exact approaches for the capacitated vehicle routing problem. *Discrete Applied Mathematics*, 123:487–512, 2002.

479. M. A. Trick. A dynamic programming approach for consistency and propagation for knapsack constraints. In C. Gervet and M. Wallace, editors, *Proceedings, Integration of AI and OR Techniques in Constraint Programming for Combinatorial Optimization Problems (CPAIOR 2001)*, pages 113–124, Ashford, U.K., 2001.

480. M. A. Trick. Formulations and reformulations in integer programming. In R. Barták and M. Milano, editors, *Integration of AI and OR Techniques in Constraint Programming for Combinatorial Optimization Problems (CPAIOR 2005)*, volume 3524 of *Lecture Notes in Computer Science*, pages 366–379, New York, 2005. Springer.

481. M. A. Trick and H. Yildiz. Benders cuts guided search for the traveling umpire problem. In P. van Hentenryck and L. Wolsey, editors, *Integration of AI and OR Techniques in Constraint Programming for Combinatorial Optimization Problems (CPAIOR 2007)*, volume 4510 of *Lecture Notes in Computer Science*, pages 332–345, New York, 2007. Springer.

482. E. Tsang. *Foundations of Constraint Satisfaction*. Academic Press, London, 1983.

483. M. Türkay and I. E. Grossmann. Disjunctive programming techniques for the optimization of process systems with discontinuous investment costs: Multiple size regions. *Industrial Engineering Chemical Research*, 35:2611–2623, 1996.

484. M. Türkay and I. E. Grossmann. Logic-based MINLP algorithms for the optimal synthesis of process networks. *Computers and Chemical Engineering*, 20:959–978, 1996.

485. W.-J. van Hoeve. A hybrid constraint programming and semidefinite programming approach for the stable set problem. In F. Rossi, editor, *Principles and Practice of Constraint Programming (CP 2003)*, volume 2833 of *Lecture Notes in Computer Science*, pages 407–421, New York, 2003. Springer.

486. W.-J. van Hoeve. A hyper-arc consistency algorithm for the soft alldifferent constraint. In M. Wallace, editor, *Principles and Practice of Constraint Programming (CP 2004)*, volume 3258 of *Lecture Notes in Computer Science*, pages 679–689, New York, 2004. Springer.

487. W.-J. van Hoeve and I. Katriel. Global constraints. In F. Rossi, P. van Beek, and T. Walsh, editors, *Handbook of Constraint Programming*, pages 169–208. Elsevier, 2006.

488. W.-J. van Hoeve, G. Pesant, L.-M. Rousseau, and A. Sabharwal. Revisiting the sequence constraint. In F. Benhamou, editor, *Principles and Practice of Constraint Programming (CP 2006)*, volume 4204 of *Lecture Notes in Computer Science*, pages 620–634, New York, 2006. Springer.

489. W.-J. van Hoeve, G. Pesant, L.-M. Rousseau, and A. Sabharwal. New filtering algorithms for combinations of among constraints. 14:273–292, 2009.

490. W.-J. van Hoeve and J.-C. Régin. Open constraints in a closed world. In J. C. Beck and B. M. Smith, editors, *Integration of AI and OR Techniques in Constraint Programming for Combinatorial Optimization Problems (CPAIOR 2006)*, volume 3990 of *Lecture Notes in Computer Science*, pages 244–257, New York, 2006. Springer.

491. R. J. Vanderbei. *Linear Programming: Foundations and Extensions.* Springer, New York, 2nd edition, 2001.

492. A. Vecchietti, S. Lee, and I. E. Grossmann. Characterization and formulation of disjunctions and their relaxations. In *Proceedings of Mercosul Congress on Process Systems Engineering (ENPROMER 2001)*, volume 1, pages 409–414, Santa Fe, Chile, 2001.

493. A. F. Veinott and G. B. Dantzig. Integral extreme points. *SIAM Review*, 10:371–372, 1968.

494. A. F. Veinott and H. Wagner. Optimal capacity scheduling I. *Operations Research*, 10:518–532, 1962.

495. N. Vempaty. Solving constraint satisfaction problems using finite state automata. In *National Conference on Artificial Intelligence (AAAI 1992)*, pages 453–458, 1992.

496. R. R. Vemuganti. Applications of set covering, set packing and set partitioning models: A survey. In D.-Z. Du and P. Pardalos, editors, *Handbook of Combinatorial Optimization*, volume 1, pages 573–746. Kluwer, Dordrecht, 1998.

497. M. Wallace, M. S. Novello, and J. Schimpf. ECLiPSe: A platform for constraint logic programming. *ICL Systems Journal*, 12:159–200, 1997.

498. H. P. Williams. Linear and integer programming applied to the propositional calculus. *International Journal of Systems Research and Information Science*, 2:81–100, 1987.

499. H. P. Williams. Duality in mathematics and linear and integer programming. *Journal of Optimization Theory and Applications*, 90:257–278, 1996.

500. H. P. Williams. *Model Building in Mathematical Programming.* John Wiley, New York, 4th edition, 1999.

501. H. P. Williams. The formulation and solution of discrete optimization models. In G. Appa, L. Pitsoulis, and H. P. Williams, editors, *Handbook on Modelling for Discrete Optimization*, pages 3–38. Springer, New York, 2006.

502. H. P. Williams. *Logic and Integer Programming.* Springer, New York, 2009.

503. H. P. Williams and H. Yan. Representations of the all_different predicate of constraint satisfaction in integer programming. *INFORMS Journal on Computing*, 13:96–103, 2001.

504. R. Williams, C. Gomes, and B. Selman. Backdoors to typical case complexity. In *International Joint Conference on Artificial Intelligence (IJCAI 2003)*, pages 1173–1178, 2003.

505. R. Williams, C. Gomes, and B. Selman. On the connections between backdoors, restarts, and heavy-tailedness in combinatorial search. In *International Conference on Theory and Applications of Satisfiability Testing (SAT 2004)*, pages 222–230, 2004.

506. L. A. Wolsey. Faces for a linear inequality in 0-1 variables. *Mathematical Programming*, 8:165–178, 1975.

507. L. A. Wolsey. The *b*-hull of an integer program. *Discrete Applied Mathematics*, 3:193–201, 1981.

508. L. A. Wolsey. Integer programming duality: Price functions and sensitivity analysis. *Mathematical Programming*, 20:173–195, 1981.

509. L. A. Wolsey. Valid inequalities for 0-1 knapsacks and MIPs with generalized upper bound constraints. *Discrete Applied Mathematics*, 29:251–261, 1990.

510. L. A. Wolsey. *Integer Programming*. John Wiley, New York, 1998.

511. L. A. Wolsey. MIP modelling of changeovers in production planning and scheduling problems. *European Journal of Operational Research*, 99:154–165, 1998.

512. Q. Xia, A. Eremin, and M. Wallace. Problem decomposition for traffic diversions. In J. C. Régin and M. Rueher, editors, *Integration of AI and OR Techniques in Constraint Programming for Combinatorial Optimization Problems (CPAIOR 2004)*, volume 3011 of *Lecture Notes in Computer Science*, pages 348–363, New York, 2004. Springer.

513. H. Yan and J. N. Hooker. Tight representations of logical constraints as cardinality rules. *Mathematical Programming*, 85:363–377, 1995.

514. B. Yehuda, J. Geiger, J. Naor, and R. M. Roth. Approximation algorithms for the vertex feedback set problem with application in constraint satisfaction and bayesian inference. In *Proceedings of 5th Annual ACM-SIAM Symposium on Discrete Algorithms*, pages 344–354, 1994.

515. T. H. Yunes. On the sum constraint: Relaxation and applications. In P. Van Hentenryck, editor, *Principles and Practice of Constraint Programming (CP 2002)*, volume 2470 of *Lecture Notes in Computer Science*, pages 80–92, New York, 2002. Springer.

516. T. H. Yunes. Software tools supporting integration. In P. van Henten-
     ryck and M. Milano, editors, *Hybrid Optimization: The Ten Years of
     CPAIOR*, pages 393–424. Springer, New York, 2011.

517. T. H. Yunes, I. Aron, and J. N. Hooker. An integrated solver for opti-
     mization problems. *Operations Research*, 58:342–356, 2010.

518. T. H. Yunes, A. V. Moura, and C. C. de Souza. Exact solutions for real
     world crew scheduling problems. Presentation at INFORMS national
     meeting, Philadelphia, 1999.

519. T. H. Yunes, A. V. Moura, and C. C. de Souza. Hybrid column gener-
     ation approaches for urban transit crew management problems. *Trans-
     portation Science*, 39:273–288, 2005.

520. J. Zhou. A constraint program for solving the job shop problem. In E. C.
     Freuder, editor, *Principles and Practice of Constraint Programming (CP
     1996)*, volume 1118 of *Lecture Notes in Computer Science*, pages 510–
     524, New York, 1996. Springer.

Index